Real-Time Signal Processing: Design and Implementation of Signal Processing Systems

John G. Ackenhusen
Advanced Information Systems Group
ERIM International, Inc.

ISBN 0-13-631771-5
90000
9 780136 317715

Prentice Hall PTR
Upper Saddle River, New Jersey 07458
http://www.phptr.com

Library of Congress Cataloging-in-Publication Data
Ackenhusen, John G.
Real-time signal processing : design and implementation of signal processing systems / John G. Ackenhusen.
p. cm.
ISBN 0-13-631771-5
1. Signal processing —Digital techniques. 2. Real-time data processing. I. Title.
TK5102.9.A25 1999
621.382'2—dc21 99-20368
CIP

Acquisitions editor: *Bernard M. Goodwin*
Editorial assistant: *Diane Spina*
Manufacturing manager: *Alan Fischer*
Marketing manager: *Lisa Konzelmann*
Cover design director: *Jerry Votta*
Cover designer: *Design Source*
Cover composition: *Anthony Gemmellaro*
Cover art: *David Chmielewski, Corbis; and Nicholas Veasey, Tony Stone Images*
Project coordinator: *Anne Trowbridge*
Compositor/Production services: *Pine Tree Composition, Inc.*

Prentice-Hall, Inc.
Upper Saddle River, New Jersey 07458

Prentice Hall books are widely used by corporations and government agencies for training, marketing, and resale.

The publisher offers discounts on this book when ordered in bulk quantities. For more information, contact:

Corporate Sales Department
Phone: 800-382-3419
Fax: 201-236-7141
E-mail: corpsales@prenhall.com

Or write:

Prentice Hall PTR
Corp. Sales Dept.
One Lake Street
Upper Saddle River, New Jersey 07458

Printed in the United States of America
10 9 8 7 6 5 4 3 2

ISBN: 0-13-631771-5

Prentice-Hall International (UK) Limited, *London*
Prentice-Hall of Australia Pty. Limited, *Sydney*
Prentice-Hall Canada, Inc., *Toronto*
Prentice-Hall Hispanoamericana, S.A., *Mexico*
Prentice-Hall of India Private Limited, *New Delhi*
Prentice-Hall of Japan, Inc., *Tokyo*
Prentice-Hall (Singapore) Pte. Ltd., *Singapore*
Editora Prentice-Hall do Brasil, Ltda., *Rio de Janeiro*

To my loving, angelic wife Kay, who has borne with good cheer her stint as a widow to this work for longer than either of us have liked.

Contents

Preface

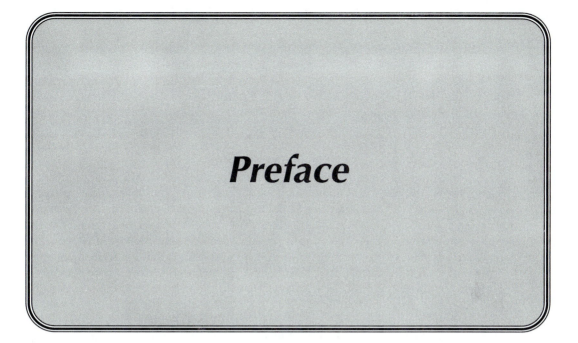

This book illuminates the methods for moving an algorithm from the general-purpose computing environment of signal processing research into the real time processing domain of signal processing applications. This application domain is characterized by a need to process at a speed that keeps up with the input and to do this on a system that is small, economical, and reliable. This is the only book to date that spans the broad range of real time implementation techniques. While other texts focus more closely on specific domains such as software code for signal processing, VLSI design of custom architectures, or real time image processing applications, this book pulls these and other topics together into a common framework, encouraging optimization of design across a wide range of approaches. Broad issues such as hardware versus software realization, space versus time tradeoffs, and off-the-shelf versus custom components are discussed here.

The intended readership includes the practicing signal processing engineer who is faced with the task of moving a signal processing algorithm that has been proven in research into a real time implementation for deployment. This book is also written for the graduating student of signal processing who seeks to quickly contribute to the practice of real time signal processing in industry. Finally, it is intended for the researcher in signal processing algorithms who requires an understanding of the process of deploying the algorithm under study into a system that must meet constraints in processing time, size, and cost.

As the field of signal processing matures, its research results are increasingly applied to the development of solutions to signal processing problems that integrate some combination of algorithms, hardware, and software into a system. Such a system must produce its results in real time, often within the limited volume and power that allow moving the system to the signal source (rather than delivering the signal to a massive immobile processor). The system also must be economical. Conversely, the set of algorithms that make up an application is developed and proven on computers with no constraints on execution time, size, or cost, since the purpose of the research environment is to provide maximum flexibility and observability to the signal processing researcher.

Preparing the algorithm for deployment requires tradeoffs and integration across these multiple domains of algorithm, hardware, and software. For example, changes to both the structure of the algorithm and the structure of the hardware and software that execute the algorithm are usually needed. Such changes can include introducing more computational steps to achieve regularity of computation, decomposing an algorithm onto a particular architecture (which may consist of multiple processors), or devising a special architecture suited for the algorithm for subsequent implementation in custom hardware. On the microscopic level, one may choose whether to perform a piece of the algorithm in hardware or in software, trading speed against flexibility. On the macroscopic level, one chooses whether to reduce the total lifecycle cost of the system, optimizing across development, production, and maintenance, or instead to minimize only one factor, such as the development cost.

This book describes the underlying principles of signal processing solution engineering, encouraging the application of these principles from one implementation to another. Its unique challenge is to articulate these techniques in a manner that is not specific to any one system, yet is based on the realities of actual implementation. Unlike the fundamentals of digital signal processing algorithms, which are the starting point for any signal processing solution and therefore permeate all implementations, the fundamentals of digital signal processing implementation enter at a later point in the design cycle. Extricating these techniques from the specifics for the implementation and discussing them in a manner that emphasizes their application across a diversity of system realizations is the purpose of this book.

Chapter 1 exposes the basic concepts that govern real time signal processing. It begins by developing a definition of real time processing, then examines how real time operation imposes constraints upon the processing of the groups of samples that comprise a digital signal. It concludes with a discussion of *approximate signal processing*, in which limited computational time is allocated to achieve the best possible partial results, often in a manner that allows later incremental improvement.

In Chapter 2, the definition of processors is built up, beginning with simple concepts of sequential and combinational logic. The discussion progresses to the general-purpose microprocessor and then moves to the programmable digital signal processor and its enhancements beyond the microprocessor. The chapter concludes with a discussion of signal processing operations that stress the capabilities of programmable digital signal processors, leading to a discussion of the design of custom signal processing architectures.

Since a given set of system requirements may be met in a variety of ways, signal processing system design involves a series of choices in multiple domains. Chapter 3 is devoted to analyzing the tradeoffs that must be performed, and so it begins with a systematic way of performing a tradeoff analysis. Tradeoff in algorithm complexity and between hardware and software implementations are discussed. Chapter 4 moves tradeoffs from the general to the specific as it identifies the central concept of the exchange of space for time and then applies this concept to space-time tradeoffs in software and in hardware.

Mapping algorithm and architecture is the main purpose of signal processing system design. Basic principles of mapping both algorithms to architectures and architectures to algorithms are described in Chapter 5. Specific examples are given in Chapter 6 that include the computing architectures of both single and multiple processors, and algorithms that require both local and global communication. Detailed examples include two-dimensional convolution on both one and multiple processors, linear predictive coding of speech, image filtering, and synthetic aperture radar image formation.

Chapters 7, 8, 9, and 10 together comprise a selection of software and hardware computing elements useful in solving real-time signal processing problems. This toolbox includes basic hardware and software structures such as buses and address computation methods that span many signal processing applications. After establishing these universal components in Chapter 7, the remaining chapters follow the stages of the information extraction chain that are common to the signal processing task of converting large amounts of data to useful nuggets of information. In Chapter 8, tools for such front-end signal analysis techniques as the Fourier transform and wavelet transform are described. Chapter 9 builds upon signal analysis to address the hardware and software structures for signal compression. Once a signal is analyzed and compressed, its information content may be compared with the information content from another signal. Such comparisons are at the root of signal recognition applications. Chapter 10 discusses computing elements for the training and recognition operations of signal comparison systems.

The discipline of life cycle systems engineering is discussed in Chapter 11. This is important for signal processing systems that are being engineered for mass production and for which lifecycle cost is an important factor. The chapter begins with a survey of the steps of the system design process, then provides a description of the systems engineering process. The subsequent steps of requirements design, hardware design, software design, and system integration and testing are then described. Finally, a case study culminates the illumination of signal processing system design. The design of a real time processor for image recognition is examined. Many of the previous topics of the book, inducing system engineering tradeoffs, special processing structures, and the stages of the system design process, are described for an actual design example that was conducted.

ACKNOWLEDGMENTS

The idea for this book began in 1984, when the author was designing computers and algorithms for real-time speech processing at AT&T Bell Laboratories, now Lucent Technologies. Dr. Larry Rabiner began the author's thinking of the underlying principles of real-time signal processing through his invitation to present an invited paper on the topic of real-time speech processing. Discussions with Dr. Ronald Crochiere over a long working relationship were valuable in refining some of the concepts that are described and in developing the framework that has since been used as the outline of this book.

Involvement with the systems engineering process is based upon the author's work as head of the Signal Processing Systems Engineering Department of AT&T Bell Laboratories, working on the Enhanced Modular Signal Processor (now in production as the AN/UYS-2) of the U.S. Navy. Skills in large-scale system engineering and design for manufacture were learned from Caryl Pettijohn, Mary Albright, David Long, David Marr, Jerry Mulyk, and Bill Robinson. More recently, work on real-time image processing architectures and algorithms was added from work at the Environmental Research Institute of Michigan (now ERIM International Inc.), with significant influences by Dr. Ron Carpinella, Dr. John Gorman, Dr. Quentin Holmes, Bob Lougheed, David McCubbrey, Paul Mohan, Dr. Jeremy Salinger, Ron Swonger, Dr. Nikola Subotic, and Dr. Gregory Warhola.

Presentation of a structured compilation of real time signal processing techniques began as a special session of the Institute of Electrical and Electronics Engineering (IEEE) International Conference on Acoustics, Speech, and Signal Processing (ICASSP) that the author organized in 1995. An expanded version of that session became the subject of a special issue of the *Journal of VLSI Signal Processing*, published in 1996. The specifics of this subject were organized into a tutorial presented at the IEEE ICASSP-96 and in-house at ERIM International, Inc. The material for that course became the basis for this book. Most recently, the 1997 IEEE Workshop on Signal Processing Systems Design and Implementation by the IEEE Signal Processing Society Technical Committee on the Design and Implementation of Signal Processing Systems (DISPS), allowed inclusion of some of the most recent topics of interest in this growing field. More generally, the IEEE, its Signal Processing Society, and the DISPS Committee have all contributed greatly to this book. Dr. Dan Dudgeon of MIT Lincoln Laboratory made excellent comments in the review of a draft of this book that have improved it significantly. Any shortfalls, of course, are those of the author alone. The help and guidance of Bernard Goodwin and Karen Gettman of Prentice Hall and of Patty Donovan and the eagle-eyed copyeditor at Pine Tree Composition are gratefully acknowledged as essential to bringing this 15-year dream to reality.

Ann Arbor
May, 1999

1

Real Time Concepts

Digital signal processing applications seek to extract useful information from an incoming set of signals, represented by sequences of numbers. The algorithms, or computational prescriptions, for extracting this information are originally invented and developed in general-purpose computing environments in which the speed of modifying and retesting the algorithm is more important than the speed of executing the algorithm. In this research computing environment, potential algorithms are first tested on small amounts of data which have been chosen or simulated to mimic certain important situations. Later, ever-increasing amounts of realistic data are used to refine and test the algorithm. At some point, sufficient performance is achieved on adequate amounts of representative data, and the algorithm is declared ready for use.

At this point, the transition must be made from the slow, flexible research computing environment to the fast, enduring environment of the compact, inexpensive real time signal processor.

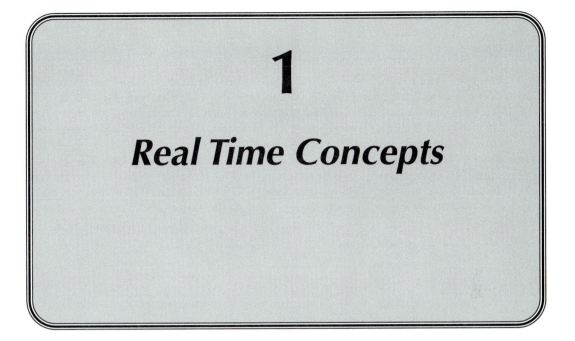

In a real time signal processing application, the samples of the digital signals arrive at the input of the processing system. These samples arrive over some interval of time, until a group of samples, known as a *frame*, have arrived that are sufficient to begin processing. Processing then begins to obtain a more information-rich version of the sample data. Performing these computations on these samples requires a certain computational time.

Real time signal processing completes the computation of the output associated with a frame of input signal samples in a time that does not exceed the duration of that frame. Moving a signal processing algorithm from the general-purpose computing environment in which it was devised to a real time, portable implementation is necessary if that algorithm is to find practical use. Although sometimes that transformation is straightforward, such as packaging and deploying the same engineering workstation as was used for developing the algorithm, most often it is difficult, requiring code rewriting, use of special hardware architectures, and life-cycle system design.

In this chapter, we build a foundation to examine the methods and structures for achieving real time signal processing. The chapter includes an examination of the time aspects of a sampled digital signal, a look at the relative timing of inputs and outputs of a signal processing transformation, and a definition of real time.

1.1 DEFINITION OF REAL TIME

We begin our discussion by examining a discrete-time signal represented by the sequence of numbers $[x(n)]$, $-\infty < n < \infty$. Here, $x(n)$ is the nth number in the sequence. In our discussions, n will correspond to uniform increments in time and will proceed in monotonically increasing fashion, from left to right in our figures (Fig. 1–1).

In our discussions, the sequence $[x(n)]$ will have been derived from sampling a continuous time-varying waveform at uniformly-spaced time increments T_x. The time of occurrence of sample n of $x(n)$ is at $t = nT_x$.

We can further define a collection of I discrete time signals as x_i, $0 \le i \le I-1$. In the most general sense, it is the intent of digital signal processing to transform the set of signals $x_i(n)$ into a second set of signals $y_j(m)$, $0 \le j \le J-1$, that is in some more desirable form (Fig. 1–2).

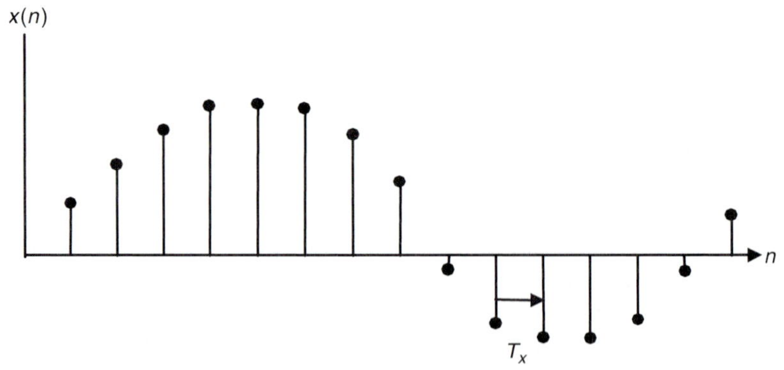

Figure 1–1 Representation of sequence $x(n)$ derived from digital sampling of signal.

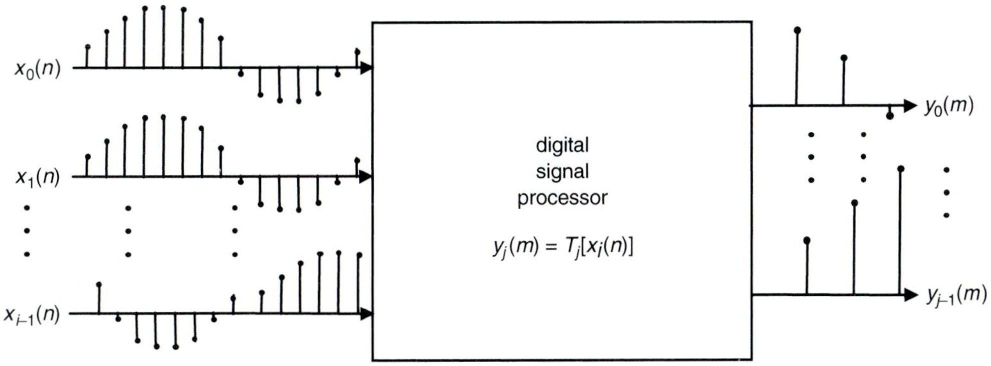

Figure 1–2 Representation of a transformation that maps the signal set $x_i(n)$ into the signal set $y_j(m)$.

This second set of signals may differ both in number of signals in the set, J, and the time spacing of its samples, T_y. The input sample period T_x is allowed to differ from the output sample period T_y. Both T_x and T_y are constant across the analysis intervals.

It is now possible to define the function of the digital signal processing system as a transformation T_j upon the set of input signals $x_i(n)$ to obtain the output signals $y_j(m)$:

$$y_j(m) = T_j[x_i(n)]. \tag{1.1}$$

The transformation T_j is indexed by j to select which output signal y_j is being produced. In the most general situation, each output sample $y_j(m)$ could depend upon every input sample of each input signal, $x_i(n)$, $i = 0, 1,..., I - 1; -\infty < n < \infty$. In practice, the dependence of $y_j(m)$ upon $x_i(n)$ is limited to some smaller range of n. This range remains constant in length and moves along $x_i(n)$ as the output signal index m advances. The time of occurrence of sample $x_i(n)$ from some reference time $t = 0$ is given by nT_x. Similarly, the time of occurrence of output sample $y_j(m)$ is given by mT_y. We can depict the idea of a limited range of n contributing to each sample $y(m)$ and show the idea of the sliding range advancing with m (here, $I = J = 1$) (Fig. 1–3).

In this figure, $y(m = 1)$ depends upon the set of L input samples $x_0, x_1, ..., x_{L-1}$. These input samples extend from time $t = 0$ to $t = (L - 1)T_x$. This set of input samples is referred to as a frame of length L samples. The period of output of $y(1)$ is shown in Fig. 1–3 as T_y, where $T_y = LT_x$.

On any real computer, to perform the computation dictated by T_j upon samples $x_0, x_1,...,x_{L-1}$ requires a certain interval of computation time, T_c, after the time of arrival of the frame's final sample x_{L-1}. If the computation time T_c is less than the time to input the samples that contribute to the computations, then $T_c < LT_x$, and the computation on samples $x_0, x_1, ..., x_{L-1}$ is completed before the arrival of

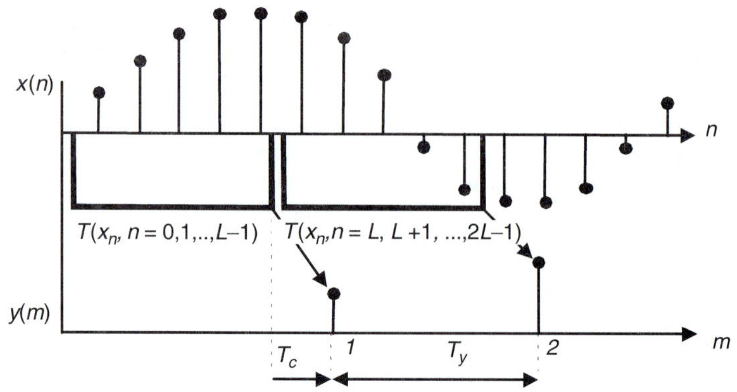

Figure 1–3 Each output sample depends upon L input samples in a manner prescribed by T.

the final sample that contributes to the next output, $y(2)$. Thus for the next frame, the computation of $T(x_L, x_{L+1},\ldots,x_{2L-1})$ may begin as soon as its final sample, x_{2L-1}, arrives. As a result, for constant frame size L and constant frame computation time T_c, the outputs $y(m)$ can keep up with the inputs indefinitely (Fig. 1–4).

Put another way, the output period equals the input analysis frame period and each output sample is produced after a computation delay, T_c, after the end of the analysis window. This system can run in a continuous manner for an indefinite period of time, because: 1) each output is based on the limited time range of adjacent input samples contained in the frame, 2) each output is computed in synchrony with delay T_c with the input, and 3) each output is emitted with a delay T_c after the end of the input frame.

Let us next examine the case where $T_c > LT_x$. In this case, the computation performed on the set of samples $x_0, x_1, \ldots,x_{L-1}$ consumes more time than the time

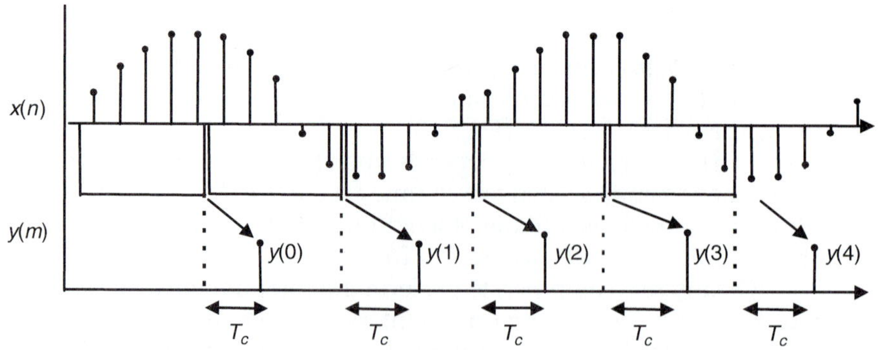

Figure 1–4 Relative timing of input and corresponding output for $T_c < LT_x$.

LT_x over which the set of inputs occurred. In this case, the computation of one frame is completed after the final input of the next frame has arrived and the outputs $y(m)$ fall increasingly behind (Fig. 1–5).

A system for which $T_c > LT_x$ cannot run continuously, because an ever-increasing backlog of input data accumulates without bound. This system is still useful for operation in a pulsed mode where the input sequence is turned off before the output lags too far behind.

The threshold of processing time $T_c = LT_x$ is a vital one to real time signal processing. Systems that have $T_c \leq LT_x$ are said to operate in real time; those with $T_c > LT_x$ operate in non-real-time (or greater than real time).

We generalize our description of real time for the case of multiple signals and formalize it with the following definition:

> For the computation to produce each output sample $y_j(m)$ for all j and m from the set of input samples $x_i(n)$ according to transformation set T_j, the condition of real time processing holds when that computing is completed in a time that does not exceed the time associated with the portion of $x_i(n)$ that contributes to the computation of $y_j(m)$.

The crucial concept to the definition of real time signal processing is the requirement that the computation that T prescribes be performed on $x_i(n)$ must be completed within a certain interval of time, namely the duration of time occupied by that set of samples $x_i(n)$. Thus, the condition of real time processing depends upon

1. the input sample period T_x,
2. the complexity of the transformation T_j,
3. the speed of the computer(s) which compute $T_j[x_i(n)]$ as measured by T_c.

In the following section, we categorize various types of real time signal processing systems on the basis of the length of the range of input samples and the sequence of computation on the input samples that produces the output samples.

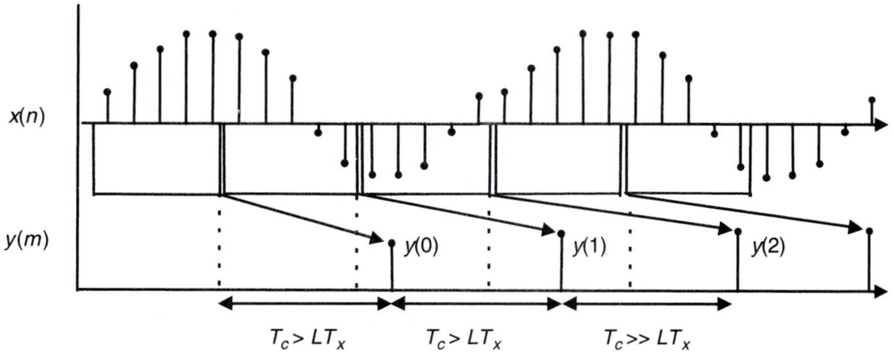

Figure 1–5 Relative timing of input and corresponding output for $T_c > LT_x$.

1.2 STRUCTURAL LEVELS OF PROCESSING

In the previous section, each sample of the set of J output signals, $y_j(m)$, depended in a manner described by the transformation T_j upon a frame of L contiguous samples of the set of I input signals, $x_i(n)$:

$$y_j(m) = T_j[x_i(n)], j = 0, 1, \ldots, J-1, i = 0, 1, \ldots I-1; \quad n = mL, mL+1, \ldots, (m+1)L-1. \tag{1.2}$$

Each output sample $y_j(m)$ is computed from a frame of L contiguous input samples. Signal processing systems can be categorized on the basis of how and when the computation of the L samples that produce $y_j(m)$ occurs relative to the time of arrival of the samples $x_i(n)$.

At one end of the spectrum, upon arrival of sample $x_i(n)$, all computations that require the use of this sample are completed and sample $x_i(n)$ is discarded before sample $x_i(n+1)$ arrives. The technique of completing all computations with one input sample before the next input sample arrives is called *stream processing*. After the stream processing of L such samples, the computation of $y_j(m)$ is then complete and $y_j(m)$ is output from the system.

At the other end of the spectrum, each input sample $x_i(n)$ is stored in memory before any processing occurs upon it. In this case, after L input samples have arrived, the entire collection of samples is processed at once to produce the output sample $y_j(m)$. This approach of performing computation on several input samples at once is known as *block processing*.

Vector processing, which may operate in either stream or block manner, is the name applied to systems with several input and/or output signals being computed at once ($I > 1$ and/or $J > 1$).

We now look at each of these types of real time systems in more detail.

1.2.1 Stream Processing

Stream processing proceeds on a sample-by-sample basis upon the input sample sequence, performing an identical set of computations upon each sample of the analysis frame and completing these computations before the next sample arrives (Fig. 1–6).

One situation in which stream processing is used is when the input sample period T_x equals the output sample period T_y. An example is given by the general difference equation used in digital filtering[1]:

$$y(n) = \sum_{k=1}^{N} a_k y(n-k) + \sum_{k=0}^{M} b_k x(n-k). \tag{1.3}$$

A block diagram of this equation indicates that for each input $x(n)$, an output $y(n)$ is produced (Fig. 1–7).

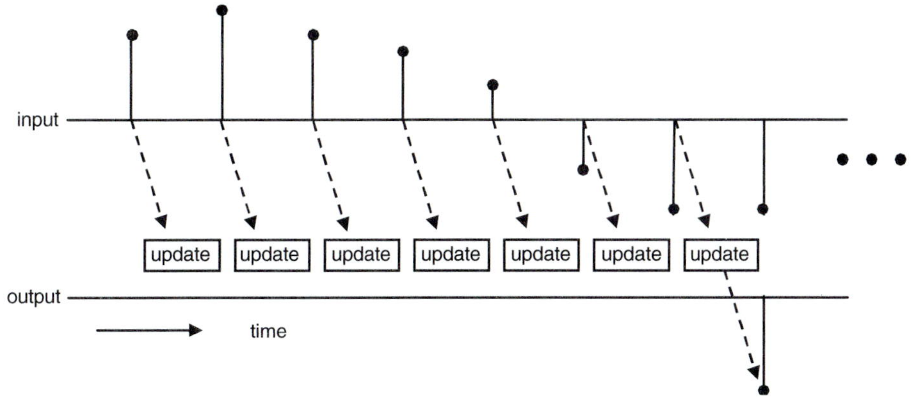

Figure 1–6 Stream processing updates the computation after the receipt of each input sample.

Storage for a set of M past input values and N output values is provided by the blocks labeled z^{-1} on the left and right side of the diagram.

We may also cast an operation that is normally represented in equations as a block processing operation into a stream processing form. An example is the computation of the sum of squares of a signal over a frame of length of N samples. This calculation provides the signal energy of the frame:

$$R = \sum_{n=0}^{N-1} x(n) \cdot x(n). \tag{1.4}$$

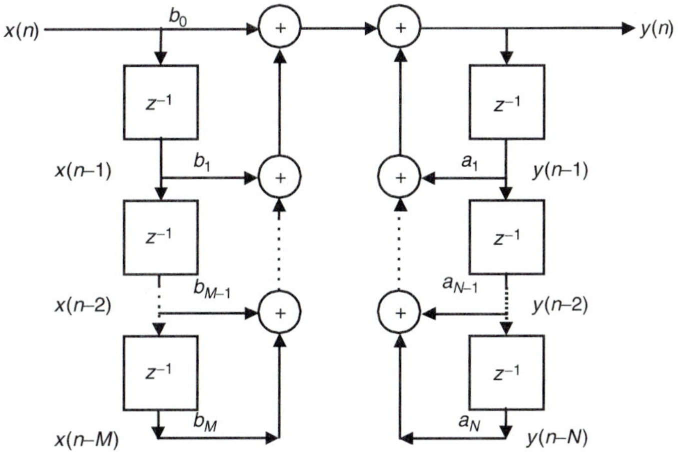

Figure 1–7 Block diagram representation of the general difference equation used in digital filtering.

Instead of waiting for all samples of the frame as is done for block processing, we instead define an intermediate accumulator variable r and update it as each sample arrives:

Initialization:

$$r = 0 \tag{1.5}$$

Do for $n = 0, 1, \ldots, N-1$:

$$r = r + x(n)\cdot x(n) \tag{1.6}$$

Finalization:

$$R = r. \tag{1.7}$$

In many applications, the accumulator register storing r is time-multiplexed, that is, used for other purposes while awaiting the next sample $x(n)$. This requires that the contents of r be written to memory and later retrieved for the next computation of the do-loop above. The writing of r to memory introduces an overhead timing penalty for reading and writing r for each update with the new sample $x(n)$.

One advantage of stream processing is that all results are kept current. Delay between an input and the output sample to which it contributes is kept to its theoretical minimum. Another advantage is that storage of inputs and outputs is kept to the theoretical minimum of the algorithm. In the general difference equation shown in (1.3), storage is required for input samples $x(n-M)$ through $x(n)$ and for output samples $y(n-N)$ through $y(n-1)$, but no additional storage is required.

A disadvantage of stream processing is that the processor must be sufficiently fast to complete all calculations before the next input sample arrives. This requirement on the processor speed is further compounded by an overhead factor in systems where $T_y > T_x$. As shown in (1.6), for each input sample, the quantity r must be retrieved from storage and then restored after the addition of $x(n)x(n)$. This adds an overhead requirement of moving data that does not appear in (1.4), in which the sums of squares are accumulated without storage.

1.2.2 Block Processing

In block processing, incoming samples are stored in groups as they arrive. After the arrival of a sufficient number of samples (as set by the nature of the implementation), processing begins on the group of samples. Because all input samples of the group, or block, are available simultaneously, processing can access the samples in a random manner rather than being restricted to sequential access.

As shown in Fig. 1–8 for a block of five samples, the first five samples are stored in a block as they arrive. After the final sample of the block has arrived, processing on the block begins. At this point, two activities proceed together: the processing of samples from block l and the input and storage of samples from block $l+1$. When processing on block l is completed, its result is output.

The case of overlapping blocks is only slightly more complicated. In this case, each input sample contributes to adjacent outputs, each output has an over-

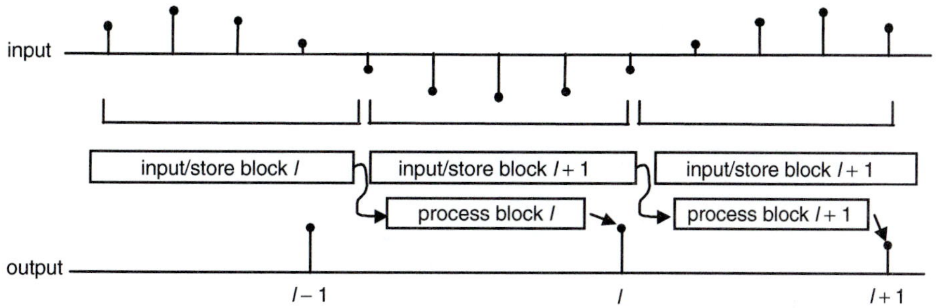

Figure 1–8 Block processing of input signal in blocks of five samples, showing timing relationships of the input, store, and process operations.

lapping block of input samples, and each input sample is in multiple blocks. In Fig. 1–9, an overlap of ½ block is shown, and each input sample contributes to two outputs. The scheme is a bit more complicated in that two input and store operations, one for block l and one for block $l + 1$, are proceeding at any one time, as are two processing operations.

Block processing techniques may be used when the output sample period T_y is much greater than the input sample period T_x, that is, for algorithms whose output sampling rate is lower than the input sampling rate. An example is the sum-of-squares computation of frame energy mentioned in the previous section:

$$R = \sum_{n=0}^{N-1} x(n) \cdot x(n). \tag{1.8}$$

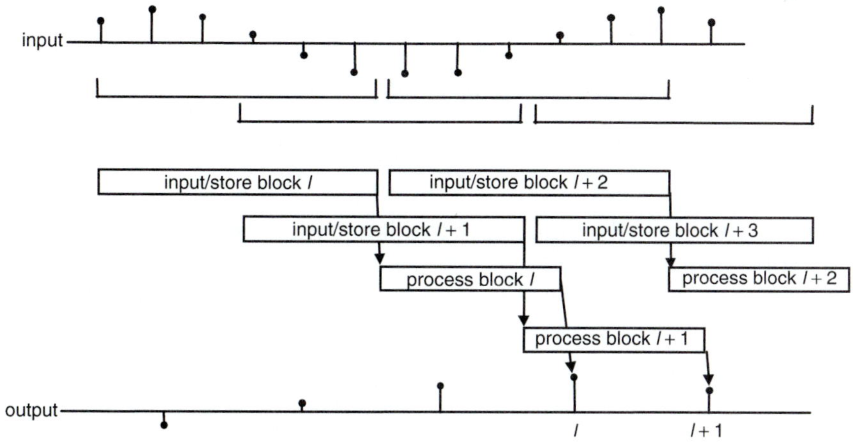

Figure 1–9 Processing for overlapping blocks assigns each input sample to multiple adjacent processing blocks, maintaining multiple simultaneous input/store and processing operations.

Casting in block form and using the style of the previous section,

Initialization:

$$r = 0 \tag{1.9}$$

Execute once:

$$r = r + \sum_{n=0}^{N-1} x(n) \cdot x(n) \tag{1.10}$$

Finalization:

$$R = r. \tag{1.11}$$

Here, the computation of the summation occurs with r kept in a local, quickly-accessed accumulator register throughout the summation. The overhead of reading and writing r from/to memory for each sample, discussed in the stream processing example above, is avoided.

Although the sum of squares can be computed in either block or streaming form, some computations are more easily computed in block form. One such example is the computation of the median signal value of the frame. Computing the median requires ordering the L samples $x(n)$ in increasing order of magnitude to form the set of numbers X_1, X_2, ...,X_L. The median is then given by $X_{(L+1)/2}$ if L is odd or by $(X_{L/2}+X_{(L/2)+1})/2$ if L is even. Since a full sort of all samples of the frame is required, the sort is much more easily performed when all samples are present than if the sorted sample list is readjusted every time a new sample arrives.

Stream processing operations can be converted to block processing form with the addition of storage for all samples of the block and with the introduction of delay of at least one frame duration used to accumulate the samples. As an example, we reexamine the general difference equation used in digital filtering, which was presented in Section 1.2.1 as a stream process. This may be converted to a block process computed in groups of samples, as discussed by Burrus.[2]

We reexamine the difference equation:

$$y(n) = \sum_{k=1}^{N} a_k y(n-k) + \sum_{k=0}^{M} b_k x(n-k). \tag{1.12}$$

We apply the z transform to this equation. The z transform, X(z), of a sequence $x(n)$, given by $X(z) = \Sigma_{n=-\infty}^{\infty} x(n)z^{-n}$, is discussed in standard references in signal processing.[1] The z transform applied to each side, with some rearranging, provides:

$$H(z) = \frac{Y(z)}{X(z)} = \frac{\sum_{k=0}^{M} b_k z^{-k}}{\sum_{k=0}^{N} a_k z^{-k}} \tag{1.13}$$

which takes the form

$$A(z)Y(z) = B(z)X(z), \tag{1.14}$$

where $A(z) = a_0 + a_1 z^{-1} + ... + a_N z^{-N}$ and $B(z) = 1 + b_1 z^{-1} + ... + b_M z^{-M}$. The convolution of a function $h(n)$ with a function $x(n)$ is formed by multiplying $h(n)$ by a reversed and linearly-shifted version of $x(n)$, then summing the values in the product:

$$h(n) * x(n) = \sum_{m=0}^{N-1} h(n) \cdot x(n - m). \tag{1.15}$$

We use the fact that the product of the z transform of two functions can be written as the convolution of the functions themselves to obtain:

$$a(k) * y(k) = b(k) * x(k) \tag{1.16}$$

For simplicity, we choose:

$$a_0 = 1 \tag{1.17}$$

$$a_k = 0; k > 3 \tag{1.18}$$

$$b = 0; k > 2. \tag{1.19}$$

The convolution may be written in matrix form:

$$\begin{bmatrix} 1 & 0 & 0 & \\ a_1 & 1 & 0 & \\ a_2 & a_1 & 1 & \cdots \\ a_3 & a_2 & a_1 & \\ 0 & a_3 & a_2 & \\ & \vdots & & \end{bmatrix} \begin{bmatrix} y_0 \\ y_1 \\ y_2 \\ y_3 \\ \vdots \end{bmatrix} = \begin{bmatrix} b_0 & 0 & 0 & \\ b_1 & b_0 & 0 & \\ b_2 & b_1 & b_0 & \cdots \\ 0 & b_2 & b_1 & \\ 0 & 0 & b_0 & \\ & \vdots & & \end{bmatrix} \begin{bmatrix} x_0 \\ x_1 \\ x_2 \\ x_3 \\ \vdots \end{bmatrix}. \tag{1.20}$$

For later simplification, we define the matrices

$$\mathbf{A} = \begin{bmatrix} 1 & 0 & 0 & \\ a_1 & 1 & 0 & \\ a_2 & a_1 & 1 & \cdots \\ a_3 & a_2 & a_1 & \\ 0 & a_3 & a_2 & \\ & \vdots & & \end{bmatrix} \text{ and } \mathbf{B} = \begin{bmatrix} b_0 & 0 & 0 & \\ b_1 & b_0 & 0 & \\ b_2 & b_1 & b_0 & \cdots \\ 0 & b_2 & b_1 & \\ 0 & 0 & b_0 & \\ & \vdots & & \end{bmatrix}.$$

We can group the input samples into blocks, or vectors, of length L samples each and define:

$$\mathbf{X}_0 = [x(0), x(1), \ldots, x(L-1)]^T; \tag{1.21}$$

$$\mathbf{X}_1 = [x(L), x(L+1), \ldots, x(2L-1)]^T. \tag{1.22}$$

Here, T in $[x]^T$ indicates the transpose of $[x]$ from a row vector to a column vector, so if

$$\mathbf{X} = [x(0), x(1), \ldots, x(L-1)], \text{ then } \mathbf{X}^T = \begin{bmatrix} x(0) \\ x(1) \\ \vdots \\ x(L-1) \end{bmatrix}.$$

Similarly, for the output samples we define

$$\mathbf{Y}_0 = [y(0), y(1), \ldots, y(L-1)]^T; \tag{1.23}$$

$$\mathbf{Y}_1 = [y(L), y(L+1), \ldots, y(2L-1)]. \tag{1.24}$$

We choose a block size $L = 3$. We can partition the **A** and **B** matrices into $L \times L$ submatrices by defining:

$$\mathbf{A}_0 = \begin{bmatrix} 1 & 0 & 0 \\ a_1 & 1 & 0 \\ a_2 & a_1 & 1 \end{bmatrix} \text{ and } \mathbf{A}_1 = \begin{bmatrix} a_3 & a_2 & a_1 \\ 0 & a_3 & a_2 \\ 0 & 0 & a_3 \end{bmatrix} \tag{1.25}$$

$$\mathbf{B}_0 = \begin{bmatrix} b_0 & 0 & 0 \\ b_1 & b_0 & 0 \\ b_2 & b_1 & b_0 \end{bmatrix} \text{ and } \mathbf{B}_1 = \begin{bmatrix} 0 & b_2 & b_1 \\ 0 & 0 & b_2 \\ 0 & 0 & 0 \end{bmatrix}. \tag{1.26}$$

We then rewrite the matrix from the convolution expression as:

$$\begin{bmatrix} \mathbf{A}_0 & 0 & 0 & \\ \mathbf{A}_1 & \mathbf{A}_0 & 0 & \\ 0 & \mathbf{A}_1 & \mathbf{A}_0 & \cdots \\ & \vdots & & \end{bmatrix} \begin{bmatrix} \mathbf{Y}_0 \\ \mathbf{Y}_1 \\ \mathbf{Y}_2 \\ \vdots \end{bmatrix} = \begin{bmatrix} \mathbf{B}_0 & 0 & 0 & \\ \mathbf{B}_1 & \mathbf{B}_0 & 0 & \\ 0 & \mathbf{B}_1 & \mathbf{B}_0 & \cdots \\ & \vdots & & \end{bmatrix} \begin{bmatrix} \mathbf{X}_0 \\ \mathbf{X}_1 \\ \mathbf{X}_2 \\ \vdots \end{bmatrix}. \tag{1.27}$$

We use the partitioned blocks in the above matrix expression to express a recursive equation:

$$\mathbf{A}_1\mathbf{Y}_0 + \mathbf{A}_0\mathbf{Y}_1 = \mathbf{B}_0\mathbf{X}_1 + \mathbf{B}_1\mathbf{X}_0 \tag{1.28}$$

or more generally:

$$\mathbf{A}_0\mathbf{Y}_{K+1} = -\mathbf{A}_1\mathbf{Y}_K + \mathbf{B}_0\mathbf{X}_{K+1} + \mathbf{B}_1\mathbf{X}_K. \tag{1.29}$$

The matrix $\mathbf{A}_0$, defined above, has an inverse $\mathbf{A}_0^{-1}$.

We rewrite the general recursion above as a recursive expression for $\mathbf{Y}_{K+1}$:

$$\mathbf{Y}_{K+1} = \mathbf{K}\mathbf{Y}_K + \mathbf{H}_0\mathbf{X}_{K+1} + \mathbf{H}_1\mathbf{X}_K \tag{1.30}$$

where

$$\mathbf{K} = -\mathbf{A}_0^{-1}\mathbf{A}_1 \tag{1.31}$$

$$\mathbf{H}_0 = \mathbf{A}_0^{-1}\mathbf{B}_0 \tag{1.32}$$

$$\mathbf{H}_1 = \mathbf{A}_0^{-1}\mathbf{B}_1. \tag{1.33}$$

The term $\mathbf{H}_0\mathbf{X}_{K+1}$ can be shown to be the $L \times L$ convolution of the first block of the impulse response of the filter:[2]

$$\mathbf{H}_0 = \begin{bmatrix} h_0 & 0 & 0 & \cdots & 0 \\ h_1 & h_0 & 0 & \cdots & 0 \\ h_2 & h_1 & h_0 & \cdots & 0 \\ \vdots & \vdots & \vdots & \vdots & \vdots \\ h_{L-1} & h_{L-2} & h_{L-3} & \cdots & h_0 \end{bmatrix}. \tag{1.34}$$

Equation (1.28) is a block realization of (1.12) that operates at a block rate of $1/L$ times the original data rate and with a delay of one block length L. The diagram of Fig. 1–10 shows the input sequence, blocked into L-length sequences $\mathbf{X}_0$, $\mathbf{X}_1$,..., and the output sequences occurring in blocks $\mathbf{Y}_0$, $\mathbf{Y}_1$, ...

Each output block is obtained from convolution of the associated input block with the truncated version of the impulse response, $\mathbf{H}_0$. The effect of blocking into frames is accounted for by both the recursive multiplication by $\mathbf{K}$ and by multiplying the previous input block by $\mathbf{H}_1$. For a filter with no zeros, $b_k = 0$ and $\mathbf{H}_1 = 0$. Fig. 1–10 then simplifies as shown in Fig. 1–11 and the simpler recursion

$$\mathbf{Y}_{K+1} = \mathbf{K}\mathbf{Y}_K + \mathbf{H}_0\mathbf{X}_{K+1} \tag{1.35}$$

is obtained.

Here, each block is represented by the convolution of the associated input block with the impulse response truncated to L points, with a correction for both the blocking and the truncation of the impulse response in the recursive multiplication by $\mathbf{K}$.

By operating on input and output samples in blocks of L samples at a time, one can obtain greater efficiency than with the direct streaming implementation. In the direct implementation for $N = M$, $2M$ multiplication operations are required to compute each output point, as shown by (1.12). When inputs and outputs are computed in blocks, the efficiency of the fast Fourier transform (FFT) may be used to compute the matrix terms of (1.28). For example, $\mathbf{H}_0$ and $\mathbf{A}_0^{-1}$ represent convolutions of length L that can be computed by adding L zeros and per-

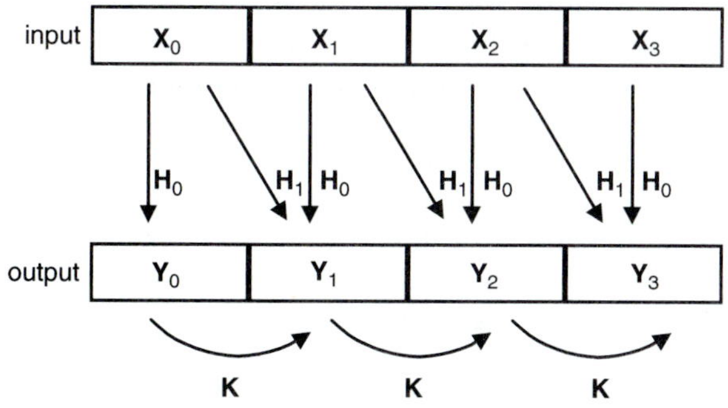

Figure 1–10 Block processing form of (1.12) with input blocks X_0, X_1,... and output blocks Y_0, Y_1,... .

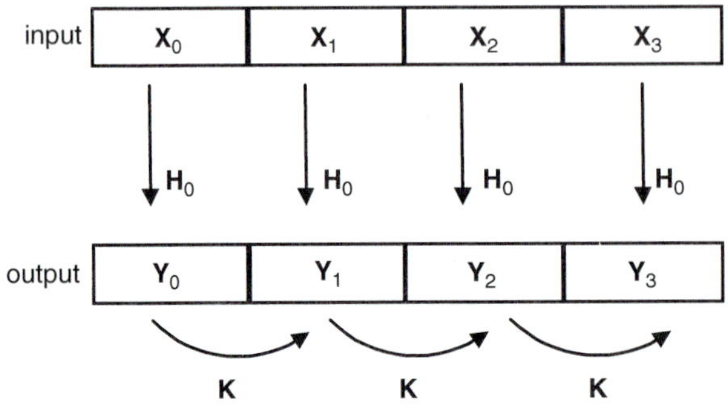

Figure 1–11 Simplification of Fig. 1–10 for filter with no zeros.

forming a 2L-length convolution using the FFT. This approach makes use of the Fourier convolution theorem (see Section 6.3.1), which states that the periodic convolution of two finite sequences can be obtained by multiplying the Fourier transforms (here, shown as FT) of the sequences, then taking the inverse Fourier transform (FT^{-1}) of the result. Thus, for $h(n) * x(n)$, and defining $H(\omega)$ and $X(\omega)$ as the Fourier transforms (FT) of $h(n)$ and $x(n)$,

$$h(n)*x(n) = \mathrm{FT}^{-1}[H(\omega)X(\omega)].$$

Given this use of the Fourier transform, the efficient structure for the Fourier transform known as the fast Fourier transform, or FFT, may be used. As described in Section 8.2.2, the FFT[3] is a block-oriented process that exploits the symmetry and periodicity that exists in the computation of the discrete Fourier transform of a sequence of length L. It reduces the amount of computation from being proportional to L^2 to being proportional to $L \log L$. Because of this efficiency it is helpful to compute $\mathbf{H}_0$, $\mathbf{A}_0^{-1}$, and $\mathbf{B}_0$ by use of the FFT.

The notion of short time stationarity of a signal is one that applies to a signal whose parameters are varying slowly with time. This signal may be decomposed into a sequence of shorter time intervals, a block of analysis provided for each such shorter sequence, and for each block, the parameters are assumed fixed in time. This leads naturally to block processing. For many signals (e.g., speech), useful models are constructed by segmenting the signal into blocks over which the signal spectrum is assumed to be constant. The entire signal is then modeled by a sequence of spectral models resulting from the sequence of blocks.

The advantages of block processing take several forms. Some algorithms (e.g., computing the median sample) require random access to all samples of the block and are difficult or impossible to execute in a stream manner. An efficiency in processing may result by keeping all numbers in the local fast storage for the duration of the block, thereby reducing the overhead of frequent read/write operations to memory. This was demonstrated in the example of computing signal energy (1.4). Another type of efficiency associated with block algorithms such as the FFT may be applied if samples are grouped into blocks.

Another advantage of block processing is that it allows a slower processor to be used. If the available processor is not sufficiently fast to keep up with input samples, the samples may be buffered up and used in computation after all input has ceased. Indeed, most digital signal processing algorithm development proceeds in this manner by using general purpose computers programmed in high-level languages to compute upon files of signal samples stored as a batch on disk.

In this example of batch processing, a major disadvantage of the approach is delay. Indeed, since by definition block processing operates on groups of samples in a block, a minimum delay of one block interval is required to allow all samples of the block to arrive. This requirement exists even if the processor is sufficiently fast to complete all operations on one block in less than one block time. Associated with block processing is the need for sufficient memory to store all the samples of the block.

Another disadvantage occurs for signals of an infinite duration that are modeled as a sequence of short-time spectra, such as speech. To perform block-oriented computation upon these signals while they are in progress requires that the processor have the ability to perform two tasks simultaneously—compute upon one block while receiving samples that make up the next. This requires either two separate processors (one for input and one for computation) or else the switching of a processor between tasks. In Section 6.4, a detailed example is given of meshing a program that computes upon input samples with a program computing output samples, both running on the same processor.

1.2.3 Vector Processing

The term "vector processing" takes on many meanings. In general, it applies to processes that operate on many signals at once which are combined as a vector. If these samples derive from the same signal, we obtain the situation of block processing that was discussed in the previous section (Fig. 1–12).

More generally, and for the purposes of this discussion, vector processing consists of operating on multiple simultaneously-arriving samples from several signals. As for a single signal channel, this processing may either occur in a stream mode (Fig. 1–13) or a block mode (Fig. 1–14).

Output may take either the form of a multi-signal vector (Fig. 1–13, Fig. 1–14) or of a scalar signal (Fig. 1–12).

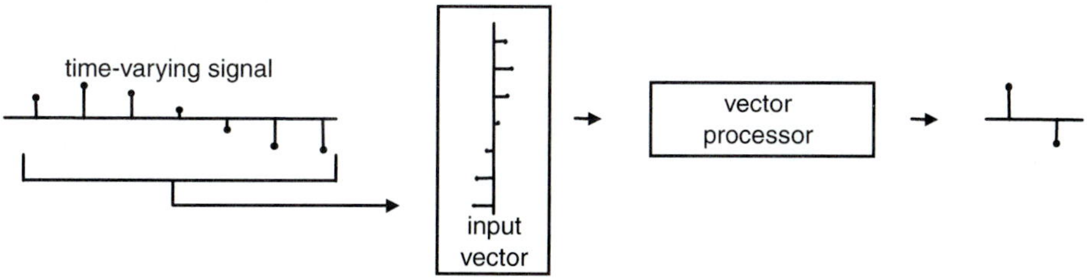

Figure 1–12 Vector processing interpretation of block processing of samples from a single signal.

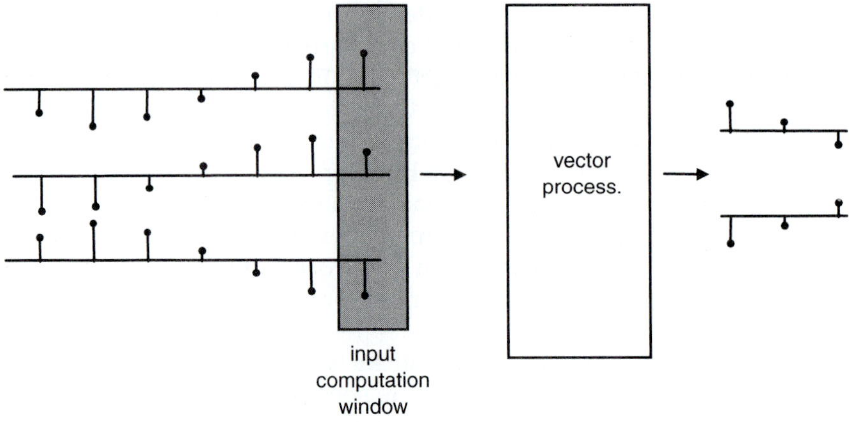

Figure 1–13 Vector processing applied to multiple signal channels, in which input samples from each channel are individually processed in stream mode (shaded processing window).

The signals that provide the multiple inputs for a vector process may be entirely uncorrelated. Indeed, one process to which vector techniques may be applied is to compute the degree of correlation between two signals. Alternatively, the signals may be interrelated or derived from the same source. One example of a vector of interrelated signals is the array of outputs from a bank of parallel bandpass filters used to perform spectral analysis of a single waveform. Each output signal derives from the original signal after it passes through a bandpass filter of a different frequency range. An example of an input signal vector is the set of signals obtained

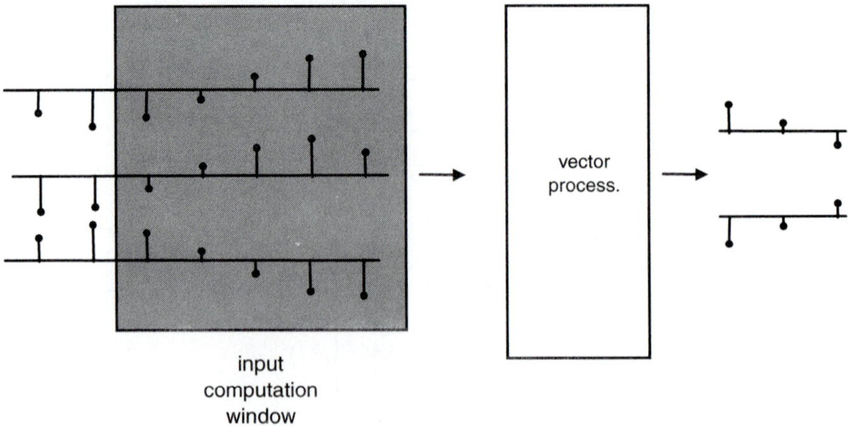

Figure 1–14 Vector processing applied to multiple signal channels (block mode).

from an array of sensors placed in different positions that record the temporal and spatial response to a geophysical impulse (e.g., explosion).

A more subtle interrelationship is obtained for signal vectors that represent a spectral model. For example, as will be discussed in Section 6.4, a vector of $p+1$ autocorrelation coefficients may be used as input to a vector process that computes a set of p linear predictive coding (LPC) coefficients. Here, the input vector represents a spectral model of the signal, yet each vector element may not be separated into a readily identifiable quantity such as the energy in a frequency range.

If we examine the case of a vector processed by block methods (Fig. 1–14), we can identify a matrix in which the rows correspond to separate signal channels and the columns correspond to units of time set by the sampling period. The conversion of an autocorrelation signal representation to an LPC representation is an example of the general case of matrix inversion, where a matrix of input signals is used to compute an inverse matrix.

The vector process may perform identical operations on each signal (row) of the input. For example in the case of parallel bandpass filters, it is common to compute the energy within each frequency channel by taking the absolute value, low pass filtering it, and then decimating (i.e., lowering the sampling rate of) each channel (Fig. 1–15).

In this case, the identical operations of rectification, low pass filtering, and decimation are performed upon each channel. If separate processors are used for each signal channel, they may be controlled by the same instruction stream.

This is an example of a class of parallel machines called Single Instruction Multiple Data (SIMD) that will be discussed more fully in Section 4.2.1. It may be distinguished from the Multiple Instruction Multiple Data (MIMD) processor that operates differently upon each channel.

Although our discussion so far has been restricted to computing structures operating on the data stream, vector processing may be applied to the instruction control as well. Certain compiler computer programs identify loops of the form:

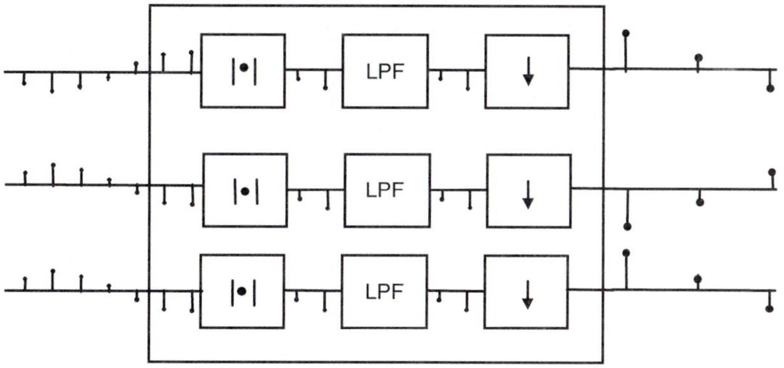

Figure 1–15 Vector processing used to determine energy for each channel of a filter bank of bandpass filters.

```
for (I=0; I<=N; I++)
      {r(I) = f(a1(I), a2(I), …, ak(I))
      }.
```

These loops apply the same sequence of operations on different data, and the vectorizing compiler maps these to parallel tasks, perhaps running on different processing units within the same computer.

The notion of vector instructions is another manner in which vector processing enters the world of instruction control. Each vector instruction initiates many arithmetic operations. Each arithmetic operation may be applied to every element of each of several vectors, such as when computing the sum, difference, or scalar multiplication of vectors. Alternatively, a single instruction can initiate the transformation of a vector from one form to another. An example of this is the instruction `LPC(P,R,A)`. This instruction performs the sequence of operations required to compute the vector of p-th order LPC coefficients from the autocorrelation vector **R** and stores the results in **A**. These vector instructions are often captured by subroutines with vector input and output.

1.3 APPROXIMATE PROCESSING[a]

Real time signal processing is characterized by the constraint of limited time in which the processing can be conducted. Processing time can be measured in number of operations or number of clock cycles available, set by such constraints as the interval between new input samples, the buffer size available to accumulate samples for processing, or the need for a timely result. In traditional approaches, if computation time runs out before the computation is completed, the result is lost, since no provision exists to output partial results and save these for later refinement. We introduce the concept of the *quality* of an answer. The quality of an answer ranges from 0 (no answer at all) to 1, for an answer in which all input data that could contribute to the answer has been fully used. Intermediate levels of quality may provide partial answers based on only a portion of the input data. This portion can correspond to a subset of the samples, or all the samples but at a reduced precision, or other subsets of the input information.

Approximate processing[4] is emerging as a way to obtain a lower-quality answer that can be reached in less time than required for the full answer. Fig. 1–16a shows the quality of answer versus time for such a computation, indicating that the full-quality answer appears at time t_{done}, but no answer is available before that time. In Fig. 1–16b, the performance curve indicates that for an approximate processor, a result of lesser quality is available earlier than the time of completion, and the quality of result improves as time proceeds, ultimately arriving at

[a]Portions adapted, with permission, from S. Hamid Nawab and Erkan Dorken, "A Framework for Quality vs. Efficiency Tradeoffs in STFT Analysis" *IEEE Trans. Acoustics, Speech, and Signal Processing,* 43 (April 1995), 998-1001. © 1995 IEEE.

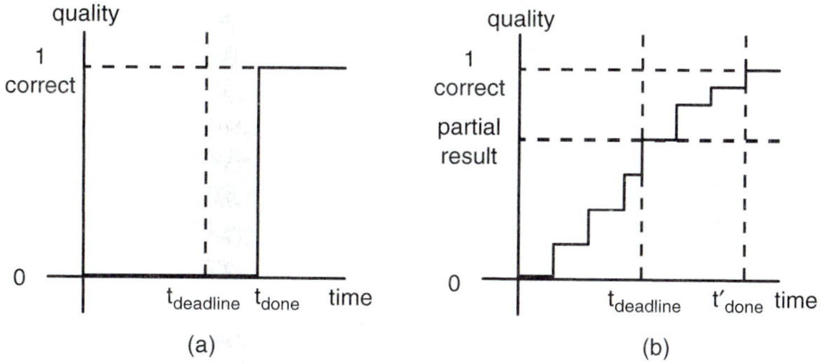

Figure 1–16 Plot of quality of result versus time for (a) traditional processing algorithm and (b) incremental processing algorithm.

the correct result. The time of completion for the approximate processing algorithm may increase over that of a traditional processing algorithm as a result of the additional operations required to compute the answer in an evolutionary manner (t'_{done} in Fig. 1–16b is greater than t_{done} in Fig. 1–16a).

1.3.1 Incremental versus Deadline-Based Processing

Approximate processing, then, provides a quality of output that increases in proportion to the amount of execution time. Two types of approximate processing may be distinguished based on whether the amount of computation time is predetermined or not. In the first case, known as *deadline-based* approximate processing, a "complete" result is provided at the expiration of time, but the result is of reduced quality. Quality measures include signal-to-noise ratio, spectral resolution, or spectral coverage, as described below. The second case, when the computation time is not known in advance, applies when the computation is interrupted, perhaps due to the arrival of another sample or task for processing. In this case, the quality of answer is proportional to the amount of computation time that has passed. Furthermore, the computation may be resumed from the lower-quality intermediate result and processed toward the full-quality answer (rather than started from the beginning) when computational resources are again available. This case is referred to as *incremental refinement*.

1.3.2 Application Example: Short Time Fourier Transform (STFT)[b]

We examine the short time Fourier transform (STFT)[5,6] and cast it into the form of approximate processing. We will rewrite it in a manner that sequentially updates the quality of results as a function of coverage in frequency, resolution in fre-

---

[b]Portions adapted, with permission, from Joseph M. Winograd and S. Hamid Nawab, "Incremental Refinement of DFT and STFT Approximations" *IEEE Signal Processing* Letters, 2 (February 1995), 25–27. © IEEE 1995.

quency, and signal-to-noise (SNR) ratio. The computation ultimately arrives at the expected answer, but it follows a route that provides lower-quality answers with less computation along the way. Several tradeoffs of quality factors in coverage, resolution, and SNR are possible.

We start with a signal $x(n)$, and apply an analysis window $w(n)$ of length N_W to achieve the windowed signal $x_{mL}(n) = x(n)w(mL - n)$, where we calculate the mth N-point discrete Fourier transform (DFT) on a window advance interval of L as follows ($N > L$):

$$X_{mL}(k) = \sum_{n=mL-N_W+1}^{mL-N_W+N} x_{mL}(n)e^{-j(2\pi/N)kn};\ k = 0,1,\dots,N-1. \tag{1.36}$$

Because we will soon quantize the sample to B bits, we first create a new signal $g_{mL}(n)$ to be the difference between each sample $x_{mL}(n)$ and its previous sample. The difference for the first sample of the analysis window (for which the predecessor lies outside the window) is formed by subtracting the last sample of the analysis window. This circular differencing operation and quantization increases the number of zero samples (e.g., two adjacent equal samples will produce a zero difference), and the zero samples can decrease the amount of subsequent operations:

$$g_{mL}(n) = \begin{cases} x_{mL}(n) - x_{mL}(n+N+1); & n = mL - N_W + 1 \\ x_{mL}(n) - x_{mL}(n-1); & mL - N_W + 1 < n \leq mL - N_W + N. \end{cases} \tag{1.37}$$

The DFT can then be written in terms of $g_{mL}(n)$:

$$X_{mL}(k) = \sum_{n=mL-N_W+1}^{mL-N_W+N} g_{mL}(n)G_n(k);\ k = 1,\dots,\frac{N}{2} \tag{1.38}$$

where

$$G_n(k) = \left(\frac{e^{-j(2\pi/N)kn}}{1 - e^{-j(2\pi/N)k}}\right). \tag{1.39}$$

With these bounds on k, we have excluded the DC term ($k = 0$) and restricted the analysis to real signals ($k \leq N/2$).

We now represent $x_{mL}(n)$ as the sum of bit vectors $x_{mL,0}(n)$, $x_{mL,1}(n)$, $\dots, x_{mL,b}(n), \dots, x_{mL,B-1}(n)$:

$$x_{mL}(n) = \sum_{b=0}^{B-1} \beta(b)x_{mL,b}(n), \tag{1.40}$$

where

$$\begin{aligned} \beta(0) &= -1 \\ \beta(b) &= 2^{-b};\ b > 0. \end{aligned} \tag{1.41}$$

Thus, we use the bit-wise representation of $x_{mL}(n)$ from (1.40) to write (1.36) as

$$X_{mL}(k) = \sum_{b=0}^{B-1} \beta(b)\left[\sum_{n=mL-N_W+1}^{mL-N_W+N} x_{mL,b}(n)e^{-j\frac{2\pi}{N}kn}\right]. \tag{1.42}$$

We note that this is a linear combination of B distinct STFTs, each on a signal with two quantization levels, 0 and 1.

We rewrite the difference equation as:

$$g_b(n) = \begin{cases} x_{b,mL}(n) - x_{b,mL}(n+N-1); & n = mL - N_W + 1 \\ x_{b,mL}(n) - x_{b,mL}(n-1); & mL - N_W + 1 < n \leq nL - N_W + N \end{cases} \tag{1.43}$$

and

$$G_{n,b}(k) = \begin{cases} -\dfrac{e^{-j(2\pi/N)kn}}{1-e^{-j(2\pi/N)kn}}; & b = 0 \\ 2^{-b}\dfrac{e^{-j(2\pi/N)kn}}{1-e^{-j(2\pi/N)kn}}; & 1 \leq b \leq B-1. \end{cases} \tag{1.44}$$

The ith approximation to the DFT is now given by

$$\hat{X}_i(k) = \sum_{b=0}^{v_i-1}\sum_{n=0}^{r_i-1} g_b(n)G_{n,b}(k);\ 1 \leq k \leq c_i. \tag{1.45}$$

Here, the indexing bounds c_i, r_i, and v_i set the bounds in frequency coverage, resolution, and signal to noise ratio, respectively. The quality of $\hat{X}_i(k)$ with respect to frequency coverage is given by $2\pi c_i/N$ rad/sample, with respect to frequency resolution is $2\pi/r_i$ rad/sample, and with respect to SNR is approximately $6v_i$dB. Full quality in frequency coverage is reached when $c_i = N/2$, in frequency coverage when $r_i = mL + N_W + N$, and in SNR when the number of bits v_i increases to B.

In a deadline-constrained processing situation, a fixed number of operations is available to compute the DFT, which we represent by P. Eq. (1.45) can be viewed as a vector summation operation of the column vectors, each of which contains the multiplicative coefficients $G_{n,b}(k)$ for a particular n, multiplying the difference signal $g_b(n)$, which has been quantized to a total of $2^{v_i} + 1$ levels.

If more than P vector summations are required to complete the full calculation of $X_{mL}(k)$, then certain vector summations may be dropped in accordance with the bounds c_i and r_i. In this manner, the number of computations may be limited to P, as imposed by the deadline constraint, yet a complete answer, of reduced quality, is produced.

Fig. 1–17 illuminates the idea of eliminating certain vector summations. The set of dots represent all vector sums, while the regions labeled a) and b) narrow either the time (decreasing frequency resolution) or frequency (decreasing frequency coverage). Region c) shows a frequency-dependent narrowing of time, resulting in full frequency coverage but reduced frequency resolution at high frequencies.

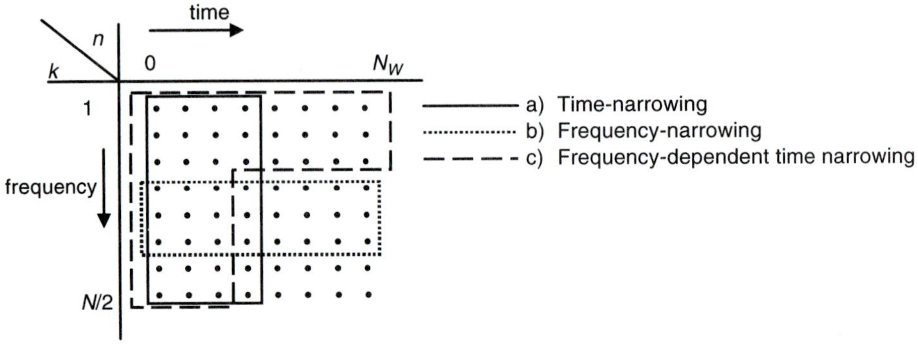

Figure 1–17 Alternative means to select a subset of time-frequency space to reduce STFT computation to constraints of deadline. © 1995 IEEE.

To implement incremental refinement, we write a recursive form for the ith refinement of $\hat{X}_i(k)$, as

$$\hat{X}_i(k) = \begin{cases} \hat{X}_{i-1}(k) + C_i(k); & c_{i-1} < k < c_i \\ \hat{X}_{i-1}(k) + R_i(k) + V_i(k); & 1 \leq k \leq c_{i-1} \end{cases} \tag{1.46}$$

where $C_i(k)$ is the coverage update,

$$C_i(k) = \sum_{b=0}^{v_i} \sum_{n=0}^{r_i} g_b(n) G_{n,b}(k) \tag{1.47}$$

$R_i(k)$ is the resolution update,

$$R_i(k) = \sum_{b=0}^{v_i} \sum_{n=r_{i-1}+1}^{r_i} g_n(n) G_{n,b}(k) \tag{1.48}$$

and $V_i(k)$ is the SNR update,

$$V_i(k) = \sum_{b=v_{i-1}+1}^{v_i} \sum_{n=0}^{r_i-1} g_b(n) G_{n,b}(k). \tag{1.49}$$

The initial conditions are $c_0 = r_0 = v_0 = 0$; $\hat{X}_0(k) = 0$ for all k.

For each i, a representation of the STFT results, and as i increases, the coverage, resolution, and SNR improve. Within the confines of real time computing, a limited amount of computation resources would advance the computation to refinement level i_0. The computational resources would then be directed to other processing, and when that processing was complete, the STFT processing could be resumed at stage i_0 and advanced to a higher level of quality i_i, $i_0 < i_i$, building on the results of the computation so far rather than starting over.

Fig. 1–18 shows an example of successive refinement applied to a two-second violin recording that plays a sequence of two notes. Stage 1 requires 11 percent of the computational resources required for the final result and provides an initial approximation with reduced frequency coverage (highest frequencies are missing), reduced resolution (valleys between peaks and valleys are irregular and approach the height of peaks), and reduced SNR (noise seen between the two tones, at around 1 sec). Stage 4, which is 94 percent of full result, shows more high frequency content, higher peaks, and less noise.

Approximate processing has been reviewed for the STFT, but it can be applied to other signal processing operations. Most simply, for signal processing tasks that consist of a sequence of discrete Fourier transforms, the approximations described above can be applied to each Fourier transform. One example is the process of forming a synthetic aperture radar image from a sequence of complex-valued pulses corresponding to the reflections of radar pulses (Section 6.6). The operation contains two-dimensional Fourier transforms, interspersed with transposition of matrix rows and columns, and a crude image may be produced for rapid analysis. The introduction of a decision element, by which such a crude image is examined and a decision made as to whether the crude image has enough features to merit the expenditure of further computing resources, leads to *decision-directed incremental refinement*, where the decision can be manual (i.e., by human observation) or automatic.[7] Finally, recent work has described a means to apply incremental refinement to operations beyond the STFT, including binary hypothesis testing for signal presence using the FFT, image decoding with the discrete cosine transform, and frequency-selective filtering.[8]

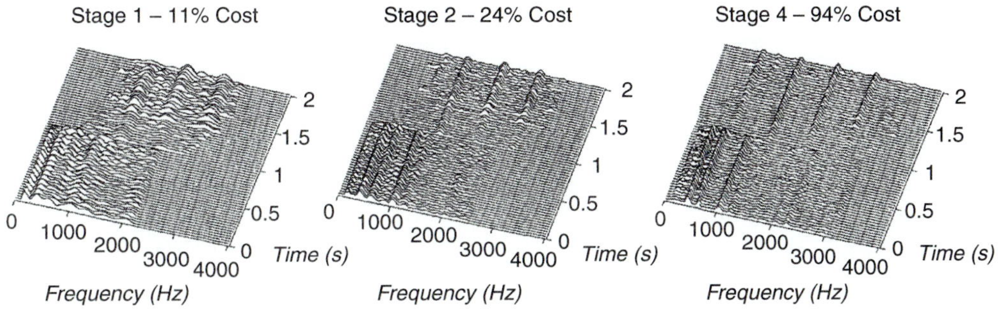

Figure 1–18 Incremental refinement of a four-stage STFT of a two-second violin passage consisting of two tones. At 11 percent of total computational cost, the structure of the spectrum is apparent, and the results approach the final quality at subsequent stages.[6] © 1995 IEEE.

REFERENCES

1. Alan V. Oppenheim and Ronald W. Schafer, *Digital Signal Processing* (Englewood Cliffs, NJ : Prentice Hall, 1975).
2. Charles S. Burrus, "Block Realization of Digital Filters" *IEEE Trans. Audio and Electroacoustics*, AU-20, (October 1972), 230–235.
3. J. W. Cooley and J. W. Tukey, "An Algorithm for the Machine Calculation of Complex Fourier Series," *Math. Computation*, 19 (1965), 297–301.
4. A. V. Oppenheim, S. H. Nawab, G. C. Verghese, and G. W. Wornell, "Algorithms for Signal Processing" *Proc. First Annual RASSP Conference*, Arlington, VA, August 1994, pp. 146–153.
5. S. Hamid Nawab and Erkan Dorken, "A Framework for Quality vs. Efficiency Tradeoffs in STFT Analysis" *IEEE Trans. Acoustics, Speech, and Signal Processing*, 43 (April 1995), 998–1001.
6. Joseph M. Winograd and S. Hamid Nawab, "Incremental Refinement of DFT and STFT Approximations," *IEEE Signal Processing Letters*, 2 (February 1995), 25–27.
7. Nikola S. Subotic and Brian J. Thelen, "Sequential Processing of SAR Phase History Data for Rapid Detection" *Proc. IEEE Inter. Conf. Image Processing*, Washington, DC, October 1995, vol. I, pp. 144–146.
8. Joseph M. Winograd and S. Hamid Nawab, "Probabilistic Complexity Analysis for a Class of Approximate DFT Algorithms," *J. VLSI Signal Processing Systems for Signal, Image, and Video Technology*, 14 (November 1996), 193–205.

2

Processors

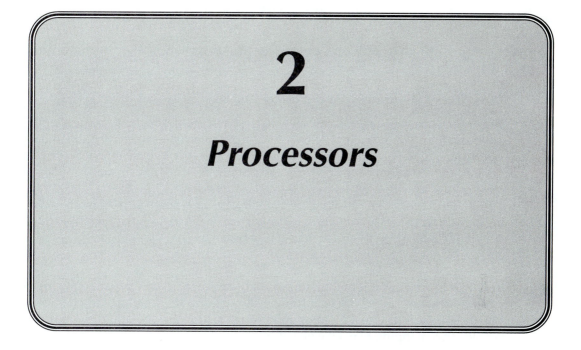

Signal processing is carried out by performing arithmetic operations on representations of the signal according to the prescriptions of an algorithm. In this chapter, we step through an evolutionary description that begins with simple logic circuits and builds up to the description of a general purpose processor. We examine the general-purpose processor in terms of the requirements of signal processing algorithms and identify the bottlenecks of the general-purpose architecture to performing the tasks of real time signal processing. In doing that, we develop the description of the programmable signal processor. We conclude with an examination of a few common signal processing operations that are difficult even for the programmable signal processor, arriving at a discussion of specialized signal processing components that are realized in custom very large scale integrated circuits (VLSI).

2.1 DEFINITION OF PROCESSOR

Here, we concentrate on digital signal processing, where the signals are represented by sequences of numbers in a binary representation of 0's and 1's and where the processing is performed by digital logic elements (hardware). After briefly exploring the methods of representing numbers in binary form, we examine a hierarchy of increasingly complex structures to perform processing. We begin with the combinatorial logic element, then progress to the sequential logic element, and finally arrive at a generalized description of the programmable processor.

2.1.1 Forms of Binary Representation of Numbers

Digital (or switched) logic uses two values of voltage (or current) at each input and output wire of every logic element (or bit)—one represented by 1, the other represented by 0. Therefore, we use binary representations consisting of a set of B bits to represent numbers to perform the arithmetic of digital signal processing. For a set of B bits, 2^B different numbers can be represented.

In our discussions, we often refer to the same number in two different ways, one *logical* and the other *arithmetical*. The number x in binary notation is made up of a set of B bits, given by the vector relationship

$$x = (x_{B-1}, x_{B-2}, \ldots, x_0). \tag{2.1}$$

This is the logical representation, because it refers to the bit patterns within the vector that make up x. The arithmetical representation requires a scheme to map the bit patterns into a number upon which arithmetic operations can be performed. This mapping may take one of several forms, as we now describe.

In *fixed point representation*, the 2^B values that B bits can represent may all be positive (called *unsigned integer representation*). In this scheme, the value of x is given by

$$x = \sum_{b=0}^{B-1} 2^b x_b. \tag{2.2}$$

Alternatively, the numbers may be centered around zero, consisting of (nearly) equal numbers of positive and negative numbers. The most common representation, *two's complement*, represents 0 by 0000 (for $B = 4$), counts up in radix-2 fashion to a maximum value of $2^{B-1} - 1$ in the positive direction, and counts in the negative direction by forming the two's complement of the corresponding positive number. The two's complement of a number is formed by *complementing* (switching) all 0's to 1's, all 1's to 0's, and then adding one to the result.

Another method of representing numbers in fixed point is *sign-magnitude notation*. Here, the leading (leftmost) bit is 0 for positive numbers and 1 for negative numbers. The positive representation of a number is the same as for two's complement, and each negative number is formed by creating the corresponding number, then turning on the leading bit. Table 2–1 shows the decimal, unsigned integer, two's complement, and sign-magnitude representations of numbers available for $b = 4$. Two values for 0 are provided for several forms of representation—one as 0 and one as –0. The two values reflect the direction of approach to 0 for those schemes that can represent both positive and negative values—in either case, both the naturally occurring value of 0 in the series {2, 1, 0} and of –0 in {–2, –1, –0} is used.

Table 2–1 Three methods of allocating four bits to represent integers.

decimal	unsigned integer	two's complement	sign magnitude
–7		1001	1111
–6		1010	1110
–5		1011	1101
–4		1100	1100
–3		1101	1011
–2		1110	1010
–1		1111	1001
–0		1000	1000
0	0000	0000	0000
1	0001	0001	0001
2	0010	0010	0010
3	0011	0011	0011
4	0100	0100	0100
5	0101	0101	0101
6	0110	0110	0110
7	0111	0111	0111
8	1000		
9	1001		
10	1010		
11	1011		
12	1100		
13	1101		
14	1110		
15	1111		

In practice, most fixed point representations of speech, image, and audio-bandwidth signals range from 8 bits to 24 bits, with 16 bits being the most common. For $B = 16$, 65,536 values, ranging from –32,768 through 32,767 inclusive, may be represented in two's complement notation.

Numbers that are fractions are represented in fixed point notation by multiplying them by a constant value. If this value is a power of two, for example 2^F, this multiplication by a constant is analogous to defining a binary "decimal point" (or more properly, radix) and allocating the right F bits to represent the fraction. Thus, if $F = 14$ and $B = 16$, the range of values may be represented in the form *xx.xxxxxx xxxxxxxx*, covering a range in two's complement from 10.000000 00000000 (–2.000000 decimal) to 01.000000 00000000 (1.999939 decimal).

Hexadecimal notation is frequently used to abbreviate the lengthy binary representation of a number to a shorter base-16 version by replacing the 16 values of a four-bit binary field. Hence, (0000, 0001, ..., 1110, 1111) is replaced with (0, 1, ... *e*, *f*). Thus, for $B = 16$, the range of values represented in two's complement is from 8000*H* to 7*fff**H*, where *H* indicates hexadecimal representation.

In fixed point representation, the decimal point position remains fixed within the bit field. Thus, for our example of $F = 14$ and $B = 16$, it is impossible to represent any number larger than 2.000. Therefore, adding two valid numbers, for example $1.5 + 1.5 = 3.0$, results in an invalid number that is beyond the range of this fixed point representation.

Floating point representation in effect moves the position of the binary decimal point to change the representation as a function of the number being represented, using a few of the B bits to define where the binary point has been placed. In floating point representation, the B-bit field is divided into three subfields: sign (s) of width b_x, mantissa (m) of width b_m bits, and characteristic (c) of width b_c bits, where $b_s + b_m + b_c = B$. Thus, a number n is represented by

$$n = (-1)^s \cdot m \cdot 2^c. \tag{2.3}$$

The mantissa is normalized to lie within a certain range, usually $1 \leq m < 2$.

The format defined by the floating point standard IEEE-754[1] provides an example of floating point representation. This standard is used in most floating point representations on general purpose computers. It is also used in many floating point digital signal processor integrated circuits. Many other floating point digital signal processors, while using their own internal representation of floating point numbers, provide conversions to and from their own format to the IEEE standard. In the IEEE-754 representation, $B = 32$, $b_m = 23$, $b_s = 1$, and $b_c = 8$ (Fig. 2–1).

Unsigned integer representation is used for the mantissa (the sign bit is placed to the far-left position) and two's complement representation is used for the characteristic. Because it is required that $1 \leq m < 2$, only the fractional part of m is represented in its 23-bit field. The characteristic c is represented in two's complement fixed point notation.

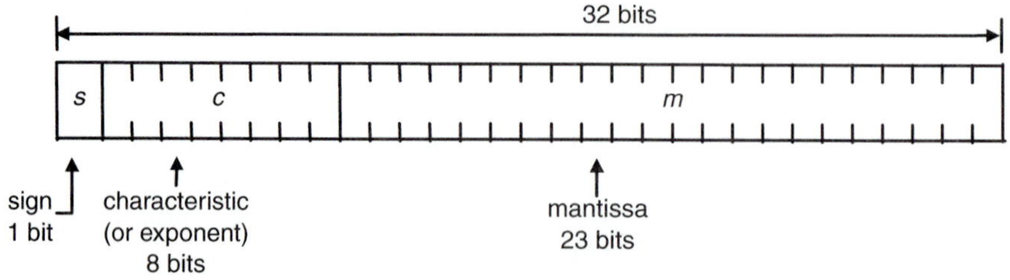

Figure 2–1 Allocation of bit fields within the IEEE-754 floating point representation.

If m and c are expressed as unsigned integers, the decimal number n can be computed from s, c, and m as:

$$n = (-1)^s \cdot (2^c - 127) \cdot [(1 + m) \cdot 2^{-23}]. \quad (2.4)$$

For this representation, n must lie in the range $-1.701 \times 10^{38} \leq n < 1.701 \times 10^{38}$ ($m = \pm 10^{23} - 1$ and $c = 255$). The smallest fraction that can be represented is $\pm 7.006 \times 10^{-46}$ ($m = 000\ldots01$ and $c = 0$).

Another method of representing numbers, known as the *residue number system*, is convenient for applying parallel processing to basic arithmetic instructions. The residue number system is described more fully in Section 4.1.3.

2.1.2 Combinational Logic

We recall the general description of a signal processing system that receives a set of signals $x_i(n)$ and transforms them to output another set of signals $y_j(m)$:

$$y_j(m) = T_j[x_i(n)]; \; 0 \leq i \leq I - 1; \; 0 \leq j \leq J - 1. \quad (2.5)$$

We look at the specific case of one input and one output signal ($I = J = 1$), suppress the subscripts i and j, and we specify that the input sampling period, T_x, equal the output sampling period T_y. We describe input and output with the common variable t:

$$y(t) = T[x(t)]; \; 0 \leq t < \infty. \quad (2.6)$$

We further specify that we look only at causal systems, that is, systems for which for each time t_c the output $y(t_c)$ depends only upon the present input and previous time's inputs ($t \leq t_c$). (The output of a non-causal system could depend upon the entire range of t, both $t \leq t_c$ and $t > t_c$.) We can write (2.6) to show this dependence by separating t into the current time, t_c, and all previous times $[0, t_c - 1]$:

$$\begin{aligned} y(t_c) &= T[x(t)]; \; 0 \leq t < \infty, \\ &= T[x(t_c), x(t_c - 1, \ldots, x(1), x(0)]. \\ &= T[x(t_c), F(t < t_c)]. \end{aligned} \quad (2.7)$$

We further identify the special case of (2.7) for which $F(t < t_c) = 0$, that is, the outputs $y(t_c)$ depends only upon the current inputs $x(t_c)$. These systems are known as *combinational* systems, and they have their outputs determined by a combination of input signals. The dependence of an output upon only the current input requires that the combinational system has no logical feedback loops within it; in tracing a signal from input to output, one cannot pass the same internal point more than once. We depict a combinational system with I inputs and J outputs as shown in Fig. 2–2.

A logic function may be specified for a simple logic element consisting of two inputs, x_0 and x_1, and one output, y_0 (Fig. 2–3).

There are $2^2 = 4$ possible combinations of input x_0 and x_1, and each input combination yields one output value. Thus, we can define a logic function by list-

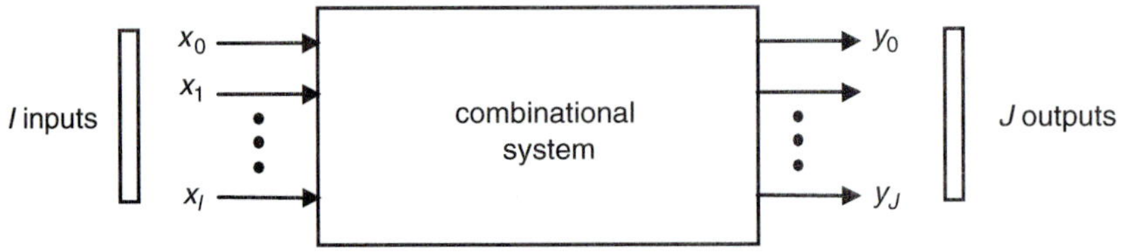

Figure 2–2 Combinational logic system.

ing the output of the logic element that results from each of the four input combinations. For this set of four input combinations, there are sixteen possible ways that the output may respond. This is because there are sixteen ways that 0's and 1's (the allowed output value) may be distributed into four bins (the four input combinations). A *truth table* may be constructed to present an exhaustive list of all possible logic mappings (Fig. 2–4).

Six of these sixteen mappings are of special significance and are given the names AND, XOR, OR, NOR, EQUIV, and NAND. We extract these particular combinations from Fig. 2–4 and place them in Fig. 2–5, where in addition we also place the one-input, one-output NOT operation. At the bottom of Fig. 2–5, we show the standard logic design symbol used to represent each function.

Note that the logic operations may be expressed by the arithmetic-like switching expressions as indicated in the figure. The most commonly used are $-x_0$ (for NOT x_0), $x_0 \cdot x_1$ (for x_0 AND x_1), and $x_0 + x_1$ (for x_0 OR x_1). These logic gates are the fundamental building blocks from which more complicated logic modules are built.

2.1.3 Sequential Logic

We return to the general description of a signal processing system T with input signal $x(n)$ and output signal $y(t)$, recalling the segmenting of t into the current time t_c and all previous time samples $[t = 0, 1, \ldots, t_c - 1]$:

$$y(t_c) = T[x(t_c), F(t < t_c)]. \tag{2.8}$$

For the combinational logic system, $F(t < t_c)$ was 0, causing the current output at time t_c to depend only upon the current input $x(t_c)$. In a *sequential* logic system, the output depends not only on the current input, but on past inputs, so $F(t < t_c) \neq 0$ for sequential systems.

Because the output of a sequential system depends upon a history of inputs, the results of past inputs must be stored somewhere. The need for storage introduces the need for a storage element. In a sequential circuit, a *register* serves as the storage unit. The register is a memory device that holds a vector of one or

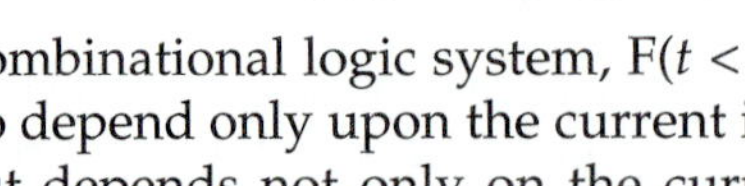

Figure 2–3 Combinational logic element with two inputs and one output.

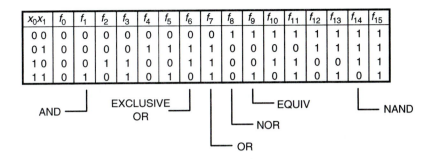

x_0x_1	f_0	f_1	f_2	f_3	f_4	f_5	f_6	f_7	f_8	f_9	f_{10}	f_{11}	f_{12}	f_{13}	f_{14}	f_{15}
0 0	0	0	0	0	0	0	0	0	1	1	1	1	1	1	1	1
0 1	0	0	0	0	1	1	1	1	0	0	0	0	1	1	1	1
1 0	0	0	1	1	0	0	1	1	0	0	1	1	0	0	1	1
1 1	0	1	0	1	0	1	0	1	0	1	0	1	0	1	0	1

AND (f_1), EXCLUSIVE OR (f_6), OR (f_7), NOR (f_8), EQUIV (f_9), NAND (f_{14})

Figure 2–4 The 16 possible mapping functions of two input variables, with designation of those used as logical building blocks.

more bits until they are changed by new register inputs. In Fig. 2–6, the input register holds the bits that enter the combinational logic module, and the output register holds the results. Each register has a control signal c that causes whatever logic level (0 or 1) that is present on the input of the register at the time that control signal c_{in} or c_{out} is asserted to be frozen on the register output.

In typical use, a series of combinational circuits sequentially applies several stages of combinational logic, with registers separating each combinational block (Fig. 2–7).

At each clock cycle, the output of a combinational logic block is clocked into a register and presented as input to the next combinational logic block. Sequential logic circuits fold this diagram onto itself and introduce a feedback path, interrupted with a clocked register set, from the output to the input of a combinational logic circuit (Fig. 2–8).

Most generally, the output of a sequential logic circuit depends upon all past and present inputs. However, in practical situations, it is not necessary to memorize the complete history of $x(t)$. Instead, values of $x(t)$ can be grouped into classes so that all $x(n)$ sequences that have the same effect on the output $y(t)$ at time t are in the

Input x_0	*Input* x_1	AND $x_0 \cdot x_1$	EXOR $x_0 \oplus x_1$	OR $x_0 + x_1$	NOR $-(x_0 + x_1)$	EQUIV $x_0 = x_1$	NAND $-(x_0 \cdot x_1)$	NOT $-x_0$
0	*0*	0	0	0	1	1	1	1
0	*1*	0	1	1	0	0	1	1
1	*0*	1	1	1	0	0	1	0
1	*1*	1	0	1	0	1	0	0

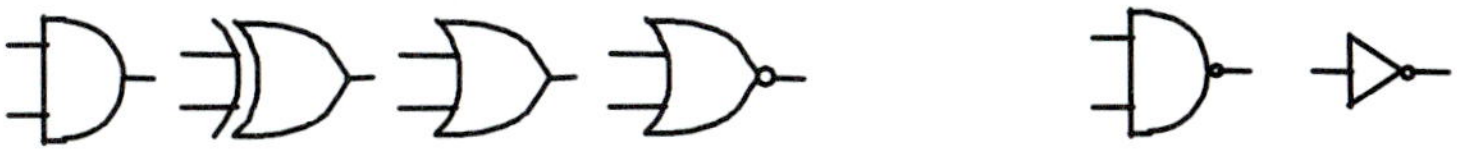

Figure 2–5 Truth tables for standard logic building blocks, extracted from Fig. 2–4.

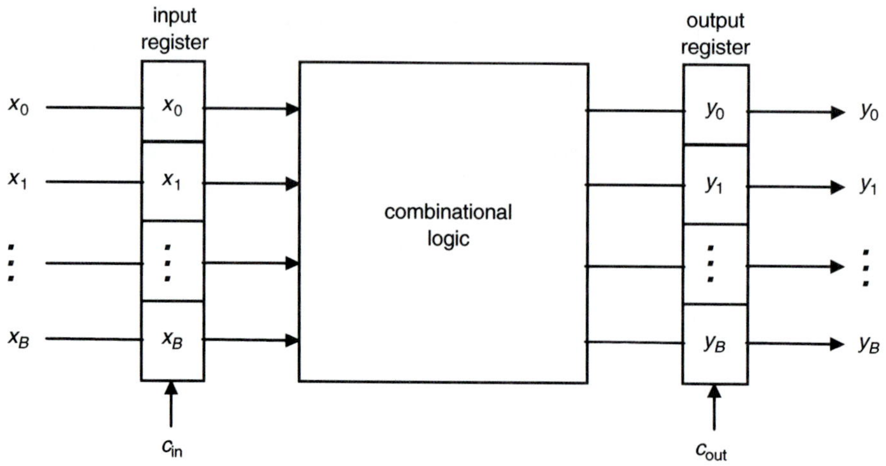

Figure 2–6 Sequential logic consists of combinational logic module preceded and followed by clocked registers.

same class. The class is called the *state* and is represented by the variable s. Using the state, the output at time t depends on the state at time t and the input at time t:

$$y(t) = T[s(t),x(t)]. \tag{2.9}$$

The state changes based on its current value and the value of the input $x(t)$. A *state transition function* may be defined to provide the next state, based on the present state and present input:

$$s(t + 1) = G[s(t),x(t)]. \tag{2.10}$$

Figure 2–9a contrasts the general dependence of a sequential logic circuit upon the complete history of input signal $x(t)$ to the simplification offered by the use of a state $s(t)$ to capture the history of $x(t)$ (Fig. 2–9b). As shown in Fig. 2–9b,

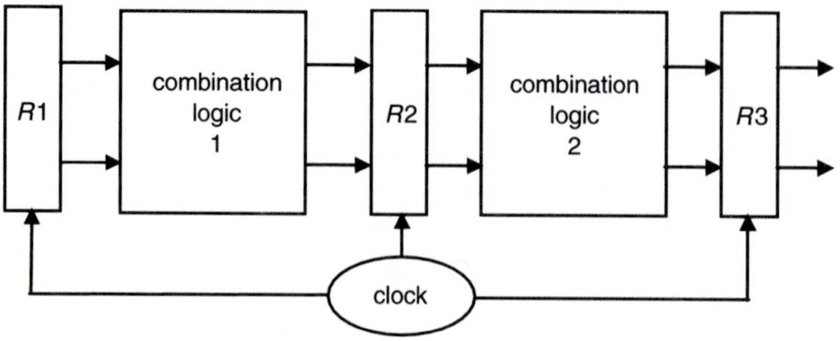

Figure 2–7 Sequential circuit built from sequence of combinational logic units, separated by registers R.

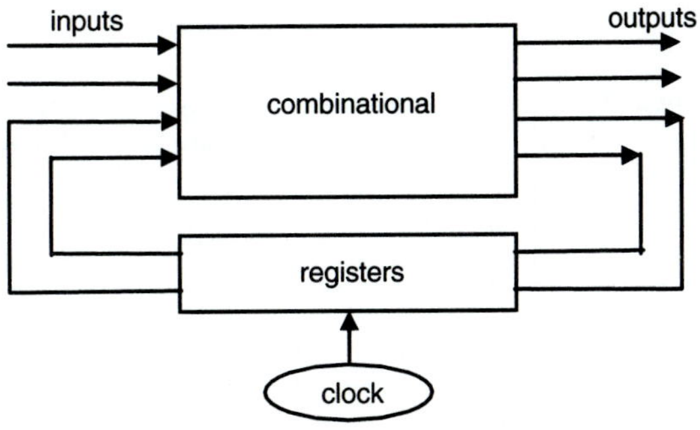

Figure 2–8 Sequential logic folds diagram of Fig. 2–7 upon itself and inserts feedback path.

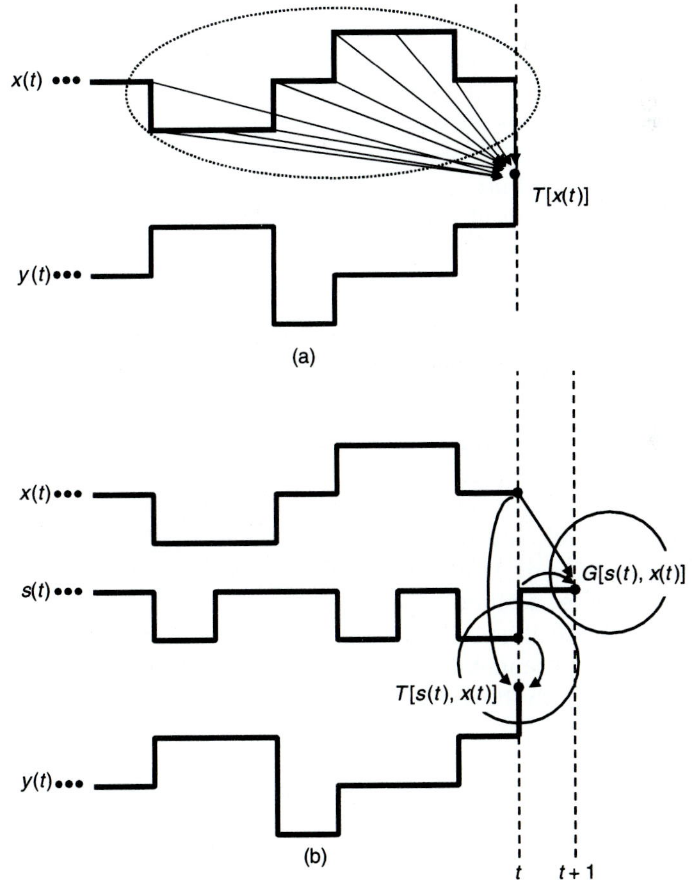

Figure 2–9 (a) The general dependence of sequential logic circuit output y on inputs x at all times; (b) capturing past history of x in state s using state transition function G and output function T.

Table 2–2 Table specifies finite-state system as next state $s(t+1) = s_{ij}$ and output $y(t) = y_{ij}$ as function of input state $s(t)$ and input value $x(t)$.

	$s(t+1), y(t)$		
	$x(t) = x_0$	$x(t) = x_1$	...
$s(t) = s_0$	s_{00}, y_{00}	s_{01}, y_{01}	...
$s(t) = s_1$	s_{10}, y_{10}	s_{11}, y_{11}	...
.	.	.	
.	.	.	
.	.	.	

the state transition function $G[s(t), x(t)]$ is updated to a value at time $t + 1$ as a function of $s(t)$, its state at t, and $x(t)$, the input at time t.

The number of states in realizable systems is finite; hence these are called *finite-state* systems. We can specify the properties of a finite-state system in several ways. We use the input sample value $x(t)$ and the present state value $s(t)$ to produce a sample output $y(t)$ and a next state $s(t + 1)$. One method of specification uses a table that presents $y(t)$ and $s(t + 1)$ for all combinations of $x(t)$ and $s(t)$ (Table 2–2).

The next state and output table entries s_{ij}, y_{ij} are subscripted with the current state i and current input j to represent their dependence on these parameters. Possible values of $x(t)$ are given by $x_0, x_1, \ldots$. To provide a specific example, we choose the example of a circuit that has a single input x that can take one of two values, 0 or 1. The output is a single signal that can also take on values of 0 or 1. The circuit begins operation at time $t = 0$. The circuit output at time t is 1 if the number of inputs of value 1 is even, and it is 0 otherwise. The circuit has two states: EVEN, if the number of 1's that has arrived is even, and ODD otherwise. The circuit need not memorize the entire history of inputs to determine whether an even number of 1's has arrived; it merely records that information through the value of its state. This circuit is represented by an instance of Table 2–3.

Another method to specify a sequential system is through the use of a *state diagram*. Here, the nodes (circles) correspond to the states $s_0, s_1, \ldots$. For each input value to a present state, an arc is drawn to show the next state, and the arc is labeled with both the input value x and resulting output value y (Fig. 2–10).

Table 2–3 Table specification of transition function and output for example circuit.

	$s(t+1), y(t)$	
	$x(t) = 0$	$x(t) = 1$
$s(t)$ = EVEN	EVEN, 1	ODD, 0
$s(t)$ = ODD	ODD, 0	EVEN, 1

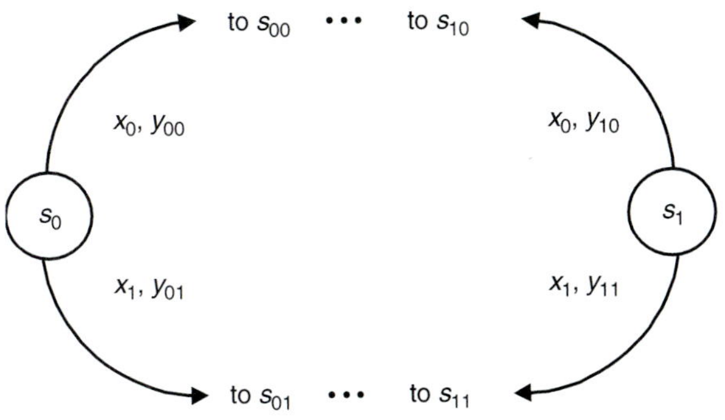

Figure 2–10 State diagram representation of sequential system: For a specific state diagram, the free arrows "to s_{ij}" would be attached to the next state specified in the state table for current state s_i and current input j as shown in the example in the next figure.

For the example of the even/odd circuit, the state diagram representation is shown in Fig. 2–11. The input to a state and the resulting output symbol are displayed in the format *output/input.* Thus, a state that emits a 1 upon receiving a 0 is labeled 1/0.

2.2 GENERAL PURPOSE MICROPROCESSOR

The program-controlled logic module, more generally called the programmable processor, is an extension of the combinatoric and sequential logic modules discussed in the previous section. To migrate to a program-controlled module, the sequential logic circuit (Fig. 2–8) can be modified by:

- supplementing some of the registers with much larger memory;
- replacing the clock with a control circuit, known as the control unit;
- injecting the signals from this control unit into each of the blocks that comprise the circuit and linking an output from memory to the control unit.

These changes provide the schematic of a programmable processor (Fig. 2–12).

The control points provide the control unit with direct input into the regions of the processor. The control point for the combinational logic unit, for ex-

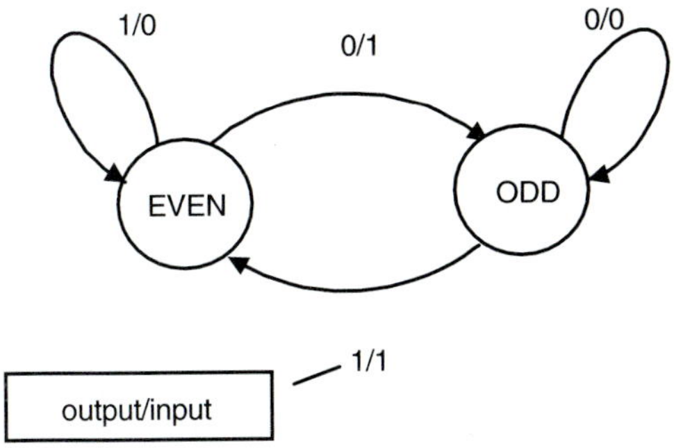

Figure 2–11 State diagram representation of EVEN/ODD example described in text.

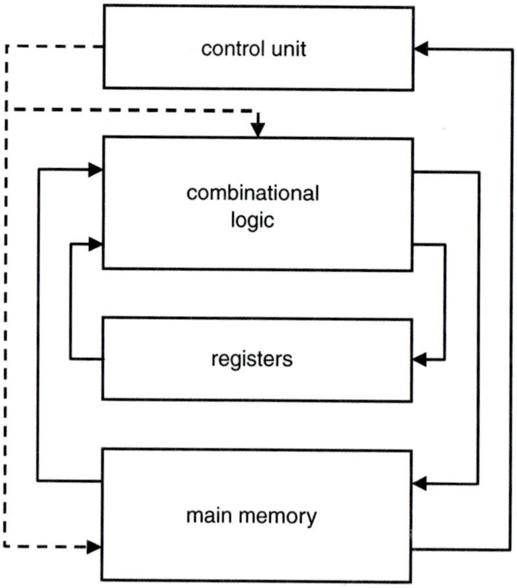

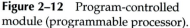

Figure 2–12 Program-controlled module (programmable processor).

ample, provides the ability to apply input data from registers or memory to one of a variety of combinational circuits, producing a choice of combinationally-implemented arithmetic operations. Thus, the simple free-running clock signal of the sequential logic circuit has now been replaced by a multi-signal, multi-destination control unit.

2.2.1 Description

The programmable processor is used for systems that are too complicated for implementing with combinational or sequential circuits. This complexity is either due to the numbers of inputs and outputs, the number of states, or the complexity in representing their functions. For the programmable processor:

- the state may be decomposed into a control component and a data component;
- the data transformation, being too complex to perform in one step, is decomposed into a sequence of simpler operations, with the sequence controlled by the control component.

Both the control and the processing units are state machines. A fuzzy boundary exists between the control and processing units. Processing units have a greater number of internal states with a greater regularity; control units have a lesser number of states that are more diverse and less regular.

Fig. 2–13 expands the processing unit of Fig. 2–12 into the structure typical of processors. The processing unit has been divided into a logical unit and an

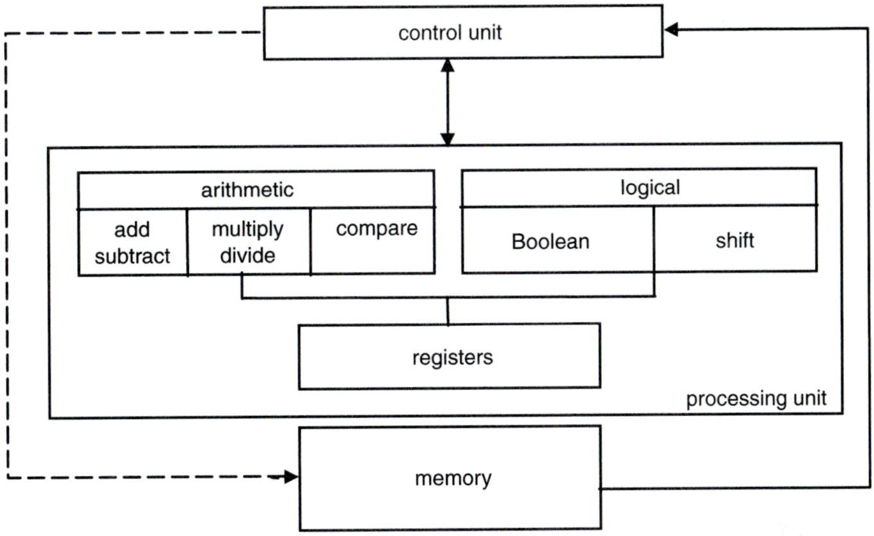

Figure 2–13 Basic structure of programmable processor.

arithmetic unit. The logical unit performs Boolean logic on individual bits of a data word and executes shifts upon the entire data word, and the arithmetic unit performs arithmetic on the word-based data values presented. The two units are often combined into the single name of arithmetic logic unit, or ALU.

The instructions that control the programmable processor require multiple steps to execute. First, the address of the next instruction is accessed, and the contents of program memory at that address are read and decoded into signals that control the processor logic elements. In the next step, one or more operands are then fetched to provide data to the instruction, and instruction execution then begins in subsequent steps. Finally, the results of the instruction are stored and the address of the next instruction is computed. Because each instruction may take several clock cycles, we must distinguish between a clock cycle (the period of the incoming clock) and the instruction cycle (duration of the shortest instruction, which is usually several clock cycles). Arithmetic instructions may require a varying number of clock cycles to complete. Simple integer addition or subtraction and compare operations are among the fastest to execute. Floating point arithmetic, particularly additions and subtractions, as well as both integer and floating point multiplication and division, can require multiple instruction cycles to complete. Similarly, while Boolean operations occupy one instruction cycle, left or right shifting of a word by multiple bits usually occurs at one instruction cycle per bit, demanding multiple cycles for multiple shifts.

The arithmetic unit may be supplemented by an accelerator to perform some of its operations more quickly. For example, some microprocessors include a floating point arithmetic unit to speed floating point operations, and in a programmable signal processor, a hardware multiplier is added to the arithmetic unit to perform rapid multiplications. Such accelerators are placed in parallel with the arithmetic unit; instructions that require accelerator functions are passed, with the appropriate data values, to the accelerator by the arithmetic unit and the results are then passed back.

Two types of registers remain close to the arithmetic and logic unit. Dedicated registers include:

- *program counter:* points to the memory location of the next instruction to be executed, used to control program flow;
- *flag register:* consists of several one-bit flags that indicate whether the result was negative, zero, overflowed, etc., used to control execution by enabling such data-dependent program tests as Jump on Zero;
- *stack:* stores the memory location of a first-in, last-out memory array that is used to store the machine state (i.e., the contents of all registers) before jumping to a new location of execution, allows resumption of computation in mid-stream when the routine that was jumped to is completed.

In addition, several general purpose registers provide and receive data from the ALU. Because of their proximity (usually on the same chip), communication with these registers is faster than with the much larger amounts of memory off chip. The registers, ranging from 4 to 32 in number, are fully functional and capable of storing data, address, arithmetic, and logical operations, and a register may be specified as an operand in an instruction, speeding execution.

Memory is allocated to program and to data. Program memory contains the instructions that control the execution of the processor. These instructions may include data values (for operands) and memory addresses (for program jumps or data access). One type of processor instruction set, known as the Complex Instruction Set Computer or CISC, mixes simple single-cycle instructions with complicated multi-cycle instructions (such as the divide operation). More recently, Reduced Instruction Set Computers (RISC) have been used which optimize the speed of a much smaller set of single-cycle instructions and use these to build up operations for the more complicated instructions.[2] As more units, such as multiple processor units and accelerators, are added, the number of bits in an instruction increases to provide the necessary number of control signals. Very Long Instruction Word (VLIW) machines perform many operations in parallel.

The *Von Neumann* architecture, typical of general purpose processors, uses the same memory for both program and data. Program and data information are usually segregated into different address regions of memory. The data memory is both read and written with values during execution. The contents of program memory, except for the rare exception of self-modifying code, is only read, not written, during program execution.

Values in data memory are accessed in a variety of addressing modes:

- *immediate:* data value is contained as part of the instruction (from program memory);
- *direct:* the address of the data value is contained in the instruction;
- *indirect:* the instruction indicates another location, usually a register, that contains the address of the data.

Other more specialized addressing modes, in which an address is calculated relative to the value contained in a page register, are typical of some machines.

Processor memory is arranged in a hierarchical manner, from small amounts of memory that are accessed in less than one instruction cycle, to large amounts of memory that require more time for access. In increasing order of size and decreasing order of speed, processor memory can be contained in:

- *registers:* up to about 32 locations directly available to the ALU;
- *cache:* local high-speed memory containing the most recently accessed instructions and data from main memory, exploits *locality of reference*, by which most instructions (such as loops) are executed repeatedly and can be kept close and quickly available;
- *on-chip memory:* general purpose memory for program or data on chip;
- *off-chip memory:* much larger memory not constrained by chip capacity, but requiring a full multi-clock-cycle memory access to obtain;
- *bulk memory:* electromechanical storage such as magnetic disk or CD-ROM that contains orders of magnitude more memory than the off-chip memory array.

With these general concepts of the microprocessor architecture in mind, we proceed to a discussion of the Intel x86 family of microprocessors as a sample architecture to illuminate these concepts.

2.2.2 Design Example: Intel 8086 Microprocessor

The Intel 8086, one of the first 16-bit microprocessors, is the first of a family of software-compatible microprocessors of increasing speed that has since evolved. The 8086 was introduced with a full set of hardware and software support. Its alternate configuration, known as the 8088, became the basis of the IBM-compatible personal computer family. We will use the 8086 and the rest the members of the Intel x86 family to

- map the microprocessor architectural features discussed above to specific architectural elements of a common microprocessor;
- describe the features of the Intel microprocessor family at a high level;

- follow the evolution of the Intel x86 microprocessor to exemplify how the speed of a computer architecture may be accelerated through careful design and the advances in speed and density of semiconductor technology.

The Intel 8086 microprocessor was introduced in 1978. Fig. 2–14 shows the block diagram of the 8086, and Fig. 2–15 indicates its use in a system, where it is augmented with a clock chip (the Intel 8284) and address bus latches (8282). The

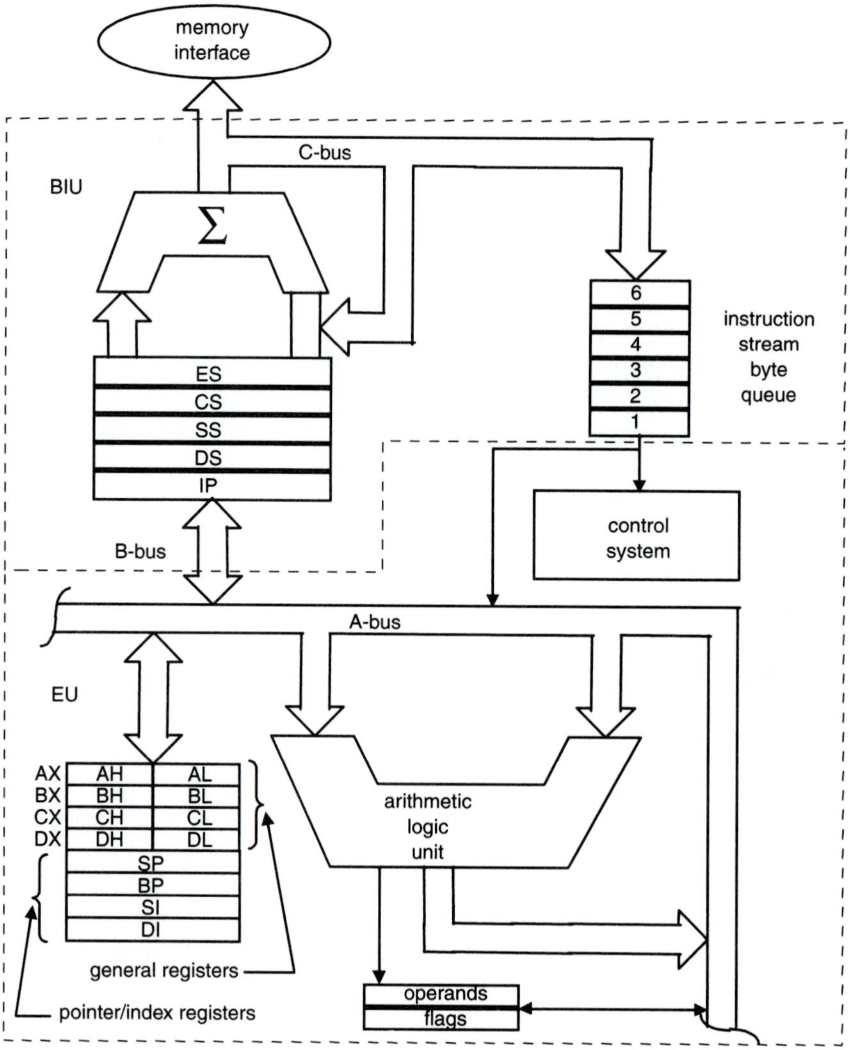

Figure 2–14 Block diagram of Intel 8086 16-bit microprocessor.[3] © 1979 Intel Corporation; used with permission.

processor consists of two processing units, the Bus Interface Unit (BIU) and the Execution Unit (EU). The BIU, which generally operates separately from the EU, performs bus operations and arithmetic to fetch operands and instructions for the EU. It provides bus control signals by means of its C-bus. The address bus of the 8086 consists of 20 bits, allowing access to 1 Mbyte of memory locations. However, the internal registers of the 8086 are only 16 bits wide. Full 20-bit addresses are calculated as the sum of a *segment register*, which is applied to the upper 16 bits of the 20-bit field, and an *offset* of 16 bits aligned to the lower 16 address bits. The segment registers are known as Extra Segment (ES), Code Segment (CS), Stack Segment (SS), and Data Segment (DS). For example, the address of the next instruction is computed from the CS register, left-shifted by four bits, plus the instruction pointer (IP) register, resulting in an address of [16 × (contents of CS)] + IP (Fig. 2–16). This is represented by CS:IP.

The BIU also prefetches the next instruction while the current instruction is executing. The multiple bytes of the prefetched instruction are stored in the instruction byte queue. In this manner, the BIU fulfills the function of the control unit shown at the top of Fig. 2–13.

The Execution Unit (EU) corresponds to the Processing Unit of Fig. 2–13, providing both the arithmetic and the logic functions of the units described in that figure. The EU contains 10 registers that are in three groups. The first is the general group, consisting of AX (accumulator), BX (base), CX (count), and DX (data). These names indicate the processing use of each register made by the in-

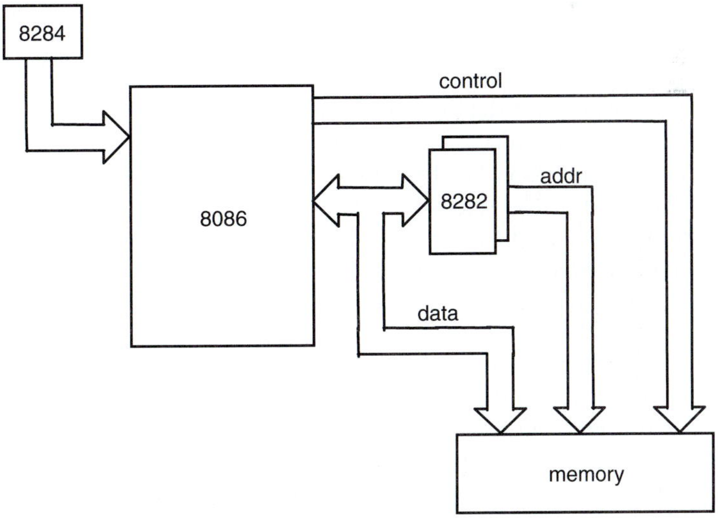

Figure 2–15 Microprocessor system combines the 8086 with clock (8284) and address latches (8282).[3] © 1979 Intel Corporation; used with permission.

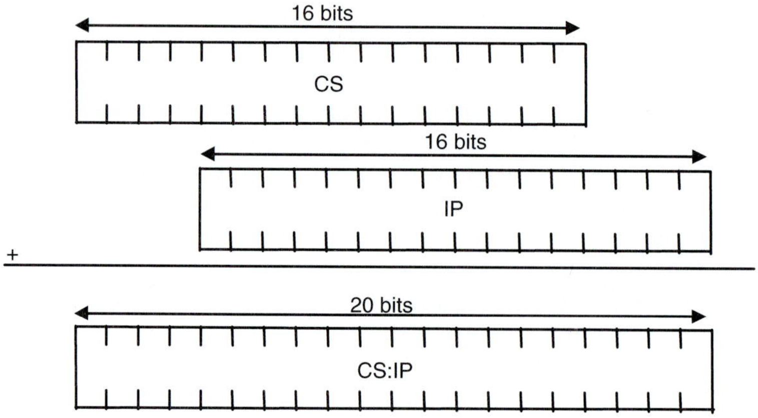

Figure 2–16 To accommodate 20 bits of address within a 16-bit architecture, the 8086 combines a 16-bit segment address (CS) as the upper 16-bits of address, with a lower 16-bit portion (IP).

struction set: accumulating a sum, providing a base address for index calculations, counting the iterations through a loop, and storing data, respectively. Each of the general registers is 16 bits wide, but the high byte and low byte of each can be addressed separately. The pointer/index registers make up the second group. The stack pointer (SP) stores the address of a first-in, last-out segment of consecutive memory locations used to store machine information (such as the return address) upon jumping to a subroutine. The base pointer (BP) is used in address calculations for data values, and the source index (SI) and destination index (DI) provide source and destination addresses for moves of large segments of consecutive memory locations from a source to a destination area of memory. Finally, the flag register serves as the third type of register.

The 16-bit BIU segment registers are combined with the 16-bit EU registers to provide full 20-bit addresses. Several addressing modes are possible to access operands in memory, made up of combinations of one segment register ("seg"), a displacement value contained in an instruction ("disp"), and one or more EU registers (Table 2–4).

The EU performs operations on data accessed via the above addressing modes by means of the arithmetic logic unit, or ALU. The ALU provides the standard arithmetic operations of add, subtract, multiply, and divide. It provides logical operations of one operand (e.g., NOT, left or right shift) and of two operands (e.g., AND, OR, XOR). The ALU also provides string manipulation operations that allow moving the contents of consecutive memory locations from one block of memory to another, or that translates one byte value for another by means of a lookup table. Finally, instructions are provided that direct the flow of execution, such as subroutine calls and returns, and jumps, conditional jumps, iteration con-

Table 2–4 Example of addressing modes that combine segment, displacement, and EU registers.[4]

Mode	Example
Base index	BX + SI* + disp + seg
Index	SI + disp +seg
Base	BP + disp + seg
Base index (no displacement)	BX + SI + seg
Indirect	SI + seg
Relative	disp + IP
Direct	address + seg

* DI may be substituted for SI throughout

trol, and interrupt handling. A few processor control instructions allow setting flag register values to cause the processor to wait.

2.2.3 The Intel x86 Microprocessor Family

The 8086 instruction set was designed to obtain maximum power and flexibility from a constrained number of transistors, 29,000, which was at the forefront of integrated circuit technology in 1978. In the ensuing years, Intel Corporation has introduced a sequence of microprocessor designs that have increased speed by more than 4,000 times while maintaining full software compatibility with the 8086. Fig. 2–17 shows the progress in speed of the Intel x86 architecture as function of time, beginning with the 8086 in 1978. The original 8086 operated at 0.33 million instructions per second (MIPS) at a clock rate of 5 MHz; the Pentium II ('686 in the numerical series, although Intel abandoned this numbering scheme in favor of names) operates at an effective 1400 MIPS, over 4,000 times that of the 8086. The fastest Pentium II runs at a 450 MHz clock rate, or 90 times that of the 8086. It requires over 19 million transistors of logic. In the future, the Merced (P7, or '786) processor moves from 32-bit to 64-bit architecture and provides another factor of four speedup.

To understand the architectural techniques available to accelerate execution speed, it is instructive to note that the 4100-fold increase in throughput of this example is made up of a 90-fold increase in clock speed and a 45-fold factor that arises from architectural enhancements. Table 2–5 describes the x86 family of processors, comparing the speed of each to that of the 8086 and dividing the speed into a clock speed factor and an architectural factor. The clock speed factor reflects the simple increase in clock speed—a processor running at a 10 MHz clock rate operates two times faster than the same one operating at 5 MHz. The architectural factor indicates changes that are independent of clock speed, such as adding two execution units instead of one to increase throughput. The product of

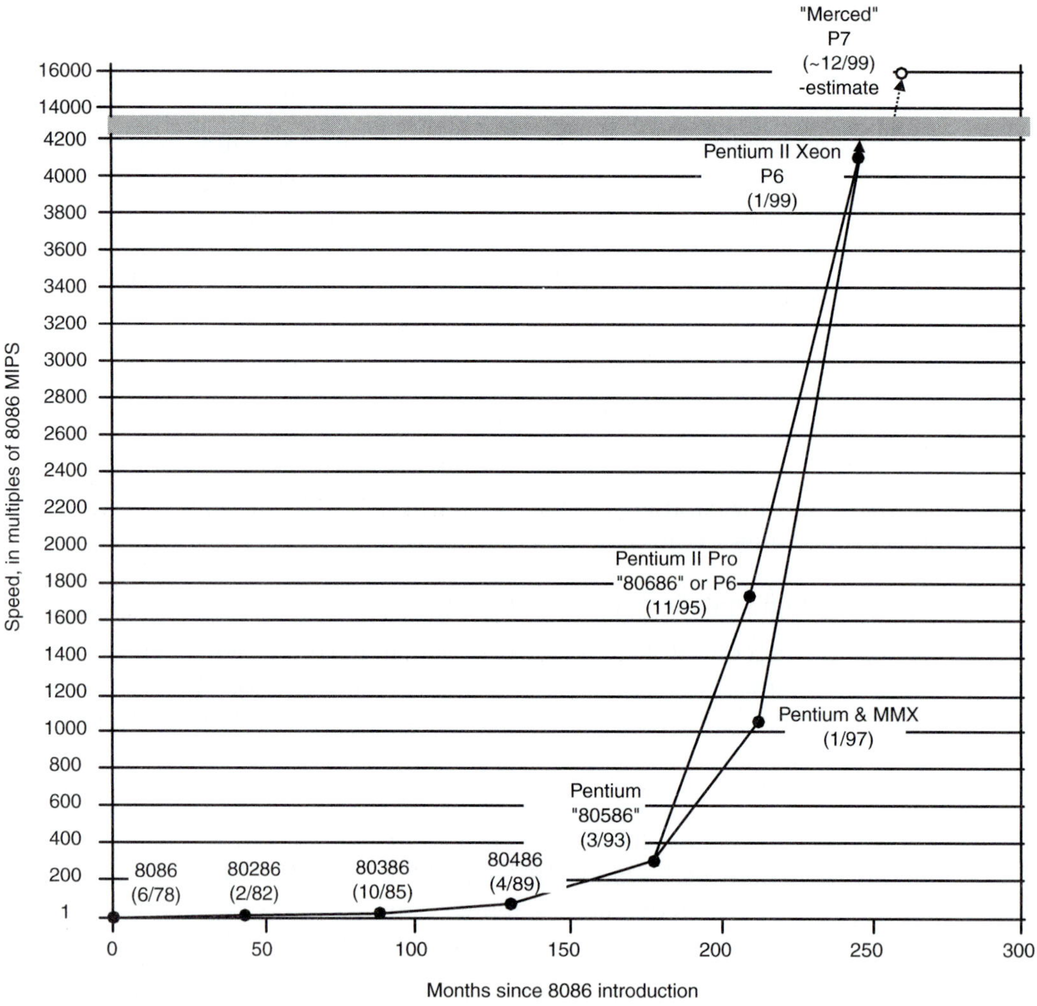

Figure 2–17 Speed of Intel x86 family of microprocessors vs. months since introduction of 8086.

the clock factor and architectural factor yields the overall speed increase. The far-right column indicates the architectural enhancements leading to major increases in speed.

The architectural enhancements of Table 2–5 may be grouped into several categories for discussion. The acceleration techniques, exemplified here for the x86 architecture, are generally applicable to other designs, including programmable digital signal processors and custom-designed integrated circuits.

Table 2–5 Architectural factors contributing to performance evolution of x86 processor family.

Processor (date of introduction)	Clock (MHz)	Speed (vs.5 MHz 8086)	Clock acceleration (vs. 5 MHz)	Architecture acceleration (vs. 5 MHz 8086)	Basis for speedup
8086 (6/78)	5	1.0	1.0	1.0	• 1 Mbyte memory address space • 29K transistors
80286 (2/82)	6	2.7	1.2	2.3	• virtual memory support • 16 Mbyte memory space • 134K transistors
386DX (10/85)	16	17	3.2	5.2	• 32 bit microprocessor with 16 bit external bus • pipelined instruction execution • 4 Gbyte memory space
486DX (4/89)	25	61	5.0	12	• on-board cache (8 Kbyte) • integrated floating point unit • effective 1-cycle execution for many instructions
486DX4 (3/94)	100	210	20	11	• doubler technology - microprocessor runs 4X bus speed
Pentium (3/93)	60	300	12	25	• 64 bit external, 32 bit internal bus • superscalar architecture (< 1 clock cycle per instruction) • separate on chip caches for code & data • more hardwired instructions • branch prediction
Pentium II "Pro" (11/95)	150	990	30	33	• 256 Kbyte secondary cache in dual cavity package with CPU • 5 parallel execution units • 14 stage pipeline • speculative and out-of-order execution • CISC/RISC hybrid

(continued)

Table 2–5 *Continued*

Processor (date of introduction)	Clock (MHz)	Speed (vs.5 MHz 8086)	Clock acceleration (vs. 5 MHz)	Architecture acceleration (vs. 5 MHz 8086)	Basis for speedup
Pentium with MMX (1/97)	200	1040	40	26	• multimedia instructions • 8 byte-addressable subfields in 8 64-bit registers • new instructions to support SIMD
Pentium II (5/97)	266	1750	53	33	• MMX capability added to Pentium Pro (P6) • cache placed on processor chip • dual independent bus architecture—dedicated bus for secondary cache • dynamic execution—multiple branch prediction, data flow analysis, and speculative execution
Pentium II "Xeon" (1/99)	450	4100	90	46	• on-chip secondary cache (vs. dual cavity package) • clock speedup
Merced (12/99 – est.)		*~16000*			• *compiler identification of parallelism relieves burden on hardware* • *predication of branches—execute all branches, then choose answer* • *speculative data fetching* • *support for massive parallelism*

2.2.3.1 Clock Speed. Processor throughput may be increased by increasing the clock speed of overall operation. Maximum clock speed is limited by propagation delay, integrated circuit layout geometries, and signal-to-signal relative timing within the integrated circuit. Some increase in maximum clock speed may be possible by testing and selecting components from the production batch, which will have a distribution of clock speeds across its components. Generally, a redesign of the integrated circuit layout, usually in a technology with smaller design rules, is necessary to reduce the dimensions and associated signal delays, thereby increasing the maximum clock speed. Signal delays on the integrated circuit chip can impose a limit. The emergence of *clock skew*, by which signal transitions that are simultaneous at one place on the chip become out of synchronization elsewhere on or off the chip, limits clock rate over longer transit distances, particularly for chip to chip on a circuit board. Thus, the more function that can be placed within the chip, the greater speed that is possible. To mitigate chip-to-chip clock skew, Intel introduced its *clock doubler* technology in its 486DX2 and 486DX4, which doubled (or quadrupled) the on-chip clock speed while keeping the off-chip clock speed lower to match board-level timing constraints. The concept of adding another chip within the same package, and dedicating a separate full-speed bus for access to it, is exemplified by the first version of the Pentium II processor, known as the Pentium Pro. It has a dual-cavity package that includes both the CPU and a 256 Kbyte cache, as described below. The dual-cavity package allows the application of major speed enhancement circuitry that could not be fit upon the CPU chip, yet could be kept sufficiently close to achieve similar speed.

2.2.3.2 Parallel Signals. Placing more signals in parallel speeds execution. With the 386DX, data is handled 32 bits at a time, rather than in the 16-bit word width of earlier processors. In similar manner, the address bus increased from 20 to 24 bits in the 80286, and a 64-bit external bus greatly increased data throughput on the Pentium. In the emerging Merced processor the internal data processing path is extended to 64 bits.

2.2.3.3 Internal Parallel Execution. To execute each instruction, the microprocessor executes multiple steps (prefetch, instruction decode, operand fetch, execute, writeback). These steps can be cascaded in assembly line fashion by the use of *pipelining*, by which separate execution units, cascaded in series, perform consecutive stages of instruction execution upon consecutive instructions at the same time. Thus, while one instruction is being fetched, the earlier instruction, fetched previously, is being decoded, and its predecessor is being supplied with operands. The concept of pipelining is developed more fully in Section 2.3.4. If each instruction stage requires one clock cycle, a sequence of five-stage instructions can be completed by a pipeline of five execution units at a rate of one instruction per clock cycle, after a five-cycle delay to fill the pipeline. If a change in execution sequence is required due to a branch, the pipeline must be emptied of the old sequence and refilled with the new, introducing a delay upon program jump. While the 8086 included pipelining of the multiple bytes of a single instruction, the 386DX introduced multi-instruction pipelining. With the 486DX, pipelining depth became sufficient to achieve single-cycle instruction throughput on many instructions.

The Pentium gained further speed through two parallel pipelined execution units, called a *superscalar architecture,* to achieve pipelined instruction throughput of less than one clock cycle per instruction. The problem of breaking the pipeline as the result of a jump is alleviated in the Pentium by *branch prediction,* which postulates that the most likely choice taken upon a branch instruction is the choice taken when that branch was executed previously; a hardware flag is used to keep track, and the pipeline is filled on the basis of this prediction of the result of the branch. When the prediction is incorrect, the pipeline is emptied and refilled as before. Parallel execution is further facilitated in the Pentium II by *speculative execution,* by which the predicted path that follows a branch is executed in advance. The Pentium II also provides a *transaction-based bus* that communicates with main memory. This bus structure allows execution of a subsequent instruction to proceed while the processor awaits the completion of a relatively long access to main memory. In fact, the results of the lengthy memory access can be injected in the midst of a pipeline to catch up with its instruction. Such ability to allow rapid instructions to execute while slow instructions are completing leads naturally to *out-of-order execution,* which is supported by the Pentium II. The CPU eventually restores order to the results and places them in proper sequence.

In the Merced processor, branch prediction is replaced by *branch predication,* by which all possible branches are executed to the extent allowed by the parallel resources of the processor. The correct result is selected upon resolution of the branch decision and the other results are abandoned.

2.2.3.4 Cache. Processor memory systems are designed with smaller amounts of quickly-accessible memory placed close to the processor, and larger amounts of memory placed farther away and requiring more access time. While an instruction may take 10 nsec or less to execute (1 clock cycle on a 100-MHz machine), an access to main memory dynamic RAM may take 60 nsec. Thus, the instruction stream for execution cannot be filled from main memory as quickly as it can be emptied. A small amount of fast memory, if placed between the processor and main memory, can greatly speed execution as a result of the concept of locality of reference. This concept states that most memory accesses will occur within a small neighborhood of addresses. By copying that sector of memory to faster static RAM, perhaps 90 percent of the memory accesses can occur three to four times faster than from main memory.

In a cache-based memory system, each memory access must first be sought in cache. If the contents of the desired memory location are present in cache (a "cache hit"), the data is quickly accessed from cache. Otherwise, if the contents are not in cache (a "cache miss"), the data is obtained from main memory. *Cache coherency*, which checks for such cache hits and maintains agreement between cache contents and main memory in the face of changing memory contents, introduces additional complexity. Two cache reading strategies are used. A *look-aside cache* is placed beside the connection between the CPU and main memory and terminates the long bus cycle in the event of a cache hit. A *look-through cache*, which is in series with the connection between CPU and main memory, will itself start the cycle to main memory in the event of a cache miss; otherwise it provides the data directly and quickly to the CPU. A cache miss results in the data from main memory being stored in cache, along with its address in main memory, as the data proceeds to the processor for use. Cache write methods are also of two types. A *write-back cache* receives the results of computation that are output by the processor, terminates the memory write cycle, and itself writes the data back to main memory whenever bus time is available. A *write-through cache* immediately writes all CPU-written data to main memory.

Several types of cache organization schemes trade off cache complexity against speed. A *fully-associative* cache allows any main memory location to be stored at any cache location: Each cache storage location has associated with it another storage location, contained in a companion tag RAM, that indicates the main memory address from which the data in that cache location came. In a fully associative cache, each processor memory request address must be compared with every location in the tag RAM (in a time short compared to the processor instruction cycle) to determine whether a cache hit is possible. A *direct map* cache (also called a one-way set associative cache) divides main memory into cache pages of size equal to the cache. The cache contains an image of the sequential memory locations associated with the current cache page. The direct map cache can only store a specific line of memory from a particular cache page image in memory into the same line of cache. Thus, memory address 1 on page 2 could only be stored in cache location 1. The direct map requires only one comparison of a memory address to determine whether its contents are in cache, since it merely computes whether the page of that memory address corresponds to the page of memory currently imaged in the cache. This single comparison is faster than searching each cache address value as required for the fully-associative cache.

Midway between the fully-associative and the direct map cache in speed and complexity are the two- or four-way *set associative* caches. These schemes divide the cache into two (or four) equal sections, or *ways*, and treat each way like a small direct map cache. Thus, disjoint images of consecutive memory segments may exist in cache, one per way. Often, memory addresses occur in one small region for program fetches (such as a repetitive loop) and in another for the data on which those operations occur (such as a vector). Thus, two localized but separate memory neighborhoods are in use. The Pentium provides two separate 8-Kbyte caches, one for program and another for data. Both are on chip and are connected to the processor by a 256-bit bus.

Multiple caches may be placed in series, with the smallest and fastest (known as Level 1, or L1) placed closest to the CPU, and a larger, slightly slower cache (L2) placed next, with the largest storage, main memory, placed furthest away. While the Pentium provides for an external L2 cache to supplement its two L1 on-chip caches, the Pentium Pro places an L2 cache as a second integrated circuit chip within the same package as the microprocessor, connecting the two with a 64-bit dedicated bus. While both the program and data caches in the Pentium are two-way set associative, the Pentium Pro enhances the data cache to four-way set associative. The Pentium II places the L2 cache back on the processor die, realizing a 10 percent gain in performance from that step alone.

2.2.3.5 Hardware Acceleration. Hardware acceleration of multi-cycle operations performed by microcode in the 8086 has been added in later processors to allow single- or reduced-cycle execution. An on-board floating point unit provides two floating point operations per clock cycle for the Pentium and later processors. Several other instructions are moved from microcode to hardwired operation in the Pentium, and the Pentium Pro speeds instruction decoding with three parallel instruction decoders—two for single-cycle instructions and one for one-to-four cycle instructions. Instructions longer than four cycles resort to microcode.

2.2.3.6 Single Instruction Multiple Data Extension. Integrated circuit technology allows ever-wider data paths, yet the data itself need only be sampled at a fixed data width. As the personal computer evolved from processing general-purpose numeric calculation to simultaneously handling CD-ROM, video, audio, and the Internet, multiple real time data streams became necessary. To this end, features were introduced to allow the wide data path, formerly used for a single 32- or 64-bit number, to be divided into parallel fields to handle multiple samples of the lower-bit width signals of multimedia processing. Multimedia extensions (MMX) allow placing multiple data values within one word. To exploit parallel processing of multiple data values, the architecture is enhanced with new instructions that allow subword parallel execution (same instruction operating on multiple data values, known as single instruction, multiple data, or SIMD), the packing and unpacking of multiple data words into a long word, and the data-dependent conditional selection of certain data values based within the long data word. In addition, certain arithmetic operations important to multimedia operation, such as the sum of absolute differences, used in the compression of frames of video data, have been added.

2.2.3.7 Dynamic Execution. *Dynamic execution*, or "instruction fission," bridges the gap between the complicated instructions that originated with the 8086 code and its minimal number of transistors, and the simpler microinstructions into which a complicated instruction is decomposed for execution on parallel architectures such as the superscalar. The complex instruction set computer (CISC) is kept in later x86 designs by the need to maintain software compatibility with the 8086 code. Regularity of execution, indicated by microinstructions of identical length and single-cycle execution times, is the goal of the recent trend toward reduced instruction set computers (RISC). RISC-like instructions are used in the Pentium and Pentium II to support such advanced features as superscalar design, super pipelining, and out-of-order execution. Original 8086 instructions are decomposed ("instruction fission") into one or more microoperations upon receipt. In this manner, the Pentium and Pentium II combine the features of the CISC and RISC designs into a hybrid system.

2.2.3.8 Further Enhancements. Further enhancements to the Intel x86 processor include natural extensions of the enhancements occurring to date. These include

- smaller design rules and resulting greater chip density;
- incorporation of more memory for cache on chip;
- ever-increasing clock speeds;
- more parallel pipelines in the superscalar architecture.

Enhancements that extend beyond extrapolations include[5]:

- migration from a hybrid CISC/RISC architecture to a RISC/CISC emphasis, where software compatibility with the 8086 is maintained with CISC instructions, but a new non-compatible, more efficient RISC language is made available to the user as well;
- 64-bit architectures, packing several instructions into a long instruction word for faster processing;
- use of very long instruction word (VLIW) instead of the RISC emphasis,[6] in which the compiler combines multiple simple instructions into a very long instruction. This technique moves the responsibility for allocating instruction microoperations to parallel resources from processor hardware to computer software;
- use of speculative loading of operands, by which loads from memory are initiated earlier in the instruction stream;
- compiler-based analysis and identification of parallelism at execution time, thereby reducing the load on processor resources demanded by the complexities of identifying parallel and out-of-order execution;
- better support of greatly parallel multiprocessing, providing more ability for one processor to read and write to registers of another in support of greater bandwidth between individual processor components (such as registers) and collaboration on massively parallel tasks.

The techniques described here have accelerated the execution speed of the Intel x86 processor family by four orders of magnitude over that of the 8086. A portion of this increase has resulted from an increase in clock speed, made possible by integrated circuit technology that allows more dense transistor placement, shorter propagation times, and the ability to move more logic on chip that was once off chip. The other portion of the performance gain has arisen from architectural enhancements, also enabled by advances in on-chip logic density. These advances allow the simultaneous processing of multiple instructions. They move relevant data and instructions closer to the processor, where they may be more quickly accessed. The increase in complexity associated with improved performance is indicated by the number of transistors by a factor of 700 from the 29 K-transistor 8086 to the 19 M-transistor Pentium II and still higher for the Merced processor.

2.3 DIGITAL SIGNAL PROCESSOR

The general purpose microprocessor is arranged to handle the most general forms of processing, including wide-ranging mixtures of operations and control flow that can be data dependent, making large jumps from one area of program memory to another. In contrast, traditional digital signal processing emphasizes repetitive multiply-and-add operations that step sequentially through data values stored in consecutive memory locations. Furthermore, digital signal processing depends far less frequently on data-dependent execution.

2.3.1 Architecture Requirements of Digital Signal Processing Algorithms

Although the general-purpose microprocessor can perform the repetitive multiply-and-add operations of signal processing, a specially-structured microprocessor, called the *digital signal processor* (DSP), can perform them faster because it introduces architectural features to accelerate the more specialized operations of signal processing.

This discussion centers upon the repetitive multiply-and-add operation that is common to much of signal processing, described as follows:

$$\sum_{n=0}^{N-1} x(n)y(n);\ n = 0,1,\dots,N-1. \tag{2.11}$$

Here, a sequence of N signal values $x(n)$ is multiplied by a corresponding sequence of signal values $y(n)$, for $n = 0, 1,\dots, N-1$. Alternatively, the signal $y(n)$ may be replaced by coefficients a_m, and for so notational convenience we at times represent the same structure by $\Sigma_{m=0}^{M-1} a_m x(n)$.

This structure becomes apparent in several common signal processing operations. The *correlation* between two digital signals, $x(n)$ and $y(n)$, is given by

$$R_{xy}(r) = \frac{1}{N-r} \sum_{n=0}^{N-1-r} x(n)y(n+r). \tag{2.12}$$

The *autocorrelation* of a signal x (i.e., the correlation with itself) is given by an expression that substitutes x for y:

$$R_{xx}(r) = \frac{1}{N - r} \sum_{n=0}^{N-1-r} x(n)x(n + r). \tag{2.13}$$

In either case, $R(r)$ provides a measure of the similarity of the two signals when one is displaced by r samples from the other. Maximum correlation is searched for by computing $R(r)$ for multiple values of the lag r and choosing the lag providing the maximum value of $R(r)$ as the lag of maximum correlation. The operation of *convolution* is similar, except that one of the signals is reversed in time:

$$R_{x*y}(r) = \frac{1}{N - r} \sum_{n=0}^{N-1-r} x(n)y(r - n). \tag{2.14}$$

Convolution is used to compute the output of a linear system as a function of its impulse response, y, when driven by signal x.

In all cases, the multiply-add kernel is apparent. Another common use of the multiply-add kernel is in the finite impulse response filter (FIR) as shown in Fig. 2–18. The output $y(n)$ results from the multiply-accumulation of $x(n)$ with the filter coefficient a_m:

$$y(n) = \sum_{m=0}^{M-1} a_m x(n - m). \tag{2.15}$$

Fig. 2–18 shows the use of a series of one-sample delays, z^{-1}, that store a sequence of m samples, $x(n)$, $x(n - 1)$, ..., $x(n - m - 1)$, and presents the appropriate sample to the corresponding coefficient for summation into output $y(n)$.

Computing each of the M stages requires the following steps:

1. Fetch instruction from program memory.
2. Fetch two operands, a_m and $x(n - m)$, from memory.
3. Multiply $a_m\,x(n - m)$.
4. Add to the accumulating sum.
5. Store the result in the delay line.

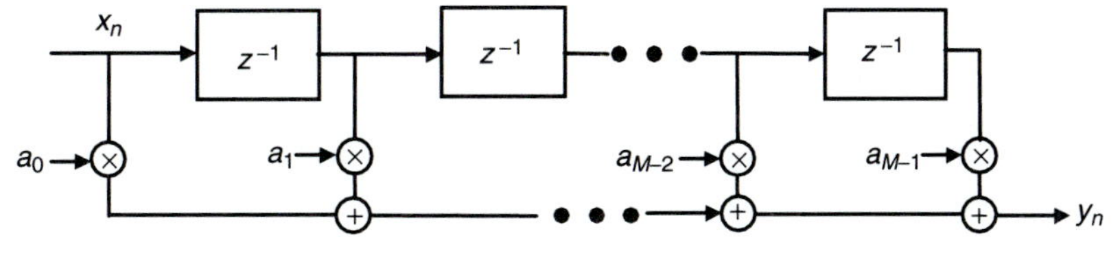

Figure 2–18 Finite impulse response filter of length M.

The general purpose microprocessor would take many instructions to perform these steps. A goal of real time signal processing is to perform these operations as quickly as possible, preferably in a single step.

2.3.2 Enhancing the Microprocessor for Digital Signal Processing

We now examine several architectural enhancements that can be made to achieve single-step operation of the multiply-add computation, thereby migrating from the general purpose microprocessor to the programmable digital signal processor. We will again express the multiply-add computation as $\Sigma_{n=0}^{N-1} x(n)y(n)$and use it to motivate a series of enhancements to the general purpose processor that speed its execution. We will first examine enhancements to perform the multiplication of $x(n)$ and $y(n)$, then discuss changes to the instruction circuitry that commands the repetitive multiply and add be performed, and finally look at improvements to access the $x(n)$ and $y(n)$ operands themselves.[7]

Multiplication operations may require several clock cycles on a traditional microprocessor, where they are performed in a serial manner that requires repetitive shift-and-add operations (see Section 3.3.2). To achieve the speed required by the multiply-intensive digital signal processing operations, a fully parallel multiplier is used that is capable of forming the product of its two inputs within one clock cycle (see Fig. 2–19, point a). Furthermore, the multiplier is placed directly within the ALU data path, rather than placed to one side as in the typical microprocessor (Fig. 2–19b). This assures that all data flowing into the ALU is immediately presented to the multiplier inputs.

The multiplier is immediately followed by an adder (Fig. 2–19c), so that the add operation is performed in the immediate next clock cycle after the multiplication.

It is apparent that cascading the processing elements of multiplier and adder in series allows the simultaneous operation of both, with the multiplier performing its work at time n on the product $x(n)y(n)$, while the adder operates on the product $x(n-1)\ y(n-1)$. This technique of overlapping processing steps on adjacent processing elements is known as *pipelining* and is discussed in greater detail below (Section 2.3.4). Pipelining relies on the fact that both the addition and multiplication instructions take the same amount of time, which in our case is one clock cycle. More generally, the instruction set of the programmable signal processor is designed so that all instructions take an equal (and minimal) number of clock cycles. Equal-period instructions allow operations on separate processing units to be performed in this pipeline manner without introducing either data pileups (caused by a pipeline element requiring more processing time than its upstream neighbor) or gaps (in which a pipeline element completes its work more quickly and must wait for new data) (Fig. 2–19d).

Further speed is introduced into the multiply-accumulate pipeline by providing separate data buses for each of the two multiplier inputs. Instead of re-

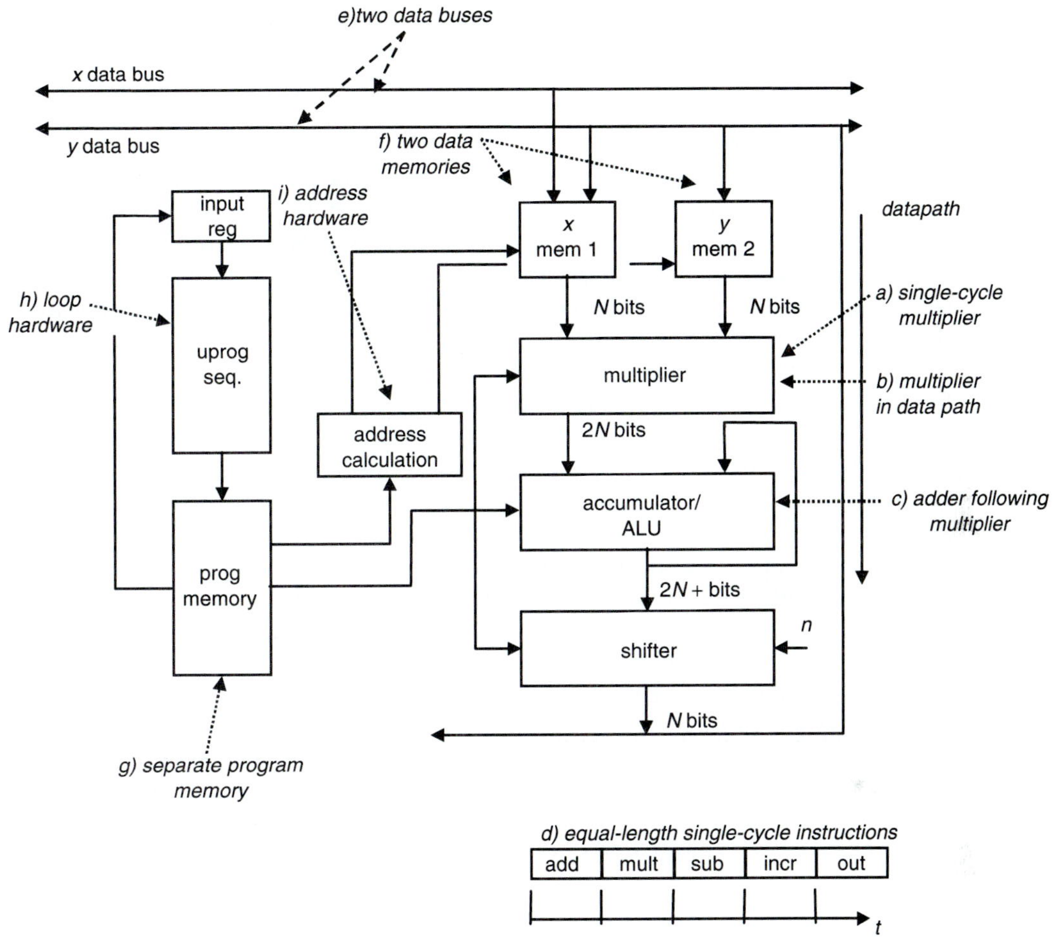

Figure 2–19 Generic digital signal processor architecture introduces enhancements to standard microprocessor architecture ($a - i$, see text).

quiring an operand fetch for $x(n)$ and another on the same data bus for $y(n)$, both may be fetched in the same single instruction cycle on two separate data buses (Fig. 2–19e). The two data buses go to two separate memories, one for x data and one for y data (or a fixed coefficient a), to avoid the conflict of two numbers entering the same memory at the same time (Fig. 2–19f). Each memory also has its own address bus which originates with an address calculation unit. A third memory and associated bus is used for storing program instructions to avoid the delays in accessing data which would be caused by the simultaneous need to fetch instructions from the program memory while fetching data from data memory (Fig. 2–19g). The separation of the processor memory space into a memory for data and another memory for program is referred to as the *Harvard architecture*, and is in contrast to the single-memory von Neumann architecture associated with general purpose processors.

The instruction module of the digital signal processor includes hardware assistance to perform iterative loops with low overhead. Instead of fetching each instruction on a separate cycle, a means is provided to repeat a single instruction or a small sequence of instructions in hardware, so that the multiple-clock-cycle sequencing of fetching and decoding each instruction is avoided on tight loops (Fig. 2–19h).

Accessing the sequence of operands $x(n)$, $y(n)$, $n = 0, 1, \ldots, N - 1$ is a regular sequential operation that is accelerated by special provisions in the signal processor. In the general microprocessor, part of each instruction field includes information on the address of the instruction operands. This information must be decoded and the operands must be fetched in subsequent operations before all data is ready for input into computations. In the special case of signal processing, these operands occur in sequential locations in memory. Therefore, the general capability to fetch operands from anywhere is augmented by the special capability to fetch operands more quickly if they are in sequence. Hardware for simple incrementing of the address pointer within the same clock cycle is provided, so that no extra clock cycles are used in accessing the necessary data (Fig. 2–19i). Further enhancements allow the access of simple regular sequences of operands, for example modulo N (i.e., $0, 1, \ldots, N - 1, 0, 1, \ldots$) with no additional delay. Auto-increment on fetch, auto-decrement on fetch, and cyclic access on fetch are all standard signal processing data sequences that are accelerated to be computed within a single clock cycle on modern signal processors.

2.3.3 Generic Digital Signal Processor Architecture

The series of signal processing enhancements to the general purpose microprocessor can be displayed in an architectural drawing of a generic digital signal processor (Fig. 2–19). At the top of the figure are the two data buses, x and y. Each is connected to a data memory, x *mem 1* or y *mem 2*, which in turn is connected to either input of a single-cycle multiplier. Dedicated address calculation circuitry performs the cyclic addressing of operands in the x and y memories discussed above. Each memory presents data of N bits width to the multiplier, and the multiplier presents a $2N$-bit wide product to the input of the accumulator. The output of the accumulator is connected to its other input, thereby providing the means to form the running sum $\text{sum}(n) = x(n)y(n) + \text{sum}(n - 1)$. Following the accumulator is a shifter that selects the most significant N bits of the accumulating sum and places the results on the y data bus for transmission to memory. A third memory is devoted to storing the program: It is addressed by a microprogram sequencer that calculates the address of the next instruction, either by simple increments, repetitive loops, or by jumping to a number placed in the input register (which itself can be contained in an instruction within the program memory).

2.3.4 Stages of Data Path

With the numerous resources available, the signal processor achieves a speed advantage over the general-purpose microprocessor by performing operations with these resources at the same time. It is useful to re-examine the basic multiply and accumulate instruction and develop a snapshot of how machine resources are applied to its execution. We first identify the datapath through which the computation passes, consisting of the five stages of instruction fetch, operand memory, multiplier, accumulator, and shifter/output, as identified in Fig. 2–20.

The multiply-accumulate equation is rewritten in an iterative form:

$$\begin{aligned} y(0) &= a_0 x(0) \\ y(n) &= a_n x(n) + y(n-1);\ n = 1, 2, \ldots, N-1. \end{aligned} \tag{2.16}$$

We write this iterative form as a program of multiple instructions that steps through the data values a_n and $x(n)$. We decompose this instruction into the individual operations that must be performed in its computation, following its

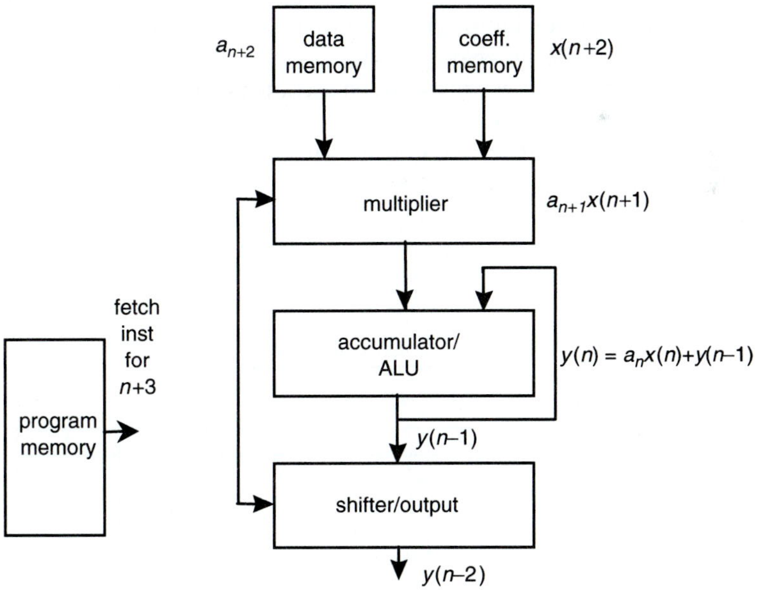

Figure 2–20 Datapath of digital signal processor consists of five stages: instruction fetch, data fetch, multiply, accumulate, and shift/output. The cascading of operations due to pipelining is shown.

progress through the digital signal processor and identifying which portion of the processor is occupied at each step toward the answer. Fig. 2–21 presents a *reservation table*, in which DSP resources are listed vertically as row headings and time slots, corresponding to instruction cycles, are listed horizontally as column headings.[8] A number in a cell indicates which instruction is being computed by a particular machine resource (given by a row) and at which time slot (given by column position).

To execute an instruction, five machine resources are used, one per clock cycle, resulting in these five stages of computation:

1. Fetch instruction $[y(n) = a_n x(n) + y(n-1)]$ from program memory;
2. Fetch the operands for the multiplication, a_n and $x(n)$;
3. Perform the multiplication $a_n\, x(n)$;
4. Add the results to $y(n-1)$, already in the accumulator;
5. Shift and output the result.

We note in Fig. 2–21 that five cycles are required to complete the operation, and ten cycles are required to complete two operations. However, only one of the five elements of the processing chain is in use at each cycle. Throughput can be increased by advancing the early stages of the next instruction to occur while later stages of the previous instruction are in progress. This technique of pipelining uses registers within the datapath simultaneously to increase throughput. Pipelining is shown in Fig. 2–22, where each clock cycle begins the execution of a new five-cycle instruction. All processor resources are active at the same time (after a five-cycle startup), and each clock cycle produces a new output (after a five-cycle lag).

Immediately after fetching instruction 1 from program memory at timeslot 1, instruction 2 is fetched at time slot 2 in the pipelined approach (rather than at timeslot 6 in the non-overlapped approach). In like manner, after fetching operands a_n and $x(n)$ during timeslot 2, operands a_{n+1} and $x(n+1)$ are fetched at timeslot 3 (rather than timeslot 7).

instruction cycles →

	1	2	3	4	5	6	7	8	9	10
prog mem	1					2				
operand mem		1					2			
multiplier			1					2		
accumulator				1					2	
shifter/output					1					2

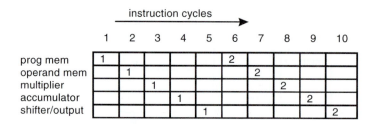

Figure 2–21 Reservation table shows time slots progressing horizontally, machine resources used ordered vertically, and instruction processed as number in cell. Case shown is without pipelining and requires 10 cycles to compute two instructions.

instruction cycles →

	1	2	3	4	5	6	7	8	9	10
prog mem	1	2	3	4	5	6				
operand mem		1	2	3	4	5	6			
multiplier			1	2	3	4	5	6		
accumulator				1	2	3	4	5	6	
shifter/output					1	2	3	4	5	6

Figure 2–22 The use of pipelining overlaps the execution of consecutive stages of each instruction.

Pipelining can present programming difficulties when the execution of an instruction must wait until an earlier result is complete. To illustrate a difficulty of pipeline programming, let us imagine that we wish to multiply the final sum of (2.16) by a gain factor, g. The multiplication by g can occur only after the sum $\Sigma_{n=0}^{N-1} a_n x(n)$ has been formed, and this sum must pass out of the accumulator, through the shifter, and out into memory, then make its way back from the memory to the multiplier for multiplication by g. Fig. 2–23 shows the reservation table of Fig. 2–22 augmented by a seventh instruction to perform the multiplication by g.

The arrow and shaded boxes indicate that the final sum, completed at time 10 by instruction 10, is not available to the operand memory until time 11, delaying the result of instruction 7 until time 14. Delaying the completion of instruction 7 until its data arrives from instruction 6 introduces gaps in the pipeline, and machine resources are idle.

Programming a pipeline machine can present a challenge, particularly if the programmer must keep track of data precedence requirements such as the one just described. One way of avoiding this detailed programming analysis is to introduce a hardware interlock. This interlock automatically senses the requirement to wait until the accumulator result passes through the output shifter and back into the data memory before continuing the execution of the instruction that

instruction cycles →

	1	2	3	4	5	6	7	8	9	10	11	12	13	14
prog mem	1	2	3	4	5	6	7	wait for data →						
operand mem		1	2	3	4	5	6				7			
multiplier			1	2	3	4	5	6			↑	7		
accumulator				1	2	3	4	5	6				7	
shifter/output					1	2	3	4	5	6				7

Figure 2–23 Reservation table for repetitive multiply-accumulate (instructions 1–6) followed by multiplication of the result by a gain factor (instruction 7).

uses this result. The hardware interlock delays the completion of execution of an instruction that must wait for data.

As shown in the reservation tables, the execution of an instruction is scattered over several time slots and across multiple machine resources. Either of two approaches may be used to avoid requiring the programmer to keep track of both dimensions of the reservation table. Each method segments the matrix in a different manner, providing a stationary snapshot of machine activity. The first of these is *time-stationary coding*, in which the program instruction of the DSP contains multiple fields, each of which defines the activity of a particular DSP resource at the same instant of time. An example is shown in Fig. 2–24, which depicts in a single instruction all the activities that occur during one instant of time, or one column, in the reservation table.

Alternatively, *data-stationary coding* applies the multiple fields of the DSP instruction to specify all operations that are performed on a single piece of data, despite the fact that they are performed at different times. As shown in Fig. 2–25, this amounts to capturing a left-to-right diagonal column in the reservation table.

2.3.5 Examples of Digital Signal Processors

The Analog Devices ADSP-2106x[9] is an example of a high-performance floating point digital signal processor. With three independent parallel floating point units in its signal processing core (multiplier, barrel shifter, and arithmetic logic unit, or ALU), three instructions are executed together in each cycle. Thus, in one version of the DSP, each unit operates at 40 million floating point operations per second, but a peak throughput of 120 million floating point operations per second (MFLOPS) is achieved. As shown in Fig. 2–26, two separate data buses, PM and

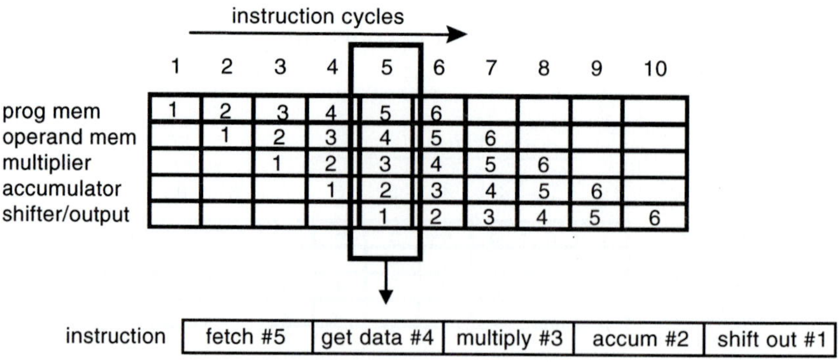

Figure 2–24 In time-stationary coding, a single DSP instruction consists of a field for each machine resource that indicates its action at the same instant of time.

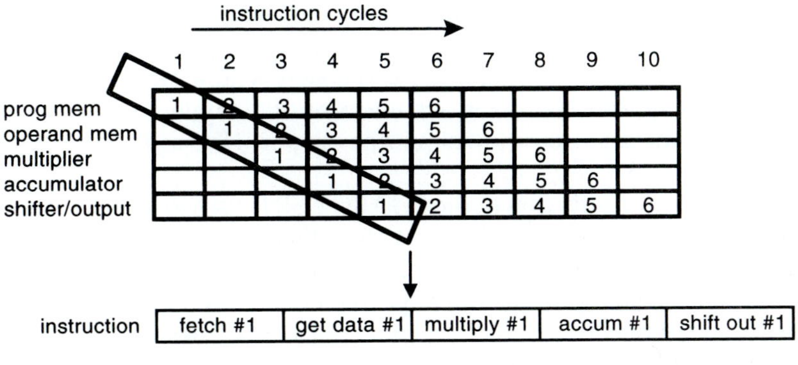

Figure 2–25 Data-stationary coding uses the fields of the DSP instruction to describe all actions on the data within an instruction, regardless of its time of execution.

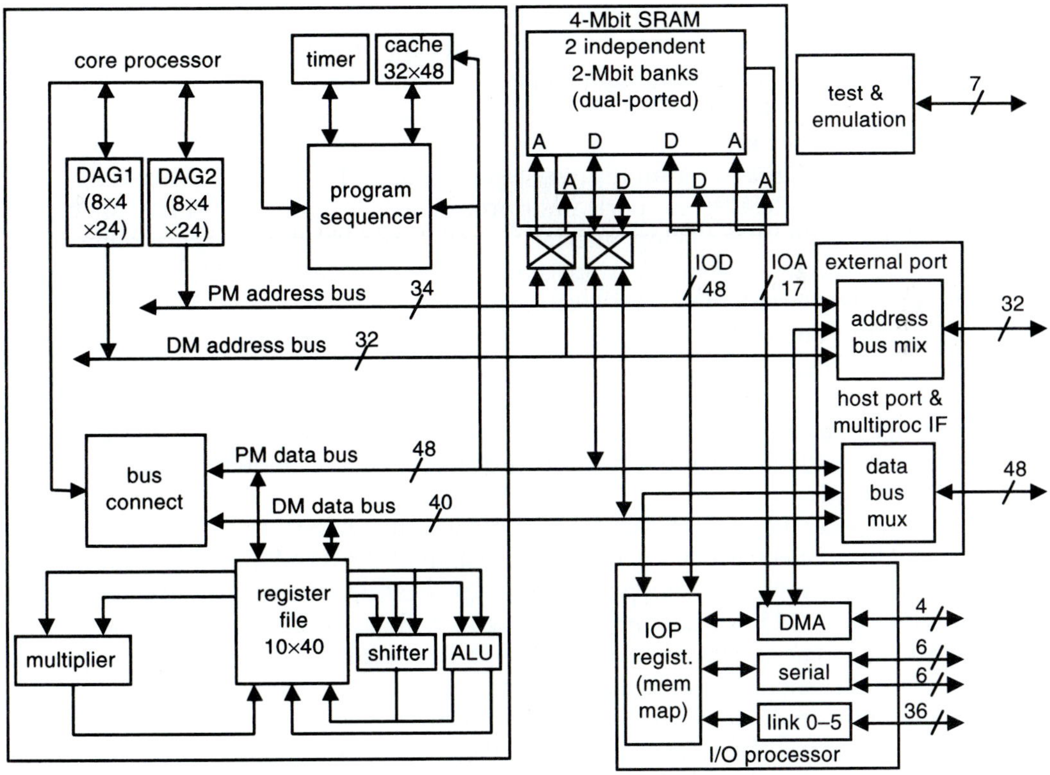

Figure 2–26 Architecture of ADSP-21060 includes a signal processing core of a multiplier, shifter, and ALU fed by a register file.[9] © 1998, Analog Devices, used with permission.

DM, provide the multiple data buses that distinguish a digital signal processor from a general purpose processor. The use of separate memories and dedicated buses for program and data are associated with a Harvard architecture (as opposed to the Von Neumann architecture of a single memory for both), and the ADSP-2106x is also known as the Super Harvard Architecture Computer (SHARC).

A major distinction of the SHARC DSP is its two large on-chip dual port static RAMs, providing the ability to store the large amounts of randomly accessed data needed for the two-dimensional fast Fourier transform used in image processing. The dual port memory speeds memory access by allowing simultaneous read and write to different memory locations through the dual ports. This is shown in Fig. 2–26 as the 4-Mbit SRAM. Each SRAM may be attached to either of the two data buses and may receive addressing from either of the two address buses, PM Address or DM Address. Each address bus has an address generator circuit, DAG1 or DAG2, that provides indirect addressing and allows implementation of circular buffers.

The computational core consists of a multiplier, barrel shifter, and ALU clustered around a 32-word × 40-bit register file. The register file transfers data among these three computation units and stores their results. Instructions are contained in a 48-bit word. An instruction cache uses three buses to perform single-cycle fetching of an instruction and four data values, allowing full-speed operation of digital filter multiply-accumulate and FFT butterfly processing.

The SHARC DSP supports multiprocessing. As many as six SHARCs can be combined with a zero-chip interface, and a unified address space allows one SHARC to access the memory space of another. Distributed on-chip bus arbitration resolves simultaneous bus access requests.

Input and output rates of up to 500 Mb/sec can be accomplished between processors using the external data port for the ADSP-21060 (subsequent versions provide faster transmission). In addition, four serial ports provide a 40-Mbit/sec transfer rate each. Direct memory access (DMA) capability and the input/output section operate independently of the processing section, allowing both to operate at full speed.

A faster version of the SHARC family, the ADSP21160, achieves a peak throughput of 600 MFLOPS, or five times that of the ADSP2106x. The speedup is gained by providing two parallel core processing sections (multiplier, barrel shifter, and ALU centered around a register file), and by increasing the overall clock rate. The 21160 is software compatible with the 21060, but it applies a single-instruction, multiple-data stream approach to operate the two cores on a common instruction stream, rather than using the single-instruction, single-data approach of the 2106x. Examples of the use of the 21060 architecture are provided in Sections 6.3 and 6.6.

The incorporation of multiple processors on a single chip is also seen in the Texas Instruments TMS 320C80 Multimedia Video Processor (MVP).[10] In this case, described in a trade study example in Section 3.3.3, four 32-bit digital signal processing cores are combined with a general-purpose processor on a single chip. These processors may be driven by independent instruction streams, thereby implementing a multiple-instruction, multiple-data (MIMD) architecture, described more fully in Section 4.2.1.

The use of digital signal processors for multimedia processing has led to the use of the very long instruction word (VLIW) architecture, as implemented by the Texas Instruments TMS 320C62 and TMS 320C67. Multimedia functions include the simultaneous or near-simultaneous operations of video compression/decompression, interactive two-dimensional and three-dimensional graphics, data/fax modem, telephony, multichannel digital audio and music synthesis, and video teleconferencing.[11] As semiconductor technology has permitted the placing of ever-wider data processors and buses on a chip, the opportunity to handle data words in parallel has arisen. The data size for pixels or signal samples in many multimedia applications is no greater than 16 bits, yet processors of 64 bits can now be built. The use of *subword parallelism* and *partitionable* ALUs can support combining multiple samples into one simultaneously processed word, speeding execution.

In addition to the growth of multimedia applications, the use of specialized instructions in the DSP set, such as those supporting the DSP-specific architectural features shown in Fig. 2–19, as well as the use of heavy pipelining requires the use of many more bits to control these additional modes and resources. In the typical DSP with an instruction width limited to 16 to 48 bits, certain of the multiple resources, such as several adders, multipliers, and memories, must be controlled implicitly, as shown in Fig. 2–27a.[12] Controlling each resource separately is necessary to achieve optimum compilation from a high-level language in the face of the specialized DSP instruction set, yet sufficient bits are not available in the instruction word of the traditional signal processor.

The VLIW architecture provides individual control of each resource, as shown in Fig. 2–27b. In the normal VLIW design, a several-hundred bit instruction word is divided into dedicated fields to control each resource separately. In an enhancement, the Texas Instruments VelociTI TMS 320C62 architecture allows the reconfiguration of these control fields to switch among resources (Fig. 2–27c).

The TMS 320C62 processor is a fixed point processor that executes up to eight instructions per cycle. The architecture consists of two sets of four functional units as shown in Fig. 2–27c, where each set includes unit D (for address calculation), M (for multiplication), S, and L (both for general ALU functions). Each of the four units for both sets are controlled by a 32-bit instruction word that provides instruction values, data values, and other information. Thus, a total instruction width of 256 bits is used. The processor achieves an instruction rate of 1600 million instructions per second. The TMS 320C6702 achieves a similar instruction rate in the floating point domain, thus surpassing the 1 GFLOP barrier.

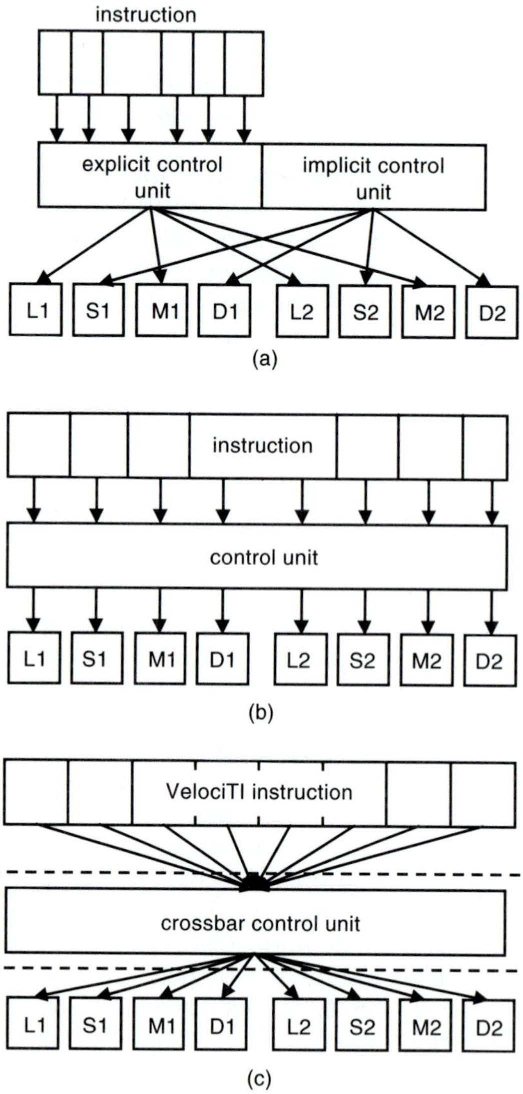

Figure 2–27 Traditional architecture (a) limits instruction width and must rely on implicit assumptions to control all resources. VLIW architecture (b) provides explicit control of each resource. VelociTI (c) allows changing the correspondence between VLIW bit fields and resources.[12] © 1997 IEEE.

2.4 SPECIALIZED PROCESSORS

Thus far, we have examined a variety of existing architectures with the intent of mapping the algorithms of interest, via programming, to fit the architecture. But what if we allow the architecture to be changed to better suit the algorithm? Both technical and economic considerations enter into a decision to pursue specialized architectures. We will discuss the technical motivations, then review both the manufacturing and the design process in brief, to provide a foundation for considering the economic factors in custom-integrated circuit design. After reviewing the economic factors, we conclude by examining the influence of integrated circuit architectures on algorithm design and discuss new directions emerging in the design of integrated systems.

2.4.1 Motivation

Careful timing analysis of a signal processing algorithm programmed on a processor may reveal bottlenecks, or small portions of code that contribute disproportionately to execution time. These bottlenecks may be repeated many times as the program progresses from start to finish. An execution profile, or plot of instruction addresses versus time, is shown in Fig. 2–28. The algorithm begins at its initial instruction (shown at the origin of the figure), progresses linearly for a small amount, then jumps to a subroutine at a higher address value. Within that subroutine, a loop is executed three times, indicated by the sawtooth ascent and drop of the execution address within the subroutine. The program then exits the subroutine, progresses a bit further within the lower-addressed space of the main routine, then again jumps to the subroutine and repeats its inner loop. This subroutine is repeated twice more before the program completes.

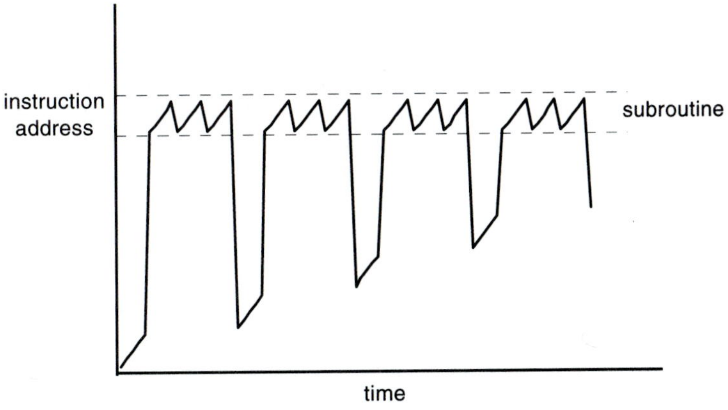

Figure 2–28 Plot of location of instruction being executed versus time shows that the vast majority of program time is spent in iterative loop within subroutine.

In this example, 75 percent of program time is spent in the subroutine loop. If the execution time of that loop could be significantly accelerated, for example by a factor of 10, then overall program execution time could be decreased, in this case to less than one third of its original duration.

A custom processor architecture can be used to effect speedups of over an order of magnitude on a difficult multi-cycle instruction sequence. As will be shown later, a particular combination of registers, logic elements, and interconnections can allow a custom architecture to achieve in 1 clock cycle what a traditional programmable architecture requires 10, or even 100, clock cycles to complete. By sacrificing the flexibility of programmability, speed of a specific kernel operation can be greatly enhanced.

In addition to speed, cost savings may motivate the use of custom processors. Just enough processing resources can be designed, and the overhead required by flexibility can be avoided if it is not needed. A carefully designed custom integrated circuit can absorb many supporting integrated circuits, reducing chip count, board area, assembly difficulty, and cost. For similar reasons, low power concerns can motivate the use of custom integrated circuits.

Other motivations to design a custom processor include:

- *improved reliability:* the challenge of producing integrated circuit designs correctly on the first attempt has driven the use of circuit simulation to the point that "what you simulate is what you get." As a result, integrated circuit performance is thoroughly simulated over a range of parameters such as voltage, temperature, and error conditions, unearthing possible performance problems, correcting them, and achieving higher reliability than is possible with a partially simulated collection of circuits mounted on circuit packs with sockets, solder, and wires.
- *meeting footprint compatibility:* systems may require selective upgrade of internal components by producing a plug-in compatible upgrade for an outdated processor, without changing the board or system design around the processor. Plug-in replacement upgrades can present a particular challenge to designers, as processor speed is upgraded while maintaining input/output compatibility of its pinout.
- *protection of intellectual property:* while software and even hardware designs can be reverse-engineered and duplicated, integrated circuits are extremely difficult to duplicate or fully understand by inspection or analysis in operation. Innovative or proprietary techniques can be encoded within an integrated circuit and protected from easy duplication.
- *fit with company's core business:* as integrated circuit technology has grown to offer ever-increasing functionality on a chip, the function of a chip has progressed from a component to a system. This has placed some semiconductor manufacturers in the business of selling systems as they sell integrated circuits. Thus, a semiconductor manufacturer may be able to recover an invest-

ment by selling a system as long as it contains the custom integrated circuits that are the core business of the company. This can motivate the inclusion of custom processors within systems to assure fit within the business arena of a particular company.

Several factors can render a computational kernel difficult for a conventional programmable processor and require the processor to expend many execution cycles. Transcendental operations (log, sin, ...) can require a Taylor series expansion and multiple cycles to execute. Even simple multiplications may require multicycle bit-at-a-time shift-and-add operations to compute. Data-dependent execution, in which the flow of execution is determined by the size or rank ordering of incoming data values, can require multiple steps on a traditional architecture. Specialized sequencing of operands, resulting in specialized operand address calculations, can consume multiple cycles to simply compute where the data is stored. Section 3.2 provides more information on factors that increase the difficulty of executing a computation.

An example of an operation requiring multiple cycles on traditional processor architectures is the kernel operation of dynamic programming, discussed further in Section 7.2.4. Dynamic programming finds the lowest-cost path through a network, computing the global minimum-cost path as a sequence of locally-minimum paths. This kernel requires the following.

1. Examine n numbers ($n \approx 3$) and choose the minimum.
2. Note the index of that minimum value (first, second, third,... of the list of n).
3. Add that minimum to another number.
4. Retrieve a value from a memory location with address based on the index of the minimum of n numbers computed in (2) above.
5. Apply possible additional constraints, such as requiring that if the first number of the list of n was chosen in the previous iteration, it cannot be chosen currently.

Other examples of signal processing kernels that are difficult (that is, require a large number of computing cycles from a standard programmable processor) include:

- codebook search for vector quantization;
- two-dimensional nearest-neighbor operations for image processing wavelet or subband coding analysis;
- Fourier transform.

These are discussed in further detail in Chapters 7 through 10.

2.4.2 Overview of Integrated Circuit Design and Manufacture Process

To understand the design process by which a specialized processor is implemented using integrated circuit technology, we first describe the integrated circuit manufacturing process.

An integrated circuit (IC) consists of a silicon chip (or *die*) that has been cut from a silicon wafer that was covered with multiple geometric patterns and layers during the manufacturing process. The wafer fabrication process consists of the series of steps by which these geometric patterns are transformed into an operating integrated circuit. A wafer is built up as a sequence of successive layers, with each layer patterned with geometrical shapes. To pattern a layer of silicon, an insulating layer of silicon dioxide is placed on the surface of the silicon wafer. Sections of the layer are selectively removed, exposing the underlying silicon, as follows. The pattern to be transformed to the wafer exists as a blank mask, or a transparent sheet covered with an opaque material. The opaque layer is selectively removed in the desired patterns, leaving a patterned mask with an arrangement of transparent and opaque regions. The silicon dioxide layer on the wafer is coated with an opaque, light-sensitive material, which is then covered by the mask and exposed to light. Silicon dioxide layer areas that are under transparent areas of the mask are exposed to light and undergo a chemical change; those under opaque regions of the mask are left unchanged. A subsequent step applies an etching chemical to the layer which selectively etches away the exposed regions, leaving behind a series of bare silicon conductive layers. (In the equivalent negative photolithography process, unexposed regions are etched away, leaving behind insulating layers of silicon dioxide.) This process is repeated on the wafer layer by layer, resulting in a structure consisting of logic elements and interconnections.

Each wafer has the same pattern applied horizontally and vertically multiple times, thereby forming a rectangular array of repetitive patterns. These arrays are separated into individual chips by scribing, then breaking the wafer, along a grid of intersecting lines. Individual die are next tested. Chips that pass are affixed within a plastic or ceramic package. Fine metal leads are bonded to metal contact pads around the edge of the die and to pads on the package that connect with external lines.

Both the wafer and the mask (as well as the process) can contain *defects*, e.g., imperfections of the atomic lattice, flaws in the mask, or dust on the surface. The chip that contains the area of the defect is flawed and is almost always non-functional.

Yield statistics can predict the likelihood of a chip containing a defect. If we assume that the defects are distributed uniformly with a density of d defects per unit density, a chip of area a has the probability P_0 of having no defect, given by:

$$P_0 = e^{-da}. \tag{2.17}$$

As shown by Fig. 2–29, a chip with area many times $1/d$ will never be found without a flaw. To maximize yield, it is necessary to minimize d and keep a less

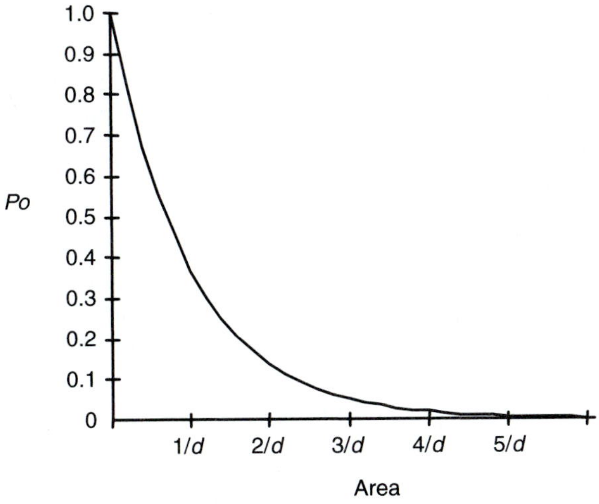

Figure 2–29 For a density of *d* defects per unit area as chip area increases (in multiples of *d*), the probability P_0 of a defect-free chip drops exponentially.

than a few times *d*. The possibility of defects mandates that each chip be tested to assure that it does not contain a defect that could impact performance.

We now turn our attention from the manufacturing process to the design process. To simplify design, it is necessary to separate the design steps from the manufacturing process. This separation assures that the manufacturing characteristics of a mask do not depend upon the nature of the pattern. The design process is kept compatible with the manufacturing capabilities by the use of a limited set of *design rules*. Design rules are constraints on mask features, including the minimum allowed line width, minimum separation of features, and similar constraints governing pattern features. These constraints are imposed to reflect the constraints set by the resolution of the manufacturing process, and they depend upon the characteristics of the particular fabrication line.

Three levels of custom design, in order of increasing complexity, are:

- programmable logic array;
- standard cell design;
- full custom design.

The programmable logic array (PLA) implements the combinational functional blocks that were placed between the registers of sequential logic elements, as was shown in Fig. 2–7. We can understand the PLA by imagining a logic block with *n* inputs and *m* outputs. For each combination of *n* input values, the *m* output signals take on a specified value. A block that implements all ways that *m* outputs could respond to *n* inputs could be a memory of 2^n *m*-bit words. The *n* inputs would correspond to an address into the memory; the desired *m* bit output would be stored as a value at that address.

The PLA operates in similar fashion to the memory, but more efficiently encodes the redundant terms. Redundant terms correspond to long sequences of input values that produce the same output value. A PLA consists of an AND plane of logic that is followed by a NOR plane. It implements logic functions that can be decomposed into the form

$$output_m = \text{NOR}[\text{AND}(input_1, input_2, \ldots, input_n)]. \tag{2.18}$$

A PLA is programmed with the desired logic values after manufacture, usually in the field. A PLA of one million transistors can only accommodate about fifty thousand logic elements. The overhead paid for easy injection of design information is a high number of transistors per logic element—about 200—as compared to the more advanced methods of standard cell or full custom, where only five or six transistors per logic element are needed.

In standard cell designs, elements of a library of optimized logic components are combined into more complex logical entities. Typical library elements include 1-bit adders and registers, which can be placed in parallel to achieve the desired number of bits of precision. Standard cells usually have fixed geometrical layouts and input/output positions, which can introduce area inefficiencies as elements are combined. Standard cell designs are more rapid and less expensive to implement than full custom designs.

In full custom designs, each logic element is designed individually, allowing greater area efficiency and operating speed. Design occurs in a hierarchical manner, with small elements being built up carefully and then replicated to achieve larger regular structures. Fig. 2–30 displays a typical integrated circuit design process. All later design steps are interspersed with simulation, which includes paths to repeat the current or earlier design stages until the design is as desired. Simulation is increasingly being used at the earliest stages of design as well, as exemplified by the executable system specification and multichip, multiboard system-level functional simulations.

The *requirements design* phase of Fig. 2–30 includes the clear specification of what the circuit is to do. It defines both normal and fault conditions, and it may take the form of a written document or an executable specification that can be simulated. *System definition* produces a quantitative description of all input and output patterns that are expected from the circuit. Its results can be simulated at the functional level. In the *module design* phase, system requirements are decomposed into individual subsystems or modules, which in turn are functionally specified. Behavioral simulation generally provides register-level, clock-cycle-accurate analysis. *Logic design* consists of implementing the functions of each module from the lowest-level components, whether they be standard cells or individual logic elements. Timing simulation examines operation at a resolution that is several times finer than the clock cycle, to search for timing glitches, signal overlaps, and skew-induced errors. The results of logic design culminate in the *layout* of the geometric patterns that will place the logic on the silicon. Associated with layout geometries are delays dependent upon signal path length, parasitic

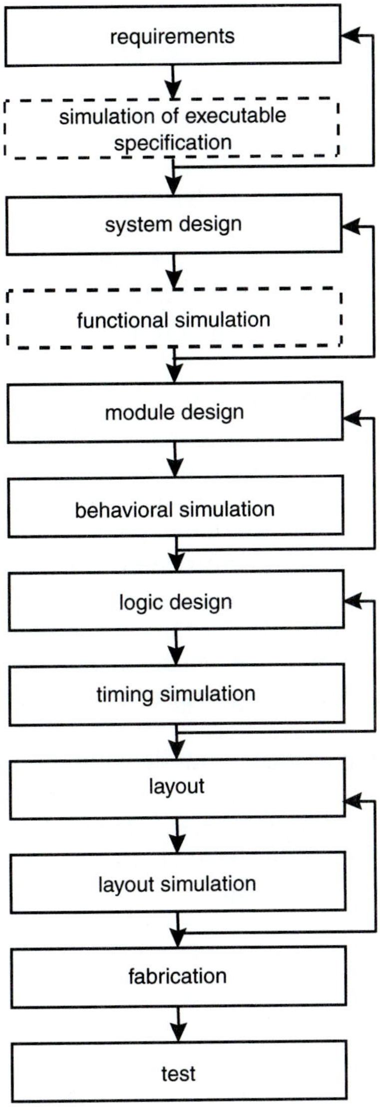

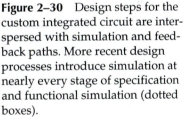

Figure 2–30 Design steps for the custom integrated circuit are interspersed with simulation and feedback paths. More recent design processes introduce simulation at nearly every stage of specification and functional simulation (dotted boxes).

coupling across nearby signals, and other geometry-related effects that are modeled in layout simulation. Finally, the wafers are fabricated, chips are excised, and their functionality is tested.

Often, a partial design, or *redesign*, is sufficient to correct a flawed design or to implement it in newer, finer-geometry technology. Finer design rules are used, and smaller areas, lower cost, and greater speed can result from a redesign. Small bug fixes can be introduced by re-entering the process at the logic design stage.

Layout updates to newer, finer-resolution technology begin at the layout stage, followed by layout timing simulation to screen for asynchronous conditions introduced by logic speedup.

2.4.3 Integrated Circuit Technology

Integrated circuit technology is subject to two trends:

1. Integrated circuit density, as measured by the number of components per unit area, is increasing exponentially as a function of time.
2. Integrated circuit cost per unit function is decreasing as a function of time.

Moore's law,[13] named for Gordon Moore of Intel Corporation, observes that the logic density of integrated circuits, as measured in bits per square inch of silicon, doubles every year. The number of transistors on a chip increases in proportion to chip area, which in turn is based upon the defect density of the manufacturing process (as well as a minor dependency on the density of transistors—memory is much denser than logic). The number of transistors on a chip increases as the square of the decrease in line width set by design rules. As will be discussed shortly, circuit design techniques are improving to address the ever-increasing complexity of chip designs. As a result, the number of transistors in a typical design doubles every two years, and the performance of the leading edge microprocessor doubles every eighteen months.

Due to these trends of increasing circuit density, the area required for a particular function on a chip decreases with time. However, die size cannot decrease below a certain size set by the length of perimeter required to place connection pads to get signals into and out of the chip. Therefore, the manufacturing cost of the die, which is proportional to its area, stops decreasing with improved technology when the chip size becomes limited by input/output (I/O) connections ("input/output-limited"). The problem of I/O is compounded because as chips take on increasing function, they usually require more I/O pins. Furthermore, I/O imposes a penalty of speed and area—drivers are needed to take signals off the chip, signals take longer to propagate the distances of chip, and such techniques as parallel-to-serial conversion of I/O, useful in reducing signal count, require more time.

New methods of packaging and interconnecting bare-die integrated circuits, generically called multichip modules (MCMs), can help overcome these limitations (Section 11.7.4). With MCM technology, multiple bare-die chips are mounted on a high-density substrate or base and are packaged as an integrated group.[14] This technique extends the high-density wiring achieved within integrated circuits to the next higher level of interconnection between integrated circuits and permits wiring densities covering up to 90 percent of a silicon substrate (as compared with the 10 percent typical of conventional circuit boards).

2.4.4 Integrated Circuit Economics

Choosing to include custom integrated circuits within the design of a signal processing system is a major decision, with significant economic and technical consequences. Therefore, it is important to supplement our review of integrated circuit technology with a review of the cost factors of integrated circuit design. The cost of an integrated circuit can be divided into its cost to create the design plus the cost to manufacture the necessary quantities. We will examine each factor in turn.

Integrated circuit design cost is proportional to the amount of engineering effort required to create the design, that is, the average number of persons on the design team multiplied by the design time. There is also a cost of delay in time of the product to market, by which any delay in system release due to designing and creating the integrated circuits delays product availability. This effect allows earlier available competing systems to gain market share (Section 11.2.4). We concentrate on the direct costs of engineering effort.

Engineering effort depends upon the difficulty of the design, which is determined by the number of new logic elements (whether low-level transistors or high-level standard cells), the effort of designing each element, integration efforts to combine these elements, and the labor-saving capability of design tools. The ability to reuse existing components from earlier designs or to include entire functional units such as digital signal processors as core macrocells can greatly reduce design costs. Tools that support the aided or automatic translation from specification language to silicon design can increase productivity. Parametrically described macrocells, which receive a limited number of parameters for input and then automatically generate the appropriate logic structure, allow a balance between the efficiency of customization and the design speed of standard cells. Input parameters can include both logic descriptors like number of bits of precision and layout descriptors like cell shape and position of leads. Engineering effort is reduced by the ability to simulate performance at even the high system level (not just at the lower gate level), and to automatically progress between high- and low-level simulation as needed. Simulation fidelity, which assures that if the circuit works in simulation it will work in reality, is crucial and has become the lifeblood and point of pride of suppliers in the simulator market. Engineering effort increases under the addition of the constraint of backward compatibility, in which the new design is required to duplicate form and function of an existing component. Finally, the degree to which technology thresholds are being pushed, whether in circuit speed, chip size, or layout density, has a strong impact on design costs.

Therefore, the ideal low-cost design consists of a few logic blocks (preferably reused from an earlier design) which are repeated in a regular fashion. Signal flow supports regular layout in an ideal design, and performance demands of the design are well below technology thresholds. Table 2–6 displays examples of estimated development costs for several microprocessors.[15] Because microprocessors are designed with the expectation of large markets, they expend high design costs to achieve low manufacturing costs, and they approach the limit of current technology, using full custom design techniques. As shown in the table, microprocessor development costs rise at about 25 percent per year; this does not include a similar rise in integrated circuit process technology costs, which also rise as design rules become finer. The cost per transistor is shown to remain approximately constant. Despite the fact that advances in design tools decrease design time, more complex circuit designs allow less internal visibility and require more time to design the clever tests and test circuitry needed. Also, to reach the market in time, larger integrated circuits require larger design teams, and more overhead time is spent communicating and coordinating team efforts.

We turn now to the integrated circuit manufacturing costs and the factors that influence them. Integrated circuit costs are divided into the cost of the silicon chip, the cost of the package, and the cost of testing. The major factor that determines chip cost is the cost of the wafer from which multiple chips are diced, divided by the number of successfully operating chips that pass final testing. Wafer cost is proportional in a roughly linear manner to the diameter of the wafer, while the number of chips obtained from the wafer is proportional to its area, or the square of its radius. Thus, larger wafers are more economical, driving a trend from 50-mm to 200-mm diameter wafers, with 300-mm wafers emerging. The number of process steps, including the number of layers of metal, also influences wafer cost. The process yield determines the fraction of chips cut from a wafer that are fully functional. As discussed previously, yield depends upon defect density and chip area in an exponential manner. Fig. 2–31 shows that as time progresses and manufacturing processes improve, the maximum area of a chip that is likely to be defect-free increases.

Testing of individual chips on the wafer is performed before dicing, and inoperative chips are noted. Once the wafer is diced, some testing may again occur to check for dicing-induced failures, but testing is usually postponed until after the chip is packaged, to allow also catching any packaging-induced failures. Postponing testing until after packaging presents difficulties for multichip module designs, which use diced but unpackaged chips. Multichip module designs require fully tested bare chips for assembly into wafer-scale integrated circuit assemblies, requiring that chip manufacturers pull up their full chip testing to a stage prior to packaging for MCM-intended die.

Table 2–6 Cost of microprocessor design increases at about 25% per year.

Year	Processor	Est. Development Cost [15]	Transistors (thousands)	Development cost per transistor
1977–79	Motorola 68000	$5M	68	$74
1989	Intel 80486	$100M	1,200	$83
1993	Intel Pentium	$250M	3,100	$80

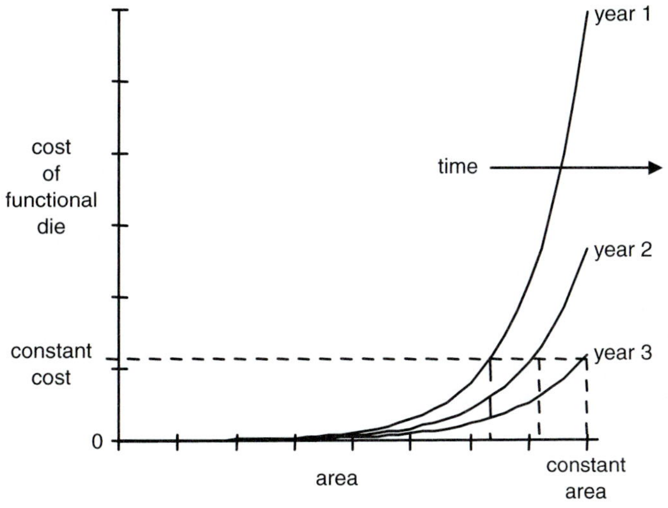

Figure. 2–31 Die area achieved for a constant cost increases with time, due to lower defect density from improved manufacturing processes.

Testing costs are proportional to testing time, both pre- and post-packaging, plus the cost of design-specific test fixtures (amortized across the production volume of the design). Testing time is determined by chip complexity, visibility, and regularity of internal structure. The adequacy of testing is expressed in terms of *fault coverage*, or the fraction of all possible faults of a particular type that, if present, could be detected by test procedures. A typical fault coverage analysis steps through each input and output to every lowest-level logic element, postulating that it is artificially forced to a fixed logic level ("stuck at"), and then determines whether the test procedures can detect that. As chip complexity increases, the complexity of tests to detect a stuck-at fault in the middle of a chip increases, and the number of signals to be tested increases as well. As described more fully in Section 11.6.3, dedicated test circuitry that allows daisy-chain connection of signals and serial readout is used to improve visibility.

Packaging costs are determined by the composition of the package (more-expensive ceramic versus less-expensive plastic), the package size, and the number of I/O leads. Special packages, such as multiple cavity packages and those with especially dense I/O signals, are more expensive.

A simple example illuminates the interplay of design cost, manufacturing cost, and market size. The design of the Intel Pentium microprocessor is estimated to have cost $250 M,[15] while its cost for quantity manufacture was $65 per unit in 1995. Since the cost to design must be amortized over the number of integrated circuits sold, the design cost becomes a smaller contributor to the total cost if the market is large. The personal computer market into which the Pentium is used is estimated to be 65 M systems in 1995, so the $250 M design cost, when spread among one year of systems, is about $4 per system, far smaller than the $65 Pentium manufacturing cost.

The number of systems sold that contain the integrated circuit depends upon the time interval that the design is current, which relates to the time required to develop and build a replacement design. Alternatively, the interval that the design is current can be the time that commercial product performance requires to catch up with the performance of the custom integrated circuit. For complex designs such as the microprocessor, the time to design ranges from two to four years. Because design costs are growing at a rate of 25 percent per year as well, economic success depends upon the difference between growth in design costs and rate of growth of the market. For integrated circuits that are less complex, a design that costs $1 M to complete of an integrated circuit that costs $40 to fabricate would need to exceed quantities of 4,000 per year over a four-year life to limit total integrated circuit cost to $100 per unit.

2.4.5 Trends in Custom Processor and Associated Algorithms

As might be suspected from the above discussion, the number of units needed to sell to achieve economic success has been increasing with time. While 50,000 units may have been adequate in the past, perhaps 250,000 units are necessary now, due to increased design costs and complexity. Often, the desired single-chip functionality can be achieved more easily by awaiting the arrival of the next generation of processor technology, which doubles in functionality every year, than by expending two to four years developing a new design for a limited application. Thus, performance levels expected for a complex custom processor should significantly exceed the capability of off-the-shelf processors expected at the time that the custom processor enters production. For the two- to four-year design cycle of the custom processor, off-the-shelf technology may progress by a factor of ten. The custom processor performance should exceed this order of magnitude. Ironically, as more logic can be placed on a chip, the complexity of function that may be implemented grows in a manner that can increase its specialization and decrease its range of application as the chip takes on increasing amounts of system-level, system-specific function. As a result, the trend is now away from the use of custom processors for all but the most generic or demanding high-performance jobs.

As the design costs for state-of-the-art custom processing elements increase, as their specialization increases, and as funding for the development of low-quantity prototype systems decreases, more use is being made of contemporary catalog processors. These processors, known as commercial off the shelf, or COTS, improve in performance with time and are enhanced by the manufacturers in response to the competitive need to keep these processors current with those from other manufacturers. In contrast, improving custom processors to keep pace with technology requires constant reinvestment from the designer of the system using the custom processor. Therefore, a growing emphasis is being placed on mapping increasingly high throughput signal processing algorithms on arrays of COTS processors. For consumer products, where quantities are high and integrated circuit costs must be kept low, a custom integrated circuit can be included at relatively minor cost and may be essential to meeting size, speed, or power requirements. Indeed, even in small-quantity, high-performance systems, custom integrated circuits may be the only method to achieve performance goals. Thus, while an emphasis on COTS components is growing, there will be situations that require custom processors, and so we continue our examination of custom processor trends.

The advent of very large scale integrated circuit technology (VLSI) encourages the use of algorithm-oriented architectures of arrays of custom processors. An algorithm is more amenable to VLSI implementation if it requires only a few different types of logic cells which are replicated as needed. Data flow that is simple and regular aids design, and algorithms that use extensive pipelining and multiprocessing aid the use of multiple copies of custom processors for their implementation.

The desirability of regular layout in VLSI may imply structuring an algorithm for decomposition into parallel computation, even at the expense of adding placeholder operations to achieve regularity. For example, in a data-dependent computation flow, both paths of a branch may be computed and either result selected, as opposed to executing only the path of a branch warranted by the data. In like manner, iterative loops may be padded with "no-op" operations to the next power to two in the quest for regularity.

Table 2–7 gives some examples of commercial custom processors. The first three are actually programmable digital signal processors that can be placed as a

Table 2–7 Examples of custom processors.

Function	Part	Description
Digital signal processing core	LSI Logic Oak Core	16-bit DSP; 25-nsec cycle time
Digital signal processing core	Texas Instruments 320C5x	16-bit DSP compatible with Texas Instruments 320C2x; 20-nsec cycle time
Enhanced digital signal processor	IBM MWave DSP	DSP with built-in support for Sound Blaster audio card for personal computers
Fast Fourier transform	Zoran ZR 38001	Fast Fourier transform (0.67 msec for 1024-point complex FFT)
Multitap filter	LSI Logic L64240	64-tap 20 MHz transversal filter; two 32-tap sections, 8-bit coefficients
Rank-value filter	LSI Logic L64220	Output chosen from a sorted list of input values

macrocell within an integrated circuit and surrounded with custom interfaces and logic elements on the same piece of silicon. The other three examples execute particular functions that are common in signal processing operations, with specialized architectures accelerating the processing to speeds not obtainable with current programmable architectures.

In the IC design process, the trend is to increase the amount of simulation available early in the design process (Fig. 2–30). The ability to simulate multiple chips operating together is becoming routine, as is the use of entire algorithms running on representative data, which are now used to supplement much simpler tests targeted to stress particular logic elements. As simulation becomes available earlier in the design, testing becomes more distributed throughout the design process. Early designers would build the article, then test it. However, present design practices include virtual testing (in the simulation domain, without actual hardware) as well as physical testing of the final results (see Section 11.6).

Testing of integrated circuits has changed as circuits have become more complex. The ratio of I/O pins to logic on chip has been decreasing as logic quantities grow, providing less visibility into internal functionality. Large chip designs now include specific circuitry to aid testing by increasing visibility. One standard method, known as JTAG,[16] places additional registers at critical points along the datapath (Section 11.6.3). These registers may be daisy-chained together and read out serially when the chip is put into a test mode, increasing internal visibility.

The recent program known as RASSP (for Rapid Prototyping of Application Specific Signal Processors), run by the U.S. government, seeks to codify many of the design trades into an end-to-end process.[17] As described in Section 11.1.3, RASSP provides:

- *virtual prototype:* executable specification of an embedded system and its response to stimuli;
- *executable requirements:* allows customer for custom integrated circuit development to execute chip functions, assure correct specification, and approve the design with full knowledge of its function prior to the start of detailed logic design;
- *connected levels of design process:* links between higher and lower steps of development (e.g., from functional level to logic level) are automated, and the progress from one stage to the next is aided by automation;
- *accelerated design interval:* the RASSP program aims to accelerate the typical design interval by a factor of four;
- *merging of hardware design with software design:* achieves a concurrency known as hardware-software codesign.

Emerging custom processor designs make increasing use of memory. As more functions fit onto the chip, a greater portion of activity becomes data dependent, using memory-intensive operations such as symbolic reasoning and table lookup. The standard cell is being extended from middle-level components (registers, ALU) to entire processor cores, and as a result, user-definable chips are emerging that can be specified with a few parameters and designed by a nonspecialist. Chip area is becoming limited less often by the area of logic and more often by the area needed for I/O signals. Advancing integrated circuit logic densities are increasing the design cost and increasing the number of units required to be sold to maintain economical use of custom VLSI. The advancing capability of standard COTS circuits is emphasizing the use of COTS over custom integrated circuits in small-quantity system designs, but the excellent value of space and economy for modest custom designs are leading to greater use of custom processors in consumer and commercial products. Finally, efforts such as the RASSP program are again striving to decrease the design time of custom processors, again reducing the cost of state-of-the-art custom processor design, and perhaps lowering the barriers for use of custom VLSI.

REFERENCES

1. *Standard for Binary Floating-Point Arithmetic,* IEEE Std 754 (Piscataway, NJ: IEEE, 1995).
2. John L. Hennesey and David A. Patterson, *Computer Architecture: A Quantitative Approach,* 2nd ed. (San Francisco, CA : Morgan Kaufman, 1996), C-1–C-26.
3. *MCS-86 User's Manual* (Santa Clara, CA : Intel Corporation, 1979).
4. Christopher A. Titus and others, *16-Bit Microprocessors* (Indianapolis, IN : Howard W. Sams & Co., 1981).
5. Tom R. Halfhill, "Intel's P6," *Byte* (April 1995), 42–58.
6. Tom R. Halfhill, "80x86 Wars," *Byte* (June 1994), 74–88.
7. Edward A. Lee, "Programmable DSP Architectures: Part I," *IEEE ASSP Magazine* (October 1988), 4–19.
8. Edward A. Lee, "Programmable DSP Architectures: Part II," *IEEE ASSP Magazine* (January 1989), 4–14.
9. Analog Devices, One Technology Way, P.O. Box 9106, Norwood, MA 02062–9106.
10. Texas Instruments, Inc., Post Office Box 1443, Houston, Texas, 77251-1443.
11. Robert E. Owen and Steven Purcell, "An Enhanced DSP Architecture for the Seven Multimedia Functions: The MPACT 2 Media Processor," in *Proc. 1997 IEEE Workshop on Signal Processing Systems, SiPS-97,* Mohammad K. Ibrahim, Peter Pirsch, and John McCanny, Eds. (Piscataway, NJ : IEEE Press, 1997), 76–85.
12. Ray Simar, Jr., "DSP Architectures, Algorithms, and Code-Generation: Fission or Fusion?" in *Proc. 1997 IEEE Workshop on Signal Processing Systems, SiPS-97,* Mohammad K. Ibrahim, Peter Pirsch, and John McCanny, Eds. (Piscataway, NJ : IEEE Press, 1997), 50–59.

13. G. Moore, "Cramming More Components onto Integrated Circuits," *Electron.* (April 1965), 114–117.
14. Rao R. Tummala, "Multichip Packaging—A Tutorial" *Proc. IEEE,* 80 (December 1992), 1924–1941.
15. Nick Tredennick, "Technology and Business: Forces Driving Microprocessor Evolution" *Proc. IEEE,* 83 (December 1995), 1641–1652.
16. *Standard Test Access Port and Boundary-Scan Architecture* IEEE Std 1149.-1a (Piscataway, NJ: IEEE, 1993). Also, Kenneth B. Parker, *The Boundary Scan Handbook* (The Netherlands: Kluwer Academic, Inc., 1992).
17. T. Egolf and others, "VHDL-Based Rapid System Prototyping," *J. VLSI Signal Processing Systems for Signal, Image, and Video Technology,* 14 (November 1996), 123–156.

3

Tradeoff Analysis

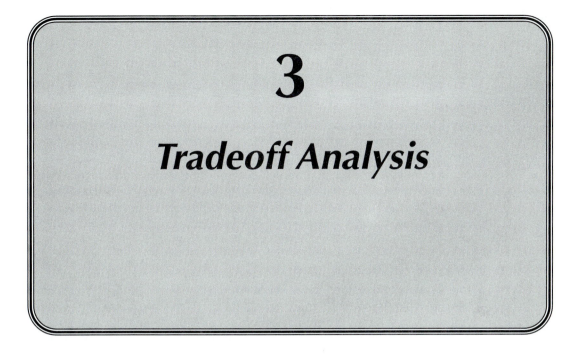

The problem of designing a signal processing system to meet the needs of a real-time processing requirement has no single correct solution. Instead, a signal processing system is crafted through a series of tradeoffs that map elements of signal processing technology into an integrated solution within constraints imposed by the requirements. The relative priority of constraints for a particular system provides the major influence for determining which of the many possible solutions is the best.

3.1 TRADE STUDY

This chapter describes the many tradeoffs that the signal processing designer makes in designing a real time digital signal processing system. It provides both framework and tools for making informed design decisions. By concentrating on the principles embodied in conducting tradeoffs, this discussion endures beyond the signal processing techniques of a single design or generation of technology and establishes a method for performing tradeoffs in a variety of design situations. Tradeoffs performed for one example are performed according to principles that extend to other systems and states of technology.

3.1.1 Types of Tradeoffs

We examine tradeoffs within several areas:

- *algorithm space:* what makes an algorithm "difficult"? Maximizing algorithm speed entails making its computational structures regular and repetitive, thereby allowing the partitioning of the work or data across multiple hardware processors that operate together.
- *software versus hardware:* operational control, by means of a sequence of instructions or program that directs the hardware resources, versus type, number, and complexity of hardware processors. Software is fairly quickly developed, easy to change, and flexible, but it executes relatively slowly, while hardware is more lengthy to design and harder to change, but is faster in execution.
- *space-time tradeoffs in software:* the use of a common software subroutine each time a function is needed saves space and programming efforts, but calls to subroutines must be made more general to accommodate all possible calling situations, and jumps from one section of software to another take time to execute. This is because the state of the machine, as specified by all register contents, must be saved by writing it to memory before jumping and then restored to the registers upon return. Saving the machine state allows resumption of the execution in progress when the jump occurred. Finally, table lookups can replace the computation of results, trading more memory space to store the table for more computation time required to compute a function without a table.
- *space-time tradeoffs in hardware:* when real-time throughput requirements outstrip the capabilities of a single processor, speed can be improved by allocating work among multiple processors, increasing space and speed. In another form of this tradeoff, a purpose-built hardware architecture is used to accelerate a particular processing application beyond what can be achieved with available existing processors. Custom processors require an additional design interval, increasing the time before the system is available in order to achieve a speed advantage.
- *mapping algorithms and architectures to each other:* the types of trades discussed above culminate in the matching of algorithms and architectures. Mappings are constructed through a combination of modifying the algorithm and modifying the architecture—structures and sequences within the algorithm are rearranged to fit them into a specific architecture for real-time execution, and new architectures based on the target algorithm may be developed to attain real-time throughput. The process for mapping proceeds from abstract to specific representations of the algorithm, as will be discussed in Chapter 5. Throughput bottlenecks are identified and alleviated by applying principles of real time execution and by inserting structures for rapid execution that are taken from a repository of experience and past designs.

3.1.2 Formal Trade Study

We describe the use of the formal trade study as a tool to guide designers through the multiple choices that must be made to achieve a design. The trade study is applied both to large system-level issues, such as architecture, and to small subsystem questions, such as choice of input/output interfaces. Thus, a design consists of a hierarchy of trades and choices.

A formal trade study consists of the following steps:

- specifying the goal of the trade study;
- imposing the necessary constraints;
- defining the evaluation criteria;
- developing an assortment of alternative approaches;
- scoring each of these options in accordance with the evaluation criteria;
- selecting a candidate solution;
- validating that the choice achieves the goal within the constraints.

The trade study begins with a set of requirements to fulfill, establishing the goal of the design. Often, the requirements are not fully specified at the start of the design and they may evolve as the design progresses. A throughput specification, such as computing a particular algorithm within a specific time, is a typical form of requirement.

Constraints set conditions that must be achieved in reaching the goal. Examples of constraints include limits on maximum cost, maximum weight, maximum power requirements, or maximum downtime due to failure.

A plot of system requirements versus a constraint can show the allowable regions within which the trade study is completed. Fig. 3–1 shows a plot of per-unit cost versus throughput for a hypothetical system. A maximum allowed cost is shown as a horizontal line, and a minimum allowed throughput, set by the requirement to operate in real time, is shown as a vertical line. The shaded area is

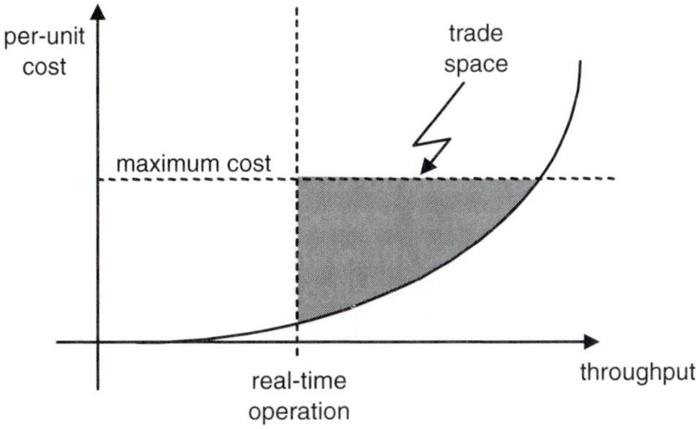

Figure 3–1 In a plot of cost versus throughput, an area for tradeoffs is bounded by constraints on minimum throughput and maximum cost.

the trade space, as set by the constraints of maximum cost and minimum but real time throughput and the functional dependence of cost upon throughput. When pairs of all such constraints and functional dependencies are mapped, a trade volume in the multi-dimensional space is defined. In system design, tradeoffs are conducted to identify the best points, or design choices, within the volume within a multi-dimensional space set by the constraints.

The concept of "best" choice is based upon evaluation criteria that are developed as a step in the trade study process. Sources of possible evaluation criteria include system requirements, goals, knowledge of the application, and experience from past designs. Often, criteria are generated and weighted in importance by a brainstorming processing involving a diverse group of experienced design engineers.

The evaluation criteria are weighted, for example, as either "must have," "try to have," or "nice to have." This is best performed by quantitative simulation of the system, providing the impact of a change in one criterion upon the performance of the total system. More typically, system-level simulation may not be available. Members of a diverse group of designers (or better still, potential customers for the system) may generate a rank ordering of available criteria. A list of candidate criteria is first generated in a brainstorming process, and duplicate criteria are combined or removed. Then, each person rates each criterion as high, medium, or low, and a collective score is computed. If several participants are performing the ranking, the top and bottom score for each criterion can be thrown out prior to combining, to achieve greater statistical robustness. The end result is a list of criteria with a numeric score weighing each in importance.

Next, a set of alternative solutions is developed. These may come from past designs (appropriately modified), from references in the literature, or from the creative efforts of design team members. These options are then scored against the evaluation criteria. In some cases, mathematical models or simulations may be available to compute quantitative scores. Otherwise, educated judgment is made by a diverse group of experienced designers, whereby each option and criterion is discussed and a consensus is sought. Based on the scores of each option against the evaluation criteria, a choice is made for the highest-scoring alternative. This choice is compared against the goal and the constraints as a final check of validity.

Fig. 3–2 presents an example of a trade study to choose between a hardware and a software approach of implementing a function. The requirement is to achieve real-time execution of that function, and evaluation criteria are grouped into "must have," major (weighted as 1), and minor (weighted as 1/2) groups. The options are two (software and hardware), and each is scored using a simple scale of –1 (fails or poor), 0 (modest influence), and 1 (meets criteria or favorable result) against each criterion. Computing a score shows a clear advantage to the software solution for this example. A final check of the software solution against throughput and constraints on maximum cost, size, and power is then made to substantiate the choice.

Criteria	Factor	Weight												Total Score	Weight Score	Choice
(must)	throughput	must														
(major)	design time/cost	1.00														
	unit cost	1.00														
	size	1.00														
	power	1.00														
	flexibility	1.00														
(minor)	memory required	0.50														
	processors required	0.50														
	risk	0.50														
	expandability	0.50														
	cost of smallest system	0.50														
Options	Software			1	–1	1	1	–1	1	1	0	1	1	1	4	3.5	X
	Hardware			–1	–1	–1	–1	1	–1	–1	0	–1	0	1	–5	–3.5	

Figure 3–2 Example of a trade study that compares a software and hardware option to achieve the requirement of real-time execution. The evaluation provides the higher-weighted score to the software option, suggesting that it be selected.

3.2 ALGORITHM COMPLEXITY

As we have seen, an algorithm may be viewed as a prescription for transforming one set of data into another. A *data set* is a structured set of data elements defined by both the range of the values of its elements and the structure among them. An *algorithm* is a particular realization of a computation, and a *computation* is a function that receives and generates data sets. The algorithm decomposes the computation into subcomputations with an associated precedence relation that determines the order in which these subcomputations are performed.[1]

The precedence relation can be described with a directed graph in which the nodes represent subcomputations and the connecting arcs define the data flow. Algorithms can be classified into sequential or parallel types according to the structure of the precedence graph. As shown in Fig. 3–3, a sequential algorithm has only one subcomputation active at a time, while a parallel algorithm has multiple subcomputations occurring at once.

Initial algorithm descriptions occur at the highest level with little or no specifics on implementation. Subsequent descriptions include more implementation details, such as number of bits of precision of data values. A *program* is a

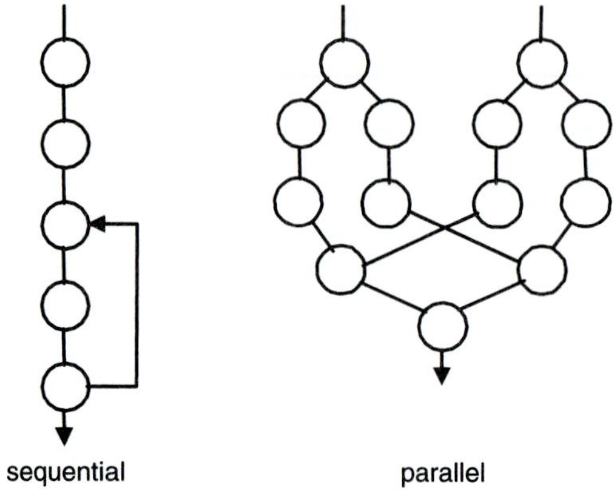

Figure 3–3 Signal flow diagram for sequential and parallel algorithms.

more detailed level of algorithm description that enforces precedence as an ordering or sequence of steps which is used to automatically direct the execution of the algorithm on a computer.

Repetitive, regular subcomputations are sought in decomposing algorithms onto architectures, leading to parallel structures which can be partitioned across multiple computing units and executed simultaneously. As will be discussed in Section 4.1.6, two types of parallelism may be achieved. *Task parallelism* allows simultaneously performing different tasks on the entire dataset on different processors, as shown on the right in Fig. 3–3. *Data parallelism* simultaneously performs the same tasks on subsets of the data.

The factors that increase algorithm complexity also decrease the amount of repetitiveness, regularity, and parallelism. Parameters of an algorithm that determine its complexity for real-time implementation include:

- throughput (number of operations, amount of data to process, and time available to process);
- range and precision of numbers;
- amount of "difficult" instructions (those requiring significantly more than the average number of instruction cycles);
- data-dependent execution, whereby the instruction sequence is influenced by the incoming data;
- precedence relations within the algorithm, as well as the lifetime of data values within the computation;
- global versus local communication of data;
- random versus regular sequencing of data addresses;
- diversity of operations.

We next look in more detail at these parameters of complexity.

3.2.1 Throughput

Real time processing, as described in Chapter 1, entails computing a certain number of operations upon a required amount of input data within a specified interval of time, set by the period over which that data arrived. Therefore, real time throughput requirements impact algorithm complexity by determining: 1) the number of operations to perform, 2) the amount of data upon which they are performed, and 3) the time available to perform these operations. These three factors affect the ability of processing within the data sample period. The faster the data input, the less time there is to process between samples. The more complex the processing, the harder it is to complete between samples. The complexity of the real time implementation increases with the data throughput and the amount of processing.

3.2.2 Numeric Range and Precision

The precision and range of numbers impact complexity. As discussed in Section 2.1.1, a number is represented with a fixed number of bits, which may be allocated according to a tradeoff between dynamic range and precision. Dynamic range is the range between the most negative and the most positive number encountered. The number of bits determines the number of numeric levels available and how precisely consecutive digital numbers sample the continuous space of numeric values. Complexity increases with the number of bits. As the number of arithmetic operations increase, the need for bits increases. For example, adding two numbers of N bits each results in a sum of $N + 1$ bits, and multiplying two N-bit numbers yields a product of $2N$ bits. Computers have a fixed number of bits in the arithmetic path, so results of operations that exceed the allowed number of bits must be reduced in size, either by scaling, truncation, or rounding. In a purpose-built (custom) architecture, increasing the number of bits increases the area, approximately as the square of the number of bits. A class of machines known as *bit serial architectures* achieves maximum area efficiency by limiting communication paths to one bit, at the expense of requiring additional time to decompose and recompose parallel numbers to and from a series of one-bit values. Floating point representations (Section 2.1.1) reserve a subfield of a parallel field of bits to serve as a scaling component, increasing dynamic range at the expense of the added complexity of floating point arithmetic.

The effect of finite word length on digital filters provides an example of the requirement for numeric precision. For the relatively simple lower-order infinite impulse response filter, finite word length effects can lead to *zero input limit cycles*, whereby after the input to the filter goes to zero, the output decays and then oscillates within a non-zero range.[2] The effects of limited precision from finite bit width is also seen in hidden Markov modeling for speech recognition, which involves computing the probability of a particular observation sequence O, given a hidden Markov model λ (Section 10.2.3). The sequence O consists of a series of observations $\{O_1, O_2, \ldots, O_T\}$, each emitted at a time $t = 1, 2, \ldots, T$, with the system in state q_t at time t. The requirements is to compute $\Pr(O_1, O_2, \ldots, O_t; i_t = q_i \mid \lambda)$ where i_t is the Markov model state at time t set to q_i. The observations O_t are either continuously variable or are selections from a codebook of perhaps 1024 possible discrete values, based on vector quantization (Section 9.2.2). Observations are made every 15 msec, and so a typical 1-sec utterance consists of 67 observations, each of which is taken from a 1024 (2^{10}) element codebook. Thus, $(2^{10})^{67}$ possible sequences exist for one second of speech, and the probability of any one sequence is on the order of the inverse of $(2^{10})^{67}$. As a result, the probability tends to zero with geometric speed, and finite word length effects will result in probabilities underflowing to zero. Scaling of these probabilities must therefore be performed in computation.[3]

3.2.3 Multi-Clock-Cycle Operations

Multi-clock-cycle operations depart from the single instruction cycle timing that is typical of simpler instructions. Multiple clock cycles are needed either because the operation is relatively uncommon (e.g., the $\tanh^{-1}$ operation) and has not been the target of speedup with dedicated hardware, or because it requires multiple approximations to compute, for example, computing a reciprocal using a Taylor series expansion. The operation of multiplication, which serves as the building block of digital signal processing, was in older designs performed in multiple cycles by repetitive shifts and adds. Even now, one may trade a smaller number of bits in parallel for more clock cycles of operation (see Section 3.3.2). The single bit communication link of bit-serial arithmetic presents the extreme case. Since chip area increases approximately as the square of the number of parallel bits used in processing, bit serial arithmetic provides the best use of operations per second per unit silicon. However, bit serial arithmetic requires multiple clock cycles to reassemble each word or to perform the arithmetic operation on a bit-by-bit basis. Of course, as the unit of work performed becomes larger, and as full algorithms are built up from basic bit-level operations, more and more clock cycles are needed.

3.2.4 Data-Dependent Computation

High-speed computing is most easily achieved for algorithms that are regular, i.e., that perform the same operations on each piece of data. Regular algorithms ease the use of multiple processors operating in parallel. Data-dependent computations and data precedence requirements for sequential execution pose obstacles to achieving task parallelism (executing multiple tasks in parallel). The requirement of global communication increases the difficulty of achieving data parallelism (performing parallel computations on subsets of the data).

Data-dependent computation introduces data-dependent timing, making it difficult to fit the various data-dependent processing operations within the fixed time interval. One must instead budget for the maximum possible time of execution and insert wait states into shorter operations to lengthen them to this maximum. Data-dependent computation also results in stochastic rather than deterministic execution paths through decision-intensive code, again resulting in stochastic execution time. Timing may depend upon the specific bit patterns within a binary number. For example, Booth's multiplication algorithm scans a bit field for zeros and ones, performing different operations for each (Section 3.3.2). Execution time may depend upon data magnitude as well, as exemplified by operations to sort lists of numbers.

3.2.5 Data Lifetime and Precedence

The lifetime of a particular piece of data in a computation contributes to complexity. Computations that use a piece of data once and then discard it (e.g., the running sum of incoming samples) are more amenable to stream processing algorithms (see Section 1.2). Stream processing algorithms require less storage, avoid the need to again find a piece of data from within a random memory array, and reduce the latency of results. In contrast, block processing algorithms, which collect all samples at once before acting upon them, require time to accumulate numbers, which introduces latency.

Closely related to data lifetime is data precedence, whereby earlier stages of a computation must be completed before later stages can begin. End-to-end signal recognition algorithms have strong precedence requirements—input data is first screened for areas of interest, then detailed computation occurs for these areas of interest. In like manner, to present the top score of a list of candidates as a recognition choice requires computing scores for the entire list first, imposing a precedence requirement of scoring before sorting. Similarly, the two-dimensional Fourier transform that is computed by two one-dimensional transforms requires that all row transforms be computed, then corner turn (matrix transpose) be completed, and then the column calculations be executed. Precedence rules restrict the ability to achieve task parallelism, thereby reducing the opportunity to speed computation by dispatching multiple processors to work on different stages of the chain in parallel.

As an example of data lifetime and precedence, Fig. 3–4 presents flow charts of two numeric operations. In the first, a running sum is computed, treating each input sample once and then discarding it. In the second, the median of a list of values is computed. This second flow chart has two complications:

1. The execution path depends upon the relative magnitude of number pairs, which are sorted into a rank order list prior to choosing the middle, or median, value.
2. The sorted list cannot be computed until all numbers have arrived, necessitating the storage of all numbers used in the computation and postponing sorting until all numbers have arrived.

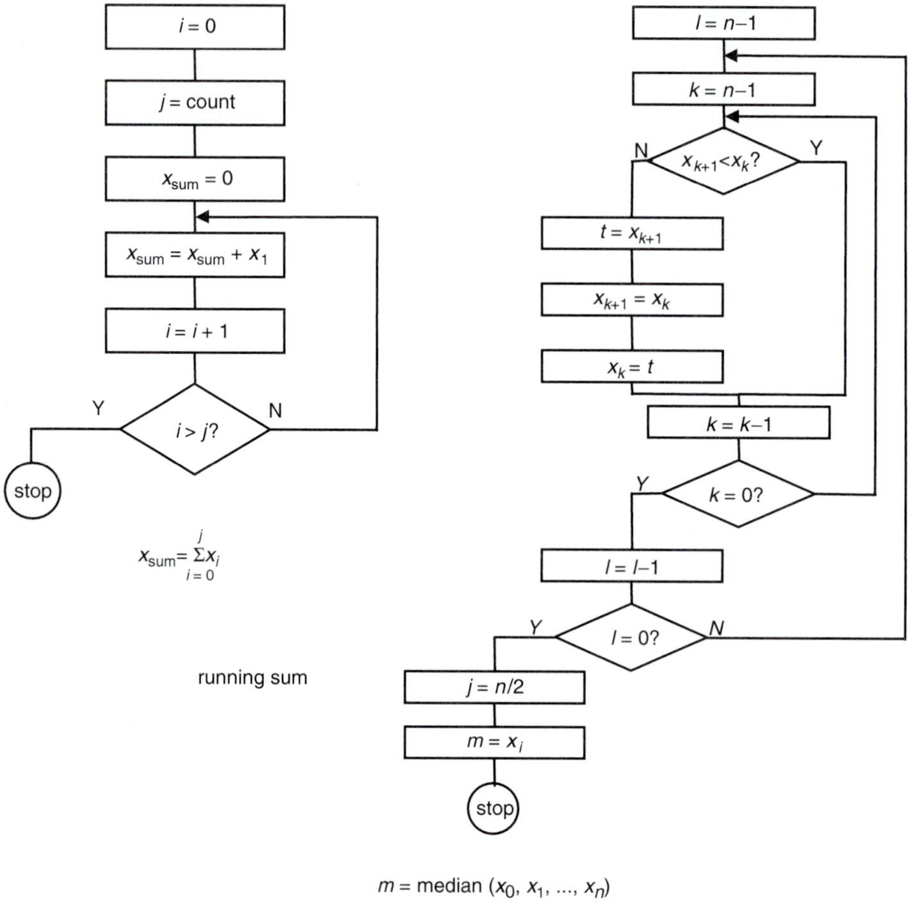

Figure 3–4 Running sum algorithm (left) uses each number x_i to update sum, while median computation (right) requires sorting of entire list of numbers before choosing the median m.

3.2.6 Global versus Local Communications

Just as the requirement for task precedence can restrict exploiting task parallelism, the requirement for global communication can restrict the ability to subdivide an input stream and process various portions in parallel (data parallelism). We consider the example of a two-dimensional array of data, such as pixels in an image, to describe the difference between local and global communication requirements.

In a computation that requires local communication, the transformation of each pixel depends only on the data from those pixels adjacent to it, plus perhaps some external coefficients. Processing occurs by mean of a two-dimensional kernel, or neighborhood, whose center pixel is stepped across the image.

The left image of Fig. 3–5 shows a neighborhood around a center pixel that is stepped across the image. This center pixel is updated as a function that depends on pixels in a neighborhood immediately around the center pixel, as well as the center pixel itself. Thus, the update of pixel $P_{x,y}(t)$ is given by a function F of surrounding pixels at the previous time interval $t-1$:

$$P_{x,y}(t) = F\left\{\begin{matrix} P_{x-1,y-1}(t-1), P_{x-1,y}(t-1), P_{x-1,y+1}(t-1), \\ P_{x,y-1}(t-1), P_{x,y}(t-1), P_{x,y+1}(t-1), \\ P_{x+1,y+1}(t-1), P_{x,y+1}(t-1), P_{x+1,y+1}(t-1) \end{matrix}\right\}. \quad (3.1)$$

The kernel neighborhood of ± 1 pixel can be expanded to ± n pixels or changed from a square shape. The communication required to update each pixel is restricted to the neighborhood of surrounding pixels. This communication is *local*—pixels in far regions of the image do not affect the update.

In contrast, for *global* communication, the computation of one data value depends upon other data throughout the array, not just nearest neighbors. As an example, we choose the two-dimensional Fourier transform, computed as the result of two one-dimensional Fourier transforms (Fig. 3–5, right). The first Fourier transform is performed on each row of the image. Each row or group of rows may be mapped to a processor within a multiprocessor array to allow parallel processing.

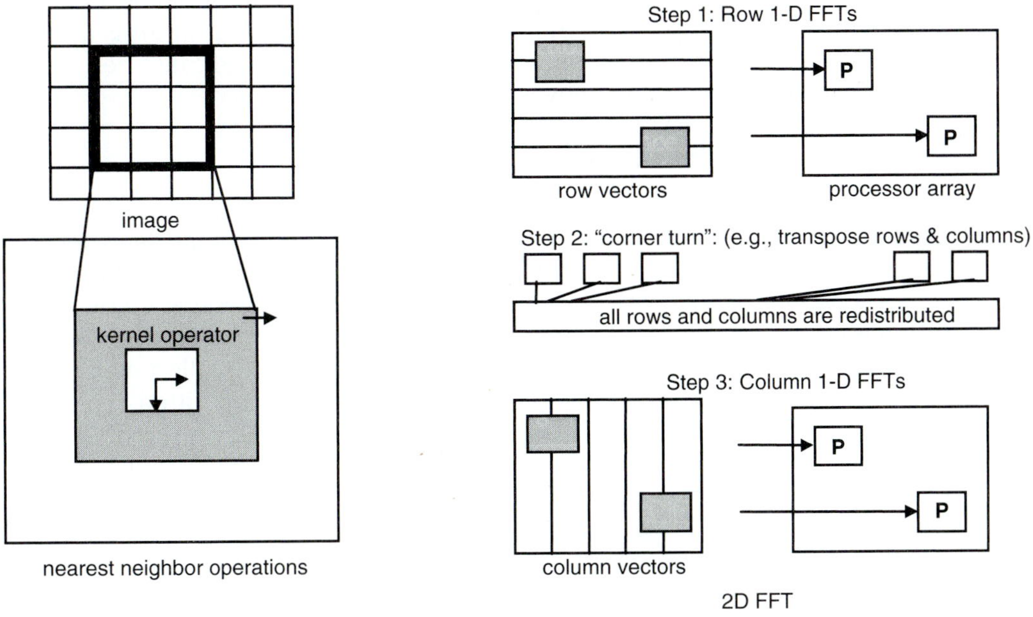

Figure 3–5 Comparison of local communication (left) with global communication (right) for processing of two-dimensional array.

After completing the Fourier transform on the rows, image rows and columns are transposed in a corner-turning operation. Finally, a Fourier transform is performed on each column, possibly with multiple processors, to complete the two-dimensional Fourier transform.

The transpose operation requires moving each pixel, and each output pixel is affected by every input pixel, necessitating global communication. The global communication makes it difficult to subdivide the image into several subimages to perform smaller two-dimensional Fourier transforms on each subpatch with parallel processors.

3.2.7 Randomness of Data Addressing

Just as data values are computed, address locations for data must also be computed. The regularity of addressing data influences the complexity of signal processing algorithms. At the simplest extreme, one coefficient used repetitively to multiply each data value (e.g., constant gain factor) requires no extra effort for address computation. On the other extreme, very complicated, irregular, nonrepetitive addressing sequences can require a separate processor devoted exclusively to computing addresses. Examples include data dependent operations like sorting and indirect addressing techniques that use a computed data value to derive an address of data in a structured list.

Fig. 3–6 shows a one-dimensional array of numbers, with arrows indicating the jumps through address space. The plots show a waveform of address values versus time. These plots range from the simple case of a single address (horizontal line) through linear and circular addressing, more complicated periodic addressing, and finally random movements through memory.

3.2.8 Diversity of Operations

One can imagine creating a coordinate system by which algorithms are characterized, in which each axis corresponds to one of the attributes that characterize an algorithm, such as the list used here in the discussion of complexity (throughput, range/precision, multi-cycle operations, . . .). If each subcomputation of an algorithm is characterized by a point in that space based on the value of its attributes, an entire algorithm will describe a volume made up by the collection of points. The size of that volume corresponds to the diversity of the algorithm.

Greater algorithm diversity results in greater algorithm complexity, as computational requirements extend beyond the capability of a single regular architecture. Although diversity opposes regularity, it is a natural consequence of the structure of advanced signal processing algorithms. Signal analysis algorithms, for example those for speech or image recognition, receive vast quantities of data at the input and derive a compact representation of that data at the output, resulting in a data rate reduction.

Fig. 3–7 shows examples of the structure of end-to-end speech and image recognition algorithms. The speech algorithm begins with preemphasis, which

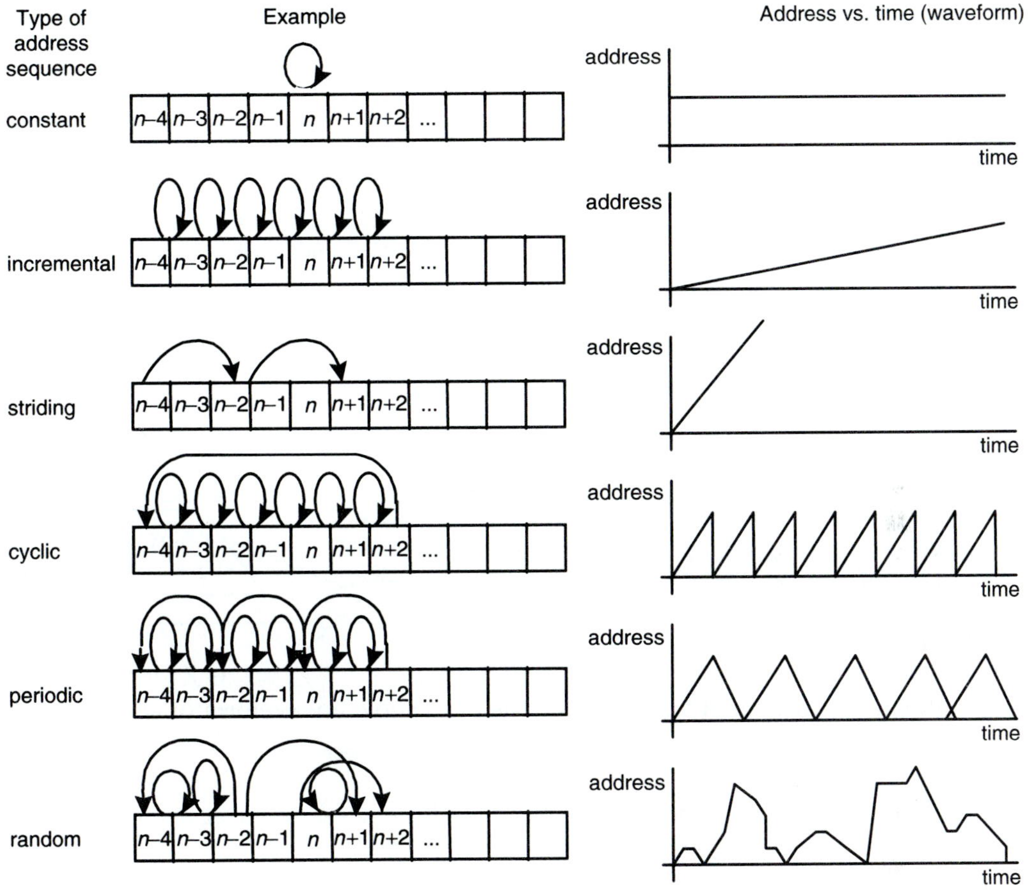

Figure 3–6 Addressing sequences occurring in signal processing, from simple to complex.

equalizes the average spectrum versus frequency, then proceeds to endpoint detection (finding the utterance within the listening window), spectral modeling, acoustic matching, matching to known components such as phoneme-like subword units, and finally the application of a higher-order language model. In like manner, the image processing begins with scanning a scene for detection of possible areas containing targets of interest (the analogue to endpoint detection), then proceeds to the computation of features, which are often based on spatial frequency. A symbolic representation is then achieved by mapping these features to those potential targets. Finally, higher-level knowledge, such as patterns of deployment and geographic information about allowable locations, is used to select the final recognition candidates.

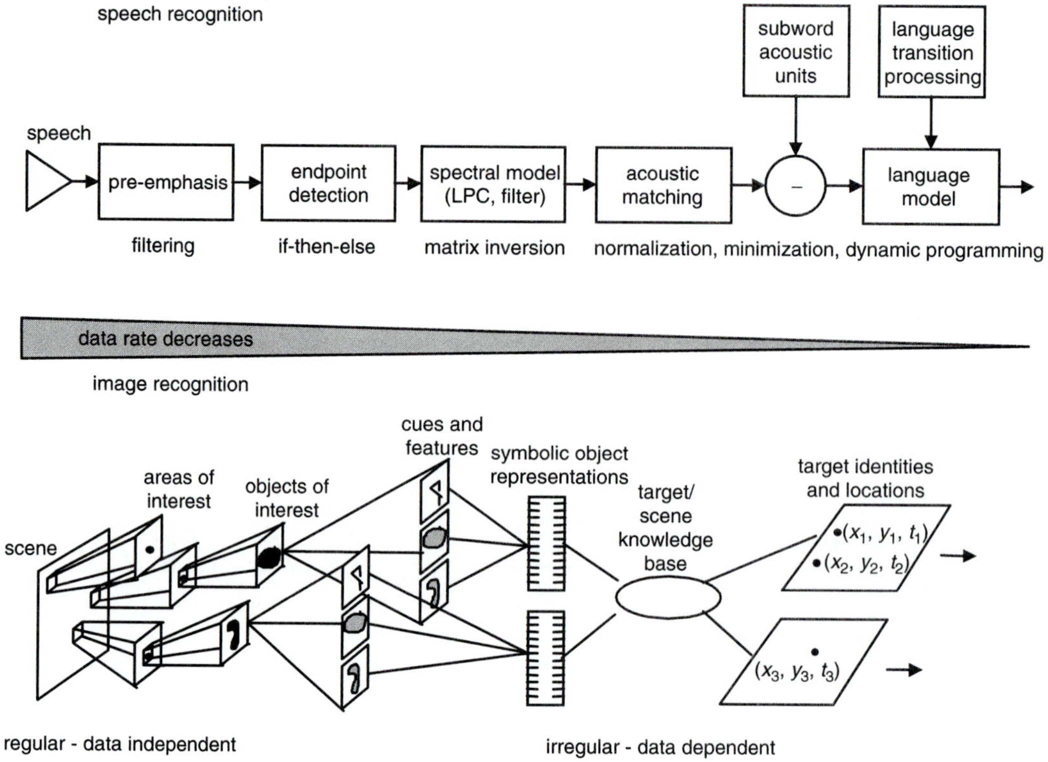

Figure 3–7 Speech and image recognition algorithms proceed from regular operations on high rates of data to irregular operations on low data rates.

Both the speech and image recognition processes begin with structured algorithms operating on all input data in a regular, repetitive manner. Data dependence then enters to complicate the processing, beginning with the exclusion of areas not containing the word or object of interest from subsequent processing. The search for objects of interest can be placed into regular form, at the cost of increased computation, by postulating an object at every possible location. While postulating a word at every frame is done for connected speech recognition,[4] it is too computationally cumbersome for image recognition algorithms to postulate an object at each pixel.

Matching and search algorithms are next reached in the recognition chain. These algorithms require data-dependent execution to perform global matching, which is computed as a sequence of local matches combined along matching paths based on the ranking of match scores. Moving through a decision tree also requires data-dependent execution.

Higher-level or contextual processing is very irregular and data dependent. For speech, these operations include applying language models, transition probabilities, and lookup tables to determine parts of speech. Synonyms and grammar rules are also used in the search. In image recognition, target scene knowledge is applied at this stage, including overlaying the image and its potential target detections with a mask of allowable areas for target location (based on effects of distance from roads, slope of terrain, obstruction of view by trees or rocks, etc.), as well as the spatial interrelationship between groups of targets.

Therefore it can be seen that recognition algorithms progress from identically processing every input sample to choosing the best model by ranking numbers, constructing global paths, and outputting a choice and a confidence score for that choice. The front end consists of structured signal processing algorithms for which the bulk of the computation is aided by regular structures. More irregular operations then follow that are not easily characterized and are not amenable to regular computing structures. The computing resources for recognition and analysis algorithms must span a full range, from regular to irregular, perhaps by using a single powerful processor that can perform some portions of the operations well and can perform the rest adequately. Another choice is a *heterogeneous* architecture that consists of several interconnected processors, each tuned to execute a portion of the computation, with parallel, regular structures at the front and fast general purpose computing at the rear.

3.3 SOFTWARE-HARDWARE TRADEOFFS

In designing a real-time digital signal processing system, a basic first choice is whether to implement a function in software or in hardware. Once asked at the system level, this question is repeated for the several modules that make up the system. Here, *software* means programs that control a set of computing hardware resources to implement an algorithm, while *hardware* consists of particular computing components that may also include custom controlling software. Therefore, the software versus hardware trade is also a trade of program control versus processing elements.

3.3.1 Types of Control

Fig. 3–8 shows the various categories of control systems classified as programs (software) and processors (hardware). The program stores its control information in one of two forms. In *procedural*, or sequential, control, the program specifies the correct sequence of operations, while in *declarative*, or event-driven control, the program specifies the state of active operations.

Procedural control emphasizes *control flow*, by which the program specifies what happens at each clock tick, and operations are executed in an order determined by the program. In declarative control, or *data flow*, operations are executed in an order and at a time determined by data interdependencies (i.e., the ar-

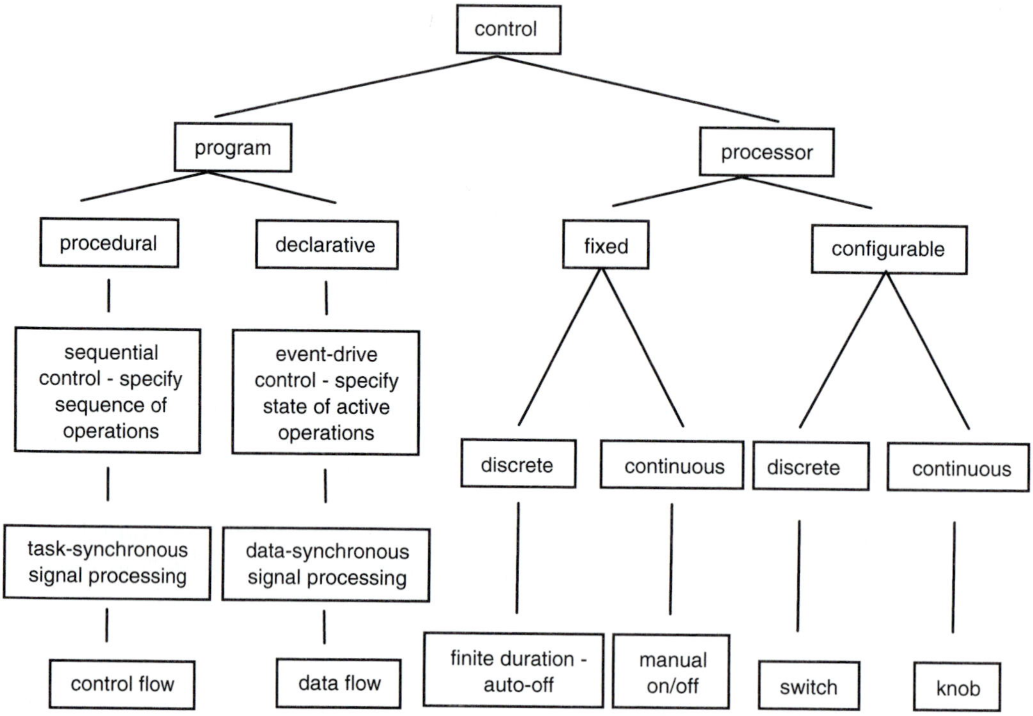

Figure 3–8 Taxonomy of software and hardware methods of control.

rival of all necessary inputs to a processing node before that node is scheduled for execution) and by resource availability (i.e., scheduling a task at the next time that an appropriate processor becomes available).

A hybrid of data flow and control flow is also possible. An example is the U.S. Navy AN/UYS-2 Enhanced Modular Signal Processor (EMSP).[5] This computer uses data flow for task scheduling on its various processors and control flow for executing tasks within each processor. The AN/UYS-2 dataflow architecture is discussed more fully in Section 4.2.

As shown in Table 3–1, the software-hardware tradeoff involves many of the same factors as the decision to design custom hardware (Section 2.4).

A programmable processor can be programmed for a variety of tasks, but as a result, for a particular task, the processor is slower than a processor that was custom-built for that task, and not all of the programmable processor capability is used. Thus, software implementations direct flexible resources on existing hardware, the capability of which extends beyond the specific problem at hand. Conversely, hardware solutions focus on the specific problem and can be stripped to include only the essentials needed. They are faster and smaller than software solutions, but they take longer to design and are harder to change.

Table 3–1 Overall characteristics of software vs. hardware design choice.

Factor	Description	Software	Hardware
Execution speed	need only achieve real time; little advantage in being faster	slower than hardware, but sufficient if real time is met	faster than software
Time to design	design interval from start to first article—software allows incremental design and phase-in of function; hardware does not	shorter	longer
Size	system size, measured by weight, volume, power	larger	smaller
Flexibility during design	ability to accommodate changes during design interval	higher	lower
Flexibility after design	ability to change function in field	higher	lower

The design options for software and hardware may be contrasted. Software design can either be machine independent or machine specific. In machine-independent design, a high level language such as C is compiled to a machine-specific language by a computer program. High-level language programs are faster to write, as each line of high-level code results in many machine-level instructions, and they may be moved from one machine to another (i.e., they are portable) by simply recompiling for a new machine. They are easier to maintain and update, and so machine-independent high-level language is the preferred method of implementation, if it can run quickly enough. Machine-dependent software requires writing the original software in the language that reflects the architecture of a specific computer. While the resulting software runs faster, it takes longer to write and must be rewritten if moved to another machine.

Hardware design, like software, can also occur at several levels of detail based on the capability of the functional building blocks that are manipulated and connected during design. At the macroscopic level, entire processors can be replicated and interconnected according to a variety of topologies. At the microscopic level, individual transistors are connected to form the logic gates that are built up into functional units.

More typically for hardware design, at the top level of the design hierarchy, one begins with an existing processor core and then customizes it by adding special input and output circuits and controlling such parameters as amount of memory, number and arrangement of processors, and distribution of memory among processors. At the middle level of design detail, existing processors are combined into a multiprocessor system, possibly designing custom-integrated circuits to provide interconnection capability and perhaps designing a custom processor as a member of the array. At the most detailed level, as described in Section 2.4, an individual processor is crafted through the interconnection of lower-level components. These components may be field-programmable gate arrays, macro logic cells of entire processor elements, standard logic cells, or transistors for a full-custom design.

3.3.2 Software/Hardware Tradeoff Example - Multiplier

The multiplier is central to digital signal processing, and so it is used here as an example to illuminate several principles of the software/hardware tradeoff. These underlying principles include:

- simple structures, given enough time, can compute nearly anything;
- techniques that humans use for computation can also be used as the basis for machine computation for in real-time operation;
- small modifications to the techniques of human computation can achieve greater efficiency;
- even greater efficiency can be obtained by algorithm changes;
- further speed can be gained by decomposing the problem into pieces with separate parallel resources.

This discussion begins with the slowest and simplest method for computing the product of two numbers, and then it proceeds to faster, more complex methods. It begins with a software technique expressed in a few lines of code whose execution time is very long and it arrives at a hardware technique requiring an array of logic elements that can produce an answer within a single clock cycle.

The simplest method that may be used for multiplying two positive integers, x and y, is to merely add x to itself y times. This requires y addition operations, and if y is represented by a B-place number of radix r (that is, $y = \Sigma_{b=0}^{B-1} j_b r^b$), the number of operations is seen to be proportional to the exponential of the number of bits. The number of operations need to compute y, or $N_y^{rep\text{-}add}$, for this repetitive-add technique is of the order of r^B, or $y = O(r^B)$. Thus, for a binary number, $r = 2$ and $y = O(2^B)$.

Another software technique implements the paper-and-pencil method that humans use to multiply and accomplishes its result in time proportional to the number of places (digits) in the radix implementation ($N_y^{pap\text{-}pen} = O(B)$). Fig. 3–9 shows the method for the multiplication of two 4-place numbers, x (represented in bits by $(x_3 x_2 x_1 x_0)$ and $y = (y_3 y_2 y_1 y_0)$.

For this paper and pencil method, all partial products are formed and shifted into proper alignment before addition begins. The shift and add method of multiplication may be written:[6]

				x_3y_0			
			x_3y_1	x_2y_1	x_2y_0		
		x_3y_2	x_2y_2	x_1y_2	x_1y_1	x_1y_0	
+	x_3y_3	x_2y_3	x_1y_3	x_0y_3	x_0y_2	x_0y_1	x_0y_0
	x_3y_3	$x_3y_2+x_2y_3$	$x_3y_1+x_2y_2+x_1y_3$	$x_3y_0+x_2y_1+x_1y_2+x_0y_3$	$x_2y_0+x_1y_1+x_0y_2$	$x_1y_0+x_0y_1$	x_0y_0
×	2^6	2^5	2^4	2^3	2^2	2^1	2^0
p_7	p_6	p_5	p_4	p_3	p_2	p_1	p_0

Figure 3–9 Multiplication of two 4-place numbers $x = (x_3\,x_2\,x_1\,x_0)$ and $y = (y_3\,y_2\,y_1\,y_0)$ forms and adds all partial products and multiplies each by the proper factor of r^b, where r, the modulus of operation, is 2 in this example.

$$x \cdot y = x \cdot \sum_{b=0}^{B-1} y_b r^b = \sum_{b=0}^{B-1} x \cdot r^b \cdot y_b \tag{3.2}$$

Here each of the B places of y requires a one-position left shift of x, followed by multiplication by the single radix-r digit y_b, requiring multiple operand capability and storage. For a computer, it is more efficient to perform the addition concurrently with the formation of partial products, thereby avoiding the need to place partial products in memory and retrieve them for addition.[7] As a result, a recursive method is used instead of the direct method of (3.2) to form a sequence of partial products:

$$p^{(j+1)} = \frac{1}{r}\,[p^{(j)} + r^{(B)} \cdot x \cdot y_j] \text{ for } j = 0, \ldots, B-1. \tag{3.3}$$

Completion of the recursion gives:

$$p^{(B)} = x \cdot y. \tag{3.4}$$

This approach requires B steps, with each step consisting of a shift, a multiply, and an add. The number r^B multiplying x merely indicates that x has to be properly aligned with the significant half of the partial product.

The above methods can be extended from positive integer multiplication to sign-magnitude multiplication by rewriting y in two's complement:

$$y' = y - 2^B \operatorname{sign}(y) = \sum_{b=0}^{B-1} y_b 2^b - 2^B \operatorname{sign}(y) \tag{3.5}$$

The multiplication is then performed recursively as before, except for the last step, which subtracts rather than adds the multiplicand in the last step. In the particular case of computing the scalar product of two vectors, $\mathbf{x} = (x_0, x_1, \ldots, x_{M-1})$ and $\mathbf{y} = (y_0, y_1, \ldots, y_{M-1})$, the product is given by $\mathbf{x} \cdot \mathbf{y} = \sum_{m=0}^{M-1} x_m y_m$. The use of *distributed arithmetic* can convert the sum of products to a series of power-of-two computations that can be looked up in a table (Section 9.3.1).

The discussion of multipliers has so far focused on software techniques, ranging from $O(2^B)$ to $O(B)$ steps. Operations can be further accelerated by looking more carefully at the bit patterns of the operands themselves, either via software or hardware.

The first technique to examine individual bit patterns is Booth's algorithm,[8] which can be applied in either hardware or software. Booth's algorithm exploits the fact that any sequence of 1's in consecutive places of a binary number can be represented by only two 1's, appropriately placed (Table 3–2).

The Booth's algorithm representation, which requires a simple shift and subtract, is then used instead of performing multiplication on each digit sequentially. Thus, if a number $y = (y_{B-1}, y_{B-2}, \ldots, y_0)$ and all digits y_{i-1} to y_{i-n} are 1, then

$$y_{i-1} \cdot 2^{i-1} + y_{i-2} \cdot 2^{i-2} + \ldots + y_{i-n} \cdot 2^{i-n} = y_i \cdot 2^i - y_{i-n} \cdot 2^{i-n} \tag{3.6}$$

and so

$$x \cdot (y_{i-1} \cdot 2^{i-1} + y_{i-2} \cdot 2^{i-2} + \ldots + y_{i-n} \cdot 2^{i-n}) = x \cdot y_i \cdot 2^i - x \cdot y_{i-n} \cdot 2^{i-n}. \tag{3.7}$$

This technique converts the one-bit-at-a-time multiply-shift-add sequence to an operation of seeking consecutive 1 bits and for them:

1. Add two's complement of x with its least significant digit on digit location $i - n$.
2. Add x in its true representation with its least significant digit on place i.
3. Do nothing for the intervening bits.

To accommodate two's complement multiplication, y is represented as before:

$$y' = \sum_{b=0}^{B-1} y_b \cdot 2^b - 2^B[\text{sign}(y)]. \tag{3.8}$$

The multiplication is again performed as before, but again, in the final step, the multiplicand is subtracted rather than added.

Table 3–2 Booth's algorithm represents consecutive 1 digits in binary numbers by the difference of two binary numbers each containing one 1 bit.

Standard representation		Booth's representation	
Binary	*Decimal*	*Binary*	*Decimal*
1	1	1	1
11	3	4–1	100–011
111	7	8–1	1000–0001
1111	15	16–1	10000–00001
•	•	•	•
•	•	•	•
•	•	•	•

As shown above, Booth's algorithm entails looking for consecutive fields of 1's. Looking for adjacent 1's and taking the appropriate steps is suited to hardware implementation, and thus, Booth's algorithm is an important means of speeding up hardware multipliers.

Another way to speed multiplication through hardware is with the fully parallel multiplier. The parallel multiplier is a combinational circuit rather than a sequential one, and thus its multiplication time is set by logic settling time rather than by multiple clock cycles. The parallel multiplier exploits the fact that the product of two one-bit binary numbers is merely the logical AND of those bits (see Fig. 2–4). The product of two multi-bit numbers is computed as the sum of appropriately shifted combinations of the partial products formed through all pairwise AND combinations of bits of the multiplier and multiplicand. Each bit-wise product is formed by a dedicated AND gate, and a two-dimensional matrix interconnection is used to achieve all possible pairwise combinations of bits and to apply the appropriate shifting. Multiple-input full adder circuits are used to accumulate sums of partial products: The number of simultaneous inputs is set by the number of partial products that are added at a particular binary digit location of the product.

Fig. 3–10 shows a parallel multiplier circuit to implement the multiplication of two 4-bit binary numbers. A diamond structure of AND gates forms the partial products that were shown in Fig. 3–9 indicated by the circles in Fig. 3–10. As shown in the blow-up view, each circle has two inputs. These inputs are AND-ed together to produce an output, and they are also passed through without AND-ing as diagonal outputs to subsequent stages. Each circle, except those on the left perimeter of the diamond, also has a horizontal input, which is passed through the circle and placed next to the AND gate output as a parallel horizontal output. This forms an output bus that increases in width by one bit for each circle that is passed. Each horizontal output bus enters a full adder that adds two, three, or four 1-bit numbers (FA2, FA3, and FA4), receiving any carry inputs and putting out the appropriate number of carry outputs (one less than the number of inputs).

3.3.3 Software versus Hardware for Infrared Processor

A pixel-level target detector for infrared (IR) imagery provides a second example of a software/hardware tradeoff.[9] The task involves two preprocessing algorithms for a medium wave IR filter, and the implementation choice consists of a TMS 320C80 Multi-media Video Processor (MVP) programmable signal processor versus a field programmable gate array-(FPGA) based reconfigurable processor known as CHAMP (Configurable-Hardware Algorithm-Mappable Preprocessor). In this work, both solutions were developed and compared using the evaluation criteria of accuracy, processing speed, processing latency, design effort, and final unit cost.

The MVP consists of four 32-bit digital signal processors in a multiple-instruction, multiple-data (MIMD) format (see Section 4.2.1 for a description of

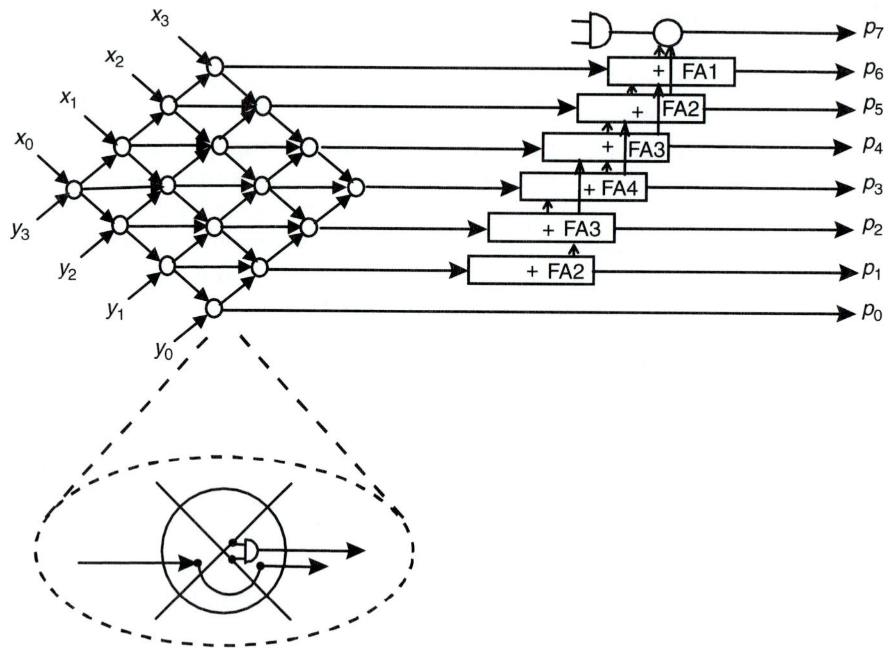

Figure 3–10 Parallel multiplier uses rhombus of AND gates to form all 1-bit partial products and full adders to sum the terms for each output bit p_n.

MIMD) and achieving a throughput of 2 billion operations/sec. The MVP also contains a reduced instruction set (RISC) general purpose processor with an integrated IEEE-754 floating point unit, which is programmed in the C programming language. A 50-Kbyte static RAM is on the chip. The four signal processors are connected with a crossbar switch operating at 1.8 Gbyte/sec (for instructions) and 2.4 Gbyte/sec (for data). The MVP chip is supported by 1M × 64 external dynamic random access memory (DRAM) and 64K × 8 external electrically programmable read only memory (EPROM) (Fig. 3–11).

The CHAMP processor (Fig. 3–12) contains eight processor elements connected with a crossbar interconnect. Each processor element is implemented with two FPGAs, each with 576 configurable logic blocks of 25 gates each. Each element also contains a 16 K × 32 bit dual port RAM and is connected with two 32-bit ports to the crossbar. Each processing element is capable of 0.5 billion operations/sec, and the crossbar supports 760 Mbyte/sec communications. A 256K × 16 three-port RAM supports the array by storing multiple frames of data.

Two algorithms were implemented on these architectures. The first, an electro-optic (E/O) algorithm, reads an incoming stream of data pixels (256 × 256 pixels at 10 frames/sec) and performs preprocessing operations such as thresholding. The second, known as the midband algorithm, is more complicated, ap-

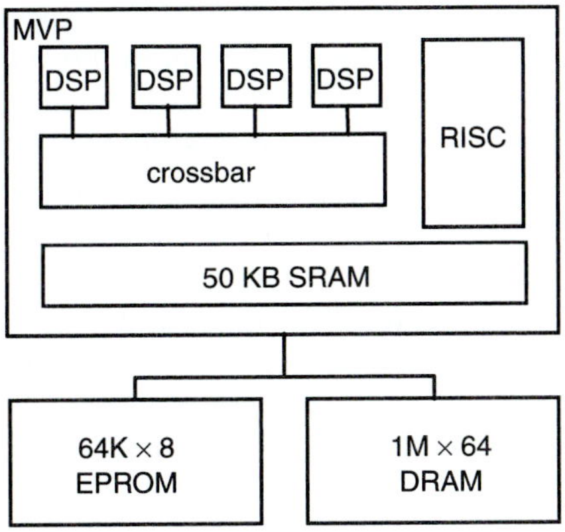

Figure 3–11 Block diagram of Multi-media Video Processor (MVP) processor board.

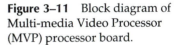

plying a pixel-adaptive spatial filter. Both algorithms detect pixels that correspond to targets in the images.

Benchmarks used for the comparison of implementations were:

- *accuracy*: based on 1) *exceedance*, or pixels declared as targets (i.e., each image pixel was subjected to a binary hypothesis test, 1 (target) versus 0 (no target). For this benchmark, comparisons were made to known truth; a pixel-by-pixel determination of correct classification, false alarm, or missed detection was made; 2) *pixel value*, or comparing intermediate results that lead to target detection, but with more bits of precision than the final binary output of exceedance; and 3) *threshold*, that is, the value of threshold computed to define exceedance;

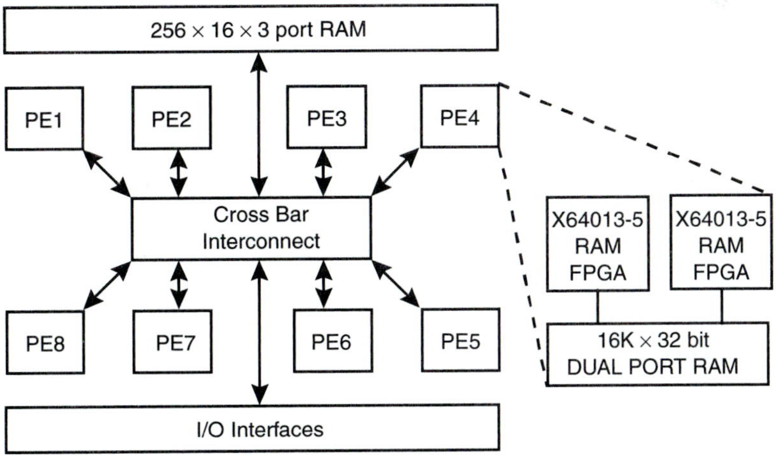

Figure 3–12 Block diagram of CHAMP board, implemented as eight FPGA-based processors. © 1995 IEEE.

- *speed:* this included 1) throughput—length of time required to process one frame, measured from the start of the first pixel to the start of the first pixel of the next frame; 2) latency—difference in time between start of processing for a pixel of a frame and the completion of results for that pixel;
- *unit cost:* for both the MVP, cost given for next-generation version of hardware. For the CHAMP, the next-generation enhancements include multichip module packaging (see Section 11.7.4), a threefold increase in throughput per board, and improved automatic reconfigurability. For the MVP, the new hardware includes two MVP processor chips (each with four DSPs and one RISC processor) and a wider selection of interfaces.
- *design time:* the time required to design the CHAMP and the MVP implementations of each algorithm.

The results of the comparison are shown in Table 3–3.

As shown in Table 3–3, the software (MVP) approach requires significantly less design time and has a unit cost that is significantly less than the hardware (CHAMP) approach, but it runs about 25 times more slowly (for the midband algorithm) or four times more slowly (for the E/O algorithm). Moreover, the software approach provides more accurate results on the midband algorithm.

3.3.4 Conclusions

The above case study exemplifies a systematic approach to performing the software/hardware tradeoff at the system level:

1. Determine whether a solution developed by software programming existing hardware will meet the speed and space requirement. This determination can be quite difficult, as one must answer the question "Can this

Table 3–3 Comparison of hardware ("CHAMP") vs. software ("MVP") approach to real-time implementation of IR array preprocessing algorithm.

Criteria		CHAMP	MVP
Accuracy (EO/midband)	Exceedance Pixel Threshold	100% / 99.57% 100% / 99.5% 100% / 98.6%	100% / 100% 100% / 100% 100% / 100%
Speed (EO/midband)	Throughput (10 fr) Latency	32.8 msec / 32.8 msec 618 µsec / 68.1 µsec	135 msec / 135 msec 13.8 msec / 82.4 msec
Cost (per unit, in next generation HW)		$40K	$15K
Design time (EO/ midband)		28 days / 44 days	9 days / 28 days

processor run this algorithm at this speed?" The topic of software/hardware performance estimation is treated further in Section 4.1 and 4.2.

2. If software meets the performance specification with reasonable reserve for program growth, use the software approach. One definition of "reasonable reserve" is used for developments for the U.S. government and specifies a reserve of 33 percent in real time and 50 percent in program memory.
3. If software on an existing processor is inadequate, apply a hardware solution that requires a minimum amount of new design. Such a minimal approach may involve the designing of a multiprocessor system that cascades several existing processors;
4. If small multiprocessor designs are inadequate, then the design of a customized processor must be undertaken.

Even for the full custom hardware design, some programmability should be incorporated to accommodate late changes that can occur in field application. These changes might include changing gain or setting minimum confidence thresholds for pattern recognition algorithms.

A trend has been observed that within a class of signal processing applications, increasingly complex custom hardware solutions are developed to attain a certain level of capability, such as is sufficient to attain real time throughput on a major type of problem. After that point, a plateau in hardware is reached, and subsequent improvement occurs in software and algorithm efficiency. As stated in Section 2.4, choice of a custom hardware design should be made only if hardware will achieve more than a factor of 10 increase in performance as compared with available software solutions, because during the design interval for custom hardware, existing processors will increase significantly in performance.

REFERENCES

1. Milos D. Ercgovac and Tomas Lang, *Digital Systems and Hardware/Firmware Algorithms* (New York: John W. Wiley and Sons, 1985).
2. Alan V. Oppenheim and Ronald W. Schafer, *Digital Signal Processing* (Englewood Cliffs, NJ : Prentice-Hall, 1975).
3. L. R. Rabiner and B. H. Juang, "An Introduction to Hidden Markov Models," *IEEE ASSP Magazine* (January, 1986), 4–16.
4. J. S. Bridle, M. D. Brown, and R. M. Chamberlain, "An Algorithm for Connected Word Recognition" *Proc. IEEE Inter. Conf. Acoustics, Speech, Signal Processing*, Paris, France, May 1982, pp. 899–902.
5. J. D. Seals and R. R. Shively, "EMSP: A Data Flow Computer for Signal Processing Applications," in *VLSI Signal Processing*, Peter R. Cappello and others, Eds. (Piscataway, NJ : IEEE Press, 1986) pp. 421–426.

6. Donald E Knuth, *Seminumerical Algorithms, The Art of Computer Programming, Vol. II* 2nd ed. (Reading, Mass: Addison Wesley, 1989), p. 253.
7. Ercgovac and Lang, *Digital Systems and Hardware/Firmware Algorithms*, p. 688.
8. A. D. Booth, "A Signed Binary Multiplication Technique," *Q. J. Mech. Appl. Math.*, 4, part 2 (1951), 236–240.
9. Kerry L. Hill and others, "Real-Time Signal Preprocessor Trade-Off Study," *Proc. IEEE National Aerospace Electronics Conf. (NAECON)*, Dayton, OH, May 22–26, 1995, pp. 328–335.

4

Space-Time Tradeoffs

The concept of trading space against time occurs throughout the design and implementation of signal processing systems. In the software domain, code size (corresponding to memory space) is traded against execution speed (time). In hardware design, multiple processors in parallel (requiring larger physical space) accelerate program execution (time). For both software and hardware, increased design time and effort usually lead to smaller software or hardware modules for a given throughput.

4.1 SPACE-TIME TRADEOFFS IN SOFTWARE

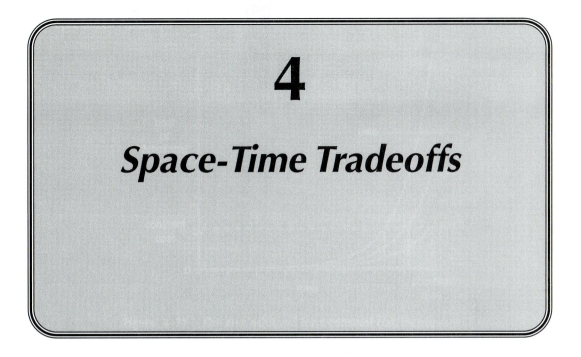

The primary goal of real time software development is to apply programming techniques that minimize the time of computation for an algorithm to the point that it achieves real time throughput. Real time processing was defined in Section 1.1.1 as the condition under which the processing upon a group of input samples required less time than the input period over which those samples were gathered. A secondary goal of real time programming is to use programming techniques upon real time software to minimize the use of various processing resources, such as space of code or data in memory, or number of clock cycles, while still maintaining real time execution. Real time software serves two goals that are sometimes contradictory: to produce an algorithm implementation that is easy to understand, code, debug, and run on a variety of different platforms, and to provide a program that makes most efficient use of the unique computational resources of a particular platform, thereby running as fast as possible.

The running time of a program is central to the concept of real-time signal processing. A program's running time depends on:

1. data input to the program;
2. quality of code generated by a compiler to create the machine-specific object program;
3. the type and speed of the instruction set on the platform upon which the program is run; and
4. the complexity of the algorithm that underlies the program.

Because the program running time can depend on the particular data that is input to the program, one can have both a worst-case running time and an average running time. The worst-case value is easier to compute, but it is often several times the average running time. The average running time is more difficult to determine and depends upon the statistical distribution of input values, which is often not known.

The absolute running time of a program, measured in seconds, depends upon the compiler, the machine, and possibly the data, all of which must be specified to specify the running time. More common and useful is the *dependence* of run time upon the number of input values, n. The "big O" notation ("O" for order) is used to specify how the execution time of a program scales with n, and programs are compared based on run time factors, listed here in increasing order:

$$O(1) < O[\log(n)] < O(n) < O(n^2) < O(n^3) < \ldots < O(e^n) < O(n!). \tag{4.1}$$

A program with a high growth rate can actually execute faster than a lower-growth-rate program for very low values of n, but as n increases, its execution will grow to outstrip that of the low-growth-rate programs.

A set of rules governs how the running time of a program is based on the run times of its individual modules. A read or write operation is generally $O(1)$. The run time of a sequence of statements is the largest running time of any statement in the sequence, to within a constant. The run time of a conditional branch, such as an IF statement, is the run time of the longest branch, plus time to evaluate the branch statement (generally $O(1)$). The time to execute a loop is the product of the number of times through the loop times the largest possible execution time of one pass through the loop. Because of this multiplicative effect, the instructions within a frequently executed loop are often the best place to begin optimizing execution time, since a small decrease in loop execution time is multiplied by the number of times through the loop.

With the preliminaries on running time now in hand, the concept of a space-time tradeoff to decrease running time will be discussed. This will be followed by a discussion of the effects of tradeoffs in the language used to specify the algorithm and program it on the machine. A discussion of software architectures then follows, and the section concludes with a discussion of the information content of a program, which is preserved as space-time tradeoffs are executed.

4.1.1 Space-Time Tradeoff

For software, the space-time tradeoff involves trading program memory space for execution speed. Decreasing execution time can be accomplished by precomputing as much information as possible and storing it in memory. As an extreme example to emphasize the concept, one can imagine combining the several input values for a processing module into a single large address, accessing a very large memory based on that address, and outputting the correct responses that correspond with those inputs. Such a table lookup could be accomplished far more quickly than performing any arithmetic or logic on the inputs. However, an extremely large memory would be required (for 100 8-bit inputs that result in a unique set of outputs, a memory of $(2^8)^{100} = 2^{800}$ memory locations, each containing the full set of bits to describe the multiple output signals for that input, would be needed).

More reasonable tradeoffs of space versus time may be made to precompute results. For example, the evaluation of the function $\sin(x)$ is usually implemented as a Taylor series expansion, using subtracts, adds, and multiplies from a typical computer instruction set to iteratively compute $\sin(x)$ to the desired precision. Alternatively, if x is an 8-bit number, a simple table lookup from a modest memory array of $2^8 = 256$ elements can provide the value of $\sin(x)$ more quickly. The use of several dozen instruction cycles to compute the Taylor series expansion of $\sin(x)$ is thus replaced by a few hundred memory locations to provide a result in a single memory access time (1–2 instruction cycles). For more complicated operators or for more bits of precision, a hybrid of table lookup and computation can be used. For example, finding the reciprocal of a number x can be decomposed into two stages (Section 7.2.1.1):

- a table lookup that finds the reciprocal of another number x' that results from the most significant bits of x, followed by
- a linear interpolation or Taylor series expansion about the point $1/x'$ that uses the remaining lower significant bits of x to arrive at a more precise value of the reciprocal.

Another example of a software space-time tradeoff is provided by the use of subroutines. In writing a sequence of code, one may need to perform a particular complicated operation at multiple times as the program progresses from beginning to end. One means, consistent with good software design, is to perform that operation by jumping to a subroutine in a different part of memory each time the operation is needed. Since the same subroutine provides code for every time that the function is needed, memory space is saved. If the computation of the function needs to be changed, changing the one place where the subroutine is executed will update all computations of that function throughout the program.

However, associated with each subroutine call is some time-consuming overhead. A pointer must be set so that when the subroutine computation is completed, progress may be resumed in the main program at the point it was left when the subroutine was called. In addition, input parameters must be passed to the subroutine, either by copying the input values from locations associated with the main program to locations known by the subroutine, or by computing and passing pointer address values associated with these inputs. Finally, because the subroutine will be called from a variety of points in the main program and under a variety of conditions, inputs must be specified for all possible conditions. For example, a general-purpose subroutine might accommodate integer, floating point, and complex values—the input values would require augmentation to specify which data type was used. In contrast to subroutines, in-line code places the essential parts of the subroutine within the execution sequence of the main program. This speeds operation, as it avoids the overhead of calling the subroutine, but it increases memory usage, as a variant of the subroutine is written in the sequence of code each time that it needs the function results (Fig. 4–1).

Space is traded against time in the choice of the level of language used to program the machine with the algorithm. As discussed in the next section, signal processor programming may be viewed as cleverly translating algorithm concepts

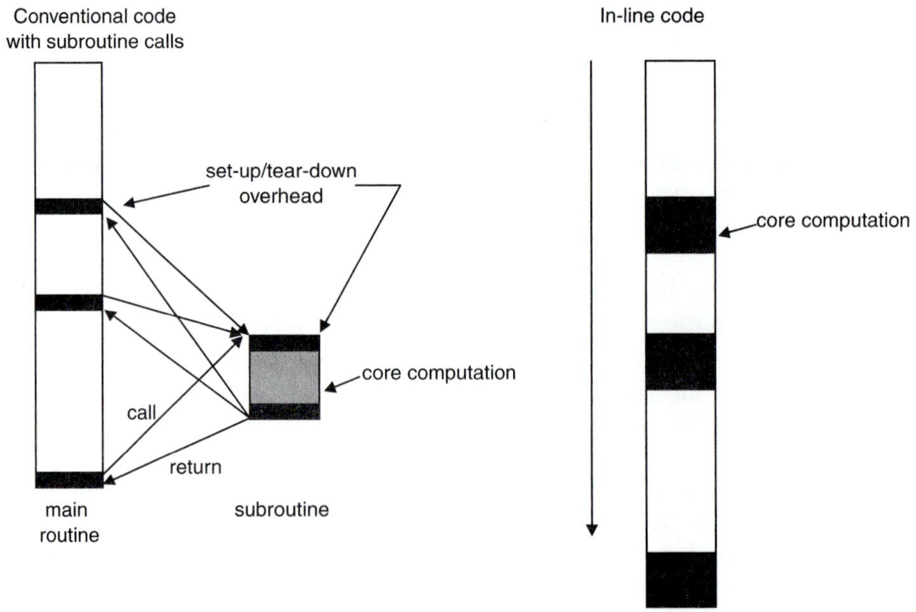

Figure 4–1 Proper use of subroutines (left) can introduce timing delays due to calling overhead, which may be alleviated with in-line code (right).

from the domain of invention (equations) to the domain of execution (program). A high-level language such as C or C++ insulates the programmer from the specifics of the machine and more closely approaches the equation domain of algorithm development. Compilers provide automatic translation from a high-level language to a machine-specific language. Higher-order languages are easier to encode (i.e., it takes less time to implement a specific function). They can be more easily moved from one computing platform to another (since the machine-specific language is generated by the machine's compiler program rather than by hand), and they follow the underlying algorithm more directly. However, the automatic translation to machine code offered by a compiler is not generally as efficient as coding by hand, since particular clever uses of registers and the timing and reuse of intermediate results are better identified and exploited by a programmer. Hand-generated machine code is generally faster and smaller, but it takes longer to develop. A mixture of higher-order language and assembly is most effective. In an example of programming an MPEG-1 audio encoder/decoder onto a Texas Instruments TMS 320C3x digital signal processor,[1] a comparison of C versus machine assembler was made. It found that for those parts of the MPEG algorithm that made good use of the specialized functions of the chip (e.g., linear filtering, fast Fourier transforms), the hand-coded machine-specific application library for the TMS320 was most efficient. For the MPEG-1 operations with little structural regularity such as encoder bit assignment and application of a perceptual model, the execution time was determined by branches and accesses to data structures, and the C compiler produced code as efficient as hand-coding, but with less effort.

A tradeoff of space versus time occurs when developing software for a single processor versus many processors. A multiple processor architecture can require separate programs for each processor, assigning a portion of the overall signal processing task to each. The collection of processors provides greater throughput than a single processor, but it requires more software development time and more memory resources to store the several programs. Multiple processors, while running faster, take longer to encode. This is because there are multiple ways to partition the software across multiple processors, with congestion and communication bottlenecks that must be avoided. With a single processor, it is clear which tasks "each" processor must execute, as the one processor must execute all tasks. Thus, while a single processor is slower in execution than multiple processors, code development is usually faster. However, this is not always true—the single-instruction multiple-data (SIMD) class of machine applies exactly the same instruction stream to multiple processors, each of which is operating on a subset of the data. In this case, a single program serves all processors as readily as it would serve one, and the speed of multiple processors comes at no added cost of memory space.

4.1.2 Algorithm Development and Programming Language

The languages used both to develop and to program the desired signal processing application exist in a continuum that trades space for time. Furthermore, a tradeoff of development time versus execution time occurs—spending time to optimize a program will make it run faster. The task of an algorithm development language is to describe algorithms with sufficient accuracy to allow understanding and manipulation by researchers and developers. A programming language states the algorithm in a manner that precisely defines its execution by machines. The algorithm specification, for use by humans, is easily understood, observed, and changed. The implementation language, for use by computers, is fast and efficient, but it is not particularly flexible to the restructuring and incremental extensions that characterize algorithm development. In the signal processing development sequence, translation must occur from the algorithm specification language to the machine implementation language for execution. This translation can occur by manual or by automatic means, or by a mixture of both. Some algorithm specification techniques seek also to serve as implementation languages or to provide completely automatic translation from the language of researchers to the language of computers.

This discussion divides the continuum between development and specification languages into five stages (Fig. 4–2):

1. *equations:* this first stage reflects the fact that most algorithm manipulation occurs using equations and the laws of mathematics;
2. *diagrams:* signal processing block diagrams that emphasize the flow of information from one processing block to the next, as well as the definition of the processing blocks;
3. *signal processing languages:* text description of operations, in the form of function names, and associated operands or data that are assembled into a program;
4. *general purpose high-order languages:* signal processing operations are described in a general purpose language, not specifically designed for signal processing, but used for the complete range of computer programming tasks. As part of the general purpose high order language, a subroutine library, suited for signal processing operations, is included;
5. *machine language:* this language reflects the specific architecture of a particular computer and controls the individual machine resources.

Equations are used for manual manipulation of signal processing operations, and they are transformed according to the rules of mathematics into progressively more useful representations. The tool known as Mathcad®[2] is an example of a language that uses equations as input in a manner similar to the way they are written in a program. As equations are written in sequence, Mathcad provides the live update of calculations. Computation flows down the page, from top to bottom and left to right. Mathcad supports symbolic processing, used in algebraic manipulation to solve an equation for a variable, as well as the more standard numerical computation. It allows the integration of equations, graphics, and supporting text on the same page, and it supports two-dimensional and three-dimensional graphics as

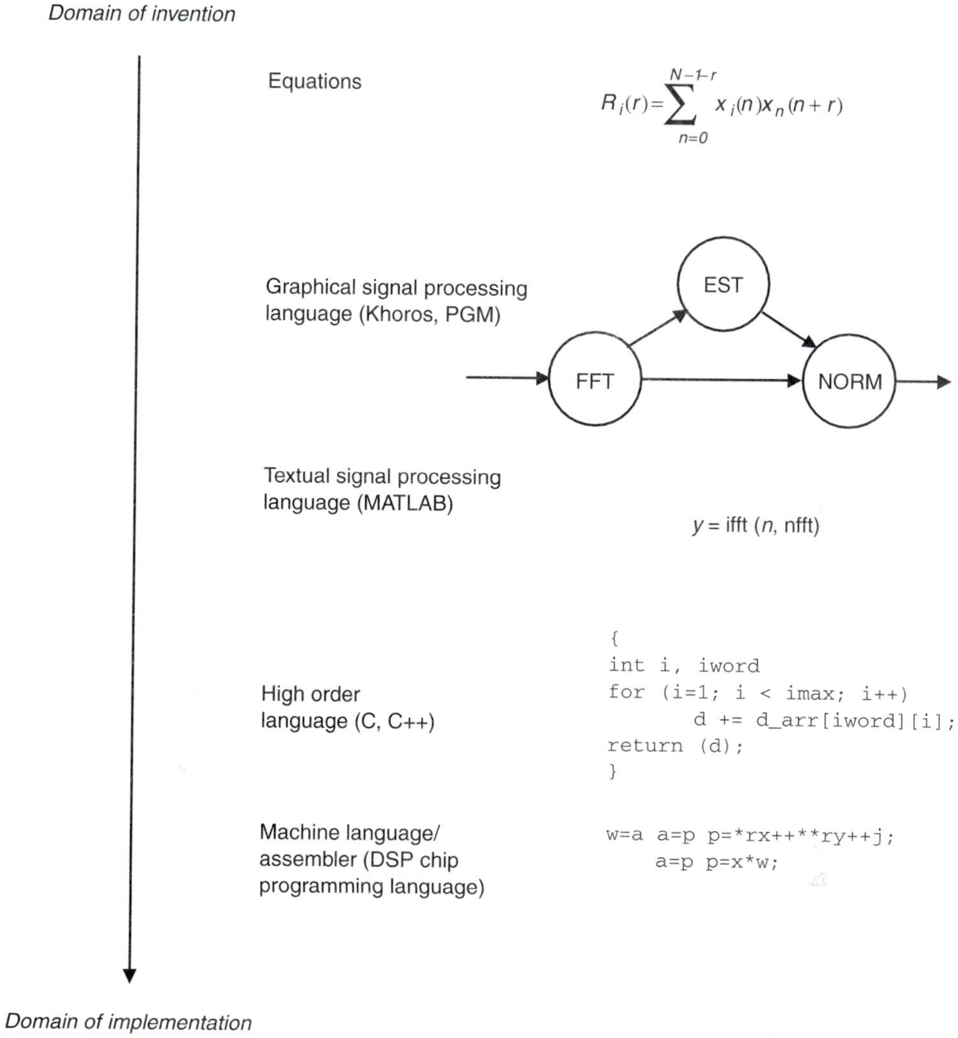

Figure 4–2 Signal processing languages proceed from those intended for invention (equations, at top) to those suited for implementation—compilers provide automatic translation down the hierarchy.

well as animation. Mathcad "Electronic Books" add the ability to include domain-specific formulas, data, and diagrams. Mathcad also offers the ability to incorporate custom functions that the user supplies in C or C++ general purpose computing languages. Fig. 4–3 shows a sample screen output of Mathcad.

Graphical signal processing uses block diagrams and connecting arrows to follow the conceptual flow of signal processing algorithms. Diagrams are pro-

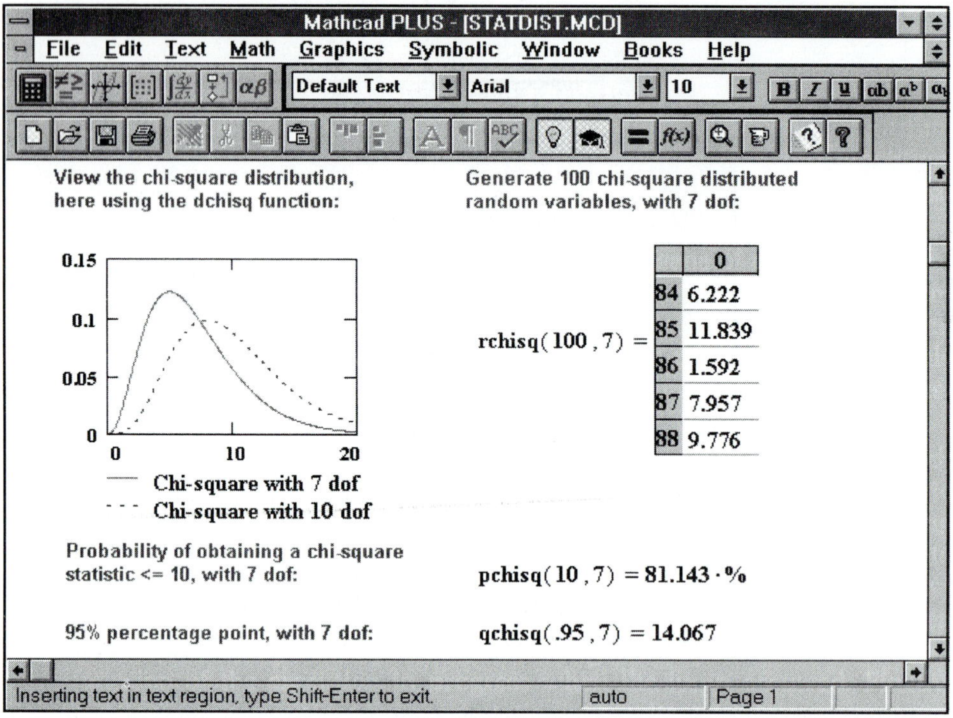

Figure 4–3 Screen interface of Mathcad®.

duced with a mouse; the resulting signal flow chart is compiled to machine-specific code (or to C or some other general computer language).

One such graphical signal processing language is Khoros.[3] Khoros provides a visual data flow programming language to support the build-test-refine iterations typical of scientific programming. The visual data flow language is augmented by support for conditionals, iterations, and subprocedures. A Khoros program is written by placing elements of the signal processing block diagram, known as *glyphs*, on the work surface by selecting the appropriate routines from menu entries. Glyphs are connected together with arrows, and tap-off points to view intermediate results may be placed between glyphs. Multiple glyphs can be combined into a single workspace, which can then be selected in a higher-level diagram as a single glyph. All Khoros data is of one type, which facilitates its exchange among glyphs. Data is represented as two-dimensional multi-band arrays, where each point has a pixel that may have a single- or multiple-coefficient vector of values. Multiple values are associated with multiple bands. Tools for image processing include elements for arithmetic, image processing, and image analysis (Table 4–1).

Table 4–1 Khoros tools for signal and image processing.[3]

Class of operations	Type of Operations	Examples
Arithmetic	Unary	scale, normalize, invert, clip
	Binary	add, subtract, multiply, blend
	Logical	AND, OR, XOR, shift
Image processing	Spatial filters	Sobel, median, 2D convolve, edge extract
	Morphology	erosion, dilation, skeletonization
	Transforms	FFT, Hadamard
	Filters	LPF, BPF, HPF, band reject, inverse, Wiener
	Geometric manipulation	shrink, rotate, transpose, interactive image warp
	Subregion	extract, insert, pad
Image analysis	Segmentation	threshold, medial axis
	Feature extraction	shape analysis, region analysis, fractal analysis, texture extraction
	Classification	k-means labeling, LRF classifier

While Khoros is designed for graphical iterative programming of image processing algorithms, the Processing Graph Method (PGM)[4] graphical language, developed by the U.S. Navy, is intended for time-domain signal processing program development and implementation. Originally developed for the U.S. Navy AN/UYS-1 and AN/UYS-2 parallel signal processors and later expanded to other platforms, PGM is intended for efficient implementation of acoustic signal processing algorithms on a multiprocessor architecture. A PGM application consists of signal processing graphs and associated command programs. A graphical workstation may be used to input the algorithm graphically, and it produces a text file description of the graph. PGM library parameters include function classes of vector operations, matrix operations, vector comparison, data conversion, data conditioning, data generation, basic signal processing, and special-purpose signal processing. The basic signal processing library includes FFT, one-dimensional IIR and FIR filters, complex demodulation, and beamforming. The special purpose signal processing class contains algorithms for sonar signal processing. In short, PGM seeks to provide an iconic description of signal processing algorithms for parallel multiprocessors that insulates the programmer from the architectural specifics of a particular multiprocessor. More information on PGM is provided in Section 11.3.1.

Not all signal processing languages are graphical. MATLAB is a textual language for high performance computing that is frequently used in signal processing.[5] The basic data unit in MATLAB is the matrix (the name MATLAB is derived from Matrix Laboratory). MATLAB is a high-order matrix language with provisions for control flow statements, functions, and object-oriented programming. It provides a working environment for managing variables, importing/exporting data, and developing applications. It includes mathematical and signal processing libraries, and it allows the linking of custom C and FORTRAN programs. SIMULINK, an adjunct to MATLAB, allows the use of block diagrams as input to MATLAB and the automatic generation of C code from block diagrams.

Before the use of equation, graphical, or textual signal processing languages, general purpose programming languages such as C and FORTRAN were used to implement signal processing algorithms. C and its object-oriented descendent, C++, are commonly used for compact, production-quality signal processing. Several factors have caused C and C++ to be used across the spectrum of signal processing development, from exploratory development to the final implementation of signal processing functions. These include the near-bit-level manipulation capability of C, the ability to handle arrays of objects afforded by C++, and the fact that the large majority of computing platforms, including many digital signal processors, allow programming in C. For those situations where machine-specific optimization is needed, most computers include C-linkable signal processing libraries that speed execution.

At the bottom level of the language hierarchy, machine-specific assembly language provides the fastest execution and most compact form of the algorithm. Many signal processing integrated circuits are programmed in their native language, even when C compilers are available, to gain the speed and memory efficiency that is realized by handcrafting the program to fit the architecture. As computing hardware becomes faster, the justification for full handcrafted assembly coding decreases. This is because:

1. Manually-generated code must be rewritten if the computer is changed.
2. Assembly code is more difficult to maintain, both because so many more statements are needed to describe a particular function, and because no natural tools exist in assembly language to enforce the rules of good programming.
3. Most timing bottlenecks occur in a few places in signal processing algorithms, usually as the innermost loop in an iterative statement (Fig. 2–28), so a better compromise is coding the entire application in a high-level language, and then casting only the small percentage of time-critical bottlenecks in assembly language.

The spectrum of algorithm specification and implementation languages runs from the equation level, used to develop algorithms, to the machine-specific level, used to provide maximum execution speed and minimum memory usage. Automated tools exist to translate any level down to the machine level, and in some cases, tools can convert higher level descriptions to C language output, providing a common lower-level description that stops short of becoming machine dependent.

4.1.3 Software Architectures

A dictionary defines architecture as ". . . the style and method of design and construction... ."[6] Software architecture is thus the style and method of design of software. Like any other architecture, it describes the hierarchical decomposition of an object. At the top of the hierarchy, the object is viewed from above as an opaque box with inputs and outputs. At subsequent levels, the box is decomposed into interconnected smaller boxes, followed by the subsequent decomposition of those boxes into lower levels of interconnected elements, and so forth. At its lowest level, the architecture includes provision for the representation of numbers, as was described in Section 2.1.1.

Before exploring the topic of software structures and conservation of program information content across program transformations, we revisit the representation of numeric quantities. As an alternative to integer and floating point operations, the *residue* number system (RNS) has for several decades received attention as a method to ease machine computation.[7] A residue representation breaks a number into the remainders that result by division of that number by mutually prime divisors, $m_1, m_2,\ldots,m_r$. It results in the representation of a number $U = (u_1, u_2,\ldots,u_r)$ as follows:

$$U \rightarrow U \bmod m_1 \equiv u_1, U \bmod m_2 \equiv u_2, \ldots , U \bmod m_r \equiv u_r. \tag{4.2}$$

According to the Chinese Remainder Theorem, any number less than the product $m_1m_2\ldots m_r$ may be uniquely represented in the form $(u_1, u_2,\ldots,u_r)$ and decoded to and from that form. This parallel representation allows arithmetic operations such as add and multiply to be performed as parallel operations on each residue with the final result being decoded at the end.

The arithmetic operations on two numbers, $U = (u_1, u_2,\ldots,u_r)$ and $V = (v_1, v_2,\ldots,v_r)$ become the following in residue arithmetic:

$$\begin{aligned} U + V &= (u_1,u_2, \ldots , u_r) + (v_1,v_2, \ldots , v_r) \\ &= [(u_1 + v_1) \bmod m_1, (u_2 + v_2) \bmod m_2, \ldots , (u_r + v_r) \bmod m_r]; \end{aligned} \tag{4.3}$$

$$\begin{aligned} U - V &= (u_1u_2, \ldots , u_r) - (v_1v_2, \ldots , v_r) \\ &= [(u_1 - v_1) \bmod m_1, (u_2 - v_2) \bmod m_2, \ldots , (u_r - v_r) \bmod m_r]; \end{aligned} \tag{4.4}$$

$$\begin{aligned} U \cdot V &= (u_1,u_2, \ldots , u_r)\cdot(v_1,v_2, \ldots , v_r) \\ &= [(u_1 \cdot v_1) \bmod m_1, (u_2 \cdot v_2) \bmod m_2, \ldots , (u_r \cdot v_r) \bmod m_r]. \end{aligned} \tag{4.5}$$

Residue arithmetic has been successfully applied to more comprehensive signal processing routines such as sum-of-product kernels in FIR filters.[8]

Residue arithmetic has several advantages:

- data path widths can be restricted to just the number needed to represent the residues;
- multiplication operations occur in time $O(n)$, where n is the number of bits in the operand, rather than in $O(n^2)$ of traditional multiplication methods;

- execution can be speeded by parallel processing of each residue—no carry propagation or other communication between residues is needed until final decoding.

Several disadvantages have kept residue number processing from receiving full acceptance:

- hard to test whether a number is positive or negative;
- hard to compare one number to another;
- hard to determine whether an overflow condition has occurred as a result of arithmetic: This can be alleviated by carefully tracking the number of bits of the maximum number that can result from each computation; hence, the sum of m products each composed of n-bit numbers ($n \times n$ product) is $\log_2 m + 2n$ bits.

Nevertheless, like bit-serial architectures, residue number representations continue to form an active area of research. Efforts have concentrated on the means to rapidly encode and decode numbers to and from residue representation, and to simplify implementation by processing each residue serially on the same residue processor, rather than in parallel. These considerations of number representation, including fixed versus floating point, residue numbers, and finite word length effects, enter into the space-time tradeoff, both at the software and hardware levels.

4.1.4 Software Structures[9,a]

The elements of software architecture include types of data structures, method of control transfers, and subroutines, as well as combining of programs to run simultaneously in real time processing. These four elements are each discussed below. The diversity of software structures that occur within a program influences its complexity for real time implementation.

Several abstract data types form the structures used in software. The first of these, the *list*, consists of a sequence of n elements of the same type, placed in a particular order $x(1)$, $x(2)$, ..., $x(n)$. The structural properties involve only the order of the list. Operations that can be performed on a list include:[9]

- insert (x, p, L)—insert item x at position p in list L
- locate (x, L)—return position of x in L
- retrieve (p, L)—return the element at point p in L
- delete (p, L)—delete the element at position p of L

[a]A. Aho, J. Hopcroft, U. Ullman, *Data Structures and Algorithms*, (pages 38, 39, 53, 57, 61, 75).

- next (p, L)—returns the position of the next element, or zero if pointer is at end of list
- previous (p, L)—returns the position of the previous element, or zero if pointer is at beginning of list.

From these basic operations, several types of lists useful for signal processing can be constructed:

- *stack* or first-in, last-out (FILO, also known as last in, first out, or LIFO): list for which all inserts, deletes, and retrieves occur from one end of the list;
- *queue*, or first-in, first-out (FIFO): list for which all inserts are from one end and all deletes and retrieves are from the other
- *deque* or double-ended queue: all insertions, deletions, and retrieves are made from either end of the list.

Lists can be implemented in two ways. In the first, known as *sequential allocation*, sequential memory locations are allocated to the list, and list elements are stored in contiguous cells. The second method, known as a *linked list*, is based on pointers. Each location in the list contains both the associated list element and a pointer to the next element of the list. A doubly linked list includes a pointer to both the next and the previous element, allowing traversal of the list in either direction. Sequential allocation, also known as array implementation, requires less storage, as only the data element, not the pointer to the data element, is stored at each location. However, the maximum size of an array must be set in advance to reserve the necessary block of contiguous memory locations. Linked lists require more storage (for the pointers) but can continue into non-contiguous areas of memory.

More complicated data structures include *mappings* and *trees*. The mapping, or associative store, is a function that maps elements of one type to elements of another (or possibly the same) type. An example is a table in which each entry contains the logarithm of its address in the table. This provides a one-to-one mapping between a number and its logarithm. It can be used for table lookup as a replacement for computation of the logarithm. Trees are collections of elements known as nodes, in which one element is distinguished as the root. A relationship associated with the tree places a structure on the nodes, in which nodes closer to the root serve as parents, and lower-level nodes serve as children.

A second abstract element of software architecture is the control structure. In the mid-1960s, Böhm and Jacopini proved that any flowchart of procedural logic could be derived from combinations of three basic types of flowcharts:[10]

- sequential: First execute instruction n, then the next instruction $n + 1, \ldots$
- IF-THEN-ELSE: If a certain condition is true, then execute a particular module of code; otherwise, execute another specified module.

- DO WHILE: Test a condition and if true, execute a particular code module, then test the condition again and repeat execution if true, else exit to the end of the module.

Each of these flowcharts has a single entrance and exit point and can therefore be treated as a black box. Nowhere is the unconditional GOTO instruction included—this function can be implemented via the three basic flow operations.

A common use of conditional control is to implement the operation of two or more modules of code within the same input sample period. For example, the data rate of a signal processing module is changed from input to output. The data rate of signal processing algorithms generally decreases as one progresses from input to output. A control structure for the interface between high and low sampling rates can be implemented as two programs, one for the low rate and the other for the high rate. The low-rate processing program is executed after the high-rate processing has occurred a certain number of times. As a result of the two timescales, the program architecture consists of two programs, a sample update program that updates input feature vectors, and a frame recursion program that computes the output feature vector from the vectors of the previous frame. The frame recursion program is divided into smaller pieces that are interspersed with repeated executions of the sample update program. An example is discussed more thoroughly in Section 6.4.2.

Subroutines, discussed earlier, represent another type of control transfer. While subroutines save program memory space, they increase time of execution, not only due to the calling overhead, but because like any other sort of control transfer, they break the processing pipeline. A jump requires that the pipeline be emptied, taking several clock cycles. It then must be refilled with the instruction queue for the new location of execution, further increasing execution time. When a subroutine or function is called in C, the compiler inserts additional hidden code to set up the function's stack frame and to copy the function's input variables from the main program's stack frame to the stack frame of the function.

4.1.5 Information Content of Programs[11]

With the variety of tradeoffs discussed, it is worthwhile to consider what factors might be constant across the various trades. The *information content* of a program provides an underlying quantity that is somewhat impervious to many of these trades. To discuss the information content of a program, we examine the set of transformations that can be applied to a program and describe some measurable properties of an algorithm.

An algorithm, when it is expressed as a program in a particular language, can be translated among four forms:[11]

1. It can be *compiled,* i.e., converted from a higher-level language to a lower-level language.
2. It can be *optimized,* or restructured in a particular machine language to run more effectively on a particular machine.
3. It can be translated from a lower-level language to a higher-level language, or *decompiled.*
4. It can be expanded into a *canonical* form by unwinding loops, replacing subroutine calls with in-line code, etc., thereby recovering the results of optimization.

Fig. 4–4 shows a cyclical diagram of the transitions. Form A is the original high-level language representation, B is the machine-language version that results from the compilation of the program, B' its optimized form, and A' is the canonical form. Although ideally, there is no loss at any transition, in practice some loss occurs at each step.

The following quantities can be measured for an algorithm and can be combined as a representation of its information content:

η_1: number of unique *operators* in its implementation
η_2: number of unique *operands* in its implementation
N_1: total number of all operators appearing in the implementation
N_2: total number of all operands appearing in the implementation

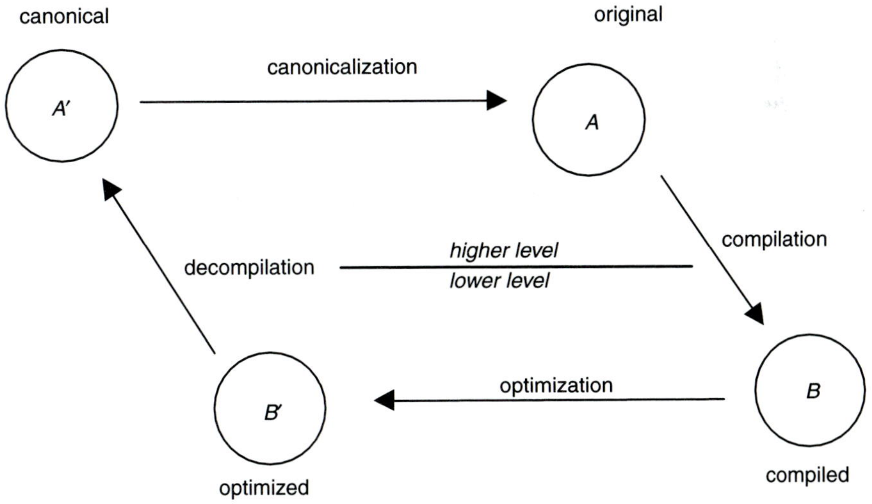

Figure 4–4 Cycle of program transformation incurs some loss at each transition.[11] (By permission of estate of Maurice H. Halstead.)

The *vocabulary* of a program may be expressed as the total dictionary of operators and operands that it uses:

$$\eta = \eta_1 + \eta_2. \tag{4.6}$$

The length of a program is its total number of operands and operators:

$$N = N_1 + N_2. \tag{4.7}$$

We can set lower and upper bounds on N, the program length, as a function of η, the vocabulary size. It seems counterintuitive that the length of a program would depend upon its vocabulary, but the reasoning proceeds as follows. The lower bound on N is simply reached by requiring each vocabulary element to appear at least once, that is, $\eta \leq N$.

Setting the upper bound requires two additional conditions that are consistent with actual code:

1. There are no identical substrings, or sequences of operators and operands, for any length η. This is reasonable, because if a particular substring appeared more than once, efficient programming practice would require that each substring should only be evaluated once, and any multiple occurring substrings would be given their own unique operand, increasing η by one. From this condition, one may compute the number of possible permutations of η items. This is seen to be η^η, as may be verified for first one object A ($\eta = 1$, $1^1 = 1$; single-class A), two objects ($\eta = 2$; $2^2 = 4$; AA, AB, BA, BB), three objects ABC ($\eta = 3$; $3^3 = 27$; AAA, AAB, AAC, ABA, ..., ABC, BAA, BAB, ..., CAA, ..., CCC) and so forth. As a result, a program could consist of at most η^η substrings of length η, giving the upper limit of $N \leq \eta^{\eta+1}$.
2. Operators and operands must alternate in occurrence. This requirement eliminates those sequences that consist of adjacent operators or operands. The vocabulary η is then divided into the number of unique operators η_1 and number of unique operands η_2 and the number of separate permutations of operators and operands are multiplied to replace the number of substrings:

$$\eta^\eta \rightarrow \eta_1^{\eta_1} \times \eta_2^{\eta_2}. \tag{4.8}$$

Thus, there are $\eta_1^{\eta_1} \times \eta_2^{\eta_2}$ allowed substrings of length η, reducing the upper limit on N to find

$$\eta \leqslant N \leqslant \eta \times \eta_1^{\eta_1} \times \eta_2^{\eta_2}. \tag{4.9}$$

This upper limit must include not only the particular sequence of N elements that comprises the program of interest, but also all possible subsets of that set. The set of all possible subsets of N is known as the power set of N and has 2^N elements. Thus,

$$2^{\hat{N}} = \eta_1^{\eta_1} + \eta_2^{\eta_2} \tag{4.10}$$

or taking the logarithm of both sides,

$$\hat{N} = \eta_1 \log_2 \eta_1 + \eta_2 \log_2 \eta_2. \tag{4.11}$$

The calculated program length is given by $\hat{N}$ to distinguish it from the observed program length N. It is assumed that the program is written in an expert manner that is well-organized and non-redundant. Analysis by Halstead of 577 published programs in Fortran and PL/1, written by experts across several different groups and published prior to this analysis, found agreement of $\hat{N}$ and N over program lengths ranging from $N = 10$ to $N = 32{,}000$ (see Fig. 4–5).

A representation of program information content is given by its potential *volume*, V^*, given by $V^* = N \log_2 \eta$. The volume is simply the number of bits required to represent the program–$\log_2 \eta$ bits are needed to represent a choice of operator or operand from a vocabulary of η_1, and there are N such choices in a program of length N. V^* is the potential minimum volume—the actual program volume V will be larger.

If an algorithm is translated from one language to another, V will change but V^* will not. The *level* of a program representation, L, is given by $L = V^*/V$. If a program is at its most succinct, $V = V^*$ and $L = 1$. Thus, $VL = V^*$ and the product VL is a constant. As the volume V of a program increases, its level L decreases.

The highest-level program ($L = 1$) would consist of a single function call and assignment statement. However, because the requirement is for a general-purpose computing language, one would need an infinite list of available procedures to allow single-function calls to anything that may be done.

Thus, the highest level language for a task has $\eta_1 = 1$, so L is proportional to $1/\eta_1$. On the other hand, the number of operands is unlimited. If an operand is repeated, that indicates a lower-level representation, so L is proportional to η_2/N_2. Thus, the level of a program implemented in language of level L may be estimated by $\hat{L}$ where $\hat{L} \propto \eta_2/\eta_1 N_2$.

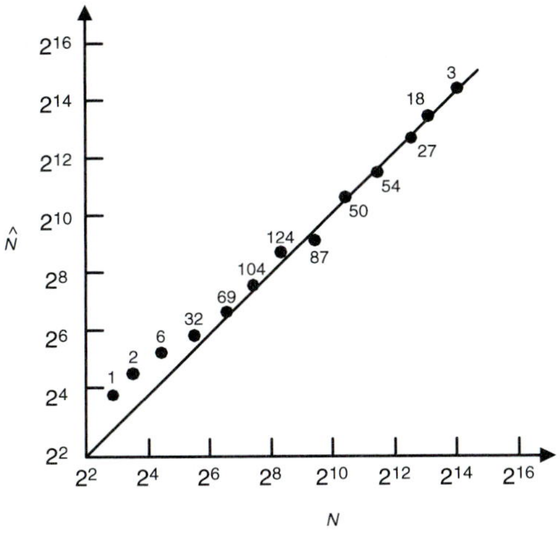

Figure 4–5 Data from 577 programs shows actual size N versus size computed as function of number of operators η_1 and operands η_2: $\hat{N} = \eta_1 \log_2 \eta_1 + \eta_2 \log_2 \eta_2$.[11] (By permission of estate of Maurice H. Halstead.)

4.1.6 Recent Techniques

Techniques of signal processing algorithm and implementation development that are moving into common use include object-oriented programming, software libraries for parallel implementation, and language preprocessors to program for a multi-processor architecture.

Object-oriented programming has been receiving strong interest because it encourages the reuse of its software elements, or objects, in future programs. Software development has become a larger expense in major systems than hardware development, and techniques such as object-oriented programming that encourage software reuse are viewed as a key element to reducing future system development costs. Object-oriented programming is exemplified by the expansion of the C programming language to C++.[12] Object-oriented programming introduces the concept of a *class*, which is a construct that defines arrays (for example, the DIMension statement of FORTRAN or the `array[n]` construct of C). It also defines *objects* as the arrays that are thus defined. In particular, an object can provide information about its contents, which are typically attributes, stored in an array, that are related. For example, an object "signal" might include the sampling rate, source, certain statistics like mean and peak values, and the signal values themselves. Thus, an object may be viewed as the more familiar array. Object-oriented programming proceeds by sending messages to objects. An object's interface consists of a set of commands, and one object directs another to perform a command by sending it a message. C++, which is a superset of C, provides the upgrades to C to obtain object-oriented programming capability.

Objects (arrays) are easy to reuse, enhance, and extend, and C++ provides the four features associated with object-oriented programming:

- *encapsulation*: the objects (arrays) encapsulate the data and functions—one cannot directly access the data in an array, but must use the array functions to access the data;
- *abstraction*: extracts the essential characteristics for a desired task—unimportant details are kept hidden from the programmer, based on the definition of relevant data structures at the start of the program;
- *inheritance*: a new class may be built from an existing class, and it inherits features of the class that built it;
- *polymorphism*: objects can be built for multiple types of data, and they will all be similar in behavior.

Software libraries for parallel implementation become important as signal processing algorithms become more computationally demanding and as they outstrip the ability of one signal processor to execute in real time. Moving to multiple processors that compute on the same problem requires changes in the methods of software development to support these processors. Efforts to develop software libraries for parallel processors have been providing sets of signal and image processing operations that are optimally or near-optimally mapped to a particular parallel architecture. Software library developments also provide tools to regenerate the signal and image processing library on new parallel architectures.

Developing high-performance signal processing software libraries involves meeting two contradictory goals: the library should be independent of parallel platform, yet it should reflect and exploit the particular architectural characteristics of a given parallel array.

An example of a parallel signal processing library is the Parallel Scalable Libraries and Algorithms for Computer Vision developed by Purdue University.[13] This program has produced a comprehensive library of image processing algorithms for a variety of parallel machines. The library has a hierarchical structure (Fig. 4–6) that consists of both architecture-independent and architecture-dependent portions, as outlined in Table 4–2.

The conflicting goals of platform independence and exploitation of architectural uniqueness are addressed by:

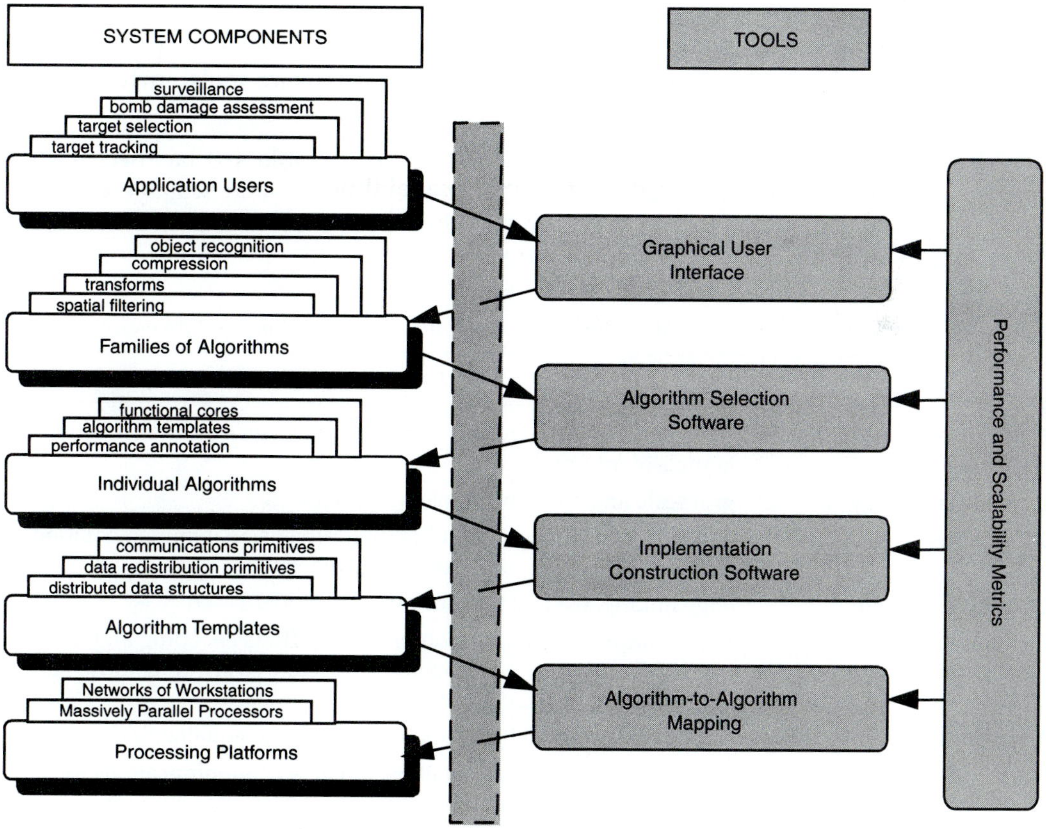

Figure 4–6 Structure of parallel scalable library for computer vision.[13]

Table 4–2 Architectural dependencies of Parallel Scalable Libraries and Algorithms for Computer Vision.

Component	Dependence upon Architecture
applications	architecture-independent
families of algorithms	architecture-dependent selection of algorithms based on algorithm and architecture parameters
individual algorithms	architecture-independent representation of individual algorithms
algorithm templates	primitives from which implementations are constructed; may be either architecture-independent (e.g., MPI standard for communications) or architecture-dependent (e.g., communications primitives optimized for a particular class of architectures

- developing and validating the C3 model (communication, computation, and congestion) to predict performance and scalability on a parallel architecture and thus support architecture-independent algorithm optimization;
- implementing fundamental algorithms and complete applications in computer vision and image processing on a variety of parallel architectures;
- developing application-driven library-based software tools, including a library-based parallel software reuse tool.

A library consisting of the elements shown in Table 4–3 has been implemented on several parallel machines, and several end-to-end scalable compression algorithms, including JPEG and MPEG-1, have been implemented using these tools.

Table 4–3 Contents of Parallel Scalable Library for Computer Vision and Image Processing.

Type	Examples
Generalized neighborhood algorithms	convolution, linear and nonlinear filtering, square and rectangular kernels of arbitrary size
Image morphology	grayscale and color morphological filters
Transforms	one- and two-dimensional FFTs; two-dimensional implementations include row-column and decimation algorithms
Object recognition	edge linking, list and contour ranking, model-based
Compression and coding	JPEG compression and decompression; MPEG encoding and decoding, JPEG and MPEG components—discrete cosine transform, Huffman coding, motion estimation
Data handling	fine-grained data redistribution algorithms to equalize I/O load
Communications	one-to-all; all-to-one; all-to-all personalized communications for coarse-grained machines; serial-to-parallel broadcasting for coarse-grained machines

In addition to libraries for signal and image processing, the ability to produce general-purpose programs on parallel machines is necessary. Work at Indiana University[14] has produced a preprocessor for C++, known as pC++, that provides an environment to allow programmers to compose distributed data structures with parallel execution semantics. This language supports data parallelism, or the execution of the same operations on multiple streams of data. Another tool supports task parallelism, or the execution of tasks in parallel, and both are combined into a preprocessor known as HPC++. These tools generate C++ code and machine-independent calls to a portable run-time system; the native C++ compiler for each of the processors then converts the results of HPC++ to the associated machine-specific code. By relying on HPC++ to generate machine-independent C++ and then using the C++ compiler supplied by the manufacturer of each processor, machine-independence of parallel processing code is obtained.

Central to the approach of pC++ is the concept of a pC++ *collection*, which is an array or set of pC++ objects. Data parallelism derives from the concurrent application of a function to each object in the collection. Each object is executed from the same instruction stream by a separate processor.

While they greatly facilitate the process of partitioning an algorithm across a parallel architecture, neither pC++ nor HPC++ provides automated mapping of the software to multiple processors. The typical pC++ development sequence is:

1. Run pC++ on pC++ code on a uniprocessor system such as a Sun workstation; partially debug parallel programs as uniprocessor tasks.
2. Run debugged pC++ on a parallel machine and debug.
3. Run that pC++ program on a second parallel machine of a different architecture. Depending upon the success of the previous two steps, this operation may involve simply recompiling without additional debugging.

A final tradeoff is that of portability versus performance. While this tradeoff is influenced by the level of implementation language, it is also impacted by hardware-specific aspects such as the communication method used between processors. At one extreme, maximum performance can be gained by programming an algorithm for a single parallel architecture; at the other, portability across all parallel architectures is the goal. A more reasonable goal is to achieve portability across a class of parallel processors—such classes are described in the next section. In the domain of language choice, peak performance will be obtained in machine-specific assembly code, while maximum portability will occur for coding entirely in a high-level language such as C. The practical compromise is programming in C code except for the time-critical modules, which are cast in form of critical subroutines and programmed in assembly language. Communication among the several processors of a parallel architecture has in the past been machine specific. A protocol known as MPI (Message Passing Interface) is now being applied as a standard portable communication method, as described in the Section 4.2.2.

4.2 SPACE-TIME TRADEOFFS IN HARDWARE

Speeding the execution of an algorithm by providing more hardware resources is often necessary to achieve real-time execution, despite the increased space and cost of these additions. Increasing speed by adding hardware nearly always means adding more processors.

An example illuminates the problems that can arise when a signal processing application grows beyond the capability of one processor. Fig. 4–7 represents a signal processing application as a flow diagram. Data from a sensor enters at the left, is split into parallel signal paths, and progresses through several signal processing operations until it reaches the output. The arrows represent data queues, which act as first-in, first-out memories (FIFOs). The circles represent nodes, which are the basic signal processing operations of the application.

These signal processing operations are executed upon a signal processor, which consists of a memory section and an arithmetic section. Executing these operations imposes some sort of load on the memory and arithmetic resources of the signal processor, as shown by the shaded levels in Fig. 4–7.

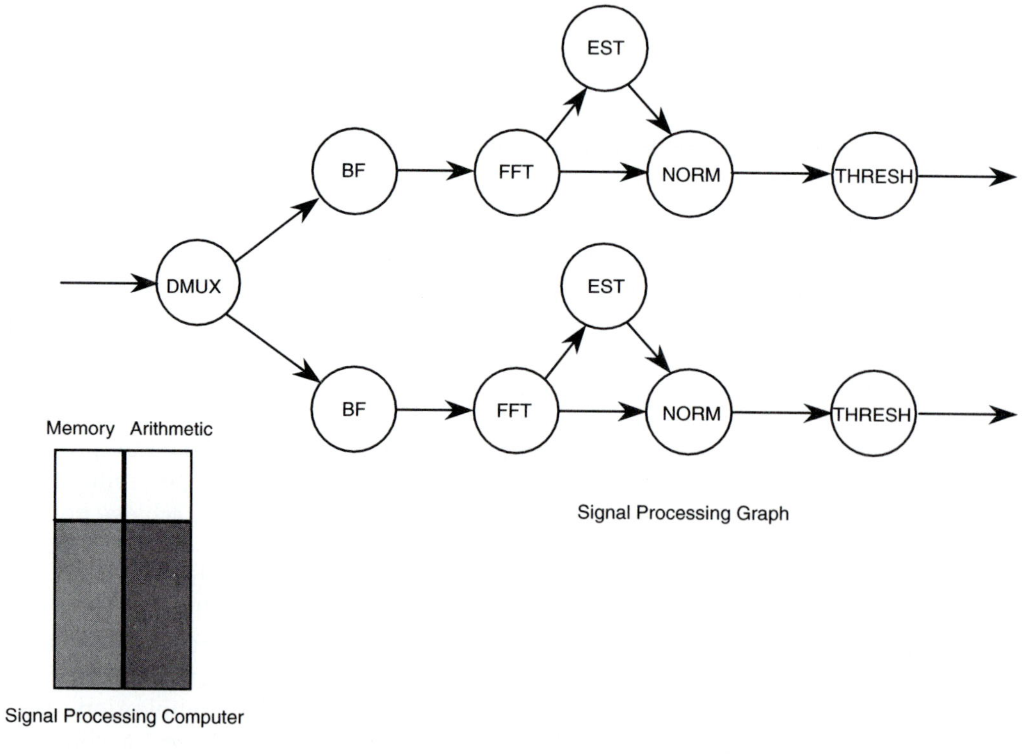

Figure 4–7 Executing processing shown by signal processing graph loads the processing and memory resources of computer as shown by shaded area.

If the complexity of the signal processing is significantly increased, it quickly exhausts the capability of the one signal processor in both memory and arithmetic. As shown in Fig. 4–8, a fourfold increase in signal processing graph complexity can be most simply accommodated by a fourfold increase in the number of processors. However, several questions then arise:

- How is the number of processors selected?
- How are portions of the signal processing flow graph allocated to the various processors?
- How are the loads upon the multiple processors balanced, given that in most applications, data rate reduction occurs so that upstream portions of the graph present a greater load than downstream portions?
- How is data communicated from processor to processor, particularly in the most general case of fully connected processors?

As shown in Fig. 4–9, these questions are addressed by:

1. Replacing the point-to-point communications of Fig. 4–8 with a *data transfer network*, or non-blocking crossbar switch, that allows simultaneous point-to-point communications between any number of pairs of computing elements;
2. Separating the arithmetic unit from the memory unit and fastening both to the data transfer network;
3. Introducing a *scheduler* element that responds to the arrival of all necessary input data for a node by scheduling that node for execution on an arithmetic unit.

To form a complete computer, elements for data input and output, plus an overall control processor, are also added.

This exercise has led to a *dataflow multiprocessor*, in which the presence of all input data needed by a node triggers the execution of that node. The dataflow multiprocessor is the basis for data flow computers, an example of which is the AN/UYS-2 Enhanced Modular Signal Processor (EMSP), used by the U.S. Navy for real-time parallel signal processing.

As this example suggests, several hardware design parameters are subject to tradeoff. These include flexibility, power, area or volume, speed, architecture of each processor element (including memory architecture, execution unit, register placement, and degree of pipelining), and implementation method (application-specific integrated circuit or ASIC, field programmable gate array or FPGA, reduced instruction set computer or RISC, programmable digital signal processor, fine versus coarse-grained architecture). One can tradeoff the number of processors versus the speed of memory. A single processor using a very fast and

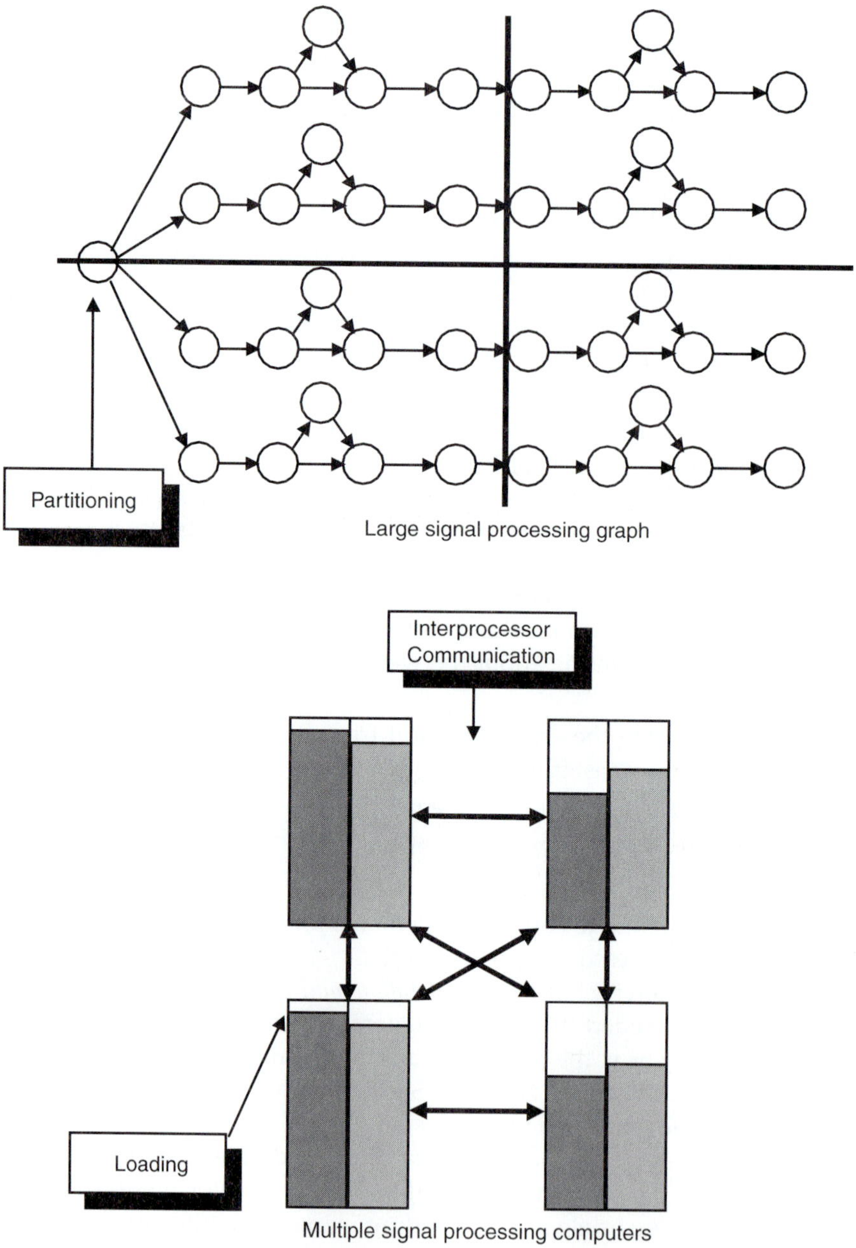

Figure 4–8 Responding to increased processing demands by simply adding more processors raises questions on partitioning, loading, and interprocessor communications.

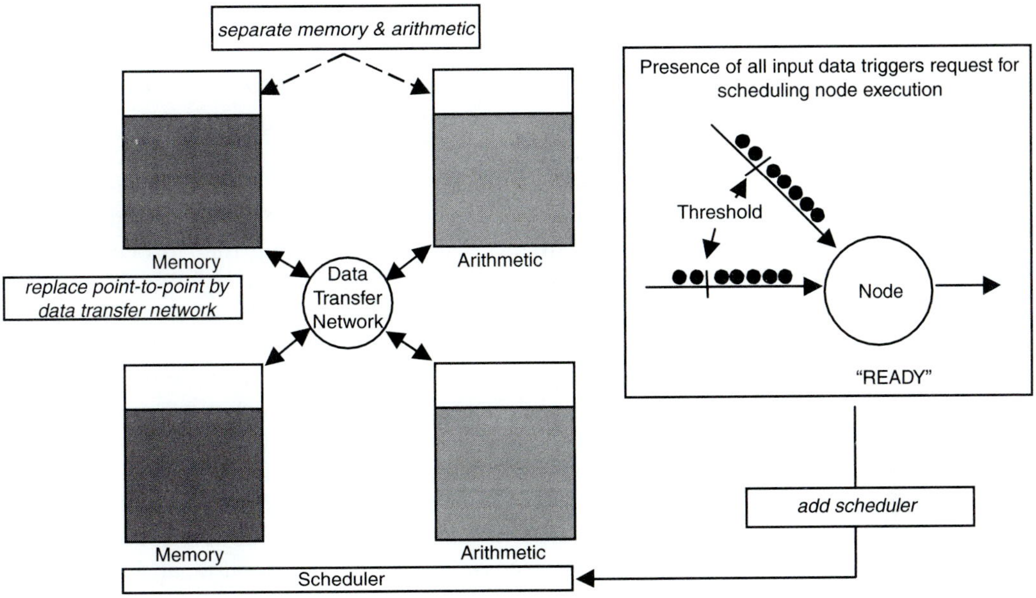

Figure 4–9 Addressing multiprocessor questions by adding architectural features to provide automatic partitioning, load-leveling, and interprocessor communications.

expensive memory may be more economically replaced with two processors and slower, less expensive memory.

The introduction of a second processor may be made in a limited domain by embedding a specialized coprocessor, suited for performing only a subset of the expected set of operations. This subset would be a class of instructions that consumes most of the execution time (see Fig. 2–28), and the main processor recognizes the type of instruction for which the coprocessor is designed, sends these instructions to the coprocessor, and awaits the return of results.

The following subsection characterizes several alternative multiprocessor architectures. Interprocessor communication is then discussed, followed by a description of control structures for multiprocessors. The section concludes with a discussion of a few examples of multiprocessors.

4.2.1 Multiprocessor Characteristics

The essential elements that distinguish various types of multiprocessors include:

- *type of processing element:* the architecture, speed, and capability of each processor that makes up the multiprocessor, and whether all processor elements are alike or of different types;

- *topology and speed of the interconnection network* that allows processor elements to communicate with each other;
- *total number of processing elements* and overall multiprocessor throughput;
- *architecture of memory*, both shared among processors and local to processors.

4.2.1.1 Flynn's Taxonomy. Multiprocessors may be characterized by how they allocate their streams of data and instructions among their processors. One may have parallel data streams, parallel execution threads, or both. Flynn's taxonomy[15] provides a widely used method of making these distinctions. The taxonomy begins with "single instruction, single data" machines, or SISD. This arrangement is really not a multiprocessor, but consists of a single processor connected to a single memory (Fig. 4–10a). The single-instruction multiple data (SIMD) machine provides multiple processors and a similar number of separate memory units. A central controller broadcasts the same instruction to all processors, and each processor operates on a different portion of the data (Fig. 4–10b). The SIMD is well suited to exploit data parallelism, as exhibited in such tasks as matrix multiplication, the vector processing of image processing, and DO-loops.

The multiple instruction, single data (MISD) multiprocessor applies the different instructions to the same stream of data (Fig. 4–10c). An example is the serial chain of processors that operate in sequential stages on data in a pipeline manner. This arrangement has also been called the "vector SIMD" under which nomenclature the SIMD of Fig. 4–10b is called the parallel SIMD.

Finally, the multiple instruction, multiple data (MIMD) arrangement provides true parallelism in both instruction and data streams. In addition to the

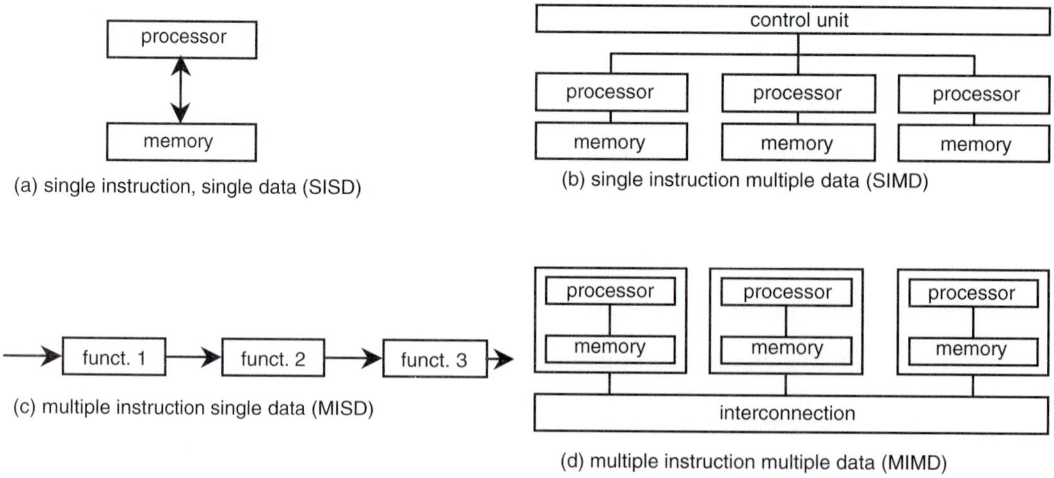

Figure 4–10 Multiprocessor types of Flynn's taxonomy.

MIMD architecture shown in Fig. 4-10d, the multiprocessor developed in our introductory example, which separates the memory from the processor (Fig. 4–9), is a MIMD machine.

An advantage of the MIMD architecture is that its processors can execute multiple jobs simultaneously, and each processor can perform any operation regardless of what other processors are doing. The flexibility of the MIMD architecture also provides disadvantages—an overhead function is required if load balancing is needed, and synchronization activities are necessary to coordinate processor timing at the end of a parallel structure, when all results must be combined.

Another distinction that is made among multiprocessor types is whether they are coarse-grained or fine-grained. In a coarse-grained architecture, the processors are relatively complex and may differ from one another, comprising a heterogeneous architecture. A crossbar switch provides a means of interconnecting coarse-grained processors. A heterogeneous coarse-grained architecture allows choosing a processor type that is best for each stage of a complex signal processing algorithm. For example, in an image processing operation (Fig. 3–7), front-end computations occur as integer operations on adjacent groups of pixels, while back-end operations are highly data dependent and diverse. A heterogeneous architecture can combine a processor suited for two-dimensional image processing with processors better suited for general purpose computing. Coarse-grained architectures are usually more difficult to program than fine-grained ones. More manual intervention is needed to map different portions of the program to different types of processors, and the programmer generally ends up with disjoint software modules that require synchronization.

Fine-grained architectures are homogeneous arrays of hundreds or thousands of small processors that are connected to their nearest neighboring processors. Since they present a uniform programming model, automatic partitioning of a task is more successful. However, fine-grain systems require highly regular, repetitive algorithms to realize their advantages.

4.2.1.2 Interconnection Topologies.[16] The two major challenges to successful multiprocessor use are the interconnecting, discussed here, and the partitioning and mapping of algorithms onto the multiple processors, discussed in Chapter 5. At the simplest level, one processor can be provided with multiple functions by switching among programs (Fig. 4–11). This technique provides a fraction of a processor per task, but allows an implementation of multitasking.

There are many ways that N processors and M memories can be interconnected. The simplest is the single common bus, shared among processors. This is an extrapolation of the SISD architecture of Fig. 4–10. Examples of a linear common bus include the VME, Ethernet, 1553, Futurebus, and PI bus. The characteristics of these buses are reviewed in Section 7.2. While a common bus is sufficient for a modest number of processors, it presents a communication bottleneck as more processors are added, since bandwidth does not grow with the number of

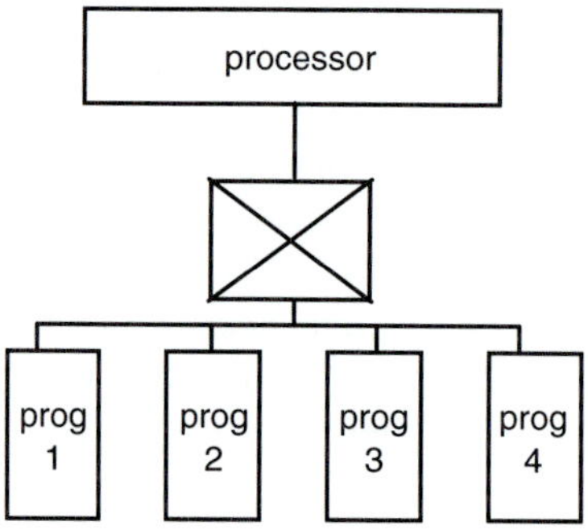

Figure 4–11 Single processor, multiple functions.

processors. The physical constraints of the shared medium limit the number of nodes that can be attached. For example, the Futurebus and PI bus are limited to 32 nodes. A hierarchical topology can be constructed by fastening linear buses to each other in either a linear or tree topology, as shown in Fig. 4–12. The linear bus is a general-purpose configuration and is not specific to one type of algorithm. It is not *fault tolerant*—the failure of any node on the bus can take down the entire bus.

Various sorts of multiconnection buses have arisen as a result of bandwidth limits on the linear bus. In general, they strive toward the goal of fully connected architectures but may seek to decrease the complexity dictated by the plague of combinatorics. Full interconnectivity of N processors requires $N(N - 1)/2$ two-way connections. In addition, the information transfer time for every processor to receive every piece of data is of $O(N)$, so the complexity of a fully connected multiprocessor scales as $O(N^3)$. Several topologies compromise between the linear bus and the fully connected arrangement. These are now examined in order of increasing complexity.

The first of these is the *star network*. A group of N processors is clustered around a central switch (Fig. 4–13). The switch may be an $N \times N$ nonblocking crossbar switch, as detailed in the figure. The central switch, while connecting each processor to all others, can serve as a communication bottleneck, but it provides higher throughput than the linear bus because it provides a separate path for each point-to-point connection, rather than requiring all point-to-point connections share a common bus bandwidth.

Serial or parallel connections (Fig. 4–14) attach processors in a manner that can be easily matched to the flow of purely serial or parallel algorithms. In the serial configuration, each processor has a pair-wise connection with its neighbors.

Joining the ends of a serial processor chain to each other produces the *ring configuration*, as shown in Fig. 4–15. The ring topology requires taking down the network to insert or remove a processor, and it introduces a latency between nonadjacent nodes. It is not fault tolerant—a single processor can bring down the network. Variants of the ring introduce either a counter-rotating or a skip-ring connection, as shown in Fig. 4–15.

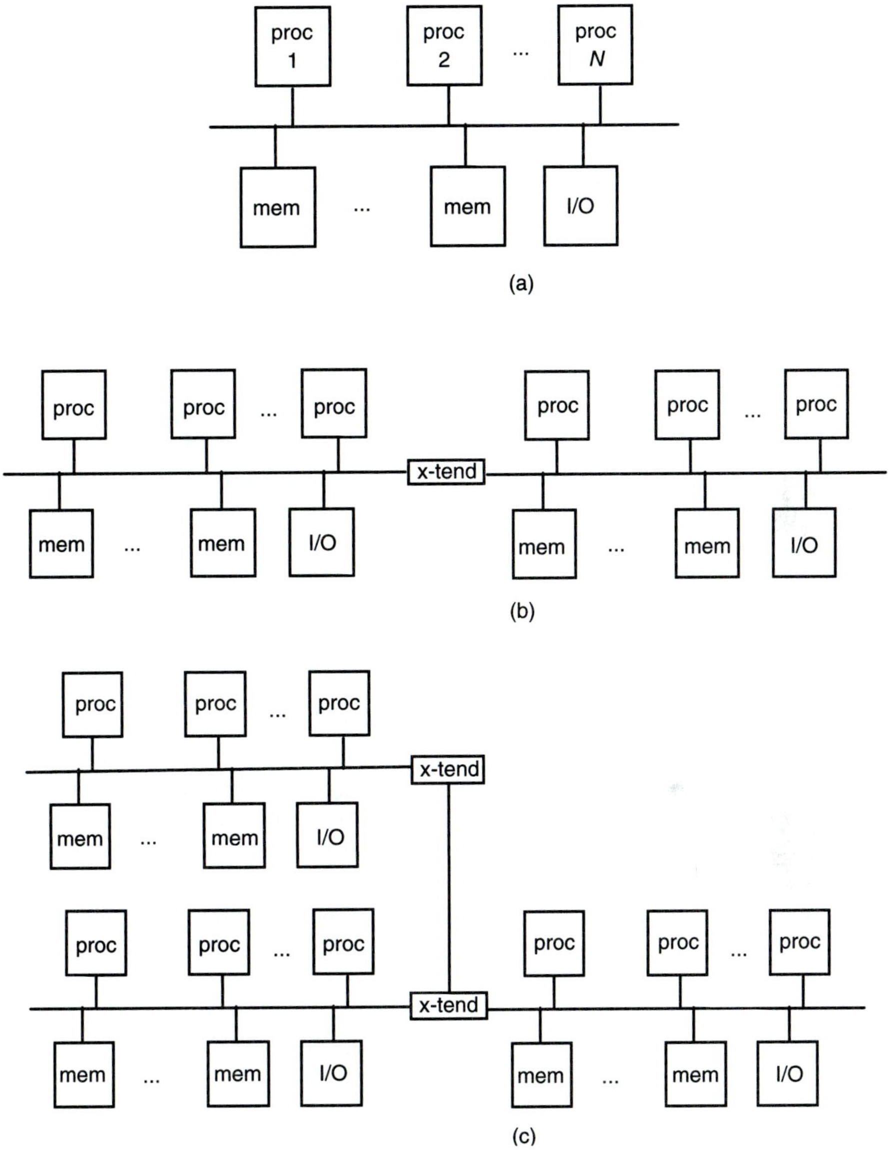

Figure 4–12 Common (linear) bus (a) and the use of bus extension units to construct larger linear (b) and tree-like (c) arrays from a linear bus.

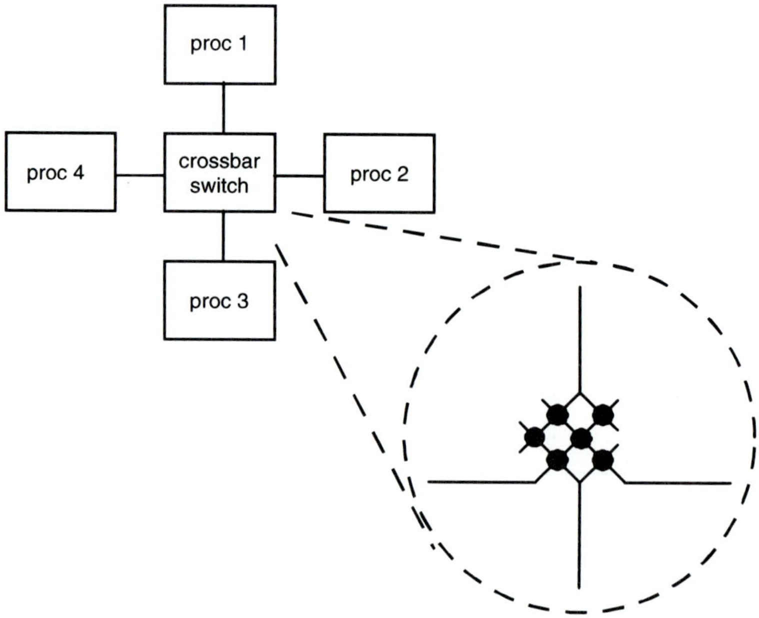

Figure 4–13 Star network with detail of 4 × 4 nonblocking crossbar switch.

Ring architectures are more scalable: Increasing system size by adding processors also adds connections and augments the bandwidth. The number of point-to-point links in a ring does not limit the number of nodes in a ring. Each link maintains a high system bandwidth as system size increases, since each processor continues to be connected on a point-to-point basis rather than a multi-drop bus. Ring protocols include the FDDI and SCSI, as described in Section 7.2.

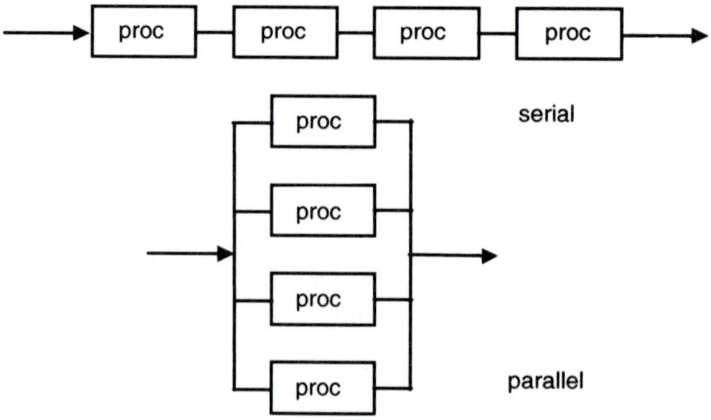

Figure 4–14 Serial and parallel topology of multiple processors.

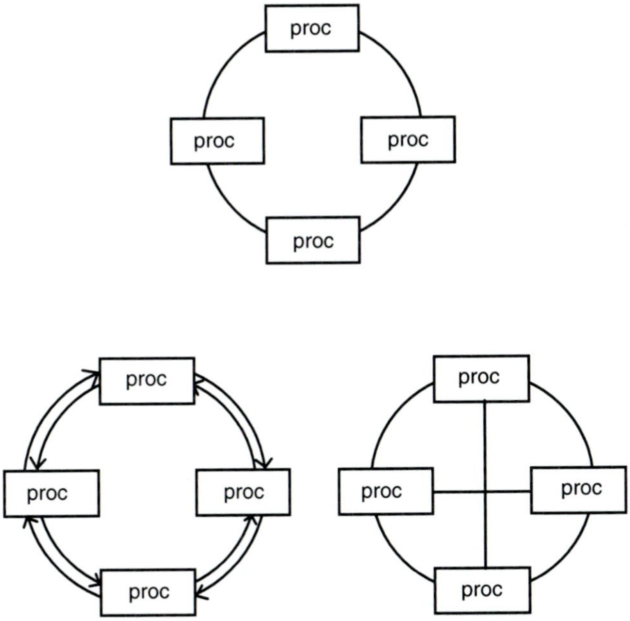

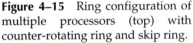

Figure 4–15 Ring configuration of multiple processors (top) with counter-rotating ring and skip ring.

A variety of interconnection topologies either approach or achieve full interconnection. These fully-connected architectures alleviate communication bottlenecks but are difficult to scale, as their size is proportional to N^2. They are often implemented with a switch, and as such are referred to as "switch-centered architectures."

Scalable switches may be built up from N-point switch building blocks (Fig. 4–16). An example is the RACEway® Interlink interconnection of Mercury Computers, in which $N = 6$. The switch building block may also be used to construct a tree network. In a tree structure, all processor are interconnected, but some connections pass through processor nodes, so processors are separated by several layers. Tree networks are suited to the hierarchical processing tasks often used in pattern recognition. The top processor of the tree performs processing that determines in which of several broad classes an unknown pattern belongs. It then sends the pattern down to the child processor devoted to processing for further discrimination within that class.

Another interconnection scheme is the *mesh* topology, which is recently receiving attention for scalable parallel processors and massively parallel processors. The mesh topology scales well—its number of links scales linearly with the number of nodes. It is slower than a fully-connected architecture, since many connections must be made through intervening nodes. A variant known as the *toroid* connects each end of a row together as well as each end of each column, allowing the use of ring-based interconnection standards such as the SCSI to be used in a mesh construction. The *weave cube* is another mesh variant in which

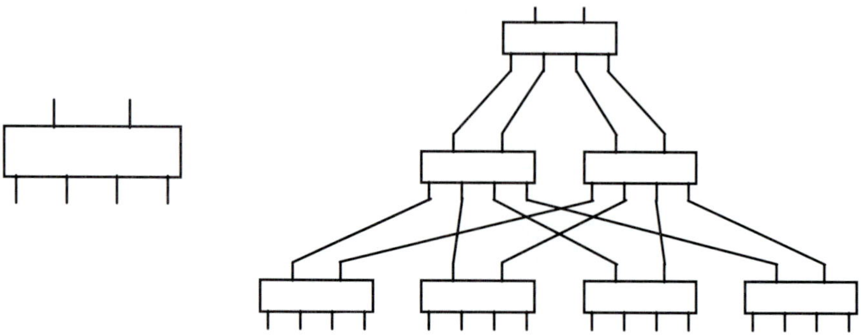

(a) (b)

Figure 4–16 Six-point switch building block (a), and fat tree constructed from these building blocks (b).

each row or column has its end connected to another row or column. The *cube*, or more generally the *hypercube*, is a series of interconnections in which each node of a 2^n processor system has n links. The advantage of a hypercube over the standard mesh is that the number of links between nodes is less. However, more links are required per node, and the system is less easily scaled. After reaching a certain size, all nodes must be upgraded with another link. Fig. 4–17 shows these variants of the mesh topology.

The *systolic* array is a special case of the mesh topology. It provides a means for local nodes to compute at speeds higher than what a local memory could support. This is achieved by using a fine-grain array of processors without memory. Each processor performs its stage of processing on a wavefront of data that passes through the mesh. As the data passes through the mesh, it is directly processed, without being stored and retrieved from local memory. While systolic arrays are successful in highly regular problems, decomposing problems to pass through sequential computation without storage to memory can be difficult. *Pseudo-systolic* arrays introduce a small amount of memory at each node to alleviate the mapping problem. However, the introduction of memory slows the flow of data, and thus a tradeoff exists between the size of the memory and communication bandwidth.

4.2.2 Interprocessor Communication

Several multiprocessor interconnection topologies have now been explored without comment on the communication that must occur over these interconnections. This subsection looks more carefully at the mechanics of communication along the lines that connect the processors in these topologies. One challenge results from the physical area of the processor that must be devoted to communication. For a word width W, the thickness of a parallel bus is proportional to W, and the bus area of a typical processor with buses running in both horizontal and vertical directions is proportional to W^2. The length of each wire also depends upon W, as

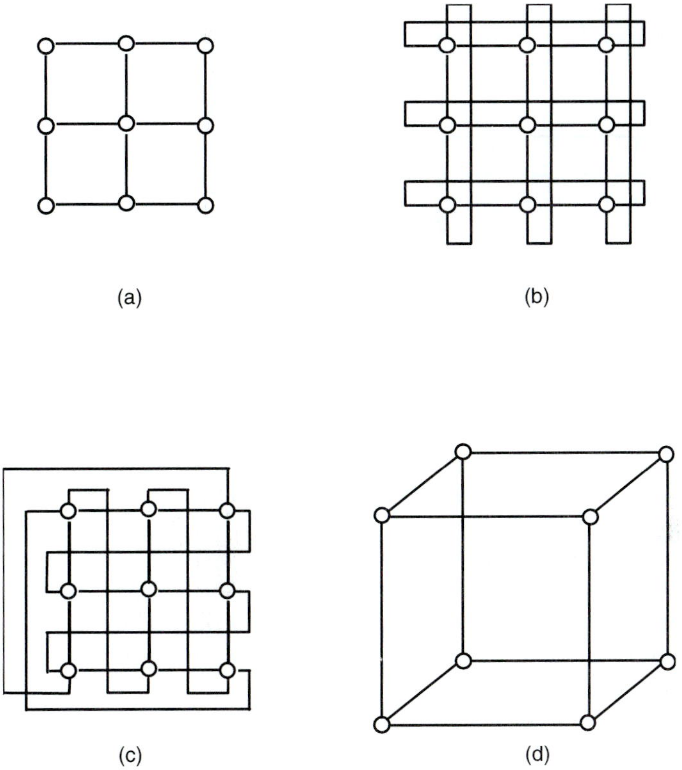

Figure 4–17 Variations on the mesh topology include (a) basic mesh, (b) toroid (c) weave, and (d) hypercube.

proportionately more communication area must be traversed, and so the signal propagation time is proportional to W. To fully connect N processors, N^2 such buses must be placed. Thus, as the number of processors to interconnect and the number of bits of interconnection increase, both the area and the delay associated with interconnection increase.

Either of two types of interconnection, shared memory or message passing, may be used for processor communication (Fig. 4–18). In a shared-memory arrangement, all processors can access memory and an interconnection network such as a switch is used. In message-passing communication, each processor has local memory (the distributed memory arrangement), and data is passed from one processor to another by a network. As shown in Fig. 4–18a, a multiprocessor architecture in which the memory modules are separate from processing modules supports shared memory communication. In a shared memory arrangement, the total memory is available to each processor as a unified address domain.

Shared memory communication is easier to program; since any memory location can be accessed by any process, the need to allocate certain data to certain

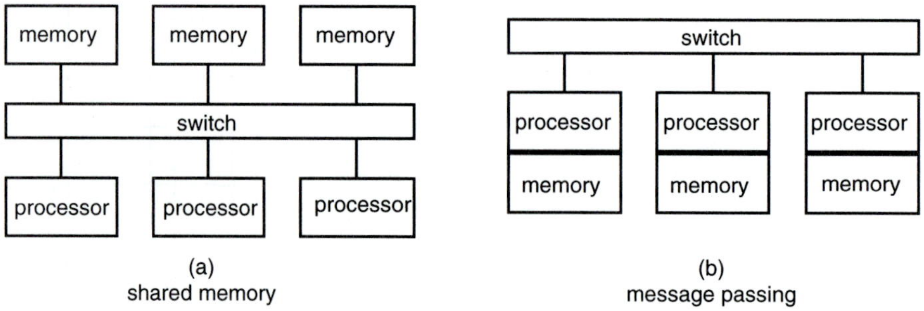

Figure 4–18 Two methods of interprocessor communication.

memory locations is avoided. However, an operating system, with its associated processing and memory overhead, is necessary to implement memory sharing.

As the number of processing and memory modules increases, shared memory operation becomes more difficult, since increasing numbers of accesses must be made over a finite-bandwidth switch. Alternatively, *message-passing communication* scales linearly as more processors are added, since each processor brings more memory. An operating system is not needed to implement message-passing communication, although many processors that use message passing also include operating systems.

Fig. 4–19 shows an example of message-passing operations. A *scheduler*, in response to the application program, issues a message to send a packet of data to a particular process. The *packet switch processor* then computes a routing path that includes the data source, the path, and the data destination. This information is added as a header to the data packet. The switch hardware and I/O circuits of each processor module then carry out the routing commands contained in the packet header and move the data packet as directed.

Message-passing operation may be either blocking or nonblocking. In the blocking form of message passing, processing stops until the send operation is complete. Message passing with blocking is easier to manage, and stopping operations until data transfer is complete means that data precedence will be honored. However, the idle time introduced as the processor awaits its data reduces the overall throughput.

Nonblocking message passing uses a buffer to permit concurrent data movement and computation. This allows communication and processing operations to proceed in parallel, providing enhanced throughput. However, communication complexity is increased as sending requires at least two operations. The first operation sends the data to the destination processor. The second operation checks to determine whether the data has been received by the other processor. If not, subsequent checks are necessary until the data is received.

An example of a standard message-passing technique is the Message Passing Interface (MPI).[17] MPI was formulated in April, 1994, with subsequent enhancements (e.g., MPI-2). It provides a widely used standard for writing message-

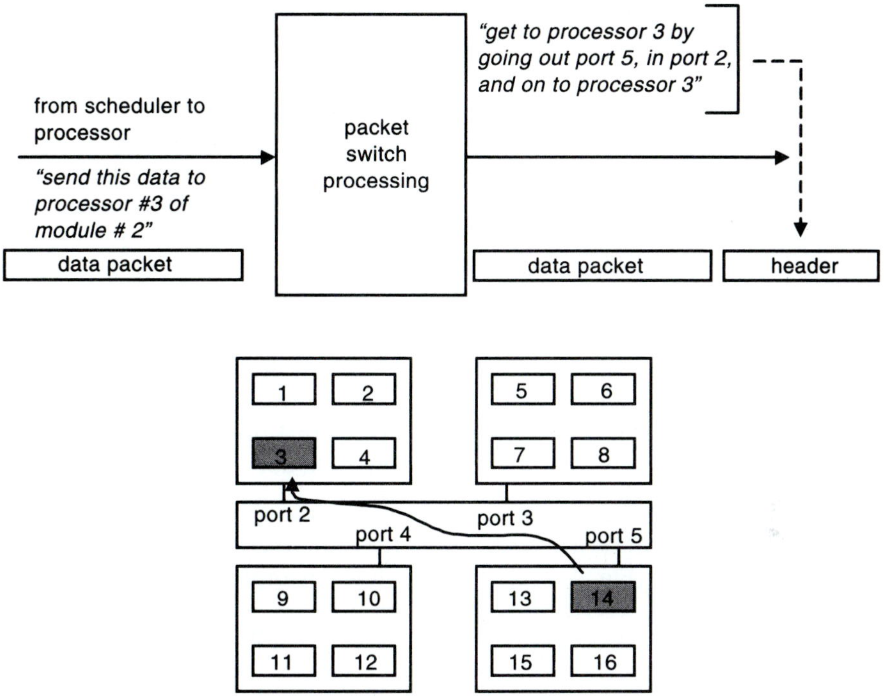

Figure 4–19 Operation of message passing movement of data from one processor to another.

passing programs. MPI supports both point-to-point and collective communications, and is intended for MIMD architectures. Because it is a standard, MPI provides an effective way to partition algorithms for a variety of multiprocessor architectures that include nearly all commercially available multiprocessors. The MPI standard sends data from one processor to another by use of the MPI_SEND command. The MPI_RECV command is called by the receiving processor. The MPI *send* and *receive* commands follow the blocking style of message passing described earlier. The full MPI library consists of about 100 commands that support point-to-point communication.

MPI has been enhanced to support its use in real time systems. This enhancement, known as MPI/RT, is part of the MPI-2 standards process, but is neither a subset nor a superset of MPI-2. MPI/RT is designed to support time-driven, event-driven, and priority-driven communication tasks, and it includes such real time features as timeouts at appropriate points.

4.2.3 Multiprocessor Control

The variety of means for multiprocessor control sets up a continuum, with *multitasking* at one end, in which multiple applications run on the same processor, and *multiprocessing* on the other, in which multiple processors work on portions of a single application. This spectrum provides two means of allocating algorithms to multiple processors:

- *control flow* assigns each task of an algorithm to a particular slice of time, based on location of the task within the program flow. It reuses processing hardware by time-sharing a single processor across tasks. The control flow approach is more appropriate for small-bandwidth communication signals (<100 kHz). Control flow leads naturally to the single-chip digital signal processor, which has a processor core for performing various arithmetic operations under the direction of a controlling program.
- *data flow* maps the algorithm directly into the hardware for an algorithm that is described as a signal flow graph. A simple but inefficient example of data flow would devote one processor to each block of the flow graph. Data flow control is less flexible than control flow, but it minimizes the control overhead associated with moving data to the correct operational units.

In a control flow architecture, processor operations are paced by a common clock and operands are written to a common memory and retrieved when needed for processing (Fig. 4–20a). Control flow for multiple processor is effective if the process can be described by DO-loops.

Data flow processing operations (Fig. 4–20b) are paced by the availability of all input data needed by a processing node. A scheduler is needed to evenly distribute the work load over the multiple processor resources, as was shown in the example of Fig. 4–9. This method allows any of multiple nodes to be executed on the next available processor of the appropriate type. Processing results flow from one stage to the next via interspersed registers. Since no program control or cen-

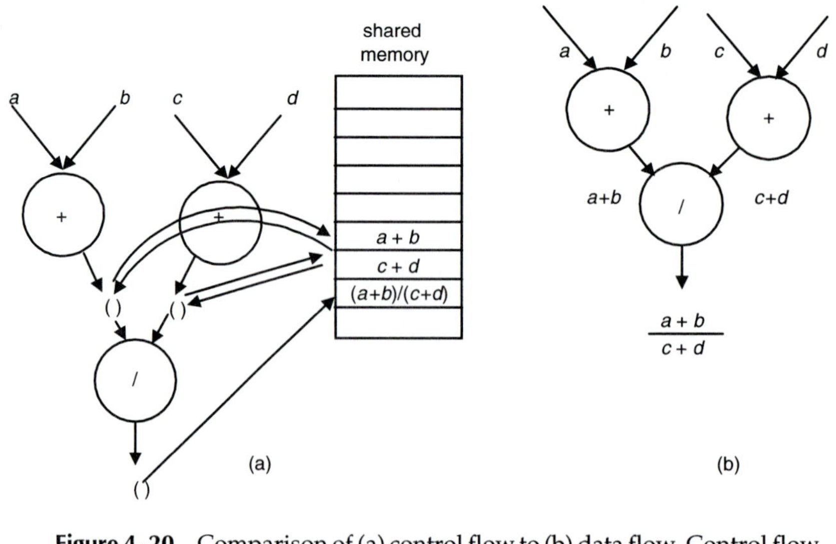

Figure 4–20 Comparison of (a) control flow to (b) data flow. Control flow involves the writing of intermediate results to memory.

tral processor orchestrates operations, and since any task can be executed as soon as its data is available, data flow supports concurrent processing.

A challenge of both data and control flow multiprocessing is interprocessor synchronization. The SIMD architecture performs processing with data-dependent timing on multiple partitions of the data, and some partitions finish before others. By adding task parallelism to data parallelism, the MIMD architecture is even more difficult to synchronize.

Two-processor synchronization provides a simplified example of the means used to synchronize separate processors. When one processor finishes, the other processor can be suspended or delayed if necessary, or the first processor can be delayed in starting its next task. If the processors are connected in series, a FIFO buffer can be used to change the timing of the data flow between the processors and allow the downstream processor to remove data at a rate different than that at which the upstream processor puts it into the buffer (Fig. 4–21, top).

A double buffer may also be used between two processors (Fig. 4–21, bottom). One side is filled by the output of processor 1 while processor 2 is removing data from the other side. When one side of the buffer has been filled and the other side emptied, the sides are switched and processor 2 begins processing on the data entered by processor 1, while processor 1 fills the other side. For systolic arrays, or for situations where all processing is completed before the next sample arrives, buffering for synchronization is not necessary.

4.2.4 Multiprocessor Examples

By examining a few commonly used multiprocessor machines, the underlying principles used in their design may be illuminated. Each machine is discussed in terms of application domain, multiprocessor architecture, characteristics of the basic processing node, and total system characteristics. Table 4–4 summarizes ex-

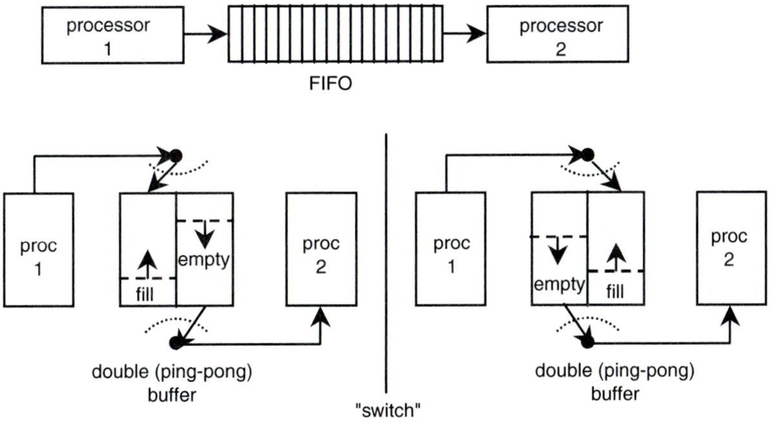

Figure 4–21 Synchronizing multiprocessors with FIFO or double buffers.

Table 4–4 Summary of common multiprocessor systems.

	Architecture			Processing Nodes			
Machine	Total MFLOPS (system)	Number nodes (system)		MFLOPS	Memory	Communication	Comment
Silicon Graphics POWER CHALLENGE XL	To 10,800	2–36	Symmetric multi-processing	300	Up to 18 GB 8-way interleaved, shared among all processors	Powerpath—1.2 GB/sec	Most widely deployed multiprocessor
Intel Paragon	150,000	2048	MIMD via parallel SIMD, fine-grained; shared memory	2 to 3 processors @ 75 MFLOPS each—Intel i860 XP (50 MHz); 2 processors—general-purpose (GP) node—one for computing, one for memory; or 3 processors—multiprocessor (MP) node—two for computing, one for either computing or messaging	16 or 32 MB shared memory for GP node; 64 or 128 shared memory for MP node	64-bit 400 MB/sec cache-coherent bus to shared memory—mesh interconnect	Used in embedded computing; militarized version available as Honeywell Touchstone

IBM SP-2	128,000	Up to 512	MIMD via parallel SIMD; distributed memory	266 MFLOPS Power 2 processor	64–512 MB memory 64-bit memory bus 64-KB data cache 32-KB instruction cache	High-performance switch—16x16; 40 MB/sec peak individual bandwidth Message Passing Interface (MPI)	Standalone general purpose scalable super-computer
Mercury RACE - SHARC	120,000	~1000	Mesh MIMD distributed array	SHARC 120 MFLOPS; also accepts PowerPC 603	542 Kbyte /processor	Mercury RACEway Interlink (ANSI/VITA 5-1994 Standard)—6-point switch building blocks, 160 MB/sec for each data path	High speed signal processing super-computer

amples of several classes of multiprocessors: the engineering workstation, embedded computer, standalone supercomputer, and multiple digital signal processing system.

4.2.4.1 Engineering Workstation: Silicon Graphics Power Challenge. A widely used multiprocessor is the Silicon Graphics Power Challenge XL workstation. Its architecture is symmetric multiprocessing, in which a common memory is shared by all processors (Fig. 4–22).

Each processor is an R8000 or R10000 reduced instruction set computer (RISC), achieving 300 MFLOPS of throughput. Each processor has 14 Mbyte of secondary cache, and from one to eighteen processors are combined in a system. A total system throughput of up to 5.4 GFLOPS is possible. A 16-Gbyte eight-way interleaved memory is shared by all processors, and the processor are linked by the Powerpath communication system, which attains a 1.2 Gbyte/sec communication rate, using a 256-bit wide data path and 40-bit wide physical address path.

4.2.4.2 Embedded Computer: Intel Paragon. The Intel Paragon is a fine-grained architecture that can contain up to 1000 nodes, with each node consisting of two to three processors each. It uses parallel SIMD paths to achieve MIMD functionality. Each node processor is the Intel i860 XP RISC, capable of 75 MFLOPS. A general purpose node uses two of these processors: one for computation that executes arithmetic, and the other for message passing that handles the message passing protocol. Each general purpose node has 16 or 32 Mbyte of shared memory. Alternatively, the multiprocessor node may include three processors, two of which are devoted to computing and the third of which is devoted either to computing or to message processing. A multiprocessor node has either 64 or 128 Mbyte of shared memory.

With its ability to cascade up to 1000 nodes, the Paragon can achieve a throughput of up to 150 GFLOPS. A 64-bit, 400-Mbyte/sec communication link arranged as a mesh provides access to shared memory.

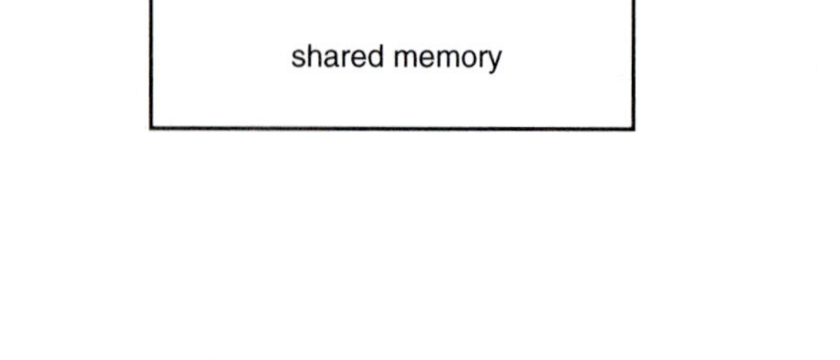

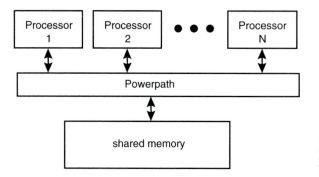

Figure 4–22 Symmetric multiprocessor architecture of Silicon Graphics Power Challenge XL.

4.2.4.3 Standalone Supercomputer: IBM SP-2. The IBM SP-2 is a general-purpose scalable supercomputer with a MIMD architecture that is achieved via parallel SIMD channels. It uses a distributed memory architecture that provides 64 to 512 Mbyte to each node. At each node is a 266 MFLOP processor attached to a 64-bit memory bus and augmented by a 64-Kbyte data cache and 32-Kbyte instruction cache. Up to 512 nodes may be combined to achieve up to 128 GFLOP of system processing capability. As shown in Fig. 4–23, the SP-2 uses a switch-centered architecture, in which each processor accesses the High Performance Switch (HPS). This switch connects up to 16 nodes with up to 16 other nodes, providing a 40 Mbyte/sec peak bandwidth between individual nodes. In addition, a slower Ethernet connection provides a second communication link, shown outside the processors in Fig. 4–23.

A MIMD message passing protocol known as MPC allows both process-to-process and global communications in the SP-2. Two communication methods are available. The *user space protocol* provides more rapid communication, but cannot be shared with other MPC processors. The slower *internet* protocol allows connectivity with other MPC processors. MPC provides C-callable communication functions (Table 4–5).

4.2.4.4 Multi-Digital Signal Processor System: Mercury SHARC RACEway®. While the previous multiprocessors have been built up of general purpose processors, the nodes of the Mercury SHARC RACE provide up to 1000 Analog Devices AD2106x SHARC (Super Harvard Architecture) digital signal processors, each capable of 120 MFLOPS (see Section 2.3.5). Alternatively, the Power PC general purpose processor may be substituted at any node. Processors are connected via the Mercury RACEway Interlink. The RACEway is a switched fabric architecture that is supported by the ANSI/VITA 5-1994 standard. It is built up from multiple six-point switch building blocks and provides a communi-

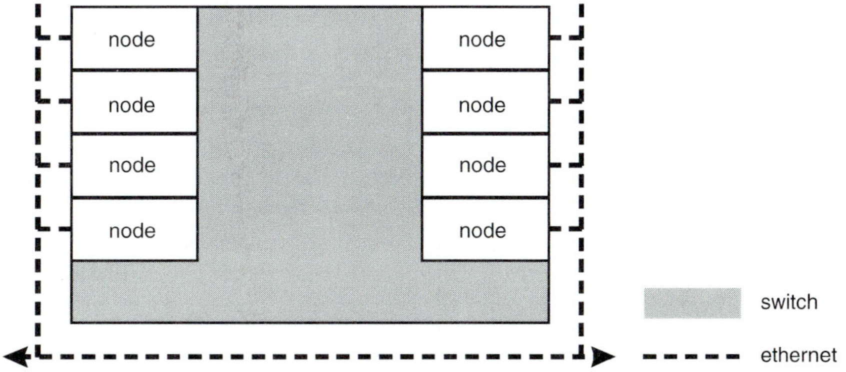

Figure 4–23 Nodes in the IBM SP-2 parallel computer are connected both by a high-speed switch and a slower Ethernet.

Table 4–5 C calls to MPC library functions.

mpc_bsend (&msg_buffer, msg_length, dest, tag
mpc_brecv (&msg_buffer, max_length, source, tag, rec_length)
where:

- msg_buffer = address of buffer containing data
- msg_length = length of buffer
- msg_length = length in bytes
- max_length = maximum length of receive buffer
- rec_length = number of bytes actually received
- dest = processor i.d. number of task to which message is sent
- source = processor task identification of source from which message is to be received
- tag = user-defined non-negative integer used to identify message transferred

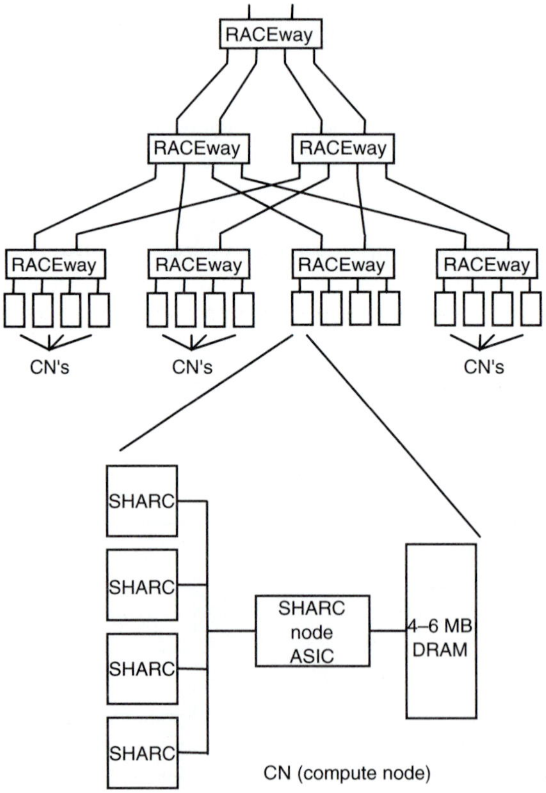

Figure 4–24 Multi-signal processor array composed of Analog Devices 2106x SHARC DSPs and the Mercury RACEway interconnect.

cation rate of 160 MB/sec for each data path. Fig. 4–24 shows the architecture of the Mercury SHARC RACE system.

REFERENCES

1. Panos Papamichalis, Jay Reimer, and Jon Rowlands, "System and Algorithm Implementation Techniques on the TMS320 Family," *Proc. IEEE Inter. Conf. Acoustics, Speech, Signal Processing,* Detroit, MI, May, 1995, pp. 2845–2848.
2. MathSoft, Inc., 101 Main Street, Cambridge, MA, 02142.
3. Konstantinos Konstantinides and John R. Rasure, "The Khoros Software Development Environment for Image and Signal Processing," *IEEE Trans. Image Processing,* 3 (May, 1994) 243–252.
4. U.S. Navy, "Processing Graph Method (PGM) Specification," ver. 2.0, 1995.
5. *MATLAB User's Guide* (Natick, MA : The MathWorks, Inc., 1992).
6. William Morris, ed., *The American Heritage Dictionary of the English Language* (Boston, MA : Houghton-Mifflin, 1980).
7. Donald E. Knuth, *Seminumerical Algorithms, The Art of Computer Programming, vol. II,* 2nd ed. (Reading, Mass: Addison Wesley, 1989) 253.
8. W. Kenneth Jenkins and Andrew J. Mansen, "Variable Word Length DSP Using Serial-By-Modulus Residue Arithmetic," in *Signal Processing Technology and Applications,* John G. Ackenhusen, Ed. (Piscataway, NJ : IEEE, 1995) 218–221.
9. Alfred V. Aho, John E. Hopcroft, and Jeffrey D. Ullman, *Data Structures and Algorithms,* (Reading, Massachusetts : Addison Wesley 1982).
10. C. Böhm and G. Jacopini, "Flow Diagrams, Turing Machines, and Languages with Only Two Formation Rules," *Communications of the ACM,* 9 (May, 1966) 366–371.
11. Maurice H. Halstead, *Elements of Software Science* (New York: Elsevier North Holland, 1977).
12. Bjarne Stroustrup, *The C++ Programming Language,* 3rd ed. (Englewood Cliffs, NJ : Prentice-Hall, 1997).
13. Leah H. Jamieson, Sussanne E. Hambrusch, and Edward J. Delp, "Parallel Scalable Libraries and Algorithms for Image Processing and Computer Vision," Defense Advanced Research Projects Agency (DARPA) High Performance Computing Initiative, 1995.
14. A. Maloney and others, "Performance Analysis of pC++: A Portable Data-Parallel Programming System for Scalable Parallel Computers" *Proc. Eighth Annual International Parallel Processing Symposium (IPPS),* Cancun, Mexico, April 1994, pp. 19940426–19940429.
15. Michael J. Flynn, "Very High Speed Computing Systems," *Proc. IEEE,* 54 (December, 1966) 1901–1909.

16. Fred Shirley, *RASSP Architecture Guide*, Document AVY-L-S-0008L-101-6, April 1995, Sanders, a Lockheed Martin Company, *http://www.dnh.mv.net/ipusers/fs/ArchGuide/ArchGuide.pdf* visited March 12, 1999.
17. Argonne National Laboratory, "MPI: A Message Passing Interface Standard" *http://www.msc.anl.gov/mpi/mpi-report/mpi-report.html*, 1994; for MPI-RT – ftp://aurora.cs.msstate.edu/pub/mpi/rt-2-26feb97.ps.

5

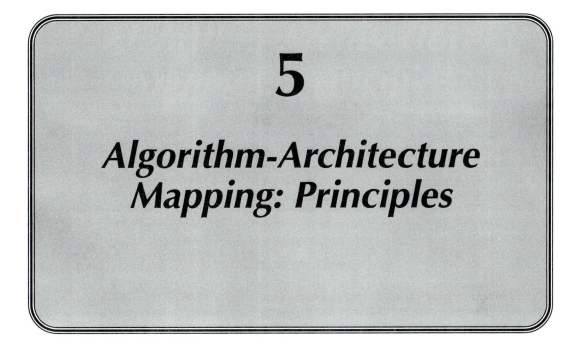

Algorithm-Architecture Mapping: Principles

Chapters 5 and 6 serve as a bridge in the description of real time signal processing systems. The basic principles discussed so far are illuminated with examples of particular tools, algorithm classes, and system synthesis. Here, the structure is built under which the tradeoffs of the previous chapters are executed: The resulting framework is then used to contain a variety of algorithm and architecture solutions as described in the remainder of this book.

The goal of this discussion is to describe the mapping of algorithm and architecture. The phrase "algorithm and architecture mapping" is used instead of "mapping algorithm to architecture." This is because both algorithm and architecture can be changed in the search for the best mapping. In practice, one is held fixed while the other is changed, but an important step to refining a mapping is iterating between holding the architecture fixed while modifying the structure of the algorithm, then holding the algorithm fixed while the architecture is improved.

5.1 INTRODUCTION

To map algorithm and architecture, a method is needed of describing each. In addition to providing a complete description, this method should allow the simulation of the algorithm running on the architecture, providing modeling of such algorithm properties as output, intermediate results, and bit-wise precision.

Similarly, the method should provide for predicting architectural properties like real time throughput, latency, and module-to-module communications bandwidth. In short, the description should support the trial and assessment of various partitions of both architecture and algorithm.

Partitioning means subdividing the algorithm or architecture into lower-level units. This discussion first examines the description and partitioning of algorithms and then looks at architecture description and partitioning. The goal of mapping algorithm and architecture is described at the conclusion of this section.

5.1.1 Algorithm Description and Partitioning

As discussed in Section 4.1.2, an algorithm may be described by equations, graphical signal processing language, general-purpose high-order language such as C or machine-specific assembly code.

A data flow graph (DFG), made up of a connection of nodes and arcs, may be used to describe and partition algorithms. A node consists of a task or computation, which includes one or more instructions, and starts with an input, performs a transformation, and produces an output. The node requires a particular execution time to complete the result. An arc represents a communication path and is characterized by a source node, destination node, bandwidth, and communication delay. Fig. 5–1 presents a picture of the data flow graph. Certain computer languages, such as PGM (Processing Graph Method) of the U.S. Navy, can directly receive such signal flow graphs as input (Section 4.1.2).

An algorithm is partitioned by subdividing it into a structure of parallel and sequential units. This involves decomposing the algorithm into chunks that fit within processing time, storage, and bandwidth of available processing resources. Multiple processing resources (or a single processing resource that is

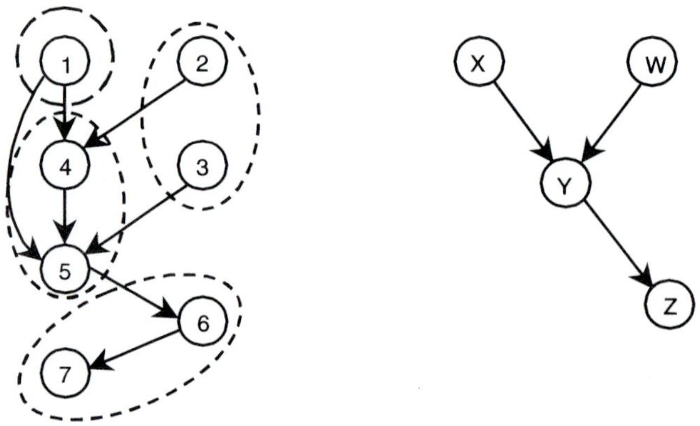

Figure 5–1 Example of data flow graph—elements of fine-grain control graph (left) can be combined into larger nodes in a coarse-grained graph (right).

multiplexed sequentially in time) are then combined to achieve real time throughput.

The process of partitioning an algorithm proceeds by partitioning its DFG. DFG partitioning steps includes:

1. Analyzing data precedence and dependence relationships;
2. Detecting parallelism;
3. Performing clustering to combine several operations into a task, and partitioning the program into tasks to increase efficiency;
4. Analyzing the communication links and timing that characterize the target architecture.

These general steps will be expanded into a detailed procedure in Section 5.2.2.

5.1.2 Architecture Description and Partitioning

Partitions of an algorithm influence the structure of the physical interconnect topology. If the interconnect topology is fixed, the best partitioning is influenced by the interconnect topology. The interconnect topology of a multiprocessor system affects system performance by introducing communication and system overhead. Overhead can be minimized through proper partitioning by:

- *data flow partitioning:* partition boundaries are located across areas of the structure diagram that have a minimum of data flow traffic (minimum bandwidth). Boundaries are often placed at the end of major data flow activities;
- *control flow partitioning:* partition boundaries are located to minimize the synchronization requirement between modules.

The parameters of an architecture performance description define:

- *resource timing*: how long each processor node requires to execute each signal processing primitive (throughput and delay, or latency); it also provides rules for combining primitives that account for any overlap in instruction execution;
- *node configuration*: the interconnection of nodes;
- *communication delay*: how long is required to move various types of data across each link?

The size of the chunks include the coarse-grained architectures as described in Section 4.2.1.1, fine-grained architectures, and elements within processors (e.g., registers, memories, ALUs, multipliers, etc.).

Several mechanisms have been developed to describe architectures. The simplest is the architecture block diagram, which is an architectural counterpart to the data flow graph used for algorithms. Several examples of architectural block diagrams were used in Section 4.2.1.2 to describe interconnection topologies. A language description for architectures, using the standard VHDL (Very high speed integrated circuit Hardware Description Language), has the ability to describe systems at multiple levels of abstraction. It is suitable for synthesis as well as simulation, and it is capable of documenting a system in an executable form to allow its simulation directly from its description language.[1]

Another class of architecture description method lies between the general block diagram and the fully specified textual description of VHDL. Exemplified by the *general-purpose memory model*,[2] the description treats both hardware latency and software overhead. It models communication time as containing both a non-recurring (setup) time and a recurring (transfer) time. Specifically:

T_d startup time—depends on software and communication protocol overhead;

τ_d transmission time for a unit of data—depends on hardware used to transmit data.

Using this model, the block communication time T to transfer m units of data is given by:

$$T = T_d + \tau_d.$$

For an architecture that has a dedicated processor for performing communications (as opposed to relying on the central processing unit), the communication delay consists of only the setup time T_d; then the main processor returns to processing while the communication processor moves the m data items. In addition, this model provides for a global communication operation, consisting of a constant value τ_g (if dedicated hardware is available) or a point-to-point time if the communication scheme uses primitives on the main processor. In this case, τ_g is $O(\log P)T_d$, where P is the number of processors.

5.1.3 Mapping

The goal of mapping is to allocate a set of operations in some optimal manner to a combination of available processing resources. Several criteria for optimization may be used. These include maximizing the throughput for a fixed set of computing resources, minimizing the amount of resources needed to achieve the specified throughput of achieving real-time operation, minimizing the system power, minimizing its cost, or minimizing its size.

A graphical view of the mapping process is provided in Fig. 5–2. Multiple algorithm segments are shown at the top, while at the bottom are multiple com-

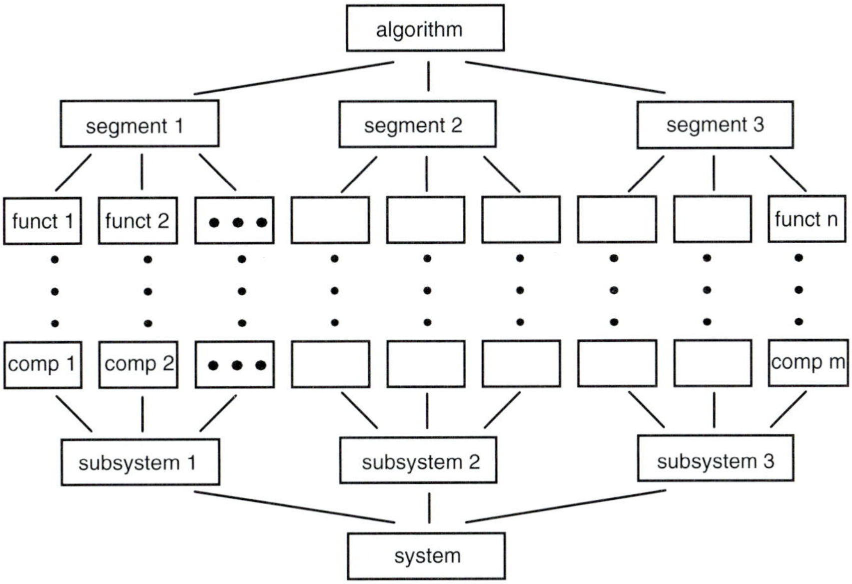

Figure 5–2 Mapping algorithm (top) and architecture (bottom) consists of decomposing each and finding allocations of one to the other at lower levels of the hierarchy.

puting resources. The mapping problem is to match computing resources with algorithm segments.

Gebotys[3] has expressed the mapping problem in a mathematical formalism that allows the use of systematic optimization techniques like integer programming. Several authors[4,5,6] have used an integer programming model coupled with a formalism like that above for simultaneously allocating and scheduling functional units of a digital signal processing architecture.

Despite the availability of mathematical tools, most algorithm-architecture mapping is performed in a heuristic manner. The partitioning of many types of algorithms can be performed by inspection, as the input data channels are independent and the sequence of processing steps are well-defined. However, transforming an algorithm into real time introduces such complexities as moving from batch input to continuous input, and overlapping the timelines of many different processing threads while simultaneously meeting data precedence and synchronization constraints. Only in simpler cases are the heuristic or experience-based inspections sufficient to achieve suitable mapping.

One measure of the optimality of an architecture/algorithm mapping is simply stated: Are all of the processing elements busy on the problem all of the time? Achieving 100 percent activity of multiple resources is challenging. As more processors are added to the problem, each may become less busy, and the total throughput may not grow linearly with the number of processors, but instead may fall below linear growth. This drop indicates the increased overhead of multiprocessing; avoiding the drop constitutes the challenge of achieving efficient multiprocessor mappings.

The following section outlines a general sequence of steps for mapping algorithms and architectures. Tools for performing each mapping step are then described in Section 5.3. Chapter 6 provides several examples of applying the process in the mapping of specific architectures.

5.2 PROCESSES AND PROCEDURES

A progression from an unmapped but proven algorithm, usually running on a general-purpose research computer, is now described, followed by a process for mapping an algorithm from the non-real-time general-purpose computing environment to the real time domain of (usually) multiprocessor systems.

5.2.1 Progression to Complete Mapping

The process of mapping architecture and algorithm is to progress from a correct algorithm, described at a level of abstraction that is both implementation-independent and timing-independent, to a system description of time-dependent, specific allocation to processing resources and the sequencing of events within those resources. The effort begins with an architecture- and timing-independent description of the algorithm and arrives at an architecture-dependent, timing-dependent description. Along the way, constraints of architecture and timing to the algorithm are successively applied. Computational results are validated before and after the application of each constraint to assure that input and output relationships have not changed. The performance is assessed periodically through the process to identify throughput bottlenecks. These bottlenecks are alleviated through iterative refinement, revising the structure and reapplying all necessary constraints.

The sequence of imposing architecture constraints usually begins with assigning the resources within one processor. This step addresses the choice of instruction, the use of the arithmetic logic unit, registers, memory, and input/output. The constraints are then extended to resource assignments across multiple processors, determining which processor is assigned to each algorithm task and determining processor-to-processor communication requirements.

The sequential addition of timing constraints accompanies the application of architecture constraints. Timing constraints include the precedence or ordering of operations, execution times, allocation to multiple processors, interprocessor communication time (both the overhead to set up a communication transfer and the per-item transfer time), the speedup factor that results from assigning parallel processors, and the needs for synchronization.

It is not generally possible to automatically solve the mapping problem in an optimal manner on the first attempt. Successful mapping requires the ability to measure the results of a particular choice of partitioning through simulation, then identifying and trying alternate partitionings, simulating performance, and iteratively optimizing.

5.2.2 A Process for Mapping

Fig. 5–3 summarizes the process for algorithm/architecture mapping that will be described here. The mapping begins with a full description of the algorithm which has already been developed and thoroughly proven capable of solving the problem within a general-purpose, non-real-time research computing environment. As described in Section 5.1.1, the algorithm is likely to be defined in a high-level language such as C and has been partitioned in the form of a main routine and several subroutines, which make calls to signal processing library functions. As part of its description, a test suite is necessary that consists of a set of known-correct outputs for a wide variety of inputs. This test suite is then used as a reference to check various stages of mapping the algorithm into real time.

The first stage of mapping involves searching for algorithm simplifications that greatly improve the execution efficiency. Algorithm simplification occurs at the highest level of description and has the potential of making the largest difference in throughput. An example technique for algorithm simplification is to limit the set of parameters from the most general case used in algorithm research to the specific set that applies to the chosen solution. Thus, a signal processing library may accept a wide range of two-dimensional filter kernels, yet a 5×5 kernel may be adequate for the task at hand. The simplification of the algorithm precedes the application of mapping and partitioning tools, because these tools will not discover simplifications at the algorithm level. Other examples of algorithm simplification are given in Section 5.3.1.

From the simplified algorithm description, a data flow graph is generated. The data flow graph introduces the constraints of operator precedence and timing. Developing a data flow graph (DFG) involves identifying trial partitions of the algorithm. Trial partitions are often based on the original partitioning of the code into subroutines and library modules. Based on the trial partitioning, modules are selected for performing the operations of each partitioned segment. In the hardware domain, these modules may be particular processor or standard logic cells; in the software domain, they are elements of a signal processing library. Behavioral simulation is performed on the partitioned version of the algorithm to assure that I/O performance matches that of the test suite. For each segment of a particular algorithm partitioning, a node is defined in the DFG, and the execution time of each node is determined, thus introducing the first set of timing information. The node-to-node communication requirements are determined in terms of the number of transfers per operation, per sample, or per second. The

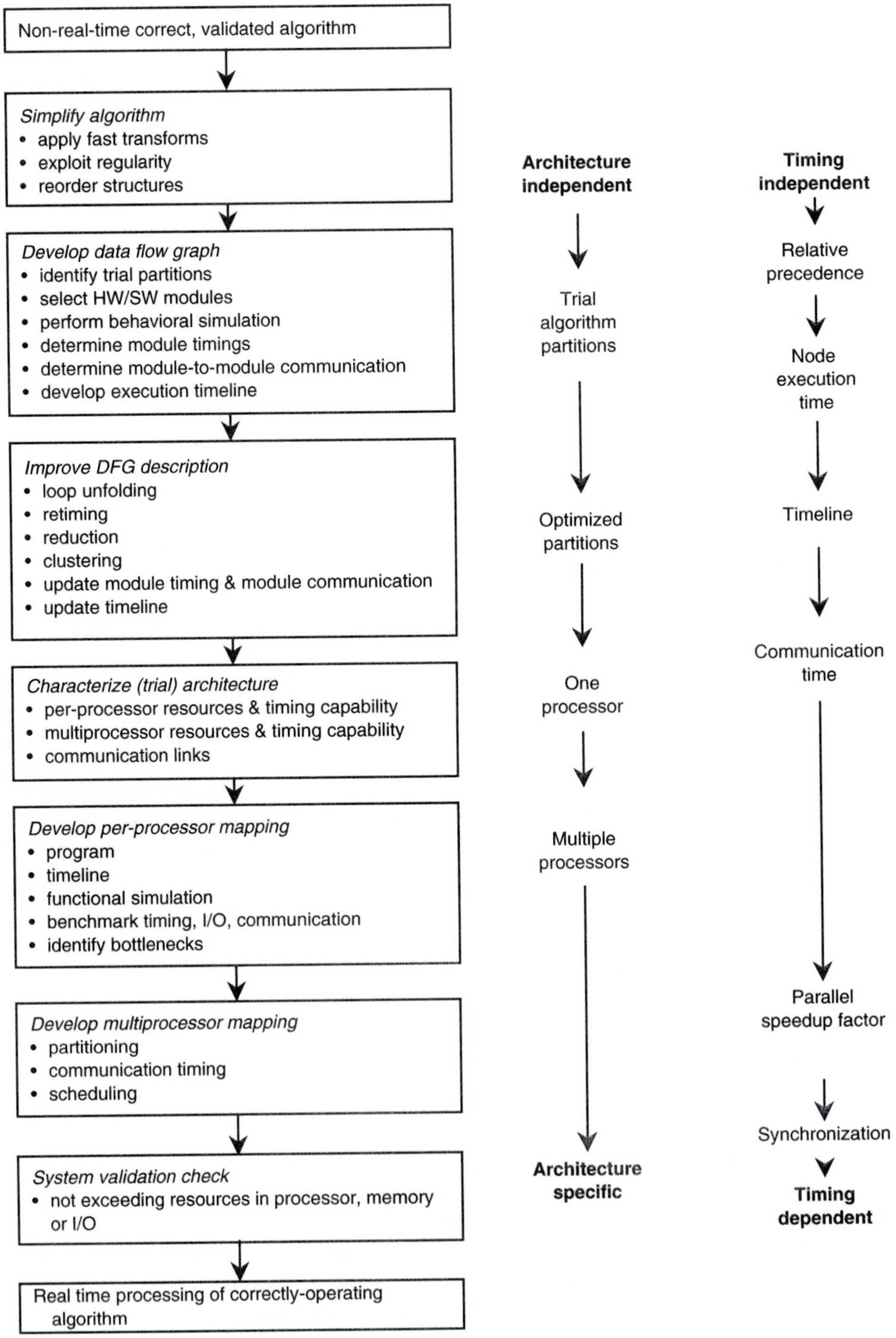

Figure 5–3 Generic process for mapping algorithm and processor architectures.

node processing times and the communication times are combined in an initial execution timeline for the algorithm.

Data flow graph improvement applies several formal methods to reduce delays and increase efficiency. With the execution timeline in hand, bottlenecks, or segments of computation time that are unusually long and pace overall throughput, can be identified. As described in the next section, DFG manipulation techniques include loop unfolding, retiming, reducing the DFG to a "perfect rate program," and clustering similar operations or sequences of operations to enable reuse of code or of intermediate results. After DFG optimization is applied, the node processing and communication times are updated, and a revised timeline is generated. Finally, the test suite is again used to validate conformance of the improved DFG.

The characteristics of a trial architecture are next introduced into the mapping. The architecture may be fixed, as it is when an algorithm is mapped to a specific processor, or it may be a parameterized multiprocessor framework under which one chooses the number and types of processors. Alternatively, the designer may have complete freedom of architecture choice. The architecture characterization, as described previously, consists of a description of per-processor resources, capabilities, and timing, multiple processor resources, capabilities, and timing, and communication links.

A per-processor mapping of the DFG is next developed. Even in a multiprocessor implementation, it may be helpful to map the entire algorithm to one processor. This single processor mapping is essential if the multiprocessor architecture consists of only this one type of processor (i.e., a homogeneous architecture). A program is developed by clever structuring of the improved DFG and the use of the native high-level-language compiler for the processor, and an execution timeline for the program is produced. Functional simulation is conducted to assure that this stage of the mapping still conforms to the test suite. For the single-processor implementation, benchmark timing, input/output, and communication requirements are determined. Timing bottlenecks are then identified, which are addressed by either iterating through the above steps or by becoming the focus of a multiprocessor partition.

Next, a multiprocessor mapping is developed (if the target architecture includes multiple processors). This can begin with all of the multiple processors performing the same operations (i.e., SIMD), then tuning the system to improve performance by reallocating operations to relieve the bottlenecks that occur. More formal techniques based on the DFG, and the use of serial instruction pipelines and parallel data partitioning, are exemplified in the following section. A choice between shared memory or message passing is made for implementing interprocessor communications. The multiprocessor execution time and communication times are determined for the mapping, and the effect of overlapping communication with processing (if the architecture has separate parallel processors for processing and communication) is computed. Memory management is arranged by which data values are placed in memory by one processor and then retrieved by another. A schedule is produced that assigns computations to specific processors at particular times, and a final execution timeline is developed.

Having completed the mapping, a final check is performed to assure that processing resources of the target architecture have not been exceeded. The test suite is again used to validate mapping conformance to the original algorithm specification. Overall throughput, which compares operations per sample and total sampling rate with the operations-per-second capability of the architecture, assures that real time is met, and design margins are introduced. Memory usage at each processor is calculated and compared against available memory. The bandwidth requirements of all processor-to-processor communication paths are characterized and compared against the bandwidth supported by the architecture to complete the validation.

5.3 TOOLS AND TECHNIQUES

This section discusses methods that may be applied to achieve the steps in the algorithm mapping process. It describes techniques for algorithm simplification, partitioning, DFG improvement, and pre-compilation improvement, attaining an initial processor sizing, communication timing, the use of integer programming for optimization, and other tools.

5.3.1 Algorithm Simplification

Regularities within a computation can be discovered and exploited to restructure it for greater efficiency. For example, such an analysis of the discrete Fourier transform led to the fast Fourier transform. Replacing the discrete Fourier transform with the fast Fourier transform replaces a computation whose complexity for N elements is proportional to N^2 with one proportional to $N \log N$. Performing several iterations of a calculation by hand with a calculator often leads to discovering patterns of symmetry that can be used for greater efficiency.

Several common examples of algorithm simplification illuminate this concept. To compute the inverse of a matrix that has been produced in autoregressive moving average (ARMA) spectral estimation, the fact that such matrix coefficients are symmetric on either side of the diagonal (i.e., the matrix is Toeplitz) can be used to compute its inverse in a more efficient recursive manner than generalized matrix inversion. In a feature measurement process, one may have a large number of features. Performing principal component analysis, which involves computing matrix eigenvalues and eigenvectors, results in a rank ordering of each feature by its importance in distinguishing the various observations. Redundant features, or features of lesser importance, may be removed from the set, thus simplifying downstream processing.

The use of quadrature mirror filters in subband coding or wavelet analysis allows the reuse of filter coefficients and partial results. Subband coding produces a bank of parallel filters that are tuned to isolate segments of the incoming bandwidth. Quadrature mirror filters have coefficients that are the same except for their sign.

For convolution operations that consist of many points, the Fourier convolution theorem can be applied to replace direct time-domain convolution with frequency-domain convolution. The circular (periodic) convolution of two finite sequences can be obtained by multiplying the discrete Fourier transform (DFT) of the two sequences and then taking the inverse DFT of the product. Once the convolution is transformed to a DFT operation, the FFT can then be used to render this approach more efficient than direct convolution.

Much of the theory of digital signal processing has been applied to simplifying the algorithm from its initial conception to a more efficient form. The leverage in easing algorithm execution in real time is perhaps the greatest at this stage of algorithm simplification.

5.3.2 Partitioning

Tools for partitioning algorithms support the development of a trial partition and its performance assessment, which is then used for refining the partition. An initial partitioning may be obtained through the use of *functional grouping.* Examining the written functional specification of the algorithm and characterizing the types of activity identifies the areas of functionality. As an example, for a real-time audio/video playback decompression algorithm such as MPEG, these blocks may include memory, control, video decoder, etc. Each block is then further divided based upon the elements of functional specification of the blocks that they serve. For example if the MPEG functional areas specification consists of a system, audio, and video segment, then the memory functional area may be further divided into video memory, audio memory, and system memory. Elements that are left over as not associated with any of the above groups are each allocated to a separate block, as exemplified by a memory that contains the present video frame to decode.

For each functional block thus obtained, a series of decomposition steps refines the partitioning. For each block, the paths through the block are listed that represent its various functions or procedures. Each path is then decomposed into a list of operations, e.g., elements of a software library. For each operation, the number of computations is estimated that are required to execute the element. When this process has been completed for all functional blocks, each path is compared to all others to identify similar routines that are used. Any similarity is then used to develop efficient, reusable design modules. For each function, the computational complexity is determined by examining the number of instructions, the type of instructions, and the execution time of each instruction. Computational bottlenecks are then identified, which may be alleviated by decomposing a slow node into multiple serial or parallel streams. Fig. 5–4 shows a possible result from such a decomposition, indicating three nodes, one of execution time τ, the next of time 8τ, and the final of execution time τ. The middle node limits the overall execution throughput and is identified as the bottleneck. Further efforts may then divide it into either serial or parallel subnodes, each of execution time τ, as shown in the figure.

Another method that may be used for partitioning is top-down decomposition. This proceeds from a high-level description of the application through an algorithm, then to an algorithm that meets the application requirements, and then a specification of the library elements that implement it. It further decomposes into the arithmetic operations of these elements. Processor-specific information is inserted at the next stage, concluding with the semiconductor technology that implements that processor (Fig. 5–5).

Top-down decomposition matches the transition from static to dynamic algorithm specification in a manner quite similar to that of functional grouping, discussed above. First, major subsystems such as data processing, test circuitry, user interface, and I/O are identified. The processing for each of these subsystems is decomposed into a list of functions, which in turn are subdivided into fundamental signal processing units, such as elements in a signal processing library. For each unit, the compute load, memory activity, and communication requirements are examined to determine speed demands. Further partitioning,

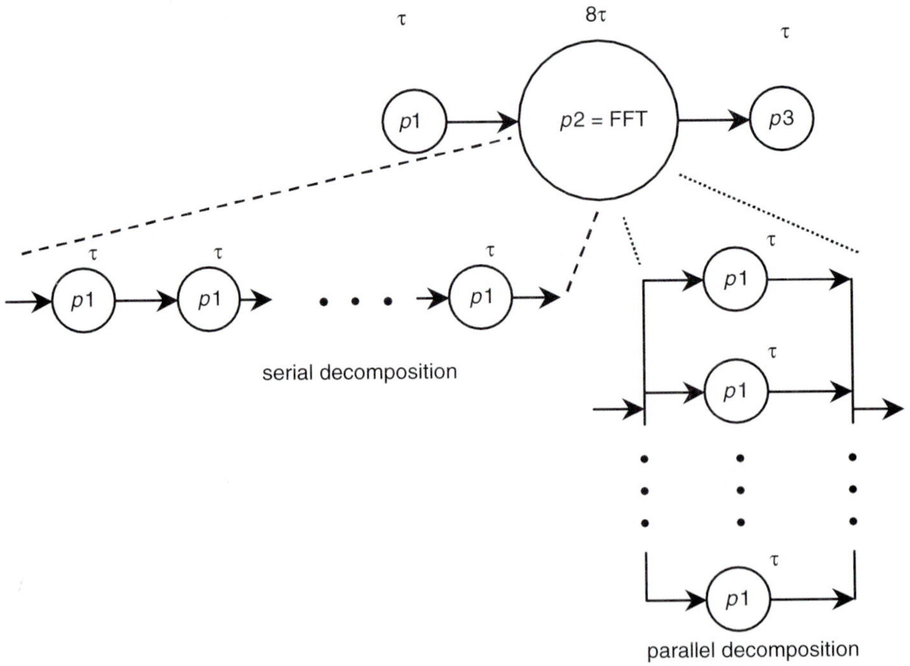

Figure 5–4 Example of identifying bottleneck node $p2$ and decomposing it into serial (at left) or parallel (at right) computations to obtain a uniform per-node execution time of τ.

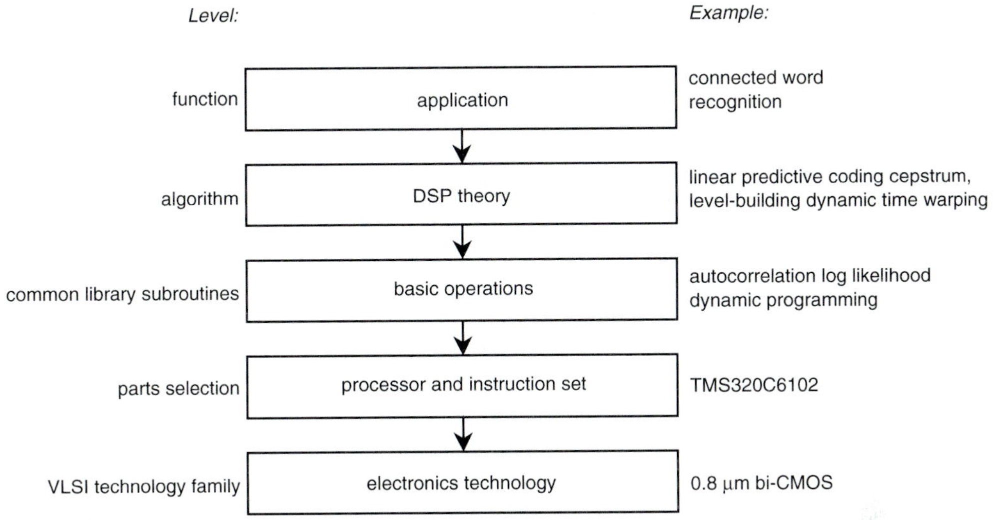

Figure 5–5 Top-down decomposition from application to semiconductor technology.

such as exemplified in Fig. 5–5, is used to partition bottlenecks into serial and parallel subcomputations. A search for subtle regularities is then made to identify common sequences of computing that occur at various places in the algorithm, regularities of memory addressing, and reuse of intermediate results.

A third variant of the partitioning process is based on the reuse of existing modules and minimizing the amount of new design that is conducted to attain real-time execution. This approach is effective when a library of optimized modules has been accumulated from past designs. In this approach, the algorithm is decomposed into segments of existing design modules. A tool to manage the accessing of past designs and production of new ones has been developed for this approach.[7] The tool's operation is shown in Fig. 5–6. When a new algorithm is presented to the system, a new data dependency graph is drawn. If the input algorithm is found to have the same data dependency structure as one of the algorithms in the library, then that architecture is retrieved directly. Otherwise, a structure for the new algorithm is built and added to the library.

5.3.3 Data Flow Graph Improvement[8,a]

The goal of data flow graph (DFG) improvement is to improve the multiprocessor scheduling of a digital signal processing algorithm by manipulating the data

---

[a]Portions adapted, with permission, from Keshab K. Parhi and David G. Messerschmitt, "Static Rate-Optimal Scheduling of Iterative Data-Flow Programs Via Optimum Unfolding," *IEEE Trans. Computers,* 40 (February, 1991), 178–195. © 1991 IEEE.

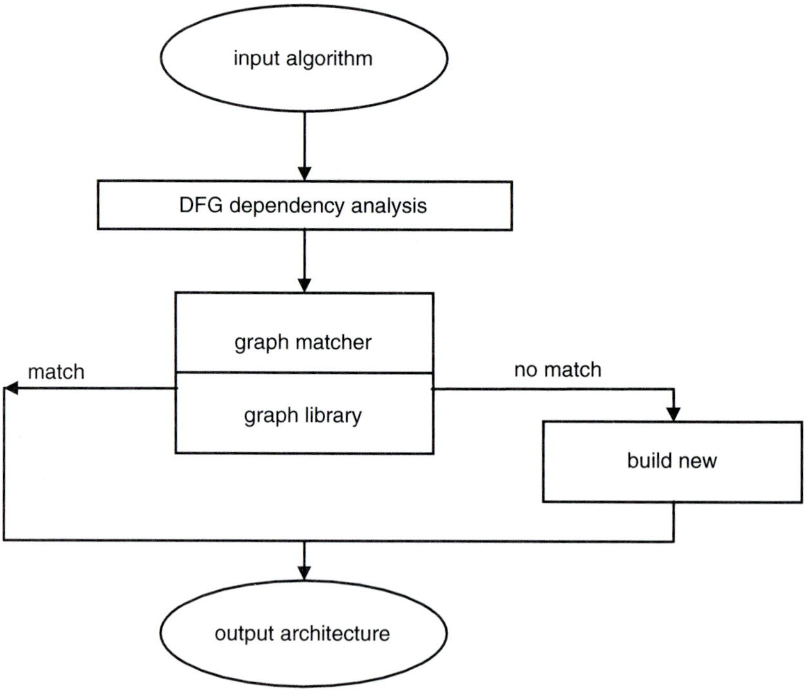

Figure 5–6 Automated aid for assessing new algorithm graph against a library of past designs.[7]

flow graph that describes it. Reordering nodes, moving around delay elements, and manipulating the graph in other ways without changing its outputs can often achieve improvement. DFG improvement is an example of the step of "canonicalization" that was one of the four forms among which an algorithm can be translated (Section 4.1.5).

5.3.3.1 Data Flow Graph Constructs. A DFG consists of nodes, which are circles that correspond to computation, and arcs, which are lines associated with communication between nodes. Fig. 5–7 shows a simple DFG and its associated program.

As shown by the listing, the program consists of a loop that is repeated without terminating, since the upper limit on n is ∞. Each iteration requires a certain interval of time, known as the iteration period.

A more complicated DFG may consist of several nodes that are connected with arcs, in which an arc can include a register that introduces a delay over one or more sample periods. The delay D causes the output to depend not only on inputs of the current sample period n but also previous periods, $n-1, n-2, \ldots$ In

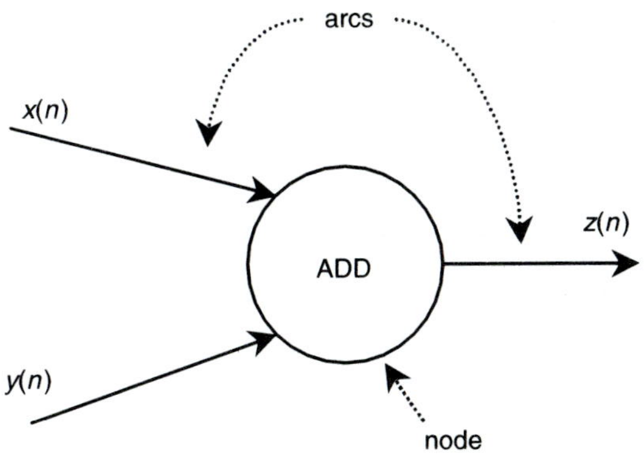


```
for {n=0 to infinity}{
        z(n)=x(n)+y(n)}
```

Figure 5–7 Simple DFG consists of nodes and arcs and represents a signal processing program.

Fig. 5–8, a delay of two sample periods between nodes *B* and *C* is shown for a slightly more complicated DFG.

Before processing begins, the contents of certain arcs are specified by initial conditions. For example the two delayed signal values required by the arc from node *B* to node *C* may be shown as $bc(-1)$ and $bc(0)$.

The notation $uv(n)$ is used to show the time series due to samples, or tokens, produced by node *U* and consumed by node *V*. A function $f_{uv}(\bullet)$ defines the function to be evaluated for the computation of $uv(n)$. Thus, the fact that a series $ab(n)$ is the result of a function f_{ab} operating on $x(n)$ may be represented as $ab(n) = f_{ab}[x(n)]$. Therefore, the program that is associated with the DFG of Fig. 5–8 may be written:

```
Initial conditions bc(-1), bc(0)
for {n=1 to ∞}{
      ab(n)=f_ab[x(n)]
      bc(n)=f_bc[ab(n)]
      cd(n)=f_cd[bc(n-2)]
      y(n)=f_y[cd(n)]}.
```

Note that $f_{cd}[bc(n-2)]$ reflects the $2D$ delay.

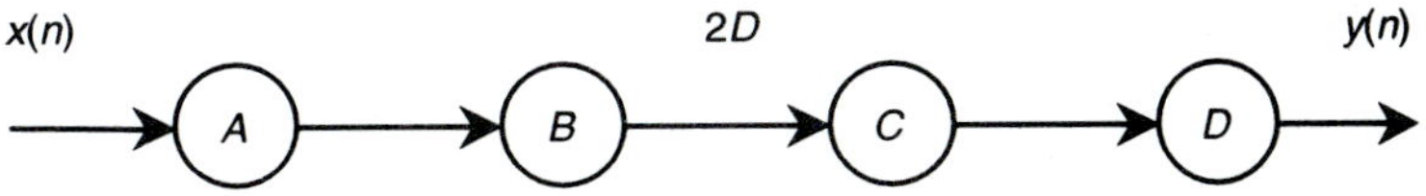

Figure 5–8 Four-node DFG with two-sample delay on arc from node *B* to node *C*.

The goal of real time signal processing is to complete all operations except the delay within one sample period. The delay is the only operation that can transcend the sample period. An arc with no registers (delays) describes the precedence within an iteration (or "intra-iteration"), while an arc with a register describes precedence between tasks of different iterations ("inter-iteration"). Optimization of the DFG will be discussed by first examining intra-iteration improvements. Further optimization will then be discussed using inter-iteration improvements.

Most realistic signal processing programs include loops, as well as delays. Two loops, labeled L_1 and L_2, form complete cycles in the DFG of Fig. 5–9.

Two types of arcs may occur in DFGs. An arc $U \rightarrow V$ is said to be transitive if there also exists (perhaps via multiple arcs) another path $U \rightarrow V$, and the number of delays of the path $U \rightarrow V$ and the arc $U \rightarrow V$ are the same. In Fig. 5–10 (top), the arc $A \rightarrow E$ is transitive because both it and the path from $A \rightarrow E$ given by $A \rightarrow B \rightarrow C \rightarrow D \rightarrow E$ have delay $2D$. Similarly, $C \rightarrow E$ is a transitive arc for $C \rightarrow D \rightarrow E$, and both have zero delay. An extended transitive arc is an arc $U \rightarrow V$ that has fewer delays than another path $U \rightarrow V$. In Fig. 5–10 (bottom), if we insert a delay between C and D, both arcs $A \rightarrow E$ and $C \rightarrow E$ become extended transitive arcs, since they have less delay than the three delays of $A \rightarrow B \rightarrow C \rightarrow D \rightarrow E$ (or $C \rightarrow D \rightarrow E$). Deletions of transitive and extended transitive arcs do not alter DFG precedence relations. Transitive and extended transitive arcs are deleted prior to DFG optimization, then are restored to the DFG after optimization is completed.

5.3.3.2 Acyclic Precedence Graph. An acyclic precedence graph (APG) is obtained from a DFG by removing all arcs from the DFG that contain delays. Extracting an APG is performed as part of intra-iteration optimization, and preserves the intra-iteration precedence of the DFG. Fig. 5–11 shows the acyclic precedence graph that results from the DFG of Fig. 5–9.

The acyclic precedence graph can be used in conjunction with the node execution times to produce a schedule for executing the DFG on multiple processors.

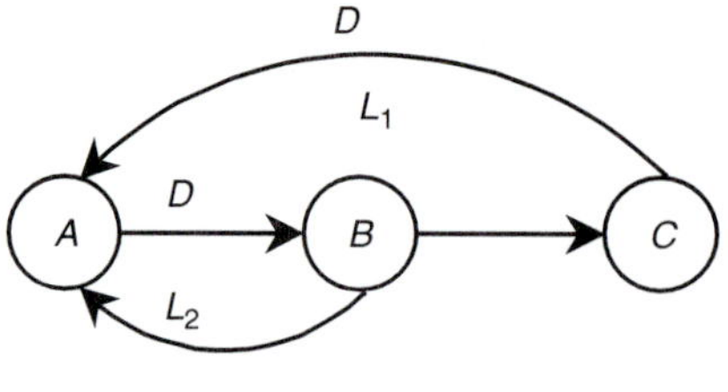


```
Initial conditions: ab(0), ca(0)
for {n=1 to ∞}{
  ba(n)=f_ba[ab(n-1)]
  bc(n)=f_bc[ab(n-1)]
  ab(n)=f_ab[ca(n-1), ba(n)]
  ca(n)=f_ca[bc(n)]}
```

Figure 5–9 DFG with loops and delays.

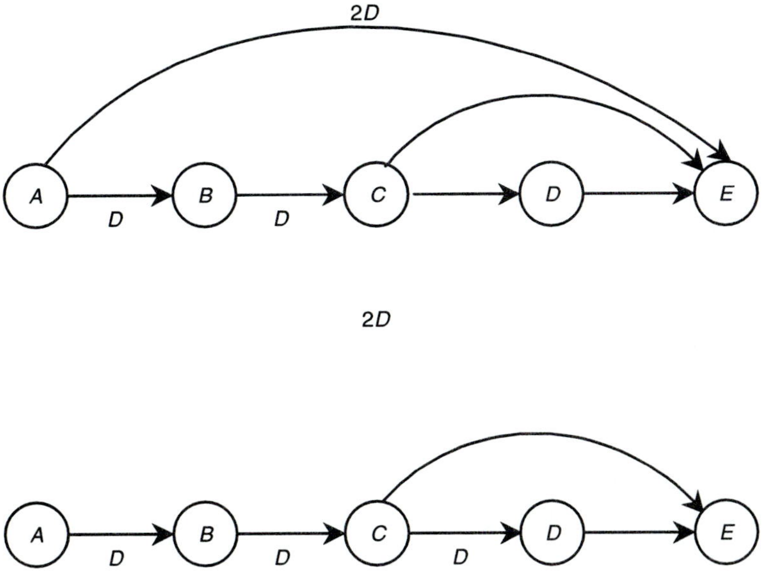

Figure 5–10 Example of transitive arc (top) and extended transitive arc (bottom).

For example, suppose the execution times of nodes *A*, *B*, and *C* in the DFG of Fig. 5–11 are:

$A = 10$ cycles
$B = 20$ cycles
$C = 40$ cycles.

The acyclic precedence graph, or APG, can be modified to show the critical path, or the path with longest execution time, using a double arrow (Fig. 5–12).

To produce a multiprocessor execution schedule from the acyclic precedence graph, assume that each processor is the same (i.e., a homogeneous archi-

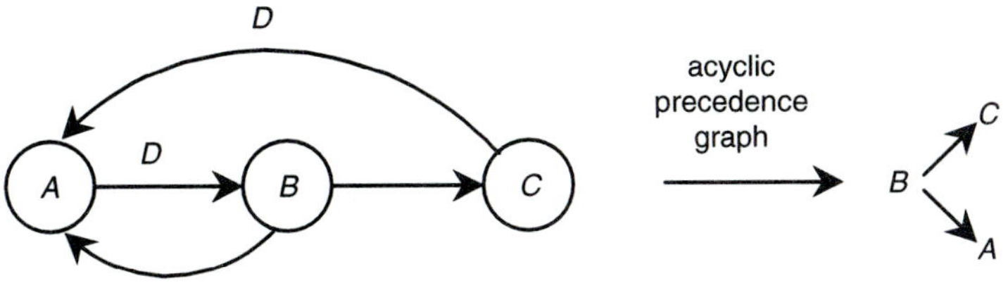

Figure 5–11 Acyclic precedence graph (right) is produced from DFG (left) by removing all arcs with delays, *D*.

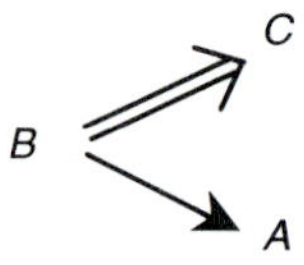

Figure 5–12 Acyclic precedence graph with node timings $A = 10$, $B = 20$, and $C = 40$ uses double arrows to show critical path.

tecture is used) and that one processor is used for each branch of the APG. Fig. 5–13 shows an execution schedule for two processors, P_1 and P_2.

The iteration period of this schedule is 60 cycles, and processor P_2 is left idle much of the time, as indicated by the shaded areas. A quick inspection shows that if nodes A and B were assigned to processor P_1 and node C to P_2, then P_2 would be fully busy and P_1 would have 10 cycles of idle time. It is the goal of the formal methods described here to systematically discover and insert such opportunities for DFG improvement.

This is an example of a non-overlapped schedule—execution of iteration $i + 1$ does not begin until all steps of iteration i have been completed. Non-overlapping scheduling optimizes the performance of a single iteration but does not optimize across iterations. The minimum iteration period of a non-overlapped iteration is equal to the execution time of the critical path. In an overlapped schedule, some tasks of iteration $i + 1$ are scheduled before all tasks of iteration i have been finished. An overlapped schedule exploits not only intra-iteration but inter-iteration precedence, and it can lead to a shorter iteration period.

A more complicated DFG can be used to demonstrate these concepts. Fig. 5–14 shows a DFG with its delays, as well as the associated program, node execution times, and acyclic precedence graph.

The scheduling method used in the previous example, which assigns a processor to each branch, produces the schedule of Fig. 5–15. It is non-overlapped with an iteration period of 3.

An overlapped schedule (Fig. 5–16) reduces the iteration period from 3 to 2 by executing B_1 on P_2 at the same time it executes D_2 on P_1, thereby overlapping iterations 1 and 2. All iterations of D and A are scheduled with a displacement of 2 on P_1; all iterations of C and B are similarly scheduled on P_2.

Another form of scheduling, known as *cyclostatic*,[9] has a processor displacement as well as a time displacement. Fig. 5–17 shows that the D, A sequence that was allocated to P_1 in Fig. 5–16 now is moved between both processors.

It is useful to determine the minimum obtainable iteration period for various types of DFGs. Knowing the minimum achievable iteration period provides a

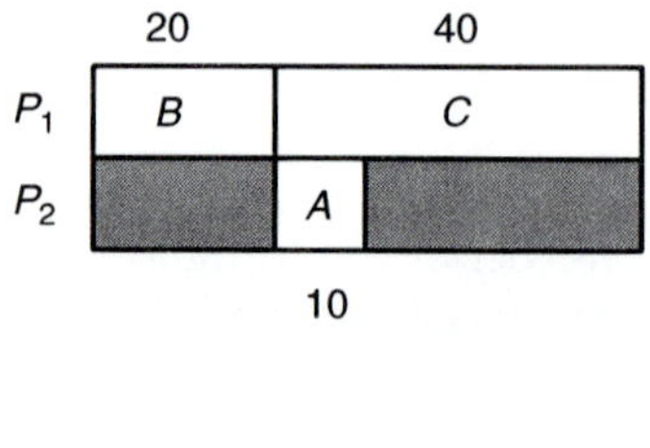

Figure 5–13 Schedule produced for DFG of Fig. 5–11 and APG of Fig. 5–12 has iteration period of 60 cycles.

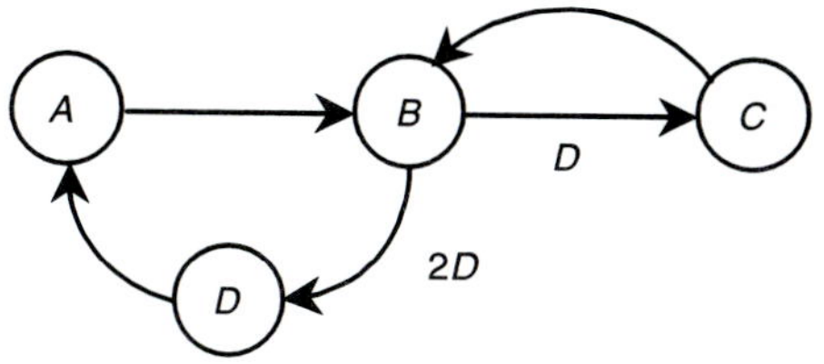

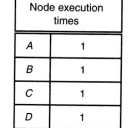

Node execution times	
A	1
B	1
C	1
D	1

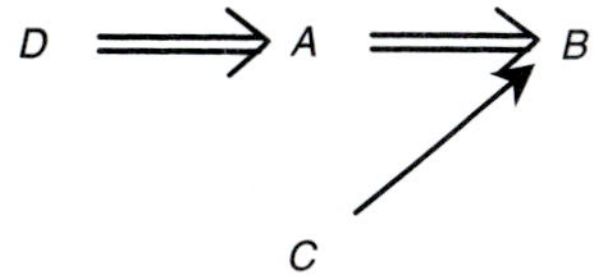


```
Initial conditions: bc(0), bd(-1), bd(0)
for {n=1 to ∞}{
   ab(n)=f_ab[da(n)]
   bd(n)=f_bd[ab(n), cb(n)]
   da(n)=f_da[bd(n-2)]
   bc(n)=f_bc[ab(n), cb(n)]
   cb(n)=f_cb[bc(n-1)}
```

Figure 5–14 Data flow graph, program, node execution times, and acyclic precedence graph.

measure of whether a particular schedule has achieved its lowest possible iteration period. Fig. 5–18 shows a DFG with no loops. The sequence of schedules below it shows that the iteration period can be decreased without limit as more processors are added.

5.3.3.3 Iteration Bound. In contrast to a loopless DFG, for a DFG with loops, an inherent lower bound is imposed on the iteration period. This limit is called the *iteration bound.* The goal in scheduling is to achieve an iteration period that is as low as the iteration bound. Such a schedule is referred to as a *rate optimal schedule.*

To find the iteration bound, each loop l is examined and the ratio of T_l, the sum of all execution times associated with loop l, to D_l, the number of delays in loop l, is computed. The iteration bound is the maximum of this ratio over all loops l, and the loop associated with it is the critical loop, l_c. Thus, where T_l = sum of all execution times associated with loop l; and D_l = number of delays (registers) in loop l.

$$T_\infty = \max_{\text{all loops } l} \left[T_l / D_l \right], \tag{5.1}$$

P_1	D_1	A_1	B_1	D_2	A_2	B_2	D_3	A_3	B_3
P_2		C_1			C_2			C_3	

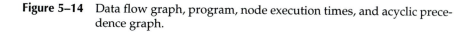

Figure 5–15 Non-overlapped schedule of period 3 to implement DFG of Fig. 5–14.

P_1	D_1	A_1	D_2	A_2	D_3	A_3	D_4
P_2		C_1	B_1	C_2	B_2	C_3	B_3

Figure 5–16 Overlapped scheduling reduces iteration period from 3 to 2.

The bound imposed on the iteration period by loop l is

$$T_l \leq D_l \cdot T_\infty. \tag{5.2}$$

Fig. 5–19 repeats the DFG of Fig. 5–11 and identifies D_l and T_l for its two loops L_1 and L_2.

The scheduling of this DFG described in Fig. 5–13 produced an iteration period of 60 cycles. Its iteration bound of 35 cycles suggests that a better schedule exists—the task is now to find it.

5.3.3.4 Achieving the Perfect Rate Data Flow Graph. The iteration bound may be decreased by reducing T_l, which can be accomplished by taking as many operations out of the loop without changing the semantics of the computation. This technique is known as *loop shrinking*. Also, as was described in Fig. 5–4, individual nodes can be decomposed into different nodes to improve scheduling by balancing the execution times of various nodes. Fig. 5–20 shows an example. This reclustering is accompanied by the danger that the new clusters will not form generally reusable operations. While nodes A and B may be commonly used arithmetic operations that can be reused widely through an algorithm, such as addition or multiplication, new nodes A' and B' may be specialized inseparable nodes that must always be adjacent and are only suited for local use. For example, A' may become an add followed by a partial product formulation for a multiplication (or an "add with the start of multiply"), while B' could become the summation of existing partial products to form a full product (a "rest of multiply" operation). Reusing A' and B' is more difficult than reusing A and B, because A' and B' correspond to very specialized functions.

While the approaches above can reduce the iteration bound, they do not define which structure can achieve that bound. Several techniques are now described to improve the schedule to approach the iteration bound without altering nodes.

Retiming of a DFG involves moving delay registers around so that the total number of registers in any loop remains the same and the input/output behavior of the DFG is preserved. Fig. 5–21 shows an example based on the DFG of Fig. 5–11; after retiming, the iteration period drops from 60 to 40 cycles when the two-processor allocation scheme is applied to the acyclic precedence graph. However, the iteration bound $T_\infty = 35$ is not yet reached.

P_1	D_1	A_1	B_1	C_2	D_3	A_3	B_3
P_2		C_1	D_2	A_2	B_2	C_3	D_4

Figure 5–17 Cyclostatic scheduling moves node execution sequences across processors.

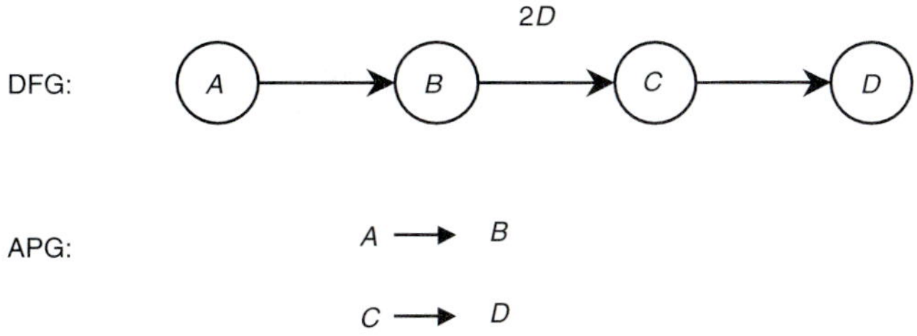

Schedule:

2 processors iteration period = 2	P_1			A_1	B_1	A_2	B_2	A_3	B_3
	P_2	C_1	D_1	C_2	D_2	C_3	D_3	C_4	D_4
4 processors iteration period = 1	P_1	A_1	A_2	A_3	A_4				
	P_2		B_1	B_2	B_3	B_4			
	P_3	C_1	C_2	C_3	C_4				
	P_4		D_1	D_2	D_3	D_4			
8 processors iteration period = 1/2	P_1	A_1	A_3						
	P_2		B_1	B_3					
	P_3		C_1	C_3					
	P_4			D_1	D_3				
	P_5	A_2	A_4						
	P_6		B_2	B_4					
	P_7		C_2	C_4					
	P_8			D_2	D_4				

Figure 5–18 For a DFG without loops, as the number of processors increases without bound, the iteration period decreases without bound.

The retiming of more realistic graphs, which are far more complex than that of this example, is more difficult, and in fact is an NP-complete problem. However, probabilistic iterative techniques can be used to achieve maximum concurrency subject to the constraint of scheduling between as soon as possible and as late as possible.[10]

Another technique to approach the iteration bound is *data program unfolding*. This involves building a DFG that describes J iterations of the original repeated DFG, where J is referred to as the *unfolding factor*. Therefore for an original DFG of N nodes and E arcs, a new DFG is constructed that contains JN nodes and JE

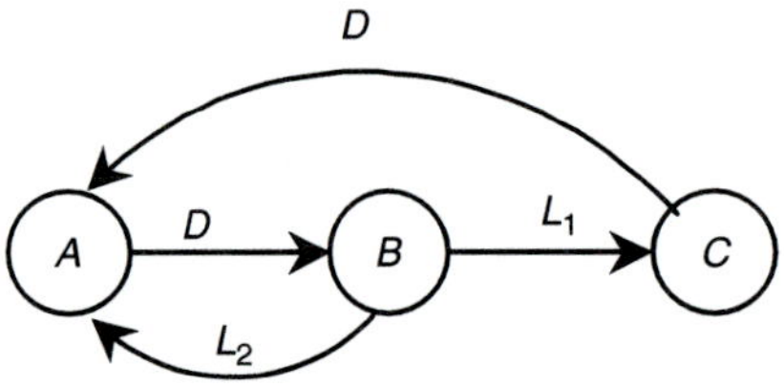

Node execution times	
A	10
B	20
C	40

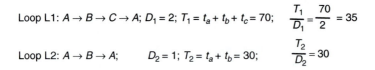

Loop L1: $A \rightarrow B \rightarrow C \rightarrow A$; $D_1 = 2$; $T_1 = t_a + t_b + t_c = 70$; $\frac{T_1}{D_1} = \frac{70}{2} = 35$

Loop L2: $A \rightarrow B \rightarrow A$; $D_2 = 1$; $T_2 = t_a + t_b = 30$; $\frac{T_2}{D_2} = 30$

Figure 5–19 DFG, node execution times, and d_l and t_l for loops 1 and 2 yield iteration bound T_∞ of 35 units.

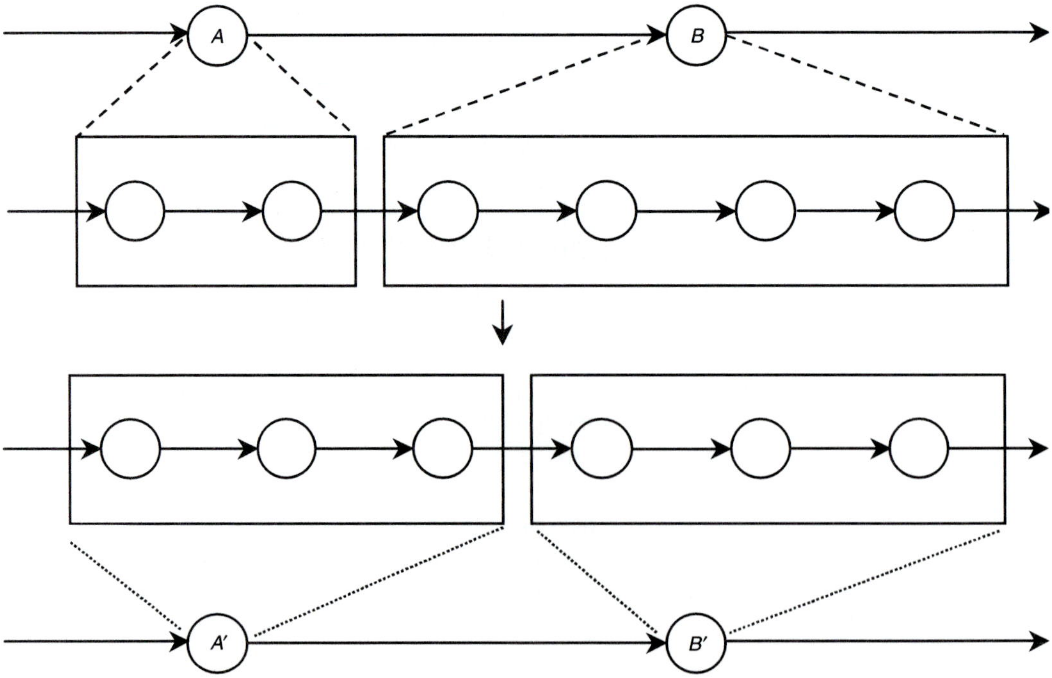

Figure 5–20 Repartitioning of nodes by decomposition into subnodes to achieve balanced execution time to improve scheduling.

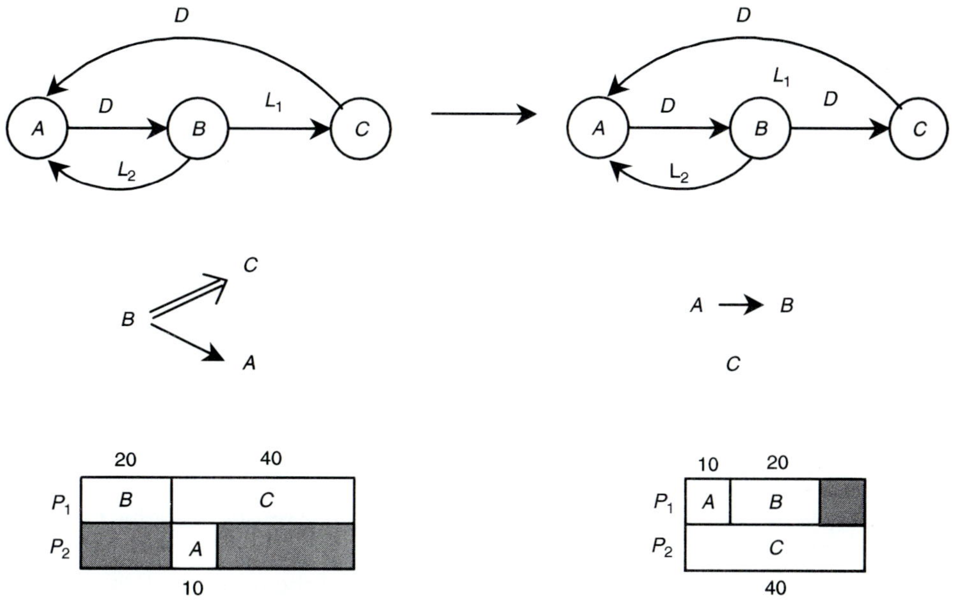

Figure 5–21 Retiming DFG to rearrange delays while keeping the delay of all loops the same allows reduction of the iteration period from 60 to 40 cycles.

arcs. Each node U in the original program is replaced by nodes $U_1, U_2, \ldots, U_J$. Node U_1 executes iterations $1, J+1, 2J+1, \ldots$ from the original DFG. An unfolded DFG is constructed from the original DFG by replacing the index n by $nJ - J + j$, where $j = 1, 2, \ldots, J$. Fig. 5–22 is an example of unfolding the example program by a factor of $J = 2$.

Original

```
Initial conditions ab(0), ca(0)
for {n=1 to ∞}{
    ba(n)=fba[ab(n-1)]
    bc(n)=fbc[ab(n-1)]
    ab(n)=fab[ca(n-1), ba(n)]
    ca(n)=fca[bc(n)]}
```

Unfolded

```
Initial conditions ab(0), ca(0)
for {n=1 to ∞}{
    ab(2n-1)=fab[ca(2n-2), ba(2n-1)]
    ba(2n-1)=fba[ab(2n-2)]
    bc(2n-1)=fbc[ab(2n-2)]
    ca(2n-1)=fca[bc(2n-1)]
    ab(2n)=fab[ca(2n-1), ba(2n)]
    ba(2n)=fba[ab(2n-1)]
    bc(2n)=fbc[ab(2n-1)]
    ca(2n)=fca[bc(2n)]}
```

Figure 5–22 Version of DFG unfolded by factor $J = 2$ is produced by substituting $nJ + j; j = 1, 2$ for n.

The procedure for unfolding a DFG by a factor J is as follows:

1. For each node U in the original DFG, draw J corresponding nodes and label them $U_1, U_2, \ldots, U_J$.
2. For each arc $U \to V$ in the original DFG containing no delay register, draw arcs $U_k \to V_k$ with no delay register.
3. For each arc $U \to V$ in the original DFG with i delay registers, do a) or b):
   - **a)** If $i < J$ (i.e., number of arc registers less than unfolding factor):
     Draw arcs $U_{q-i} \to V_q$ ($q = i + 1$ to J) with no arc register;
     Draw arcs $U_{J-i+q} \to V_q$ ($q = 1$ to i) with a single register in each arc (since U_{J-i+q} is input for B_{J+q}, which is executed by node V_q after one cycle);
   - **b)** If $i > J$:
     Draw arcs $U[i - q + 1/J]_{j-i+q} \to V_q$ with $[i - q + 1/J]$ registers ($q = 1$ to J) where the notation $\lceil x \rceil$ represents the ceiling function of x, which returns the smallest integer that is greater than or equal to x.

Applying these rules to the DFG used in the example provides the results of Fig. 5–23.

The iteration period for two iterations of the unfolded DFG is 70, which corresponds to 35 for one iteration. This is equal to the iteration bound of $T_\infty = 35$, and so the rate optimum DFG has now been achieved.

The technique of loop unrolling achieves improved efficiency because it groups more and more computations together, allowing simultaneous access to more data values. Thus, it achieves the characteristics for block processing algo-

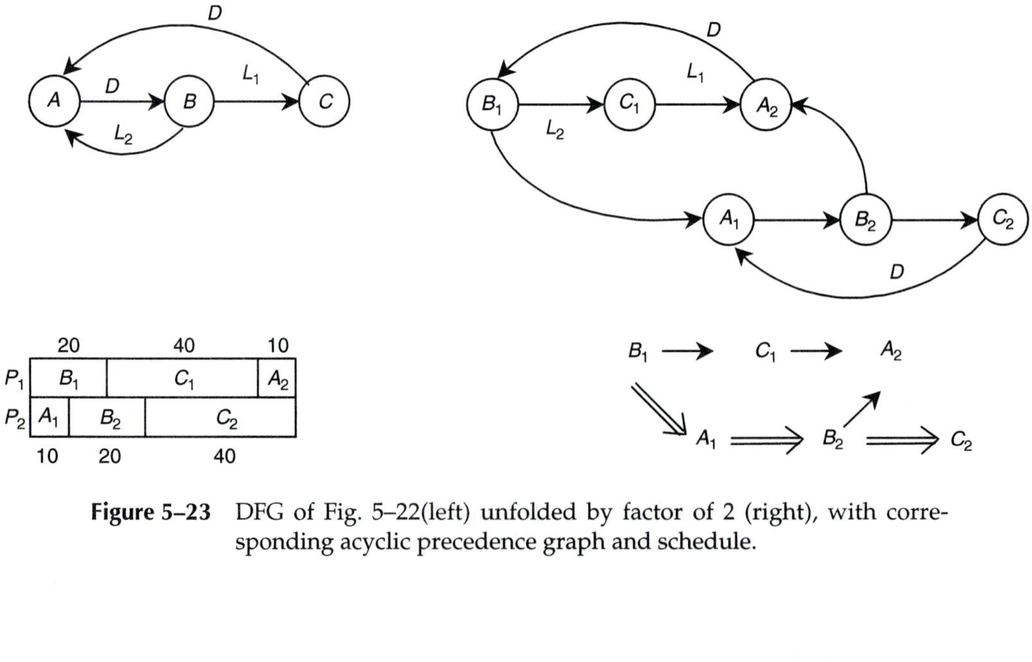

Figure 5–23 DFG of Fig. 5–22(left) unfolded by factor of 2 (right), with corresponding acyclic precedence graph and schedule.

rithms mentioned in Chapter 1. Indeed, loop unrolling may be viewed as a migration from stream to block processing.

As exemplified in Fig. 5–23, a perfect rate data flow graph has one register in each loop. The perfect rate DFG can always be scheduled in a fully static optimum manner to obtain the iteration bound. If an arbitrary DFG is unfolded by a factor equal to the least common multiple of the number of loop registers in the original DFG, the unfolded DFG is a perfect rate DFG.

A technique for scheduling the nodes in loops of a perfect rate DFG can be described, making the assumptions of an unlimited number of processors and a fully connected processor array:

1. Remove all transitive and extended transitive arcs;
2. List all loops in decreasing order of execution time, for scheduling in order of longest loop first. Also order the nodes in each loop to form a list with the node containing the loop delay register at its input at the top of the list, and the other nodes following to preserve precedence set in the DFG;
3. Assign a separate processor for the scheduling of each loop, as follows:
   a) Nodes of the critical loop are scheduled contiguously in processor P_1 (if there is more than one critical loop, choose one at random);
   b) Nodes of the next loop are scheduled in P_2 such that the schedules completed so far, on any nodes that are also in previously scheduled loops, are preserved;
4. Repeat the above until the scheduling of all loops is complete.

Fig. 5–24 uses a new DFG to illustrate the steps of this technique.

In summary, an arbitrary DFG can be transformed to a perfect rate DFG by unfolding by the least common multiple of the number of loop registers in the DFG. A perfect-rate DFG can be scheduled at a rate optimum manner (i.e., achieving the theoretical iteration bound), assuming the availability of a large number of processors and complete processor-to-processor interconnection. With the reduction of DFG optimization and scheduling to an algorithmic process, a computer program can be used to perform such optimization and scheduling. An example will be discussed in Section 6.5.

5.3.4 Pre-Compilation Optimization

The goal of pre-compilation optimization is to introduce structures and techniques at the early stage of the high-level language algorithm description to assist the subsequent compilation to improve the efficiency of its results. This optimization is both effective and necessary because compilers cannot have as full knowledge of the entire operation as the designer. Compiler code generation and optimization take place after the source code has been translated to an intermediate internal form for the compiler, and thus they cannot change the source code structure for efficiency.

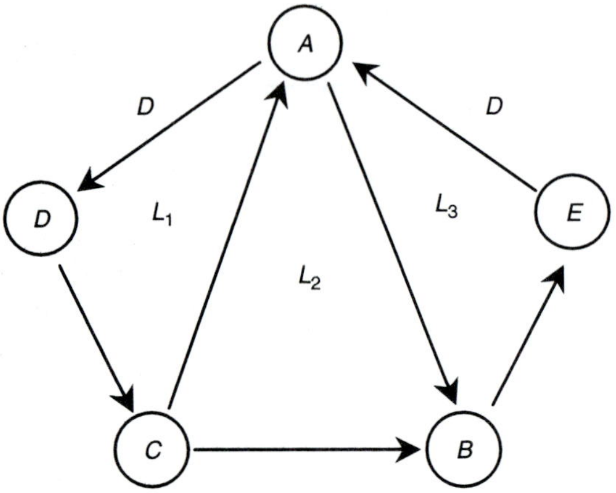

Node execution times	
A	1
B	1
C	1
D	1
E	1

Loop order	Node Order	Comment
L_1	*D*	input has delay
	C	
	A	
L_3	*(A)*	input has delay, but is already scheduled
	B	
	E	
L_2	*(C)*	already scheduled
	(A)	
	(B)	

D_1	C_1	A_1	D_2	C_2	A_2	...	...
			B_1	E_1		B_2	E_2

Figure 5–24 Use of processor scheduling procedure to develop rate-optimal schedule for DFG.

Several structural improvements can be made to improve the match between software architecture and processor architecture. These include:

- creating clearly separate modules for each function block that is of a different nature of computing, e.g., regular versus irregular, data-independent flow versus data dependent, etc.;
- excising all reused modules from in-line code and structuring them to be called as subroutines, e.g., common FFT routines;
- conversely, faster code (versus more space-efficient code) can result from embedding such common modules within in-line code—this saves on the overhead functions of calling the subroutine, returning from the subroutine, and (if the input format is customized for each use) avoiding the overhead of completely filling out a generalized input structure to the routine.

The technique of loop unrolling, discussed previously, can be applied to source code. In one example, which is described in Section 6.2,[11] loop unrolling provided a speedup of from 1.4 to 3.9 times over a range of digital signal processing operations on a commercial workstation.

Data type optimization chooses the data type (e.g., single precision integer, single precision versus double precision floating, etc.) that processes the fastest while preserving algorithm accuracy. A speedup factor of up to 2.5 times has been observed over a range of signal processing algorithms on a commercial workstation.[11]

Optimizing the order of data within tables can speed processing. In a program critical path, storing precomputed data in the expected order of access within a table that is used, or in order whose addressing is easily computed, speeds up execution by a factor of up to 1.4 over the same range of signal processing operations mentioned above.

Finally, the source code can be partitioned and reordered to reduce the traffic between cache memory and primary memory. This requires ordering computations to complete the use of data that lies within contiguous memory (or a few contiguous segments).

5.3.5 Choosing and Refining Processor Sizing

The goal of choosing an initial mix or count of multiple processors can be addressed for both homogeneous architectures and heterogeneous architectures. In a homogeneous architecture, all processors are the same, while a heterogeneous architecture contains a variety of processor types.

For the homogeneous architectures the list of processing steps per input sample is first analyzed to determine the number of operations per second to achieve real-time operation. This throughput estimate is obtained by analyzing the final data flowgraph to determine the number of operations per sample for each DFG node. This is multiplied by the number of samples per second to attain a total number of operations per second.

For each type of processing step in the DFG, an effective throughput capability is determined for the processor. While a top-level processor specification provides a peak compute speed in number of operations per second, each node will have a lower compute speed than peak, which may be determined by lower-level specifications in the processor data manual, or by direct experimental measurement in timing loops. The throughput requirement of each node is divided by the throughput capability of the processor to obtain the equivalent number of processors required for real time execution of each node. The processor counts for all nodes are then totaled, providing a total number of processors required for the application. Any additional communication requirement is then added, and a design margin is also added of 25 percent to 50 percent to provide a safety factor. The final portion of the discussion of image formation for synthetic aperture radar, discussed in Section 6.6, provides an example of this approach to partitioning.

For a heterogeneous architecture, several steps precede the above procedure for a homogeneous architecture. The algorithm description or optimized DFG is first used to determine the major sequential steps of processing. The nature of each computational segment is then compared to the descriptors of available processors. The algorithm flow diagram is inspected to look for separable high-load processing modules that are often repeated (e.g., FFT). Fig. 5–25 describes the segmentation of a computation into modules of different types, and the use of a repeated module.

For each computational module, each of the available processors is assessed as a candidate resource, and the most efficient (fastest) processor type is identified. The total algorithm is thus divided into an arrangement of sequential, parallel, and subfunction blocks with a processor type assignment to each. At this

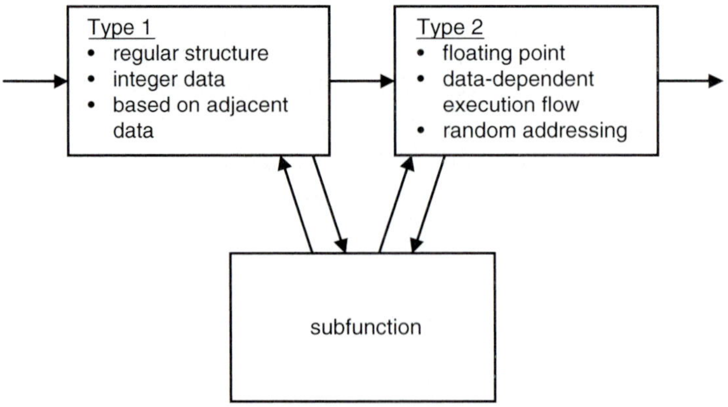

Figure 5–25 Division of large computation into adjacent blocks of different types, with common use of subfunctions.

point, each block is treated as a mapping problem to a homogeneous processor, and the processor-sizing method for homogeneous processors is then used. The mapping of subcomputations on or near the boundary between two modules, as well as other uncertain cases, is refined by performing a trial allocation to each adjacent processor module, conducting a performance estimate, and choosing the best mapping.

5.3.6 Communication Design

The design of interprocessor communication methods begins with distinguishing between local and global data communication requirements (see Section 3.2.6). For local communication, the output sample is a function of only a limited extent of input samples. For example, in the two-dimensional array associated with an image, each pixel may be processed as a function of the neighborhood of adjacent pixels. In the local case, communication requirements per output sample are independent of the size of the problem.

For the case of global communication, each output depends upon the input samples across a large range of locations. An example is the fast Fourier transform, for which the computation of each output sample depends upon all input samples in a block. Per-output global communication requirements are size-dependent—a 16-point FFT combines the results of 16 inputs into each sample, while a 1024-point FFT combines the results of 1024 inputs to form each output.

Communication design is influenced by architecture. Architecture influences the choice between shared memory and message-passing interprocessor communication schemes. If the processors can share a common set of memory modules, a shared memory approach is preferred. If each processor has its own local memory, then a message passing protocol may be considered. If message passing is chosen, the Message Passing Interface (MPI) de facto standard is often used. Because additional effort is required to add the MPI calls to the single-processor code, the use of MPI is not an automatic choice. Better performance may be more simply achieved by using the vectorizing capability of the processor native C or FORTRAN compiler. Conditions under which MPI might be justified include:

- need to port the code to a variety of parallel architectures, all supported by the MPI standard;
- code is not easily vectorized, i.e, it does not have a large number of parallel DO loops;
- code has a large serial section that is executed multiple times.

The incorporation of MPI begins with both the optimized single-processor C or FORTRAN code and the intended partitioning scheme for assigning pieces of the processing to the various processors. The MPI library calls are then added to the C code to implement cross-processor communication.

The communication bandwidth available is an overriding factor. The manner in which processors are connected determines whether communication can occur point-to-point between processors, or whether a link into and out of an intermediate processor is needed. When an architecture has communication processors that are distinct from computation processors, the timing burden of communication is reduced. An overlap between processing time and communication time can allow hiding the communication behind the processing time.

Optimizing communication timing begins by partitioning the algorithm so that communication across processor boundaries is minimized and restricted, as much as possible, to directly-connected processors. If the execution times of each node are identified first, one seeks to hide the communication time behind node execution. A tool may be used for modeling and optimizing communication times across multiple processors. The general-purpose distributed memory model of a processor, which was discussed Section 5.1.2, provides for both a communication overhead time, which is independent of transfer size, and a communication time per transfer, which depends on transfer size. A tool may be used to focus on interprocessor communication time. One such tool uses the Message Passing Interface (MPI) and a workstation to visualize the communication requirements of parallel process execution.[12] Fig. 5–26 shows the bars corresponding to communication intervals for the function types indicated by the bar number. The processor executing the functions is shown by the number on the vertical

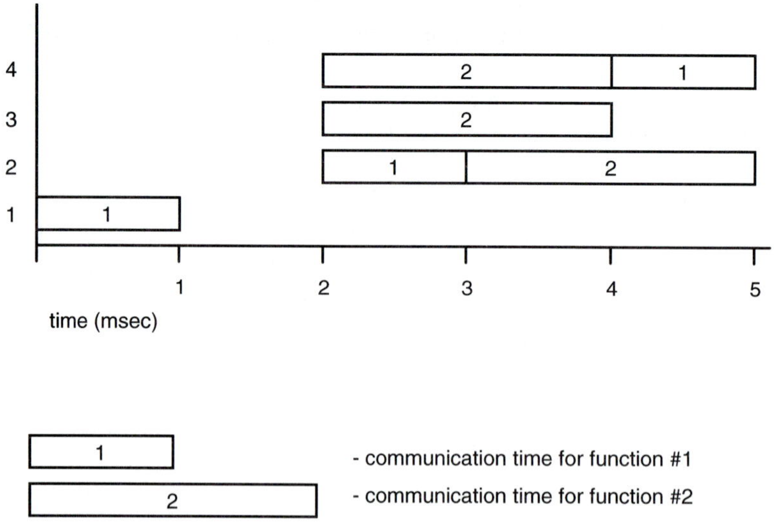

Figure 5–26 MPI-Sun[12] communication modeling tool displays a communication timeline for various functions operating on multiple processors—execution time for each node is added after communication timing is designed.

axis. After the tool is used to optimize communication paths, a prediction of execution time for each node is added and the timing is fine-tuned.

Communication analysis and modeling tools use standard communication models. One example is Myrinet,[13] a high-bandwidth, low-latency network specifically designed for workstation clusters. Myrinet consists of:

- computing elements (processing nodes such as workstations);
- network (links and switches to route data);
- network interfaces between workstations and network.

Myrinet provides typical performance of 100 μsec latency and 255 Mbit/sec throughput one way between Sun SpARC workstations. Other typical models include SCI (Scalable Coherent Interface) and MPI (Message Passing Interface).

5.3.7 Formal Integer Programming Methods

Optimum mapping of multiple sequential and parallel processing steps to multiple processors may be accomplished by integer programming methods. Integer programming maps x computing tasks to y processors while meeting constraints set by z. The notation used to describe the problem is as follows:[3]

K	= set of nodes in the DFG;
k	= a code operation of the DFG algorithm; $k \epsilon K$;
$k_1 <\cdot k_2$	= node k_1 produces output that is used by k_2 (precedence);
$x_{j,k,t}$	= these are the variables of the integer programming; $x_{j,k,t} = 1$ means that code operation k is assigned to time slot t and to function unit j;
$j_z \epsilon J(k_z)$	= code operation k_z can be implemented by functional unit of type j_z, indicating for each code operation k_z which of the range of functional units $j = 1, 2, \ldots, J$ are capable of executing k_z.
$t_z \epsilon R(k_z)$	= indicates that t_z is in a range of allowed times associated with code operation k_z; specifically asap $(k_z) \le t_z \le alap\ (k_z)$, where *asap*/*alap* mean as soon as possible/as late as possible;
I_j	= number of functional units of type j;
$j_z \epsilon op(C_z, L_z)$	= functional unit of type j_z requires C_z time steps to produce output data and can accept new data at a minimum of L steps.

Several equations set up the problem for integer programming analysis:

$$\sum_{j \in J(k)} \sum_{t \in R(k)} x_{j,k,t} = 1, \ \forall k \tag{5.3}$$

(ensures that each code operation will be assigned to an execution time and a processing unit);

$$\sum_{t_1=t}^{t_1+(L_1-1)} \sum_{k_1,j_1 \in J(k_1)} x_{j,k,t} \leq I_{j_1}, \ \forall j_1,t_1 \tag{5.4}$$

(calculates number of functional units of each type to be allocated);

$$\sum_{j_1 \in J(k_1)} \sum_{\substack{t_1 \in R(k_1) \\ t-(c_1-1)\leq t_1}} x_{j_1,k_1,t_1} + \sum_{j_2 \in J(k_2)} \sum_{t_2 \leq T, t_2 \in R(k_2)} x_{j_2,k_2,t_2} \leq 1, \tag{5.5}$$
$$\forall k_1 <\cdot\, k_2, \ \text{for } t \in R(k_2) \cap (R(k_1) + C_1 - 1).$$

(preserves data precedence; it prevents an operation k_1 from being scheduled after operation k_2 whenever $k_1 <\cdot\, k_2$ and j_1 is not a chained type functional unit).

In addition to mapping multiple operations to multiple processors, integer programming can be used to optimize a mapping with respect to a large diverse variety of system constraints. Formulas for setting up an integer programming approach to system-wide optimization have been published.[14] The goal is to minimize cost subject to the constraints and expressions of: software development cost and development time, application-specific integrated circuit (ASIC) development cost and development time, time to market, total number of programmable processors, ASIC yield, yield of field programmable gate arrays (FPGAs), system level area and power, load balancing, local memory capacity, processor utilization, interval relation, memory utilization, and processor utilization. Once the problem is set up, several generic integer programming tools are available to solve the problem.[15,16,17]

REFERENCES

1. M. Shahdad and others, "VHSIC Hardware Description Language," *IEEE Computer Magazine*, 18 (February 1985), 94–103.
2. C.-L. Wang, P. B. Bhat, and V. K. Prasanna, "High Performance Computing for Vision," *Proc. IEEE*, 84 (June, 1996), 931–946.
3. Catherine H. Gebotys and Robert J. Gebotys, "Optimal Mapping of DSP Applications to Architectures," *Proceedings 26th IEEE Hawaii International Conference on System Sciences*, Hawaii, 1993, pp. 197–204.
4. Vijay K. Madisetti, *VLSI Digital Signal Processors: An Introduction to Rapid Prototyping and Design Synthesis* (Boston: Butterworth-Heinemann/IEEE Press, 1995).
5. T. Egolf and others, "VHDL-Based Rapid System Prototyping," *J. VLSI Signal Processing Systems for Signals, Video, and Imaging Technology*, 14 (November 1996), 125–156.

6. Konstantinos Konstantinides, Ronald T. Kaneshire, and Jon R. Jani, "Task Allocation and Schedule Models for Multiprocessor Digital Signal Processing," *IEEE Trans. Acoust. Speech and Signal Processing,* 38 (December 1990), 2151–2161.
7. Juan Li and Leah H. Jamieson, "A System for Algorithm-Architecture Mapping Based on Dependence Graph Matching" in Howard S. Moskowitz, Kung Yao, and Rajeev Jain, ed. *VLSI Signal Processing IV* (Piscataway, NJ : IEEE, 1991), 157–166.
8. Keshab K. Parhi and David G. Messerschmitt, "Static Rate-Optimal Scheduling of Iterative Data-Flow Programs Via Optimum Unfolding," *IEEE Trans. Computers,* 40 (February, 1991), 178–195.
9. D. A. Schwartz and T. P. Barnwell, III, "Cyclostatic Multiprocessor Scheduling for the Optimal Implementation of Shift Invariant Flow Graphs," *Proc. IEEE Inter. Conf. Acoustics, Speech, Signal Processing,* Tampa, FL, March 1985, pp. 1384–1387.
10. Miodrag Potkonjak and Jan Rabaey, "Retiming for Scheduling," in Howard S. Moskowitz, Kung Yao, and Rajeev Jain, Eds., *VLSI Signal Processing IV* (Piscataway, NJ : IEEE, 1991), 23–32.
11. Pierpaolo Baglietto and others, "Image Processing on High-Performance RISC Systems," *Proc. IEEE,* 84 (June, 1996), 917–930.
12. Robert Cunningham, "An MPI-Based Parallel CFAR Algorithm Implementation: Implications and Recommendations for the ACP Project," *MIT Lincoln Laboratory Report,* January 1996.
13. Myricom, 325 North Santa Anita Avenue, Arcadia, CA *http://www.myri.com.*
14. Egolf, "VHDL-Based Rapid System Prototyping."
15. H. M. Salkin, *Integer Programming* (Reading, MA : Addison-Wesley, 1995).
16. R. S. Garfinkel and G. L. Nemhauser, *Integer Programming* (New York: Wiley, 1972).
17. A. Brooke, D. Kendrick, and A. Meerause, *GAMS: A Users Guide, Release 2.25,* Danvers, MA: Boyd & Fraser, 1992.

6

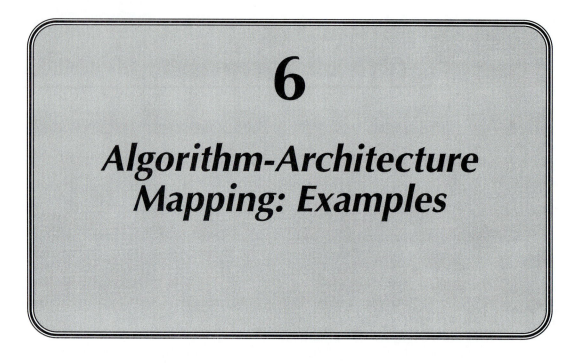

Algorithm-Architecture Mapping: Examples

The principles of mapping algorithm and architecture that were discussed previously are illuminated here by several examples. Each example differs in the means by which it applies mapping methods, and while they all accomplish the major steps shown in Figure 5.3, each accomplishes the mapping in a different way. Each example has been performed and reported by a different organization and each has resulted in a successful mapping. Indeed, it is one of the purposes of this chapter to extract and describe the underlying process for mapping by examining a diversity of successful design histories.

6.1 BASIC DESCRIPTIONS OF ARCHITECTURE AND ALGORITHM

Each of the examples is based on a different combination of two characteristics that distinguish algorithms and architectures. The first characteristic, descriptive of the architecture, is the number of processors, which may either be one or many. For mapping to one processor, optimization is performed over the individual resources of one machine. Optimization methods developed for a single processor are also used as a basis for multiprocessor implementation. This is because the algorithm allocated to each processor in an array is then optimized on that processor as a single processor mapping. Also, an early step in the mapping process is to achieve an efficient single-processor mapping and then deploy multiple processors to relieve the bottlenecks that are observed.

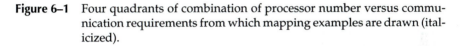

		processors	
		one	many
commun-ication	local	*convolution*, filtering	(convolution), finite impulse response filter (FIR), *infinite impulse response filter (IIR)*, dynamic programming, *convolution*
	global	discrete cosine transform (DCT), full scale block matching, Hough transform, *linear predictive coding* (LPC)	*synthetic aperture radar (SAR) image formation*, fast Fourier transform, MPEG coding, super resolution image enhancement, hidden Markov modeling

Figure 6–1 Four quadrants of combination of processor number versus communication requirements from which mapping examples are drawn (italicized).

The second characteristic, determined by the algorithm rather than the architecture, is the type of communication required among input samples to obtain the output samples. If each output sample is a function of only nearby input samples, such as a small moving window of a time-domain signal or a two-dimensional neighborhood of adjacent pixels, then the communication requirements are local. In global communication, each sample output depends on a large block of input samples, not all of which are adjacent. Global requirements affect communications, since each output may use nearly all inputs, requiring that these inputs be kept ready for processing. Global communication is typical of block processes such as transforms between time and frequency representations of data.

Fig. 6–1 displays several signal processing tasks in a matrix of processor count versus communication type. Although any of the many-processor examples could be performed by one processor at a slower speed, they are more typical of multiprocessor operations.

One example is drawn from each quadrant. In addition, one example (convolution) is examined for both a single-processor and a multiprocessor mapping and can be performed with either local communication or global communication, based on an algorithm simplification that may be used.

6.2 TWO-DIMENSIONAL CONVOLUTION ON ONE PROCESSOR[1,a]

This algorithm uses local processing to perform spatial linear filtering of an image to reduce or emphasize certain spatial frequencies for image analysis pur-

[a]Portions adapted, with permission, from Pierpaolo Baglietto and others, "Image Processing on High Performance RISC Systems," *Proc. IEEE*, 84 (June, 1996) 917–930.

poses such as edge detection. It involves sliding a matrix of coefficients over an image in a series of shifts. At each shift, a pixel of a new image is formed at the center of the matrix as the weighted average value of the pixels covered by the mesh:

$$g(x + m, y + m) = \sum_{i=-m}^{m} \sum_{j=-m}^{m} a(i,j)f(x + i,y + j). \tag{6.1}$$

The architecture to which convolution will be mapped is the Sun SpARC workstation, characterized as shown in Table 6–1. The mapping approach uses a C compiler to automatically map the algorithm to the architecture. Prior to automatic mapping, manual steps to restructure the algorithm for efficiency are performed.

In this example, the image size is 512 × 512 pixels, and the convolution mask size [$-m$ to $+m$ in (6.1)] is 5 × 5 pixels. A simple C-code routine would overrun the image edges by one-half of the size of the kernel unless the image is declared at a size that is a half mask width in excess of true image size.

C-code that is a direct expansion of the algorithm is:

Table 6–1 Description of target architecture.

Element	Characteristic	Value
CPU	Clock frequency	60 MHz
	CPU cycle time	16.6 nsec
	Data path width	32 bits
Registers	# general purpose	32 @ 32 bits each
	# floating point	32 @ 32 bits or 16 @ 64 bits
Instruction throughput	int-to-float conversion	67 nsec (4 cycles)
	integer sum	½ cycle
	integer multiply	5 cycles
	floating point sum (64 bit)	3 cycles
	floating point multiply (64 bit)	3 cycles
Memory	primary	64 MB
	RAM access time	365 nsec
Instruction cache	size	20 KB
	organization	5 way set associative
	access time	16.6 nsec
Data cache	size	16 KB
	organization	4 way set associative
	access time	16.6 nsec
Second-level cache	size	1 MB
	access time	15 nsec

```
for (x = 0; x < IM_SIZE; x++)
for (y = 0; y < IM_SIZE; y++){
        temp = 0;
        for (i = -m; i <= m; i++){
                for (j = -m; j <= m; j++){
                        temp+ = f[x+i][y+j] * a[i][j];
                }
        }
        g[x][y] = temp;
    }
```

The execution time on the Sun workstation of the above code for the 512×512 image with a 5×5 pixel mask is 1360 msec.

Several techniques discussed previously may be applied to speed execution. These are all applied to the C-source code, prior to compilation. Loop unrolling may be applied, both as external loop unrolling, by which iterations are moved from the outer loop to the inner loop, and internal loop unrolling, by which some iterations of the innermost loop are collapsed into one iteration. The technique known as data type optimization may be used to exploit the speed advantage that the Sun gives to floating point multiply (3 cycles) over integer multiply (5 cycles). This advantage is obtained by converting numbers to floating point before multiplication. This speed advantage is somewhat mitigated by a delay that is introduced in converting integer values to floating point and back (4 cycles each way).

The result of applying external loop unrolling is code that moves through the y loop in groups of N pixels at a time:

```
for (x = 0; x < IM_SIZE; x++)
    for (y = 0; y < IM_SIZE; y += N){
        temp1 = temp2 = . . . = tempN = 0;
        for (i = -m; i <= m; i++){
            for (j =-m; j <= m; j++){
                mt = a[i][j];
                temp0+ = f[x+i][y+j]*mt;
                temp1+ = f[x+i][y+j+1]*mt;
                       .      .
                       .      .
                       .      .
                tempN+ = f[x+i][y+j+N]*mt;
            }
         g[x][y] = temp0;
         g[x][y+1] = temp1;
               .      .
               .      .
               .      .
         g[x][y+N] = tempN;
    }
}
```

Internal loop unrolling collapses iterations of the innermost loop shown above into one statement to obtain:

```
for (x = 0; x,IM_SIZE;x++)
    for (y = 0; y <IM_SIZE; y++){
        g[x][y] =    f[x-m][y-m]*a[-m][-m]
                   + f[x-m][y-m+1]*a[-m][-m+1]
                   +     .         .
                         .         .
                         .         .
                   + f[x-m][y+m]+a[-m][m]
                   +     .         .
                         .         .
                         .         .
                   + f[x+m][y+m]*a[m][m];
    }
```

The results of applying external and internal loop unrolling, using a factor of 25 for loop unrolling, is to obtain a decrease in execution time from 1360 msec to 430 msec, for a speedup factor of three.

6.3 TWO-DIMENSIONAL CONVOLUTION ON MULTIPLE PROCESSORS[2]

As the size of the convolution kernel becomes larger and as the stream of input becomes infinite, real time continuous throughput is needed. This results in:

- more computation;
- the problem of partitioning across multiple processors for speed;
- achieving continuous output rather than output in discrete blocks;
- use of a digital signal processing architecture rather than a general purpose processor, resulting in less architecture flexibility (e.g., no cache for instructions or data), and a more powerful multiplication capability.

This example of convolution differs from the previous one in the following ways:

- convolution is mapped to multiple processors, rather than to one;
- the Fourier convolution theorem, as detailed in Section 6.6, is applied to replace direct convolution by multiplication in the Fourier domain.

The mapping process to be used is similar to that of the previous example. Because of the simplicity and regularity of the algorithm, data flow analysis is not applied; rather, the partitioning is applied intuitively. The steps of the process are:

1. *Define application* (equations, length of kernel, length of input, sample rate, number of bits/sample).
2. *Define computational resources*: Resources consist of a connection of nodes with input and output occurring over a bus, where the processing at each node consists of a SHARC digital signal processor.
3. *Apply algorithm simplification:* Replace continuous direct convolution with Fourier-transform-based circular convolution.
4. *Partition algorithm and/or data into blocks*: Determine how instructions and/or data can be logically divided.
5. *Map to library calls and processing instructions.*
6. *Account for system constraints* (processor, architecture).
7. *Distribute logically self-contained portions onto different processors*: Allow for any irregularities.
8. *List the major buffers needed*, and their sizes as a function of section length: Set aside additional memory for other miscellaneous use.
9. *Optimize processing parameters* (here, section length), under the constraints of a) length must equal an integer power of 2; b) length is consistent with input block size of library module; c) buffer length fits within fastest available memory.
10. *Verify that all data paths are within input/output bandwidth.*

Step 1 of describing the algorithm has been accomplished; Step 2 now motivates a description of the architecture. The basic processor is the Analog Devices 21060 Super Harvard Architecture (SHARC) described in Section 2.3.5. Each SHARC is capable of a peak compute rate of 120 MFLOPS and has 256–512 KB of static RAM. The basic node consists of four SHARCs as well as a node application-specific integrated circuit (ASIC) and 4–64 MB of dynamic RAM. The total system consists of an input card plus multiple node cards.

6.3.1 Sectioned Convolution

Step 3, algorithm simplification, introduces two structural changes. First, to accomplish a two-dimensional convolution, the problem is decomposed into a row-wise one-dimensional convolution that is followed by a column-wise one-dimensional convolution. Second, for larger convolutions, it is more efficient to replace direct convolution with a Fourier transform method that takes advantage of the convolution theorem. The Fourier convolution theorem states that the periodic convolution of two finite sequences can be obtained by multiplying the discrete Fourier transform of the sequences, then taking the inverse DFT of the result. As a result, for $y(x+m) = \Sigma_{i=-m}^{m} f(i)g(x-i)$ and defining $Y(\omega) = \text{FFT}[y(i)]$, then $Y(\omega) = F(\omega) \cdot G(\omega)$ and $y(x+m) = \text{DFT}^{-1}[Y(\omega)]$.

This example poses the problem of accommodating a continuous stream of input samples of indefinite length in real time with a minimum number of digital signal processors. The technique of *sectioned convolution* divides the continuous input data stream into sections of N points each, where each section overlaps the next by the size of the kernel, $2m + 1 = k$. Fig. 6–2 shows the sectioned convolution and indicates how Step 4, dividing the algorithm into blocks, is performed. N is further constrained to meet the needs of the FFT for an input block size that is a power of two.

In Step 5, the processing activity is mapped to calls to a library of signal processing routines. Inspection of Fig. 6–2 and perusal of the native SHARC library identifies the library cells shown in Table 6–2 for use in convolution.

Because the kernel is fixed and does not change with input data, its FFT is computed in advance and stored in a table.

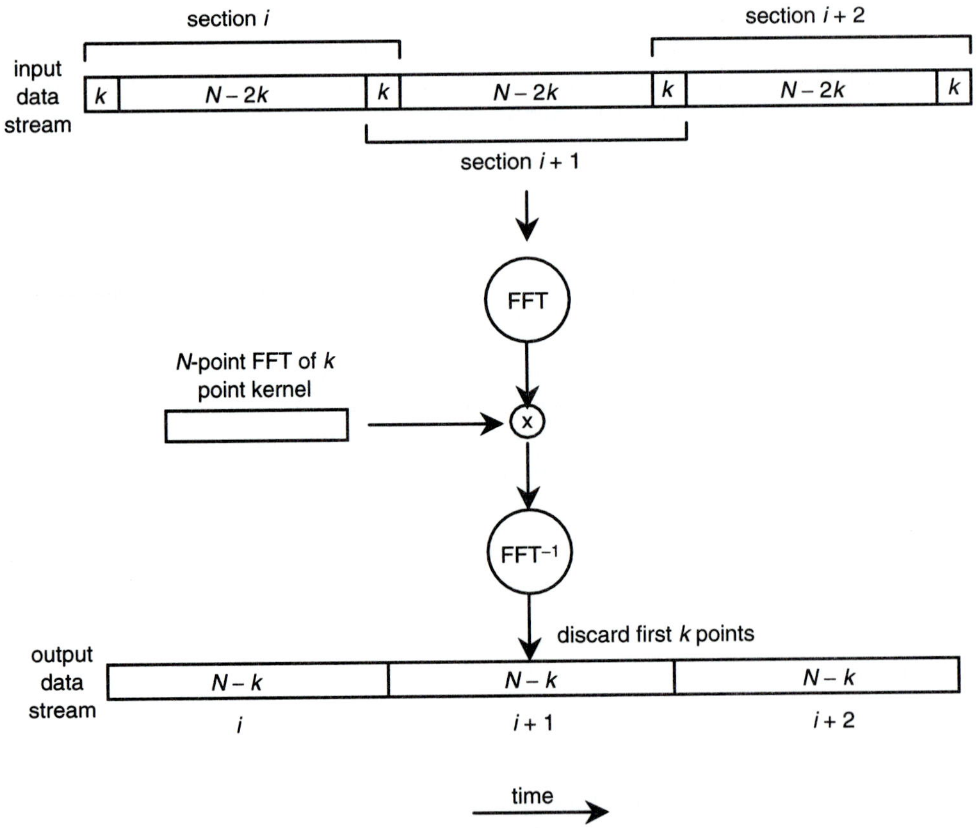

Figure 6–2 Sectioned convolution applies FFT to sections of length N, multiplies by N-point FFT of k-point kernels, then inverts the FFT and discards k points from each section to achieve continuous output.

Table 6–2 SHARC Scientific Application Library functions used in convolution.

Operation	Function
N-point int-to-float conversion	vflt()
N-point forward real FFT	rfft()
$(N/2 + 1)$-point complex multiply	cmult()
N-point inverse real FFT	rfft()

6.3.2 Including Architecture Specifics

System constraints are accommodated in Step 6. For maximum performance, each SHARC should operate only on data within its on-chip SRAM. Use of the off-chip DRAM is limited to a staging area for data either before or after loops. The SHARC is capable of receiving data into its SRAM that is written by external bus masters in parallel with conducting computation. As shown in Fig. 6–3, this

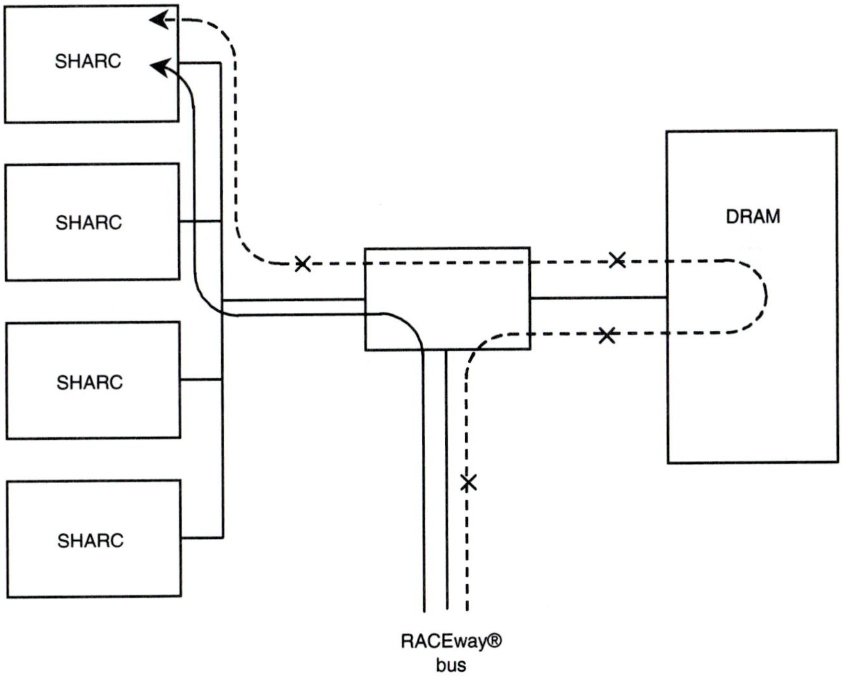

Figure 6–3 Although the large DRAM is used to accumulate output results, the input data is written directly into SHARC SRAM to gain speed.

is exploited by writing directly into SRAM over the RACEway© bus, rather than using the DRAM as an intermediary.

Data partitioning (Step 7) is accomplished by setting up input to write directly into SRAMs of the SHARCs in a round-robin manner. A total of $N - k$ elements are written into a SHARC, and the final k points are also written to the next SHARC, to accommodate the overlap region.

Buffer sizes (Step 8) are a function of the section-processing length, N. N dictates the FFT size, and the section length must be at least twice as large as the convolution kernel size, rounded up to the next power of two. Thus, for a 400-element kernel, the minimum section size is 1024.

To gain maximum speed, the FFTs are performed out of place, that is, the output data is written in memory locations that are separate from the input data. This is a direct result of the global nature of the FFT—each input sample is needed for all output samples. Thus, a double buffer is used. Table 6–3 describes the memory needed as a function of N.

The double buffer switches (or ping-pongs) between I/O and processing. Fig. 6–4 shows this repetition.

A dedicated buffer (for output) is used in conjunction with the processing half of the double buffer (for input) to perform sectioned convolution using the library calls shown in the previous table. The convolutional kernel transform buffer holds the FFT of the zero-padded convolution kernel, and the FFT weight buffer holds the twiddle factors needed for the FFTs (see Section 8.2.2 for a discussion of FFT twiddle factors).

The Analog Devices 21060 SHARC processor has 512 Kbytes of SRAM. Reserving 25 percent for miscellaneous uses, 75 percent or 384 KB is available for use as a buffer. Dividing by the 22 bytes/element computed above and rounding down to the next power of two provides a maximum section length of 16 KB.

The smallest possible section length of 1K samples favors maximum performance, because a smaller FFT requires fewer operations per sample than a larger FFT. However, with a fixed-size overlap set by the kernel length, a smaller section results in a larger per-sample penalty for the overlap as the number of out-

Table 6–3 Memory usage for convolution section of size *N*.

Buffer	Size
2 dual-use buffers (ping-pong between I/O and processing)	N 4-byte elements per buffer
1 dedicated processing buffer	N 4-byte elements
1 convolution kernel transform buffer	$N/2$ 8-byte elements
1 FFT weight buffer	$3N/4$ 8-byte elements
TOTAL	22 N bytes

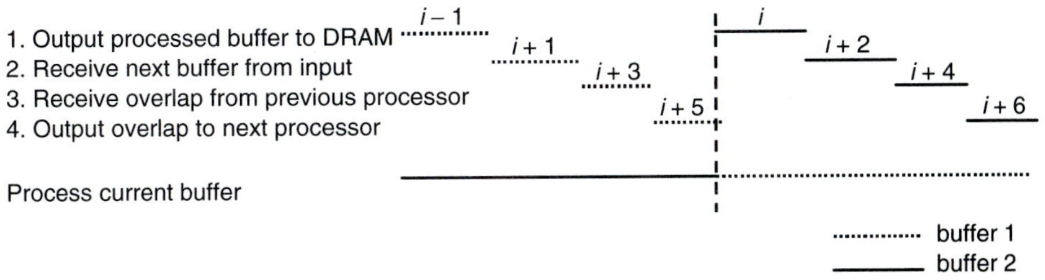

Figure 6–4 Buffer 1 and Buffer 2 timelines show switching between input/output and processing on sequential sections.

put points to amortize it over, given by $(N - k)/N$, becomes smaller. To optimize the section size for maximum throughput (Step 9), a table can be constructed that shows the processing time per step for various section lengths over the range of 1K to 16K samples. A total throughput, in samples per second, is computed by dividing the number of output points per section by the computation time per section (Table 6–4). The resulting optimum section length is 8K, which provides a maximum throughput of 1.27 Msamples/sec.

In Step 10, a final check is made to verify that processor resources have not been exceeded. Specifically, the input/output bandwidth is computed and compared with the available bandwidth. For 8K-point sections with a processing time of 6.13 msec, Table 6–5 computes the per-processor and total bandwidths.

The bandwidth requirements of writing output to the DRAM are seen to be 5.08 Mbyte/sec × 4 processors = 20.3 Mbyte/sec. Both the total bandwidth of 31.5 Mbyte/sec and the output bandwidth of 20.3 Mbyte/sec are well below the 160 Mbyte/sec offered by the RACEway, and so a four-processor node is validated as within the system I/O capability.

Table 6–4 Execution time (in msec) versus section length, and throughput versus section length.

Operation	1*K*	2*K*	4*K*	8*K*	16*K*
N-point int-to-float	.03	.05	.10	.20	.41
N-point real FFT	.26	.55	1.23	2.55	5.63
($N/2$ +1)—point complex multiply	.10	.20	.41	.82	1.64
N-point inverse FFT	.26	.55	1.23	2.55	5.63
Total processing time	.66	1.35	2.96	6.13	13.31
Output points per section	624	1648	3696	7792	15984
Throughput (Msample/sec)	.95	1.22	1.25	1.27	1.20

Table 6–5 Bandwidth requirements of sectioned convolution.

Function	Points/sec	Bytes/points	Bandwidth
Input	(8192–400)/(6.13 msec)	2	2.54 Mbyte/sec
Input of overlap	400/(6.13 msec)	2	0.13 Mbyte/sec
Output of overlap	400/(6.13 msec)	2	0.1. Mbyte/sec
Output of results to DRAM	(8192–400)/(6.13 msec)	4	5.08 Mbyte/sec
TOTAL			7.88 Mbyte/sec
x number of processors			x 4
TOTAL BANDWIDTH			31.5 Mbyte/sec

6.4 LINEAR PREDICTION CODING APPLIED TO TELEPHONE SPEECH [3,a]

Linear predictive coding (LPC), like sectioned convolution, is processed in blocks, and each input sample contributes to all outputs for the section. LPC computes an all-pole model of the short-time speech spectrum over a section length. For telephone bandwidth speech (200–3,200 Hz), eight poles are used in this example, and the processing section length is 45 msec. The short-time spectrum models the position of the vocal tract for the section interval. For each interval, the goal is to determine a set of predictor coefficients a_k, $k = 1, 2, \ldots, p$, that minimizes the sum of squared differences, E_n, between the actual speech samples $x(n)$ and estimated speech samples $x'(n)$.

6.4.1 Algorithm Description

Here, $x'(n)$ has been estimated as a linear combination of the p previous speech samples:

$$x'(n) = \sum_{k=1}^{p} a_k x(n - k). \tag{6.2}$$

From a_k, one can compute an all-pole model of the associated short time spectrum:

$$\hat{P}(\omega) = \frac{G^2}{\left|1 + \sum_{k=1}^{p} a_k e^{-j\omega k}\right|^2}; \quad G = \text{gain}. \tag{6.3}$$

In another equivalent representation, the speech is modeled by an acoustic tube consisting of p sections, the areas of which are a function of the a_k's known as *log area ratios*. The tube is excited by a random source (for unvoiced speech) or a periodic source (for voiced speech) with energy proportional to G.

In the particular instance of telephone-bandwidth speech, the signal is band-limited to 3.2 KHz and is sampled at a rate that ranges from 6,667 samples/sec, used in this example, to 8,000 samples/sec, used in standard telephone coders. The 45-msec analysis window contains 300 samples. A new analysis window is begun every 100 samples (15 msec), so the windows overlap and each sample falls in the first third of one window, the middle third of the previous window, and the final third of the second-previous window. With this bandwidth, a value of $P = 8$ value is chosen. The output of the processing for each frame consists of the logarithm of the frame energy, the zero-th through eight-order autocorrelation coefficients of the frame, and the eight a_k values. A step of gain normalization (6.8) adjusts the autocorrelation coefficients such that $R_\ell(0) = 1$, thereby using all available bits of an integer implementation of Durbin's recursion, regardless of signal size.

The following processing comprises the algorithm:

Preemphasis: $$x'(n) = x(n) - ax(n-1);\ a = 0.95 \quad (6.4)$$

Blocking into frames: $$\tilde{x}_\ell = x'(M\ell + n),\ n = 0, 1, \ldots, N-1;\quad \ell = 0,1,\ldots, L;\ M = 100;\ N = 300 \quad (6.5)$$

Windowing: $$x_\ell(m) = w(n)\tilde{x}_\ell(n);\ w(n) = 0.54 - 0.46\cos\left(\frac{2\pi n}{N}\right) \quad (6.6)$$

Autocorrelation: $$R_\ell'(m) = \sum_{n=0}^{N-1-m} x_\ell(n)x_\ell(m+n);\ m = 0,1,\ldots, P \quad (6.7)$$

Gain normalization: $$R_\ell(m) = \frac{R'_\ell(m)}{2^{r_\ell}};\ \text{where } r_\ell = \log_2 R_\ell(0) \quad (6.8)$$

Partial correlation: $E^{(0)} = R'(0)$

Durbin's recursion: for $i = 1, 2, \ldots, 8$; do $(a) - (d)$ (6.9)

(a) $$k_i = \frac{\left[R(i) - \sum_{j=1}^{i-1} \alpha_j^{(i-1)} R(i-j)\right]}{E^{(i-1)}} \quad (6.10)$$

(b) $$\alpha_i^{(i)} = k_i \quad (6.11)$$

(c) $$\alpha_j^{(i)} = \alpha_j^{(i-1)} - k_i\alpha_{i-j}^{(i-1)}\ (j = 1, 2, \ldots, i-1;\ i \neq 1) \quad (6.12)$$

(d) $$E^{(i)} = (1 - k_i^2)E^{(i-1)} \quad (6.13)$$

Extract final residual: $$E = E^{(8)} \quad (6.14)$$

Extract LPC Coefficients: $$a_j = \alpha_j^{(8)}. \quad (6.15)$$

From the autocorrelation and LPC coefficients, various transformed parameters, such as reflection coefficients and log area ratios, can be derived.

The architecture for this mapping is a single fixed-point digital signal processor, representative of the least-expensive, lowest-power part available. Its cycle time is 400 nsec, and its data path is 20 bits wide for one path, and 16 bits wide for the other. Every instruction executes in 400 nsec, requiring the four pipeline stages of decode, fetch data, accumulate output product from multiplier, and store results in memory. Data memory consists of 256 20-bit locations, and program memory is limited to 1,024 16-bits words. The small program memory size imposed the largest constraint and drove this mapping approach to minimize program memory usage at the expense of speed and data memory.

The mapping approach provided algorithm simplification through the use of several techniques, as elaborated below:

- use of Durbin's recursion to exploit the Toeplitz nature of the matrix;
- sample rate reduction;
- Taylor series approximation to raised-cosine analysis window;
- bit-level reciprocal routine to implement a divide operation;
- regularization of program by using a common piece of code for all analyses;
- data type optimization—restriction to integers.

Inversion of the autocorrelation matrix was simplified by the fact that the matrix is Toeplitz, i.e., has equal values along each diagonal. Durbin's recursion [(6.10) through (6.13)] iteratively computes the inverse more quickly than a general-purpose inversion routine.

6.4.2 Program Interleaving for Rate Reduction

To support a reduction in data rate between the input and output, a two-piece program architecture is used. Sample rate reduction is typical of transform-based coders which have an output sample rate that is less than an input sample rate. Rate reduction is addressed by a program architecture that interleaves an input-sample program with a block analysis and output program. As shown in Fig. 6–5, a conflict arises between the input time scale, one sample every 150 μsec, and the output time scale, which is 19 outputs every 15 msec. The LPC processor must process a new sample every 150 μsec, else data is lost and the output feature vector is incorrect.

The input program is interleaved with segments of the output program (Fig. 6–5). A tradeoff analysis was conducted of program memory space and processing time for inputs that were handled one, two, three, four, or more samples at a time. It was found that most efficiency was obtained by grouping the samples into four at a time. The input program updates the autocorrelation vectors every four samples by computing (6.7) in groups of four samples. An output program performs an iteration of the frame recursion given by (6.10) to (6.15), performing this every fourth sample. The timing of the interleaved input and output program is shown in Fig. 6–6.

The sample update program operates on four samples at a time each time that it is executed. This four-sample operation is a result of loop unrolling on the

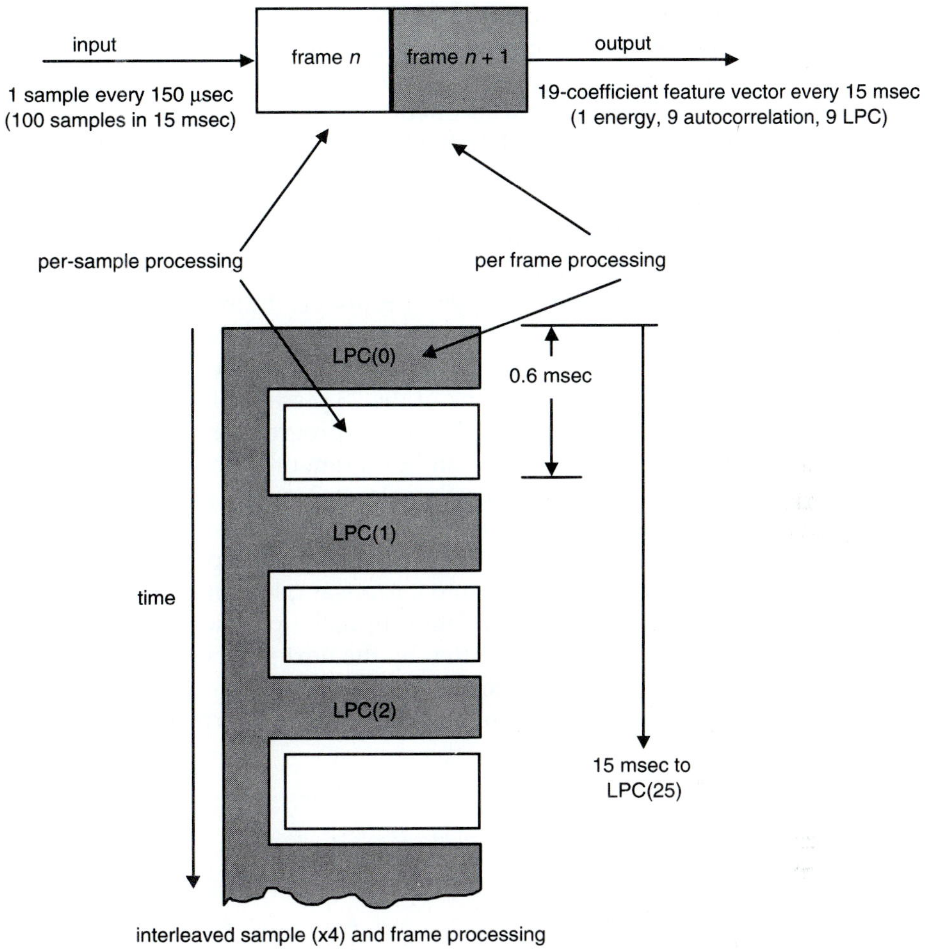

Figure 6–5 Input and output timing characteristics result in placing the spectral content of frames of 300 samples occurring every 100 samples into 19 outputs, resulting in data rate reduction. Output program LPC (shaded) is divided into 25 iteration segments, which are interleaved between four-input-sample update programs.[3] © 1985 *The AT&T Technical Journal.*

original one-sample program to speed execution. It is a compromise between fully block processing, in which autocorrelation vectors are computed on an analysis frame of 300 samples all at once, and fully stream processing, updating autocorrelations each time a sample is received. Fully block processing, while executing most quickly, requires sufficient data memory to store 300 samples, thereby exceeding the 256 data memory locations of the processor. Fully stream processing requires more operations per sample and executes too slowly to meet real-time requirements. For example, each input sample requires that address pointers must be set to access samples and autocorrelation vectors, and each autocorrelation vector must be moved through the multi-stage pipeline for each sample update. These overhead operations, which are needed whenever the autocorrelation vectors are updated, whether with one or many samples, can only be tolerated if multiple samples are processed for each update (Fig. 6–6).

Dividing the 100 sample frame period by the four sample input group requires that the output frame program be executed in 25 segments per frame period. An output operation of one value is added to each execution of the input processing program, providing 25 outputs per frame period. Nineteen of these slots contain the output data; the other six are zeros, which serve as a frame synchronization marker. As shown in Fig. 6–6, between the first and second input sample of each group of four, one of the 25 pieces of the output program is computed. Table 6–6 shows the LPC analysis program as divided into 25 parts and presents both the execution time and program memory used by each.

During the times following Samples 2, 3, and 4, portions of the autocorrelation update operation are performed. A sample is available every 150 μsec from the analog-to-digital converter and is read into an input buffer. The autocorrelation program reads the sample into a four-sample buffer at a convenient time between portions of the autocorrelation update operation, but within 150 μsec, to avoid the sample being overwritten by the next sample. Since all autocorrelation update processing occurs on four samples at once, a four-sample latency is introduced. Table 6–7 details the operations that are performed by the autocorrelation update program.

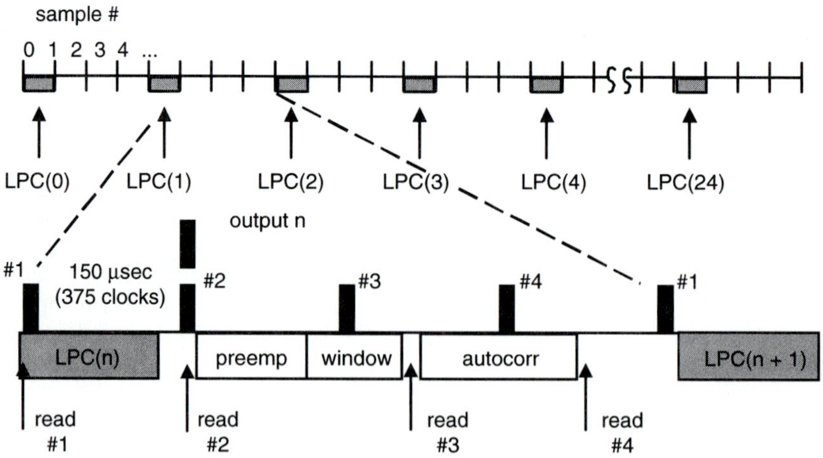

Figure 6–6 Input samples are processed in groups of four (unshaded blocks), with a segment of the LPC frame output program (shaded area) computed for each four-sample input group.[3] © 1985 *The AT&T Technical Journal.*

Table 6–6 Function, execution time, and program memory locations required for each of the 25 steps of the LPC analysis program. © 1985 *The AT&T Technical Journal.*

Label—LPC(n), where n =	Function	Execution Time (μsec)	Program Memory Locations
0	Read $R_l(m)$ to frame recursion input buffer, shift window	95	97
1	Calculate r_l	60	143
2	Calculate $R_l'(m)$; m=1,2,3,4	144[1]	50
3	Calculate $R_l'(m)$; m=5,6,7,8	144	8[2]
4	Set up for Durbin's recursion $[E^0=R_0'(0)]$	50	32
5	Calculate $1/E_{i-1}$ and Durbin's recursion (i=1)	128	226
6	Calculate $1/E_{i-1}$ and Durbin's recursion (i=2)	128	6[2]
7	Calculate $1/E_{i-1}$ and Durbin's recursion (i=3)	128	6
8	Calculate $1/E_{i-1}$ and Durbin's recursion (i=4)	128	6
9	Calculate $1/E_{i-1}$ and Durbin's recursion (i=5)	128	6
10	Calculate $1/E_{i-1}$ and Durbin's recursion (i=6)	128	6
11	Calculate $1/E_{i-1}$ and Durbin's recursion (i=7)	128	6
12	Calculate $1/E_{i-1}$ and Durbin's recursion (i=8)	128	6
13	Calculate $1/E$	128	6
14	Extract $\alpha(m)$, $m = 1,2,\ldots,8$	12	41
15 through 24	Idle	12 each	26
	Total (% used of available)	1777 (12%)	688[3] (67%)

[1] for signal 51 dB down from peak; shorter execution time for stronger signals

[2] locations include only the subroutine call; subroutine previously counted

[3] total includes 17 locations of the power-up initialization routines not included above.

The computation of the Hamming window presents a challenge. The window overlap causes each sample to lie in successive thirds of three overlapping analysis windows. After each one-sample frame overlap period, the relationships of the portion of successive windows within which each sample lies rotate cyclically. Producing the cosine-based window values and rearranging the window segments was performed by computing a Taylor series expansion of each 100-sample segment of the window about its respective midpoint, at sample 50, 150, or 250. Taylor series expansion was used because program memory was too scarce to devote 300 (or even 150) locations to window lookup values.

No instruction was available to perform the necessary divisions required by (6.10), so the reciprocal of the divisor was computed and used to multiply the dividend. An efficient reciprocal routine was used that first normalized the divisor to a value between 1 and 2 and kept track of the number and direction of shifts required

Table 6–7 Functions of four-sample autocorrelation update program. © 1985 *The AT&T Technical Journal.*

Label	Function	Execution Time (μsec)	Program Locations
Read #1	Read and convert sample	3	11
Output	Output one frame feature coefficient	6	29
Read #2	Read and convert sample	3	
Move and pre-emphasize	Shift sample buffer by four samples and preemphasize four samples	60	54
Window	Calculate window values and apply three times to four samples	111	123
Read #3	Read and convert sample	3	
Autocorrelation	Use four samples to update nine autocorrelation vectors for three overlapped functions	193	118
Read #4	Read and convert sample	3	
	TOTAL (% used of available)	382 (64%)	335 (33%)

to normalize the divisor. The number of shifts thus computed were then appended on the right with the most significant two bits after the leading 1 of the original number to provide a finer resolution value to subdivide the factors of two indicated by the count of shifts. This produced a reciprocal, and division was accomplished by multiplication by this reciprocal (Section 7.1). Program memory was further conserved by sharing the divisor normalization with the computation to determine amplitude normalization and log energy at LPC (1) (6.8). The code segment has multiple exit points, with the correct one being chosen by whether a reciprocal, amplitude normalization, or log energy was being computed. Thus, the reciprocal routine, which requires 181 program locations, shares 99 of them with the gain normalization program, saving on overall program space.

6.4.3 Achieving Iteration Independence

Another means of saving space in program memory was to develop a means to compute Durbin's recursion with a common piece of DSP code for all orders, $i = 1, 2, \ldots, 8$. The use of common code for multiple iterations is standard on a general purpose processor, but is difficult in this digital signal processor, because the DSP does not have addressing capability sufficient to handle the simultaneous arrays of α, $\mathbf{k}$, $\mathbf{E}$, and $\mathbf{R}$. Each of these four arrays would require its own index address pointer which would be incremented or decremented in units of one to step through each array, yet the DSP had only two address pointers available. Normally if program memory were sufficient, the loop would be unfolded and written as sequential code, which requires 536 program locations. A common piece of code reduces program memory usage to 119 locations and kept program memory usage within what was available.

A reordering of array contents and an adjustment of (6.10), which uses a summation for **k** with the number of terms depended on the iteration number i, resulted in simple incremental addressing restricted to two address registers and a common piece of code for all iterations.[4] This is now described in detail.

From (6.10), the summation s_i in the numerator is modified as follows:

$$\begin{aligned} s_i &= R(i) - \sum_{j=1}^{i-1} \alpha_j^{(i-1)} R(i-j) \\ &= 1 \cdot R(i) + (-\alpha_1^{(i-1)})R(i-1) + (-\alpha_2^{(i-1)})R(i-2) + \ldots (-\alpha_{i-1}^{(i-1)})R(1). \end{aligned} \tag{6.16}$$

This may be prefaced with the series $0 \times R(m) + 0 \times R(m-1) + \ldots + 0 \times R(i+1)$, since the sum of this series is zero. Thus for a model of order m, sum s_i can be formed for any $i \leq m$ by reversing the order of terms in (6.16) and calculating:

$$\begin{aligned} s_i = {} & -\alpha_{i-1}^{(i-1)} \times R(1) \\ & + \; -\alpha_{i-2}^{(i-1)} \times R(2) \\ & + \;\; . \qquad\quad . \\ & \;\;\;\; . \qquad\quad . \\ & \;\;\;\; . \qquad\quad . \\ & +1 \;\; \times \;\; R(i) \\ & +0 \;\; \times \;\; R(i+1) \\ & + \;\; . \qquad\quad . \\ & \;\;\;\; . \qquad\quad . \\ & \;\;\;\; . \qquad\quad . \\ & +0 \;\; \times \;\; R(m). \end{aligned} \tag{6.17}$$

When written vertically, this summation suggests the use of two vectors of length m consisting of sequential coefficients in memory. The first vector is $[\alpha_j^{(i-1)}; j = i-1, i-2, \ldots, 1]$, appended with the vector $[1,0,0, \ldots, 0]$ to yield a total of m elements; the second is the vector $[R(j), j = 1, 2, \ldots, m]$. With this arrangement, the summation of (6.10) becomes the regular calculation of the scalar product of vectors $[\alpha_j^{(i-1)}]$ and $[R(j)]$, regardless of iteration count i.

Fig. 6–7 diagrams the progression of two source operand address pointers $s0$ and $s1$ through the two arrays, $R(j)$ and $\alpha_j^{(i-1)}$ respectively. A destination address pointer, $d0$, is also available, and pointers $s0$, $s1$, and $d0$ can be incremented or decremented in steps of one location. Fig. 6–7 shows the case of $m = 8$ and $i = 5$ and describes how each term of (6.10) is obtained as the product of the operands pointed to by $s0$ and $s1$. Pointers $s0$ and $s1$ decrement from $i - 1$ as the computation proceeds from left to right. Calculations to the right of the heavy vertical line are from dummy terms of zero added to the sum to achieve a constant number of terms for all values of i and thus obtain regularity.

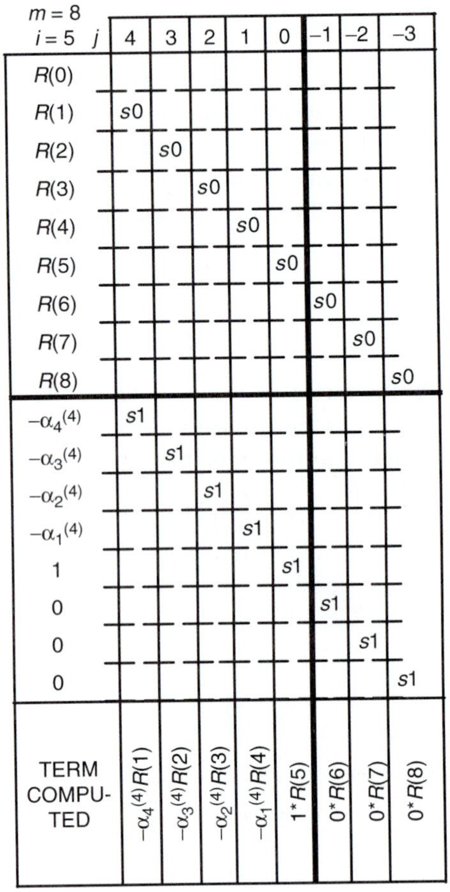

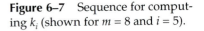

Figure 6–7 Sequence for computing k_i (shown for $m = 8$ and $i = 5$).

To cast (6.10) into the form of a scalar product of two vectors, $[\alpha_j^{(i-1)}]$ must be arranged as specified in (6.17). This means that given a vector $[\alpha_j^{(i-1)}, j = 1,2, \ldots, i{-}1]$, a vector $[\alpha_j^{(\cdot)}]$ must be computed and written in memory locations to achieve current registration with $[R(i)]$. This ordering is accomplished by computing (6.12) in a specific sequence. To demonstrate the sequence, the case of $m = 8$ and $i = 5$ is again chosen and $[\alpha_j^{(5)}]$ is computed from $[\alpha_j^{(4)}]$and k_i, proceeding in reverse order of the index j, $j = i-1, i-2, \ldots, 1$:

$$
\begin{aligned}
-\alpha_4^{(5)} &= k_5 \cdot \alpha_1^{(4)} - \alpha_4^{(4)} \\
-\alpha_3^{(5)} &= k_5 \cdot \alpha_2^{(4)} - \alpha_3^{(4)} \\
-\alpha_2^{(5)} &= k_5 \cdot \alpha_3^{(4)} - \alpha_2^{(4)} \\
-\alpha_1^{(5)} &= k_5 \cdot \alpha_4^{(4)} - \alpha_1^{(4)}.
\end{aligned}
\tag{6.18}
$$

These equations access the values of $[\alpha_j^{(i)}]$ in descending order of j on the left and access the values of $[\alpha_j^{(i-1)}]$ in ascending order of j for the first right-hand term (produced with k_5) and descending order of j on the far right. Although only

$i-1$ values of $\alpha_j^{(i)}$ are calculated, dummy calculations are appended as before to fill out the computation to m iterations and thereby attain a regular structure. This is done by augmenting the array $[\alpha_j^{(i-1)}]$ with $[0, 0, \ldots, 0]$ at the end of high j and with $[1, 0, \ldots, 0]$ at the end of low j. If the computation of (6.12) is started with one source address pointer set at $\alpha_{i-1}^{(i-1)}$ at the top of the array and the other address pointer set at $\alpha_1^{(i-1)}$ at the bottom of the array, then the computation may be conducted as iterations through (6.12) by incrementing the first pointer and decrementing the second. The offset between the starting address of the two pointers is $i-2$ and is the only portion of the recursion that changes with iteration index i (for $i = 1$, the pointers actually switch position, since $i - 2 = -1$).

In Fig. 6–8, the iterations through (6.12) are shown. Prior to computing (6.12), the value of $-k_i = \alpha_i^{(i)}$computed in Fig. 6–7 is written at the top of memory for the updated α array where it is pointed to by a constant address pointer c. The iteration then begins with $s1$ pointing to the first non-zero element, $-\alpha_4^{(4)}$. Pointer $s0$ is directed $i-2$ locations into the array, which for this case of $i = 5$, is three locations beyond $s1$. The sequential movements of $s0$ and $s1$ in opposite directions is seen as j steps from 4 to -2, and the updated elements of $[\alpha_j^{(i-1)}]$ are written sequentially as indicated by $d0$. It is clear from Fig. 6–8 that the new array $[\alpha_j^{(5)}]$ has been automatically appended by the sequence $[1, 0, \ldots, 0]$ as required to meet the alignment conditions of Fig. 6–7. To prepare for the next iteration of i, the m-element array $[\alpha_j^{(5)}]$is copied to overwrite the memory locations occupied by $[\alpha_j^{(4)}]$ when the current iteration finished.

The final technique of efficient mapping, data-type optimization, has been used by restricting all computations to take place with integer representation, which accommodates the limits of this integer DSP. At times, as exemplified in the previous discussion of reciprocal computation, normalization was performed to maintain precision of computation by scaling numbers to use all available bits.

As a result of the above tradeoffs and mappings (all driven by the need to reduce program size to fit into a 1024-word memory while maintaining real-time operation), we have succeeded in allowing a modest, inexpensive programmable signal processor to perform what in the past had required multiple processors on several chips.

6.5 TWO-DIMENSIONAL FILTER MAPPED ONTO MULTIPLE PROCESSORS[5,a]

The two-dimensional filtering of data is a local operation: Each output sample is a linear combination of previous outputs and the input values in a two-dimensional neighborhood around the sample position. In this example, data flow graph generation and optimization is conducted according to the ordered graph method.

---

[a]Portions adapted, with permission, from Winser E. Alexander, Douglas S. Reeves, and Clay S. Gloster, Jr., "Parallel Image Processing with the Block Data Parallel Architecture," *Proc. IEEE*, 84 (July 1996), 947–968. © 1996 IEEE.

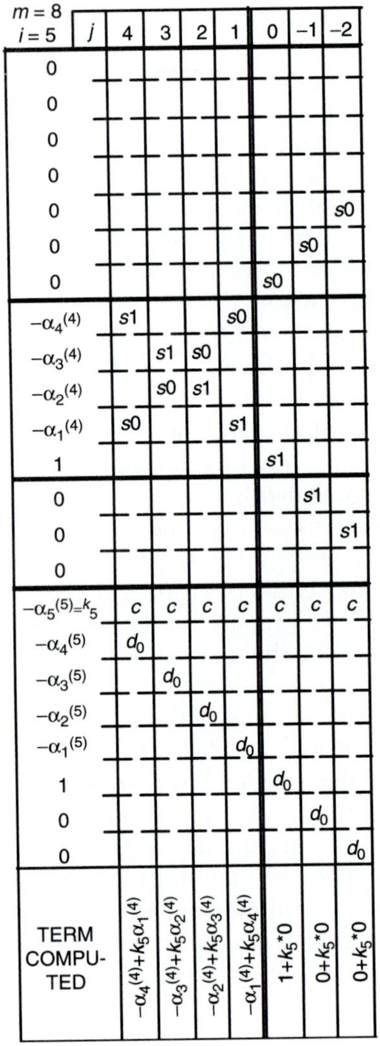

Figure 6–8 Computation sequence for $\alpha_j^{(i)}$ calculation ($m = 8, i = 5$). Iteration proceeds from left ($j = 4$) to right ($j = -2$).

The specification of the filtering algorithm is given by the following expression for a general causal two-dimensional finite difference equation:

$$g(n_1,n_2) = \sum_{h=0}^{L}\sum_{v=0}^{M} b(h,v)f(n_1 - h, n_2 - v) - \sum_{\substack{h=0 \\ h+v>0}}^{L}\sum_{v=0}^{M} a(h,v)g(n_1 - h, n_2 - v). \quad (6.19)$$

The architecture to which this equation is to be mapped is a *block data flow processor,* which is a generic representation of many types of multiprocessors. The block data flow processor has the following characteristics:

- input data is partitioned into large blocks;
- processing for a block is assigned to a single processing module; processing of multiple blocks occurs in parallel on multiple processing modules;
- a data flow approach is used to schedule block processing, i.e., processing is started as soon as an input block is available;
- each processing module provides its output as soon as it is ready;
- any exchange of data between processing modules occurs asynchronously, using point-to-point interconnection network;
- the processor connection topology is characterized as follows:
  - interprocessor communication is one-way, processor-to-processor;
  - two input buses and two output buses are provided to increase bandwidth;
  - each processor is connected directly to an input and an output bus, as well as to its upstream and downstream neighbor;
- instruction processing time consists of one cycle for an add, and two cycles for a multiply.

Mapping this algorithm to the block data flow processor consists of considering the two-dimensional input array as corresponding to an image, and partitioning it as follows:

- for two-dimensional filtering, consider the indices of the rows and columns of the matrix as an ordered pair, so that indexing along the columns of the matrix corresponds to the horizontal element of the pair, and indexing along the rows corresponds to the vertical tuple;
- schedule all computations of a single row of the image as one block (i.e., scheduled on one processor);
- schedule each row on a different processor;
- state variables that are delayed by one horizontal delay are used by the same processing module;
- state variables that are delayed by one row are transformed to the processor handling that row.

This structure supports the real-time computation on the two-dimensional matrix that results from a raster-scanned image, where the image is developed row-by-row.

Algorithm simplification results from casting the computation into a *state space* form, i.e., a form in which the outputs are expressed as a function of the inputs (as opposed to a function of both inputs and previous values of the outputs). This conversion will be described as applied both to equations and to the data flow diagram.

6.5.1 State Space Representation

The state space representation provides a form that reduces the extent of the range of dependence of output samples upon input samples, thereby minimizing the data communication requirements among processors. The state space representation converts the difference equation to a matrix vector model, where the output is represented as a linear combination of the current input and the most recently computed states. A second matrix equation computes the new states as a linear combination of the current states and current inputs.

Horizontal and vertical state variables are introduced, $Q_H(n_1, n_2)$ and $Q_V(n_1, n_2)$. The output $g(n_1, n_2)$ is expressed as a linear combination of previous horizontal state $Q_H(n_1 - 1, n_2)$, previous vertical state $Q_V(n_1, n_2 - 1)$, and the current input $f(n_1, n_2)$:

$$g(n_1,n_2) = [\mathbf{C}_H,\mathbf{C}_V]\begin{bmatrix} Q_H(n_1 - 1,n_2) \\ Q_V(n_1,n_2 - 1) \end{bmatrix} + \mathbf{D}[f(n_1,n_2)]. \tag{6.20}$$

Here, $\mathbf{C}_H$ and $\mathbf{C}_V$ are the matrix multipliers of the horizontal and vertical states, based on the definition of the states and the requirements of (6.19), and $\mathbf{D}$ is the similarly-defined matrix devoted to sample inputs, $f(n_1, n_2)$. In turn, the current state variables $Q_H(n_1, n_2)$ and $Q_V(n_1, n_2)$ are each expressed as linear combinations of current input $f(n_1, n_2)$ and the previous horizontal and vertical states, $Q_H(n_1 - 1, n_2)$ and $Q_V(n_1, n_2 - 1)$:

$$Q_H(n_1,n_2) = [\mathbf{A}_{HH},\mathbf{A}_{HV}]\begin{bmatrix} Q_H(n_1 - 1,n_2) \\ Q_V(n_1,n_2 - 1) \end{bmatrix} + \mathbf{B}_H[f(n_1,n_2)]; \tag{6.21}$$

$$Q_V(n_1,n_2) = [\mathbf{A}_{VH},\mathbf{A}_{VV}]\begin{bmatrix} Q_H(n_1 - 1,n_2) \\ Q_V(n_1,n_2 - 1) \end{bmatrix} + \mathbf{B}_V[f(n_1,n_2)]. \tag{6.22}$$

A data flow graph representation of (6.21) and (6.22) shows the two-dimensional nature of the operation (Fig. 6–9); the basic kernel that is repeated in the overall graph is also expanded (Fig. 6–10), as is the right-hand end (Fig. 6–11). The flow graphs use z_H^{-1} to represent a one-sample delay of the horizontal variable n_1, and z_V^{-1} to represent a one-sample delay in the vertical variable n_2. To achieve the state space representation of Fig. 6–9, we assign a horizontal state variable $q_{H,vL+h}(n_1, n_2)$ to the input of each horizontal delay block, where h and v are defined from the position of the delay block in the array as shown by the markers at the top and left of Fig. 6–9. Similarly, we define a vertical state variable $q_{V,v}(n_1, n_2)$ to each vertical delay block. The horizontal state variables are indexed in a raster scan format by sequential progression of the index variable $I = vL + h$:

$$Q_H(n_1,n_2) \rightarrow q_{H,vL+h}(n_1,n_2) = q_{H,I}(n_1,n_2) \tag{6.23}$$

$$\begin{gathered} Q_V(n_1,n_2) \rightarrow q_{V,v}(n_1,n_2); \\ \text{where } 0 \leqslant h \leqslant L; \\ 0 \leqslant v \leqslant M; \\ I = vL + h. \end{gathered} \tag{6.24}$$

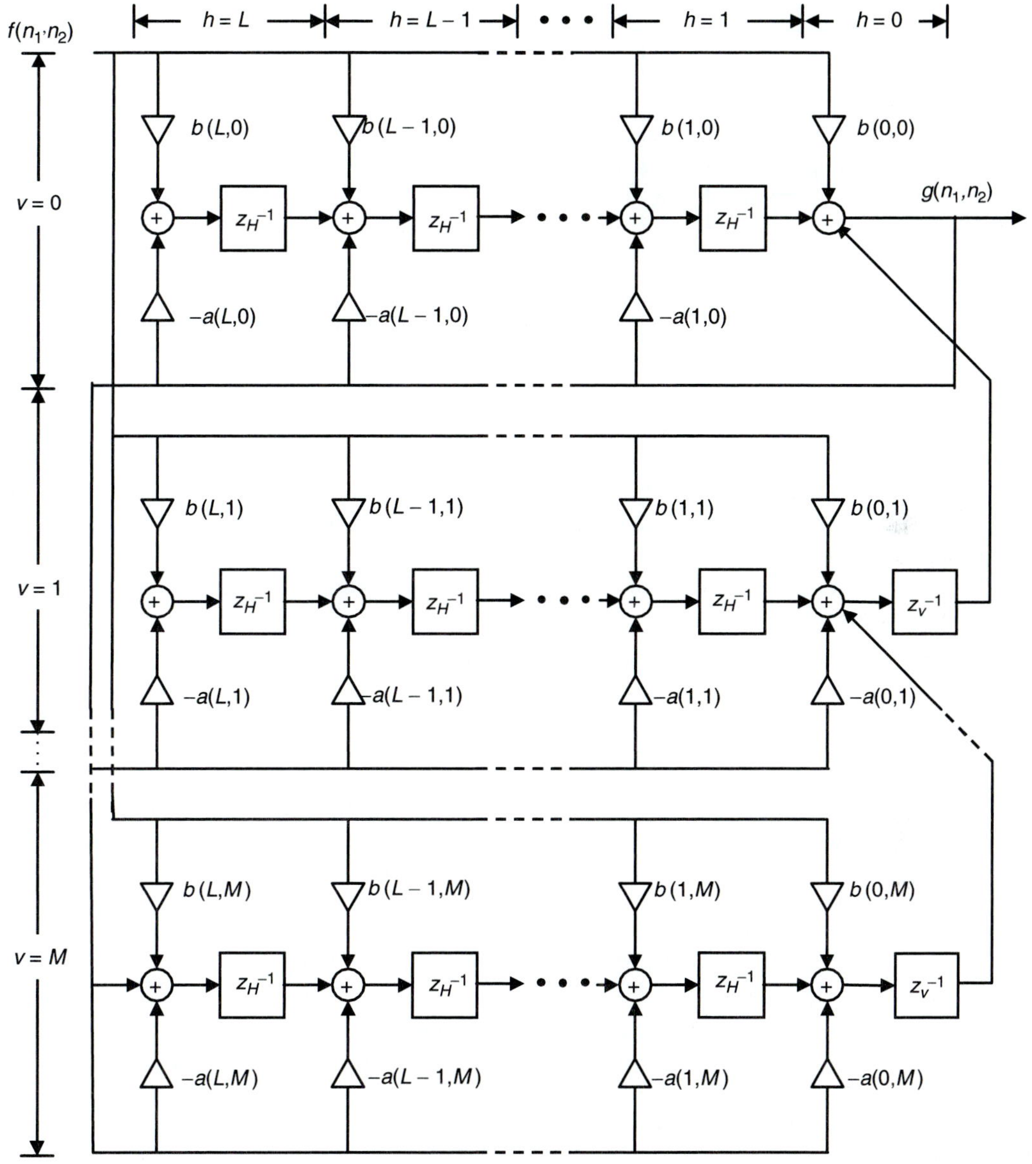

Figure 6–9 Data flow graph representation of two-dimensional filter system. Associated with each horizontal delay z_H^{-1} is a state variable $q_{H,I}(n_1, n_2)$ addressed in raster scan fashion by $I = vL + h$, and with each vertical delay z_V^{-1} is a state variable $q_{V,v}(n_1, n_2)$.[5] © 1996 IEEE, adapted with permission.

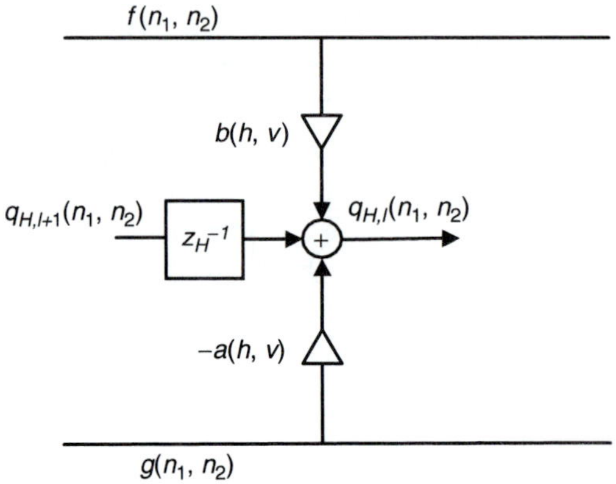

Figure 6–10 Expansion of repeated section from two-dimensional filter of Fig. 6–9.[5]

A specific value of the position index $vL + h$ is represented by I in the figures and discussion below. The repeated section that occurs throughout Fig. 6–9 is expanded in Fig. 6–10.

From the diagram of Fig. 6–10 we obtain the equation for the horizontal state variable:

$$\begin{aligned} q_{H,I}(n_1,n_2) &= b(h,v)f(n_1,n_2) - a(h,v)g(n_1,n_2) + q_{H,I+1}(n_1 - 1,n_2); \\ &\text{where } 1 \leqslant h \leqslant L - 1; \\ &0 \leqslant v \leqslant M; \\ &I = vL + h. \end{aligned} \tag{6.25}$$

Inspection of the left-hand side of Fig. 6–9 shows that the structure for computing the left-most states $q_{H,I}(n_1, n_2)$ is nearly the same as the structure for the middle states, except for the absence of the upstream state results $q_{H,I+1}(n_1, n_2)$ entering from the left:

$$\begin{aligned} q_{H,I}(n_1,n_2) &= b(L,v)f(n_1,n_2) - a(L,v)g(n_1,n_2); \\ &\text{where } 0 \leqslant v \leqslant M; \end{aligned} \tag{6.26}$$

We expand the right-hand states of Fig. 6–9 into Fig. 6–11.

From Fig. 6–11, we obtain the equation for the vertical state variable:

$$\begin{aligned} q_{V,v}(n_1,n_2) = b(0,v)f(n_1,n_2) - a(0,v)g(n_1,n_2) + q_{H,L+1}(n_1 - 1,n_2) + q_{V,v+1}(n_1,n_2 - 1); \\ \text{where } 0 \leqslant v \leqslant M. \end{aligned} \tag{6.27}$$

Finally, we obtain the output equation, as well as the final boundary condition for $q_{V,M+1}$ from Fig. 6–9:

$$g(n_1,n_2) = b(0,0)f(n_1,n_2) + q_{H,1}(n_1 - 1,n_2) + q_{V,1}(n_1,n_2 - 1). \tag{6.28}$$

$$q_{V,M+1}(n_1,n_2 - 1) = 0. \tag{6.29}$$

These equations are still defined in terms of the outputs, $g(n_1, n_2)$. To put these equations into state space form, (6.28) is substituted for $g(n_1, n_2)$ in (6.25),

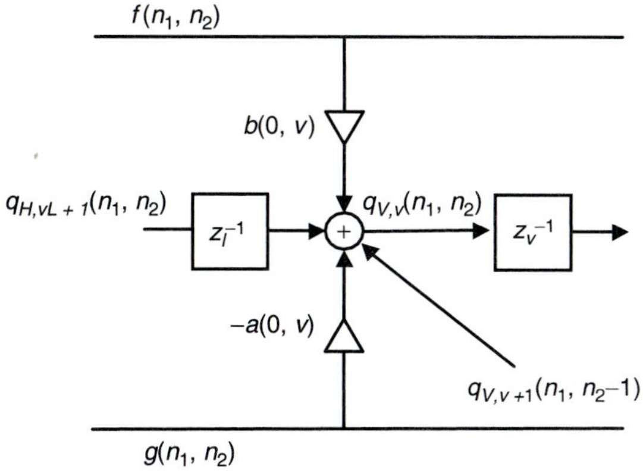

Figure 6–11 Expansion of right-hand end section ($h = 0$) from two-dimensional filter of Fig. 6–9.

(6.26), and (6.27), resulting in expressions that are restricted to terms of the current input and most recent values of state variables:

$$\begin{aligned} q_{H,I}(n_1,n_2) &= [b(h,v) - a(h,v)b(0,0)]f(n_1,n_2) \\ &\quad - a(h,v)[q_{H,1}(n_1-1,n_2) + q_{V,1}(n_1,n_2-1)] \\ &\quad + q_{H,I+1}(n_1-1,n_2); \\ &\text{where } 1 \leqslant h \leqslant L-1, \\ &\quad 0 \leqslant v \leqslant M, \\ &\quad I = vL + h; \end{aligned} \tag{6.30}$$

$$\begin{aligned} q_{H,I}(n_1,n_2) &= [b(h,v) - a(h,v)b(0,0)]f(n_1,n_2) \\ &\quad - a(h,v)[q_{H,1}(n_1-1,n_2) + q_{V,1}(n_1,n_2-1); \\ &\quad h = L; \\ &\quad 0 \leqslant v \leqslant M, \\ &\quad I = vL + h; \end{aligned} \tag{6.31}$$

$$\begin{aligned} q_{V,v}(n_1,n_2) &= [b(0,v) - a(0,v)b(0,0)]f(n_1,n_2) \\ &\quad - a(0,v)[q_{H,1}(n_1-1,n_2) + q_{V,1}(n_1,n_2-1)] \\ &\quad + q_{H,L+1}(n_1-1,n_2) + q_{V,v+1}(n_1,n_2-1); \\ &\quad 1 \leqslant v \leqslant M. \end{aligned} \tag{6.32}$$

$$q_{V,(M+1)(L+1)}(n_1,n_2-1) = 0. \tag{6.33}$$

The particular case of a second-order filter shows the above equations specialized to second order, with all equations expanded. Fig. 6–12 shows the corresponding data flow graph for the equations of the second-order section of Table 6–8. The diagram groups together all terms of Table 6–8 that are multiplied by $a\,(h, v)$.

To simplify the DFG, the delays (e.g., from $q_{H,2}(n_1, n_2)$ at the right of the DFG to $q_{H,2}(n_1 - 1, n_2)$ at the left) are not shown. More generally, each output $q_{H,j}(n_1, n_2)$ on the right is connected via a delay to $q_{H,j}(n_1 - 1, n_2)$ at the left of the diagram and $q_{V,j}(n_1, n_2)$ is connected to $q_{V,j}(n_1, n_2 - 2)$ at the bottom of the diagram.

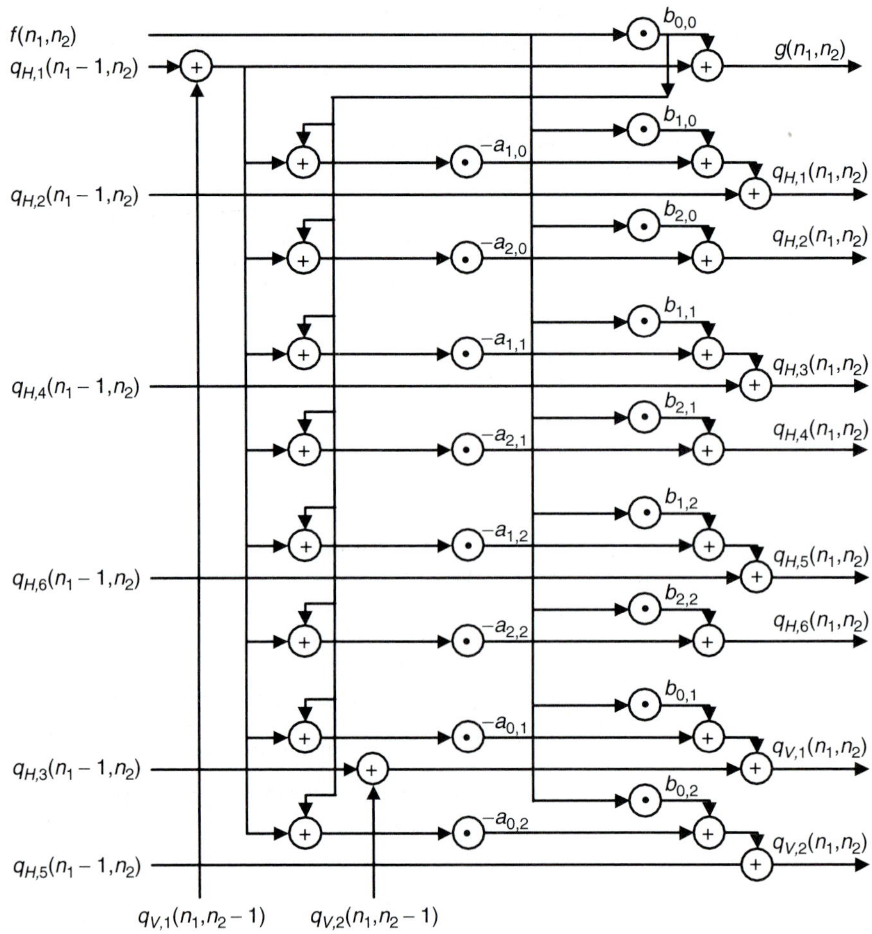

Fig. 6–12 Data flow graph for equations of Table 6–8, representing a second-order two-dimensional filter.[5] © 1996 IEEE, adapted with permission.

6.5.2 Ordered Graph Method

In improving a parallel execution schedule, the ordered-graph method uses algorithm partitioning and assumes closely coupled processors. Input to the ordered-graph method is a fully specified data flow graph, as well as the sample period that determines the delay between successive samples T_D (alternatively, the sampling rate Q may be given; $Q = 1/T_D$). An initial sanity check is performed to compare the longest-executing path time between two delay nodes to T_D. If it is greater than T_D, then the graph must be restructured until it is less than T_D. Re-

Table 6–8 Equations (6.30) through (6.33) expanded for second-order section ($L = M = 2$).

h	v	l	equation
1	0	1	$q_{H,1}(n_1,n_2)=[b(1,0)-a(1,0)b(0,0)]f(n_1,n_2)-a(1,0)[q_{H,1}(n_1-1,n_2)+q_{V,1}(n_1,n_2-1)]+q_{H,2}(n_1-1,n_2)$
1	1	4	$q_{H,4}(n_1,n_2)=[b(1,1)-a(1,1)b(0,0)]f(n_1,n_2)-a(1,1)[q_{H,1}(n_1-1,n_2)+q_{V,1}(n_1,n_2-1)]+q_{H,5}(n_1-1,n_2)$
1	2	7	$q_{H,7}(n_1,n_2)=[b(1,2)-a(1,2)b(0,0)]f(n_1,n_2)-a(1,2)[q_{H,1}(n_1-1,n_2)+q_{V,1}(n_1,n_2-1)]+q_{H,8}(n_1-1,n_2)$
2	0	2	$q_{H,2}(n_1,n_2)=[b(2,0)-a(2,0)b(0,0)]f(n_1,n_2)-a(2,0)[q_{H,1}(n_1-1,n_2)+q_{V,1}(n_1,n_2-1)]$
2	1	5	$q_{H,5}(n_1,n_2)=[b(2,1)-a(2,1)b(0,0)]f(n_1,n_2)-a(2,1)[q_{H,1}(n_1-1,n_2)+q_{V,1}(n_1,n_2-1)]$
2	2	8	$q_{H,8}(n_1,n_2)=[b(2,2)-a(2,2)b(0,0)]f(n_1,n_2)-a(2,2)[q_{H,1}(n_1-1,n_2)+q_{V,1}(n_1,n_2-1)]$
0	1	3	$q_{V,1}(n_1,n_2)=[b(0,1)-a(0,1)b(0,0)]f(n_1,n_2)-a(0,1)[q_{H,1}(n_1-1,n_2)+q_{V,1}(n_1,n_2-1)]+q_{H,3}(n_1-1,n_2)$ $+q_{V,2}(n_1,n_2-1)$
0	2	6	$q_{V,2}(n_1,n_2)=[b(0,2)-a(0,2)b(0,0)]f(n_1,n_2)-a(0,2)[q_{H,1}(n_1-1,n_2)+q_{V,1}(n_1,n_2-1)]+q_{H,5}(n_1-1,n_2)$ $+q_{V,3}(n_1,n_2-1)$
			$q_{V,3}(n_1,n_2-1)=0$

structuring may include both node reordering and the decomposition of the node into smaller, more rapid subnodes that can be moved around. Fig. 6–13 outlines the steps of the ordered graph method, which is described below.

1. *Preprocess DFG:* As was described in Section 5.3.3, a directed acyclic graph is developed that is used to optimize *intra-iteration* timing. The preprocessing steps include:
   - identify all strongly-connected components (i.e., those that contain either a single node or all nodes involved in the feedback loop); replace each

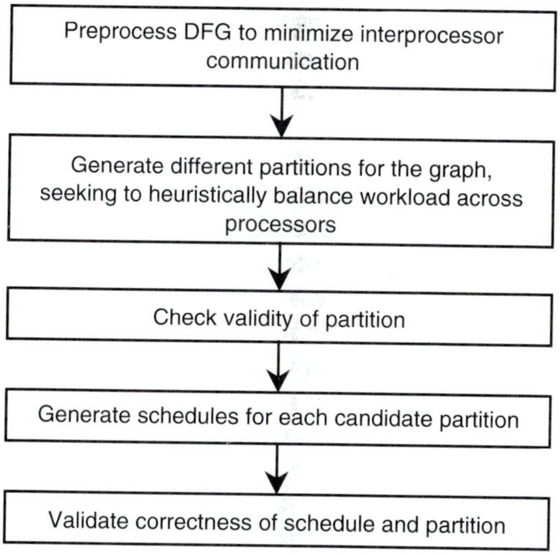

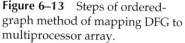

Figure 6–13 Steps of ordered-graph method of mapping DFG to multiprocessor array.

with a single node, for which the incoming edge is the union of all incoming edges for the strongly connected node, and the outgoing edge is the union of all outgoing edges of the strongly connected component;
- set computational delay to be equal to the sum of all nodes' computation delays in the strongly connected components;
- use this resulting directed acyclic graph (DAG) for partitioning (i.e., generating candidate partitions), but use the original DFG for scheduling and timing;
- check that no strongly connected component execution time exceeds T_D—if it does, convert it back to original form, remove a feedback edge, and produce a new DAG;
- obtain the DAG, which shows the precedence of operations, as the output of the above steps.

2. *Generate various candidate partitions;*
   - generate *linear extensions* of nodes in a DAG that preserve precedence (linear extensions are all possible permutations of a DAG except those that violate the arc arrows. For example, the DAG of Fig. 6–14 which has nodes $V = \{1, 2, 3, 4, 5\}$ and edges $1 \rightarrow 3$, $2 \rightarrow 3$, $2 \rightarrow 4$, $3 \rightarrow 5$, and $4 \rightarrow 5$, are [1, 2, 3, 4, 5], [2, 1, 3, 4, 5], [1, 2, 4, 3, 5], [2, 1, 4, 3, 5], and [2, 4, 1, 3, 5]);
   - process each extension with a greedy algorithm, moving from left to right. Partition p starts with the node in the ith position of the linear extension and assigns it a weight equal to the computational time of all nodes it includes;
   - partition p includes all nodes from the ith to the jth position such that the weight (sum of execution times) is less than T_D and the weight of $p + d_K$, where d_K is the delay of node K, exceeds t_D. For example, for linear extension [1, 2, 3, 4, 5], if each node has a delay of 1 and T_D is 2, the greedy partition provides [1, 2] [3, 4], [5].
   - continue for all linear extensions seeking to minimize the number of partitions and find partitions whose execution times are approximately equal;
   - repeat for all DAGs.
3. *Check validity of partition:*
   - model communication time of the partition, using model for number of words sent times the transmission time per word, minus the portion of the execution time that is overlapped by communication time;

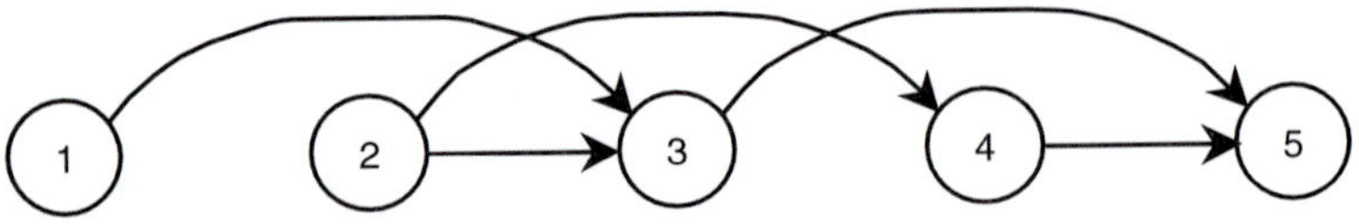

Figure 6–14 DAG for example of linear extension.

- check the network topology to assure that there is exactly one path between each source and destination processor.

4. *Scheduling:*
   - store the selected linear extension;
   - check that T_D is not exceeded by communication time plus execution time minus overlap;
   - if execution time is greater than T_D, then try a new network partition;
   - to further improve scheduling, one could try all network partitions and choose the one for which the execution time is a minimum, not simply

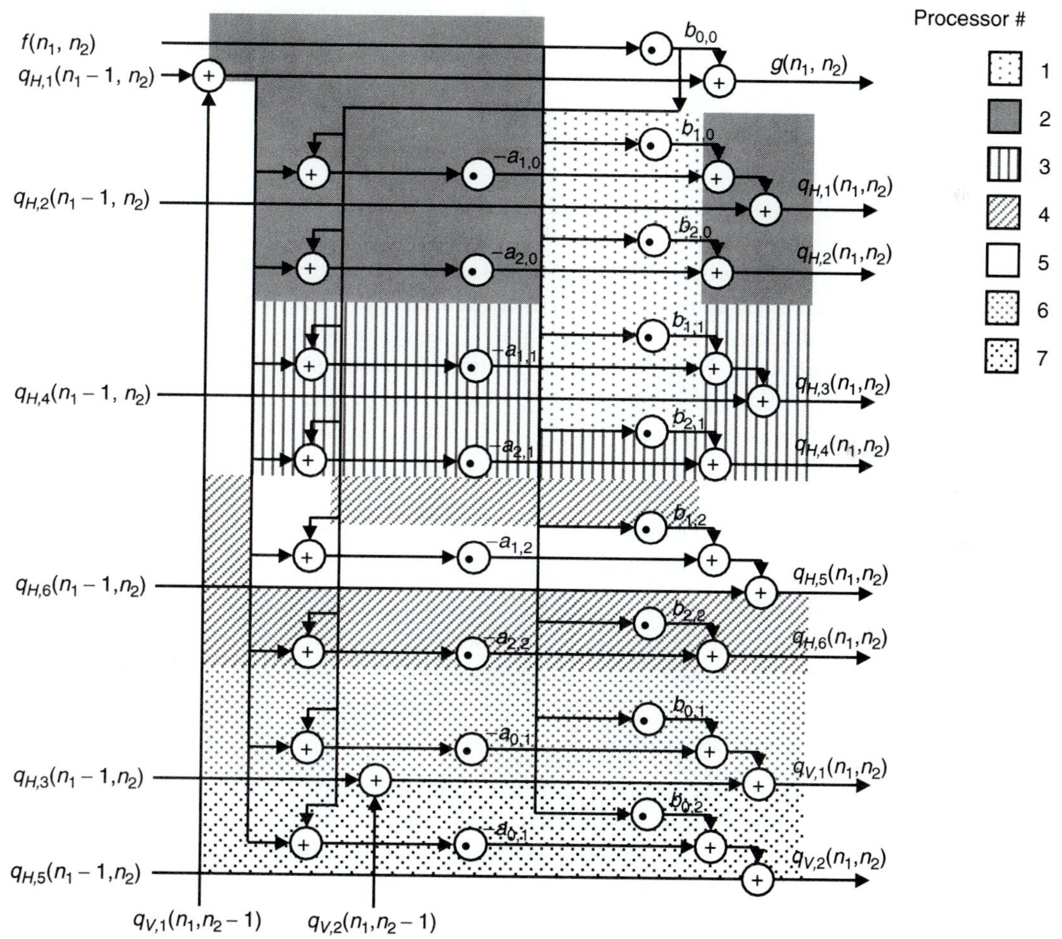

Figure 6–15 Data flow graph of Fig. 6–12, shaded to indicate assignment of processing to each of seven processors.[5] © 1996 IEEE, adapted with permission.

less than T_D (this step is optional and is only needed if the spare processor time accumulated would be used for another task).

5. *Final validation:*
   - validate correctness of schedule;
   - determine the impact of interprocessor communication;
   - compare resulting sampling rate and number of processors to the desired sampling rate and number of processors.

Fig. 6–15 shows the two-dimensional second-order infinite impulse response filter partitioned by a program that implements the above steps of the ordered-graph method, using seven block data flow processors assigned as indicated by the shading. Each processor has a multiply time of two cycles and an add time of one cycle. The array requires eight cycles per pixel, and the interconnect requirements are consistent with those of the block data flow processor array.

6.6 SYNTHETIC APERTURE RADAR IMAGE FORMATION[6]

Synthetic aperture radar (SAR) produces a high-resolution image from the analysis of the radar returns of terrain illuminated by a sequence of chirped radar pulses issued from a moving airplane. This discussion describes how to map these computations, which require anywhere from several hundred to a few thousand operations per image pixel, to an array of digital signal processors that are connected in a fat-tree topology. The real time implementation process consists of five steps:

1. Application description;
2. Static description of processing;
3. Dynamic description;
4. Determination of computational resources needed; and
5. Mapping the algorithm to the architecture.

6.6.1 Application Description

The basic concept of synthetic aperture radar is to receive both the amplitude and frequency of pulsed signals returned from objects throughout the time that they are in the beam of the moving antenna. For traditional radar, the angular beamwidth θ_B in radians is proportional to the ratio of the radar wavelength λ to the dimension of the radar antenna, L:

$$\theta_B = \frac{\lambda}{L}. \tag{6.34}$$

If that antenna is a distance R from the ground, the size of the beam footprint is approximately

$$x_B = \theta_B \cdot R, \tag{6.35}$$

where for small θ_B, the approximation $\sin \theta_B \approx \theta_B$ is made. Fig. 6–16 shows the geometry of the arrangement.

For a range of $R = 100{,}000$ m, to obtain a resolution of 1 m, the beam angle θ_B must be 10^{-5}, which for a typical wavelength $\lambda = 0.03$ m would require an antenna size L of $L = 0.03/10^{-5} = 300$ m, or larger than a football field. Such an antenna size is not practical on an airborne platform.

To increase the effective size of a smaller antenna to the size needed for good resolution, a *synthetic aperture* is created as follows. A sequence of radar pulses is issued over a time interval T, during which the antenna moves with the airplane at a velocity V to sweep out a distance $L_{eff} = VT$. L_{eff}, which is known as the synthetic aperture, can be much greater than L, since L_{eff} is determined by the motion of the plane. The result is a narrowed beamwidth of angular dimension $\delta\theta$ radians, given by

$$\delta\theta = \frac{\lambda}{2L_{eff}}. \tag{6.36}$$

Because the same spot of ground must remain in the antenna beam during the aperture time T, the effective aperture time is a function of the beam width θ_B, the range R, and the velocity of the moving antenna, V:

$$T = \frac{R\theta_B}{V} = \frac{L_{eff}}{V}; \tag{6.37}$$

$$L_{eff} = R \sin \theta_B \approx R\theta_B = \frac{R\lambda}{L}. \tag{6.38}$$

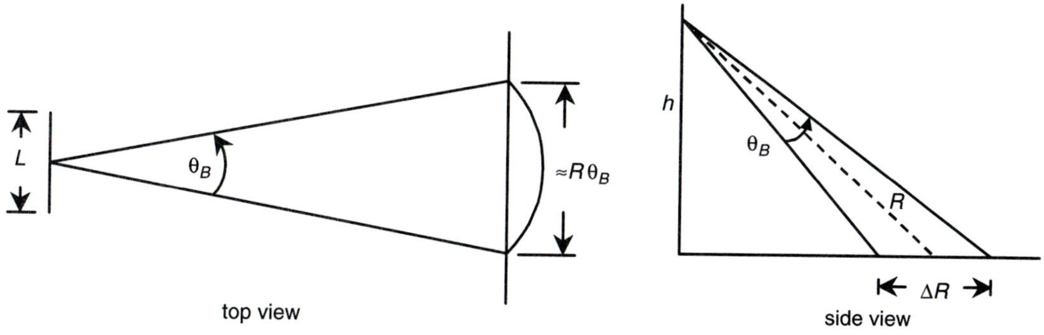

Figure 6–16 Geometry of beam of angle θ_B radians from an antenna of dimension L located at height h with average distance R from ground.

Fig. 6–17 shows how θ_B determines the maximum time T that a point A can be within the beam of an antenna moving at velocity V.

The resolution of the synthetic aperture in the direction of platform movement is given by δx, where

$$\begin{aligned}\delta x &= \delta\theta \cdot R \\ &= \frac{\lambda R}{2L_{eff}} \\ &= \frac{\lambda R}{2} \cdot \frac{L}{\lambda R} \\ &= \frac{L}{2}.\end{aligned} \tag{6.39}$$

Eq. (6.39) states that the azimuth resolution is independent of range. This is true because a target that is farther away remains in the beam longer, so the resulting increased angular resolution from its greater value of L_{eff} exactly compensates for the increased distance R. In fact, the azimuth resolution depends only upon the actual size of the antenna, which is a surprising result. The resolution in the direction perpendicular to flight path, known as range resolution, is equal to the range distance between two reflectors such that the time difference between the arrival of each of their reflected pulses is equal to the pulse width, τ. This distance is $c\tau/2$, where c is the speed of light. The pulse width τ is usually set so that range resolution and azimuth resolution are equal.

As the aircraft flies by point target A, the aircraft motion induces a Doppler frequency shift W in the return from the target that varies approximately linearly with time over the duration of the synthetic aperture time, T_x:

$$W = \frac{2V \sin \theta_B}{\lambda} \approx \frac{2V\theta_B}{\lambda} = \frac{2V^2T}{R\lambda}. \tag{6.40}$$

The time-bandwidth product is defined as the product of this frequency shift times the aperture time, T:

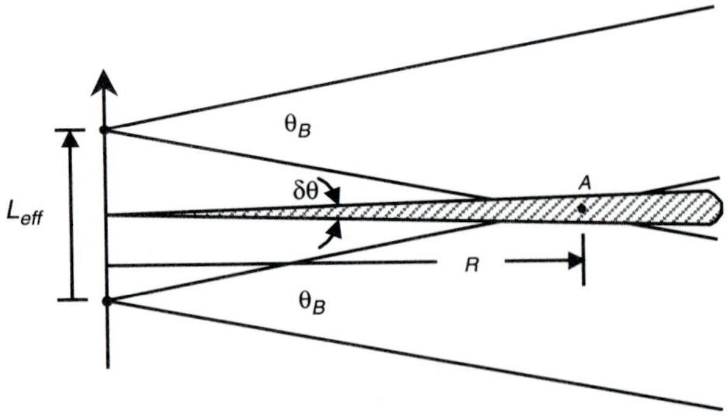

Figure 6–17 Point A at distance R remains in beam of width θ_B for the interval that it takes the platform, which is moving at velocity V, to move the distance L_{eff} between placing the beam leading edge and the trailing edge on the point. A synthetic beam narrowed to width $\delta\theta$ is the result.

$$\begin{aligned} TW &= \frac{2V^2T}{R\lambda} \cdot \frac{R\theta_B}{V} \\ &= \frac{2V^2}{R\lambda} \cdot \left(\frac{R\theta_B}{V}\right) \cdot \left(\frac{R\theta_B}{V}\right) \\ &= \frac{2R\theta_B^2}{\lambda}. \end{aligned} \tag{6.41}$$

This can be rewritten by substituting $\theta_B = \lambda/L$ and $L/2 = \delta x$ to obtain:

$$\begin{aligned} TW &= \frac{2R}{\lambda L}\left(\frac{\lambda}{L}\right)^2 \\ &= \frac{2R\lambda}{L^2} \\ &= \frac{2R\lambda}{(2\delta x)^2} \\ &= \frac{R\lambda}{2(\delta x)^2}. \end{aligned} \tag{6.42}$$

The time-bandwidth product TW can also be shown to be equal to the beam azimuth compression ratio between the real beam and the beam as narrowed by the synthetic aperture, $\theta_B/\delta\theta$.

To prevent aliasing, the pulse repetition frequency, *prf*, must equal or exceed the Doppler frequency shift W (in subsequent discussions, we will use the case of equivalence, $W = prf$):

$$\begin{aligned} prf \geqslant W &= \frac{2V^2T}{R\lambda} \\ &= \frac{2V^2}{R\lambda} \cdot \left(\frac{R\theta_B}{V}\right) \\ &= \frac{2V\theta_B}{\lambda}. \end{aligned} \tag{6.43}$$

The total number of pulses K is the product of *prf* and the aperture time, which equals TW:

$$K = T \cdot prf. \tag{6.44}$$

$$= \frac{R\lambda}{2(\delta x)^2}.$$

The return pulse is digitized after some analog processing. To preserve the resolution of δx on a pulse returned from a swath width of ΔR, the number of digital samples of each pulse, N_r, is given by the swath width, ΔR, divided by the resolution, δx:

$$N_r = \frac{\Delta R}{\delta x}. \tag{6.45}$$

The original pulse occupies a total of M_r samples; M_r, like K, is set equal to the product of the pulse width and its bandwidth:

$$M_r = \frac{c\tau}{2\delta x}. \tag{6.46}$$

The pulse width τ must be less than the time between pulses set by *prf*—it is kept as short as possible to increase bandwidth and it is typically 2μsec – 100 μsec, and $M_r = 1000 - 2000$ samples. The return of that pulse as reflected from scatterers distributed across a variety of distances along the swath ΔR spreads the pulse to N_r samples; $M_r < N_r$. Thus, to form the image, a sequence of K pulses is generated over an aperture time T, and the pulses are reflected from assorted targets that lie over a range width of ΔR and azimuth width equal to the size of the beam footprint on the ground, $\theta_B R$. The return signal is also a sequence of pulses, modified by the spatially-dependent reflection and delays due to differences in the distance.

The pulse returns are detected by the method of cross-correlating the return signal with a delayed version of the transmitted signal. As the simplest example, Fig. 6–18 shows the correlation $y(j)$ of a pulse x of N_r samples with itself, or autocorrelation, at various delays j, $j = 0, 1, \ldots, N_r - 1$:

$$y(j) = \sum_{i=0}^{N_r - j} x(i)x(i+j); \quad j = 0, 1, \ldots, N_r - 1. \tag{6.47}$$

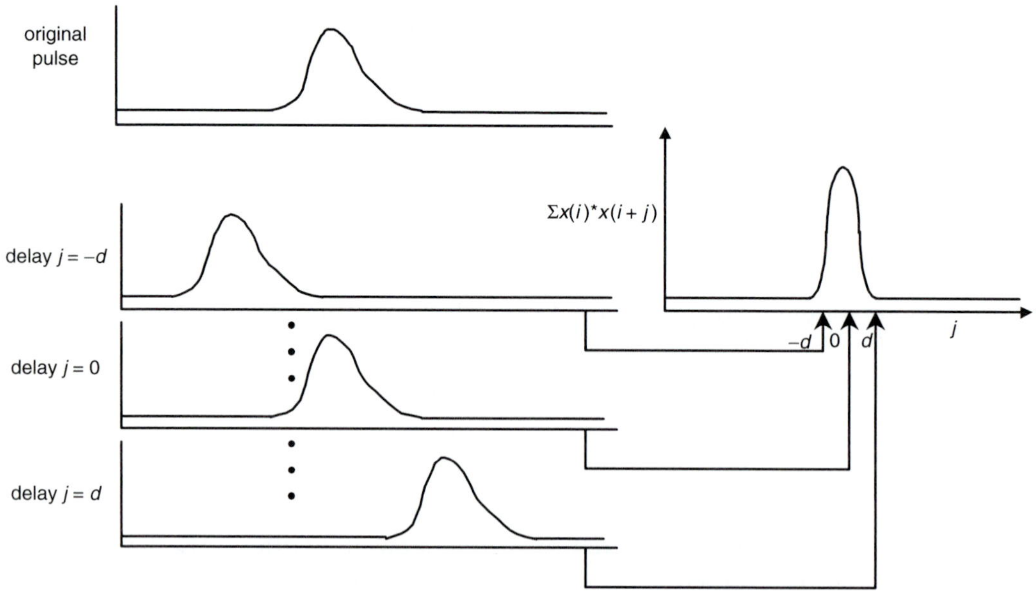

Figure 6–18 A sharp (compressed) autocorrelation pulse is obtained where the sum of products of original pulse with a delayed version of itself occurs at zero delay.

As shown in Fig. 6–18, a long pulse, when correlated with itself, produces a sharp peak at a particular time. Correlation may be used as a method of converting a long pulse, which has its energy spread out over time, to a short pulse, with most of the energy concentrated in a single pulse. This effect is termed "pulse compression." The length of the correlation interval is twice that of the original, because the correlation pulse begins when the original pulse is end-to-end with its delayed version at one end, and continues until the delayed version is at the other end of the original.

In the case of pulse detection of radar returns, a copy of the transmitted pulse is delayed and then correlated with the returned pulse. The copy of the transmitted pulse used as the range correlation kernel contains M_r samples; the returned pulse has N_r samples; the correlated signal has $M_r + N_r$ samples.

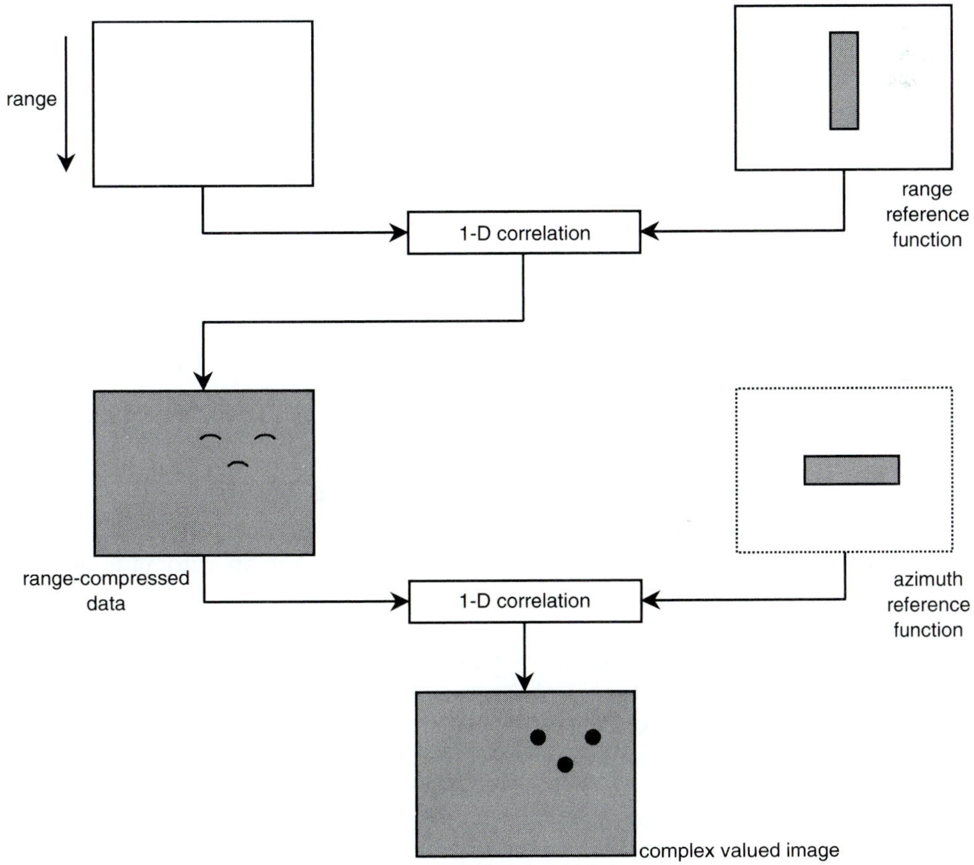

Figure 6–19 Schematic of SAR image formation process showing range compression followed by azimuth compression.

As will be shown in the upcoming description of SAR image formation, correlations in both the range and azimuth direction are performed. The range and azimuth correlations are shown in Fig. 6–19.

6.6.2 Static Description

A static description of the SAR stripmap image formation algorithm looks at the interconnection between processors, but it does not look at specifics within each processor. The steps of the image formation process are:

1. *Complex signal formation*: preserves the phase information in the signal;
2. *Range compression*: correlation processing performed for each pulse return;
3. *Corner turn*: matrix transpose that reorders data from range-sequential to pulse-sequential order;
4. *Motion compensation*: performed individually for each pulse return, for all range cells; corrects for aircraft deviation from straight-line motion over aperture time;
5. *Azimuth compression*: correlation performed individually for each range cell across all pulse returns in synthetic aperture;
6. *Image formation*: magnitude of the complex output from azimuth compression.

Several simplifying assumptions are made in this description of strip-map SAR-image formation, which include ignoring:

1. *Range walk*: change in range to target between successive pulses as a result of movement of the platform;
2. *Frequency-dependent scattering*: return from a target varies as a function of frequency, and the radar chirp contains sufficiently wide frequency variation to introduce frequency-dependent scatterer behavior;
3. *Autofocusing*: practical SAR-image formation includes a step that involves correcting for various residual phase errors through an automatic process of sharpening the image.

Also, we assume *single-look operation*, in which the SAR image is produced from a single pass over the area of the scene (in an alternative approach, the resolution can be enhanced or noise reduced by using multiple looks, with proportionately greater computation).

6.6.3 Processing Architecture

The processing element selected for this example is the Analog Devices 21060 SHARC ("Super Harvard Architecture") digital signal processor that was discussed in Section 2.3.5. It has a peak throughput of 120 MFLOPS with internal architecture that includes 512 KB of memory. Multiple SHARCs, with quantity to be determined by the upcoming analysis, are fully connected using the RACE-way® Interlink interconnection of Mercury Computer, which is capable of 160-Mbyte/sec data capacity. The arrangement of SHARCs within the fat tree offered by the RACE interconnection has been described in Section 4.2.1.2 and will be refined at the end of this analysis.

Fig. 6–20 shows the end-to-end processing chain of SAR image formation processing. The analog to digital converter (A/D) converts return pulses in bursts at a sampling rate of 50–500 MHz, placing the results into a PRF (pulse repetition frequency) buffer. The A/D sampling rate f_s is inversely proportional to the duration of a single pulse, τ, and proportional to the bandwidth of the signal being sampled, TW: $f_s = TW/\tau$. The sample buffer averages out the burst nature of samples to stretch the samples into the time interval between pulses and maintain an even rate that is equal to the average input data rate.

After conversion to floating point, a complex signal is produced from the input pulses. Fig. 6–21 shows that the input is multiplied by a reference signal in the analog domain, then filtered, oversampled, stored in a sample buffer, converted to floating point, and then filtered and decimated to produce an in-phase (I) and quadrature (Q) portion.

As shown in Fig. 6–20, two distinct compressions, or correlations, are performed. For the range correlation, the kernel is a replica of the transmitted pulse, providing a kernel of size M_r samples.

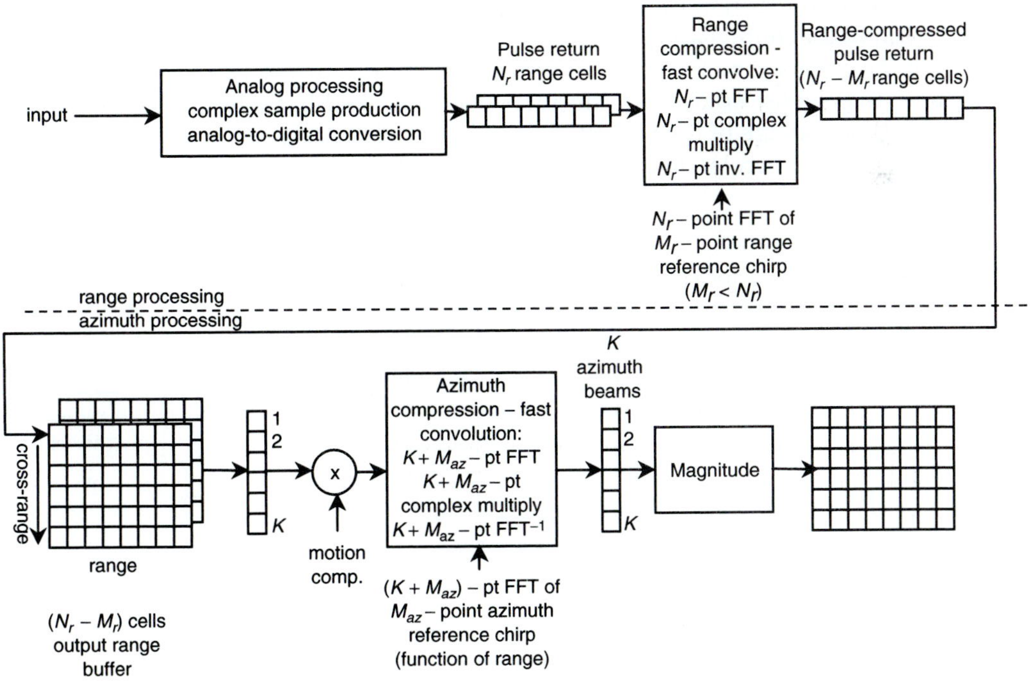

Figure 6–20 Processing chain for SAR image formation.

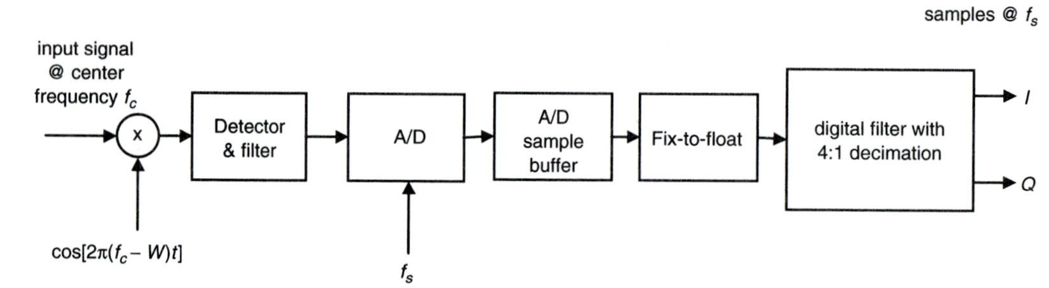

Figure 6–21 Complex samples are produced by sampling the input, centered at frequency f_c, at twice the Nyquist sampling rate $W = 2V^2T/R\lambda$.

Data is organized for SAR processing in a double buffer into which the range samples of each pulse return are placed as a row, and from which returns from all pulses at a given range cell are taken to form a column (Fig. 6–22).

Azimuth correlation occurs across all pulse returns for a given range. Its kernel size is $M_{az} = TW = R\lambda/2(\delta x)^2$, and the particular reference function used is a function of the range.

The correlation of a signal $f(i)$ with another signal $g(i)$ may also be performed as the convolution of signal $f(i)$ with signal $g(-i)$ and in turn the Fourier convolution theorem can be used to simplify the computation. The Fourier convolution theorem is applied to both the range and azimuth correlations, converting the convolutions to the sequence of fast Fourier transform, complex multiply, and inverse fast Fourier transform as was described in Section 6.3.1. This allows taking advantage of the SHARC features that were designed to speed fast Fourier transforms to accelerate convolution for large kernels.

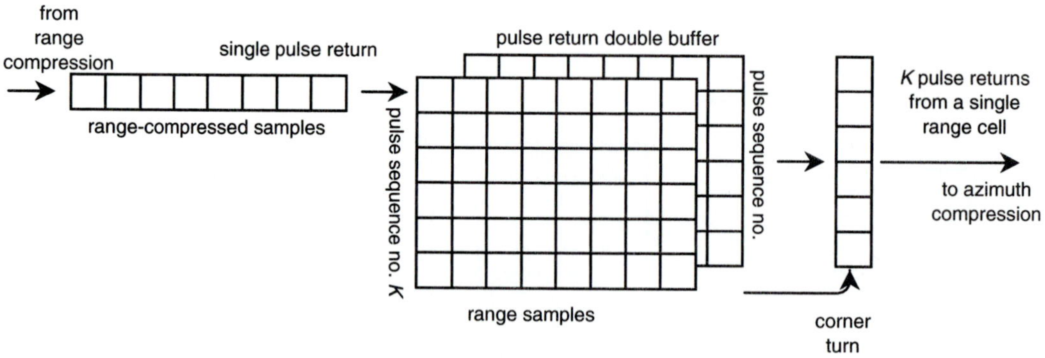

Figure 6–22 SAR pulses from the A/D converter are fed in order of range into each row of a double buffer, while samples from different pulses at constant range are pulled from each column in the other half of the buffer.

In this case, the complex signal $f(j)$, $j = 0, 1, \ldots, N_r - 1$ is convolved with the complex reference function $g^*(i)$ consisting of M_r samples, $M_r < N_r$ ($g^*(i + j)$ indicates the complex conjugate of g):

$$y(j) = \sum_{i=0}^{N-1} f(i) \cdot g^*(i + j); j = 0, 1, \ldots, N_r - M_r. \quad (6.48)$$

As written, this convolution requires $N_r (N_r - M_r + 1)$ complex adds and multiplies. Algorithm simplification via the Fourier convolution theorem provides:

$$\begin{aligned} Y(\omega) &= F(\omega) \cdot G^*(\omega), \text{ where} \\ F(\omega) &= \text{fast Fourier transform (FFT) of } f(i); \\ G(\omega) &= FFT[g(\cdot)], \text{ with } g(\cdot) \text{ padded with zeros from length } M \text{ to length } N. \end{aligned} \quad (6.49)$$

where $G(\omega)$ is the FFT of $g(\cdot)$ padded with zeros from length M to length N. All FFTs are of length $2^k = N$ and require $N \log_2 N + N$ complex multiplies and $2N \log_2 N$ complex adds. The FFT implementation is usually faster than direct convolution if $N > M$ and $M > \log_2 N$.

As in the example of Section 6.3.1, *sectioned* convolution is applied to accommodate a data sequence of indefinite length. Fig. 6–23 shows the sectioned convolution, using a section length of P samples (at this point, P and M are arbitrary—we will develop a function for them for either range or azimuth processing soon):

The values are chosen so that $P + M - 1$ is an integer power of two. Each P-sample data segment is augmented with the $M - 1$ samples of the preceding segment, and the M-sample kernel is zero-padded to achieve length $P + M - 1$. The first $M - 1$ samples are discarded from the results of each convolution.

The stage of motion compensation corrects for small deviations from linear motion by the aircraft bearing the radar antenna. These deviations are measured by an inertial navigation system and are used to correct the phase return by adding or subtracting proper values to compute the return of a straight line path for all positions. The computation amounts to a complex multiplication at each point (Fig. 6–24).

6.6.4 Dynamic Description

The dynamic description of the algorithm introduces time sequential constraints to the steps of computation, addresses the communication requirements among multiple processors, and suggests possible partition arrangements to spread the computation across multiple processors. It examines the interconnection between processors, but does not yet look inside a processor for processor-specific features.

An initial partitioning across multiple processors may be proposed by assigning each of the major processing steps to a different group of processors, which are arranged in a pipeline. This imposes a MISD (multiple instruction single data) allocation, as the same data passes through sequential stages for different types of operations. A second partitioning within these stages is gained by subdividing the data into parallel streams, achieving a MIMD (multiple instruc-

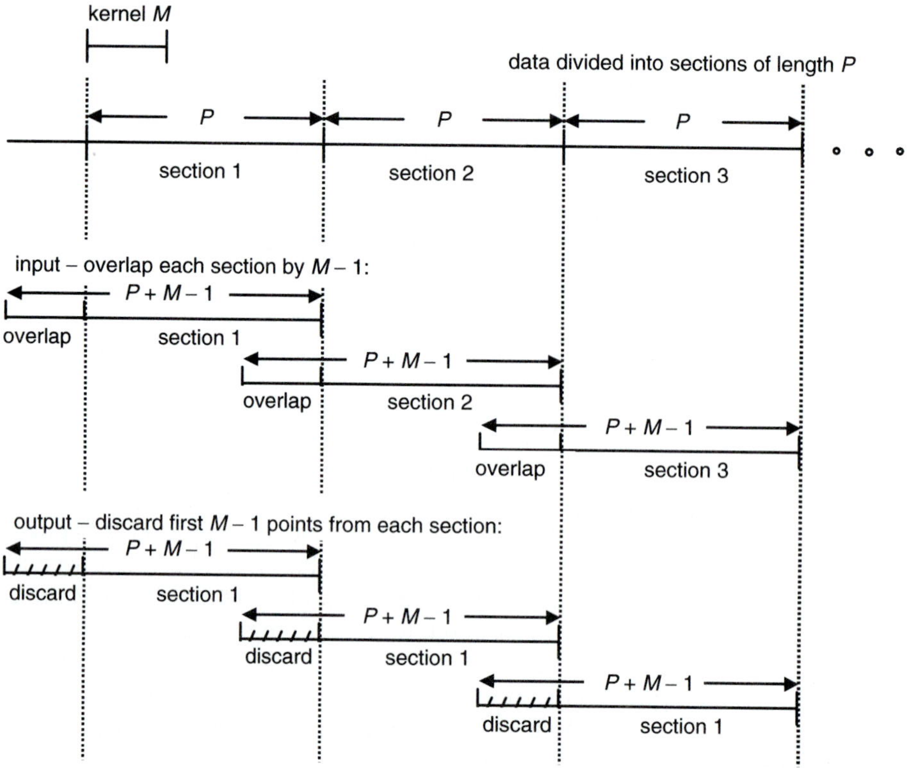

Figure 6–23 Sectioned convolution divides input stream into sections of length P and overlays them for processing by kernel of length M.

tion, multiple data) partitioning. Fig. 6–25 outlines the application of both means of partitioning.

Partitioning the data into parallel streams occurs for the three most computationally intensive blocks, as follows:

1. *Range processing*: partition by pulse return, distributing each successive pulse to a different processor in round-robin manner, returning back to the first when all others have been filled. Round robin (or for two processors, interleaved) distribution saves memory over buffering up all the pulses and then allocating a batch of consecutive pulses to each processor.
2. *Corner turn*: Distribute blocks of data, then transpose each block as described below.
3. *Azimuth processing*: Distribute across processors by range cell.

To aid this discussion, the following notation will be used. A processor is indicated by $p_{(type)}$, and an array is indicated by $p_{(type)}(index)$, where *(type)* is r for

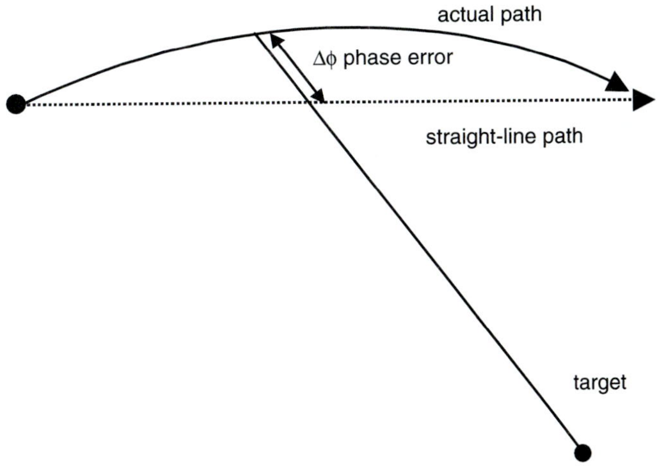

Figure 6–24 Deviation from linearity in flight path introduces phase errors $\Delta\varphi$, which are corrected by motion compensation.

range and *az* for azimuth. The processor index goes from 1 to P_r for range processors, where P_r is the total number of range processors, or from 1 to P_{az} for azimuth processors, where P_{az} is the total number of azimuth processors.

The operation of distributed corner turning, which uses the multiple p_r and p_{az} processors to transpose rows and columns, may be described using this notation for $P_r = 2$ and $P_{az} = 3$, as shown in Fig. 6–26.

The distributed corner turn first distributes a portion of the data from each range processor to all azimuth processors. It then performs a local transposition in

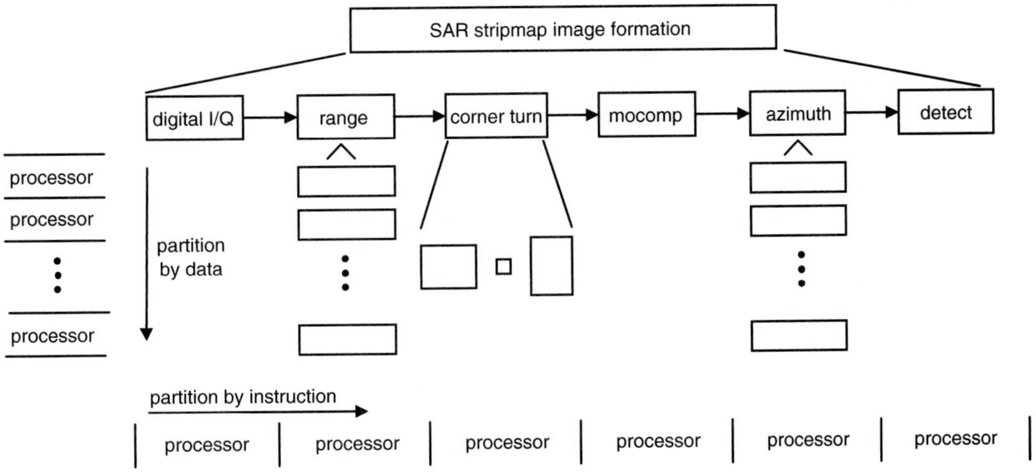

Figure 6–25 Initial partitioning of SAR image formation, first by instruction, then by data.

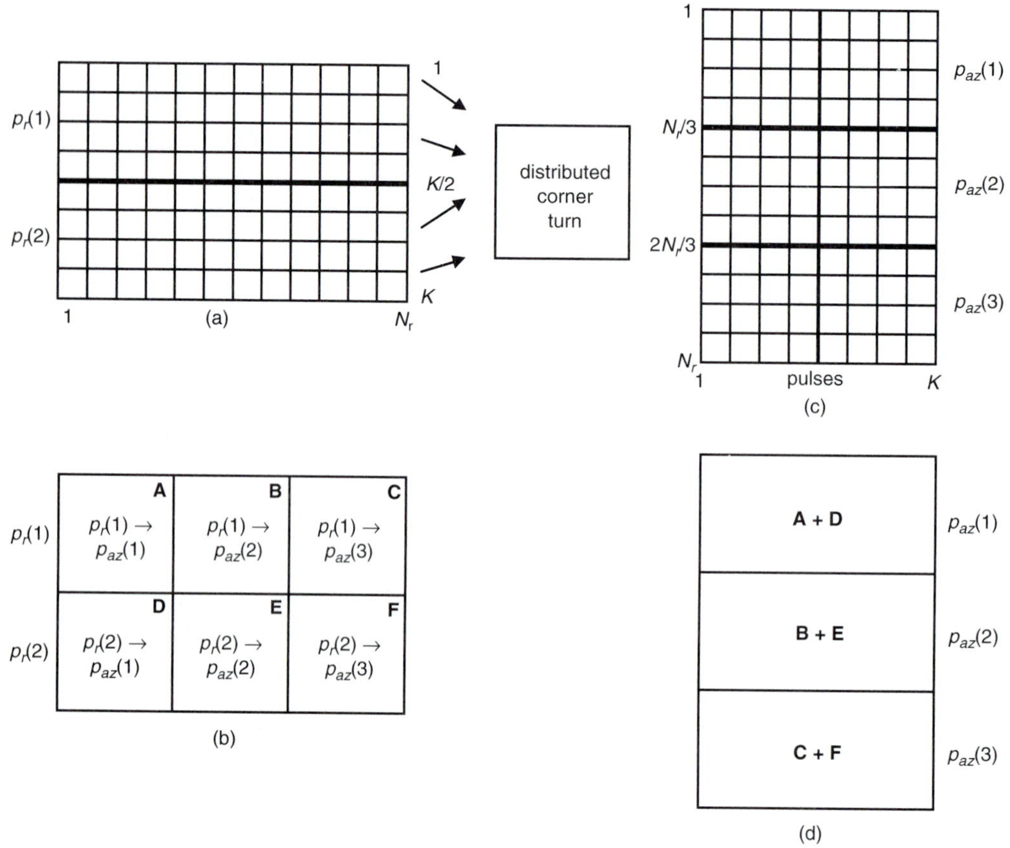

Figure 6–26 Data flow of distributed corner turn.

each azimuth processor of the resulting data. To begin the process, the data is arranged in range processors $p_r(1)$ and $p_r(2)$ as shown in Fig. 6–26a. The memory matrix contains N_r samples of each of K pulses. We process even-numbered pulses by $p_r(1)$ and odd-numbered pulses by $p_r(2)$. By alternating pulses between the two processors, both processors may be kept busy simultaneously, unlike the case in which the first half of the pulses are allocated to $p_r(1)$ and the second half of the pulses to $p_r(2)$, which causes $p_r(2)$ to wait idle until the second half of the pulses have arrived. The transfers to memories of the azimuth processors $p_{az}(1)$, $p_{az}(2)$, and $p_{az}(3)$ are shown by the columns of Fig. 6–26b. Every pulse return (each row) in each range processor must be distributed to all three azimuth processors, as follows. The first third (or $1/P_{az}$) of the samples in $p_r(1)$ memory (segment **A**) go to $p_{az}(1)$, the second third (segment **B**) goes to $p_{az}(2)$, and the final third (segment **C**) goes to $p_{az}(3)$.

Range processor 2 ($p_r(2)$) also sends one-third of the samples of each of its pulses to $p_{az}(1)$, $p_{az}(2)$, and $p_{az}(3)$, respectively. When these operations are complete, the memory content of each azimuth processor is as shown in Fig. 6–26d. Each azimuth processor performs a local transpose of rows and columns, resulting in a transpose of the total $K \times N_r$ matrix samples.

As a result of the data distribution in memory after the distributed corner turn, the data is ready for azimuth processing. The trial processor partitioning is adjusted to move the corner turn operation into the same processors that perform azimuth processing, resulting in a two-stage instruction pipeline that is further partitioned across processors by data. Fig. 6–27 shows the two banks of processors in the final partitioning and indicates the pulse distribution between them. The round robin distribution of pulses at the input and the fully connected processors between the range and the azimuth sections distinguish this mapping.

A processing timeline for the range processors (Fig. 6–28) shows pulse i is input into processor $p_r(1)$ at the upper left, and range processing commences. While pulse i is being processed by $p_r(1)$, pulse $i+1$ is input to $p_r(2)$. The round-robin assignment and processing continues until all P_r range processors are filled, and $p_r(1)$ is then revisited for feeding the next pulse. Also during the input period of pulse $i+1$ to $p_r(2)$, $p_r(1)$ is using its parallel output capability to output pulse $i - P_r$, which has been completed from the previous input cycle.

The azimuth processing includes the distributed corner turn described above, receiving the output from the range processors. Here, K pulses are included in an azimuth section. As described in Section 6.3.1, sectioned convolution on continuous data, using overlapping sections, provides identical results to performing the convolution directly on the unending data sequence.

The top third of Fig. 6–29 shows how the pulse output from each range processor is allocated sequentially across P_{az} azimuth processors, so that each az-

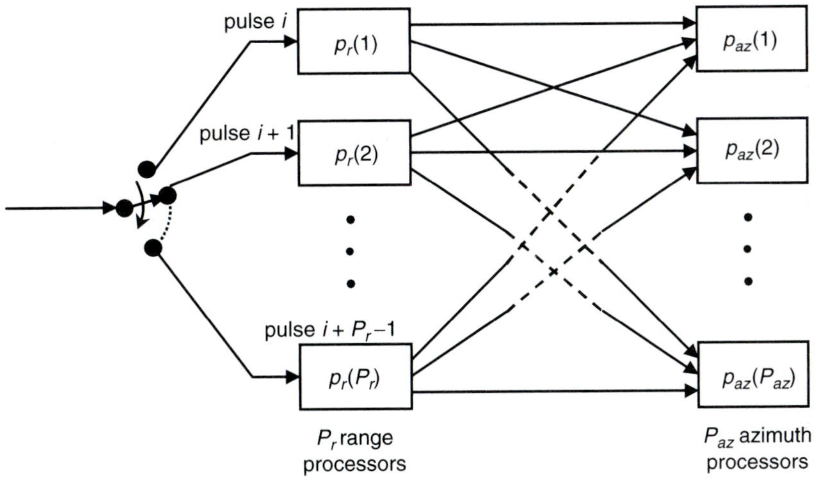

Figure 6–27 Generalized range-to-azimuth processor pulse distribution.

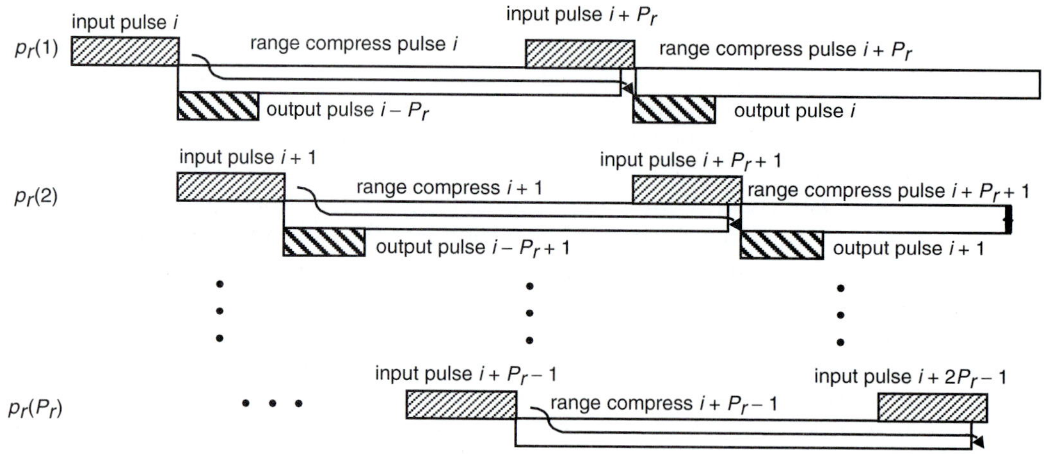

Figure 6–28 Timeline of range processing shows round-robin distribution of pulses to each processor and the overlap of input/output with processing.

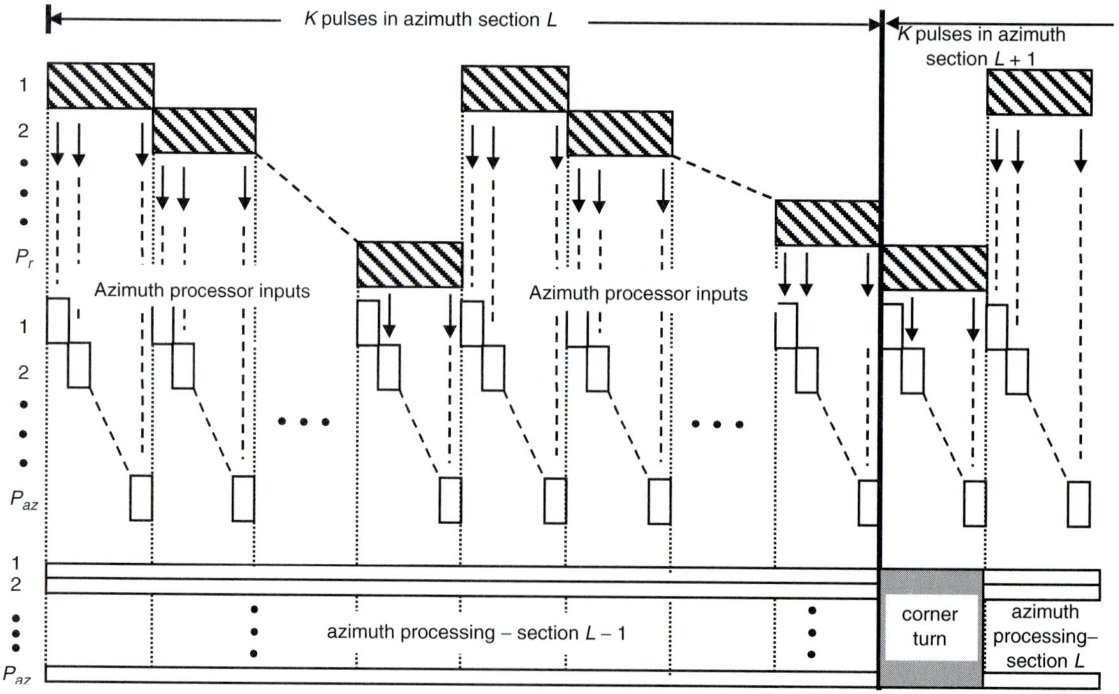

Figure 6–29 Azimuth processing distributes the same segment from each range pulse to an azimuth processor until K range pulses are distributed, then proceeds with corner turn and azimuth compression.

imuth processor ends up with a section of the compressed range pulse. The doling out of range pulses continues until K range pulses are received, and then the corner turn begins as shown at the right area of the lower third of the figure. After the distributed corner turn (shaded), the azimuth processing begins as shown at the far right. Input of range pulse segments to azimuth processors for section L proceeds in parallel with the azimuth processing of section $L - 1$.

To describe the azimuth processing, it is useful to sketch the memory buffer usage of a single azimuth processor (Fig. 6–30). An input circular buffer has space sufficient for its fraction of azimuth samples, given by N_r/P_{az}, times the total number of pulses in an azimuth section, K. It also includes an additional interval, Δ, sufficient to receive the range pulse segments that continue to arrive during the time of the corner turn operation. As range-compressed pulses are input from the range processing section, they are written sequentially in successive locations in the input buffer, and when the end of the input buffer is reached, writing continues circularly into the top. Thus, the input buffer contains $(K + \Delta) \cdot (N_r/P_{az})$ locations. When an azimuth section of K pulses is accumulated, the section is corner-turned from the input buffer to the azimuth working buffer. The size of the working buffer is set by the size of the azimuth FFT, given by K_{az}, where $K_{az} = K + M_{az}$, with M_{az}, the azimuth kernel size, given by $M_{az} = TW$ as discussed earlier. The azimuth FFT buffer is divided among P_{az} processors which accommodate N_r complex samples, so the buffer size is $K_{az}\,(N_j/P_{az})$.

6.6.5 Resource Requirements

The resource requirements are determined for processing, memory, I/O, and for processor-to-processor communication bandwidth. The processing requirement

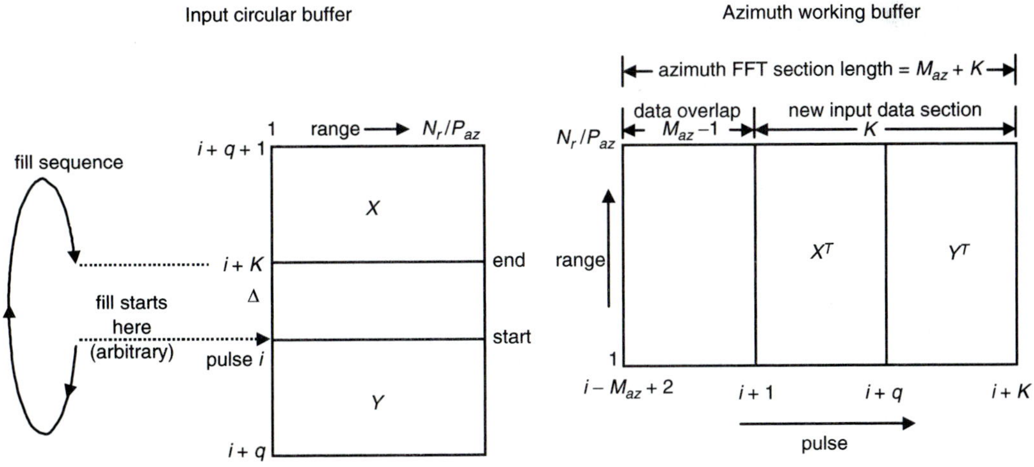

Figure 6–30 Input buffer (left) and working buffer (right) for corner turn and azimuth processing for one azimuth processor.

is computed in floating point operations (FLOP) required per sample, based on the sequence of steps applied to each input sample. A total throughput in millions of FLOP/sec, or MFLOPS, is then computed as the product:

$$\text{total rate} = \text{FLOP/sample} \times \text{samples/pulse} \times \text{pulses/sec}.$$

The sectioned convolution is examined first, because of the importance of sectioned convolution to both range and azimuth processing and its computation by the FFT. Its computations are measured in terms of *complex* adds and multiplies—these are then converted to effective real operations as shown in Table 6–9.

The butterfly operation of the FFT requires one complex multiply and two complex adds (see Section 8.2.2), for an effective load of ten real operations. The load for sectioned convolution of length N is given by Table 6–10.

A parameter tradeoff is used to determine the most efficient FFT size. Parameters include M (kernel size), N (size of FFT), a multiplier k that is the ratio of FFT size to kernel size ($k = N/M$), and P, the size of the sectioned convolution. The variables N and P, are generic for this discussion for both range and azimuth convolution. When range and azimuth convolution are considered separately, subscripts of range r and azimuth az will be used on N and M.

The tradeoff arises from the opposing facts that 1) increasing the size of the FFT decreases the percentage of the fixed M-sample overlap that the kernel requires between adjacent data sections, spreading its load over more samples and decreasing the number of operations per sample, and 2) increasing the FFT size increases the number of operations required by a factor of log N. An optimum size exists at which these conflicting effects balance each other—this size is determined by computing the load per sample versus N. The N-point convolution places a load $S \equiv 10\, N \log_2 N + 6N$ real operations. The data is sectioned into segments of length P, where $P \geq 2M$, so $N = P + M - 1 \approx P + M$. Since an N-point FFT is performed for every data section of length P, the computation per sample X is given by

$$X = \frac{S}{P} = \frac{10(P + M - 1)}{P} \log_2 (P + M - 1) + 6\frac{(P + M - 1)}{P}. \qquad (6.50)$$

The ratio of the convolution size to the kernel size given by $k = N/M$, can be used to simplify (6.49), using the approximation that $P \approx N - M = (k - 1)M$:

$$X = 10\frac{k}{k - 1}[\log_2 k + \log_2 M + 0.6]. \qquad (6.51)$$

Table 6–9 Conversion of complex operations to effective real operations.

Floating Point Operation	Example	Real Operations	Total
Add	$(x_1 + jy_1) + (x_2 + jy_2) = x_1 + x_2 + j(y_1 + y_2)$	2 adds	2
Multiply	$(x_1 + jy_1) - (x_2 + jy_2) = x_1x_2 - y_1y_2 + j(x_1y_2 + x_2y_1)$	4 multiplies and 2 adds	6

Table 6–10 Computation load of sectioned convolution.

Function	Complex Operations	Real Operations
FFT	$(N/2)\log_2 N$ butterflies	$10(N/2)\log_2 N$
N-point complex multiply	N complex multiplies	$6N$
IFFT	$(N/2)\log_2 N$ butterflies	$10(N/2)\log_2 N$
TOTAL		$10N\log_2 N + 6N$

The computational load is minimized for some value of k, but as will be shown, the memory requirement increases in proportion to k. Fig. 6–31 plots the computational load per sample given by (6.50) versus k for various values of the product TW.

As shown in Fig. 6–31, the optimum FFT size for best speed depends on the product TW and is in the range of $k = 10$. However, memory requirements increase with FFT size, and so another tradeoff of processing speed versus memory size may be made. As will be shown in Section 6.6.6, azimuth processing memory requirements are several orders of magnitude greater than range processing memory requirements. For the azimuth processing with its very large memory requirement, a fivefold reduction in memory can be gained for a 50 percent increase in computation per sample, if $k = 2$ is selected instead of the optimum $k \approx 10$. For range processing, the speed-optimum value $k = 10$ is used because its memory requirements are more modest.

We now specialize the generic discussion of fast convolution to separate range and azimuth situations. M_r denotes the range kernel size and M_{az} represents the azimuth kernel size. With section size of $P = 10\,M_r$ for range and $P = 2M_{az}$ for azimuth now selected, the computation per sample for each correlation operation is calculated (Table 6–11 and Table 6–12). The simpler input operations of format and

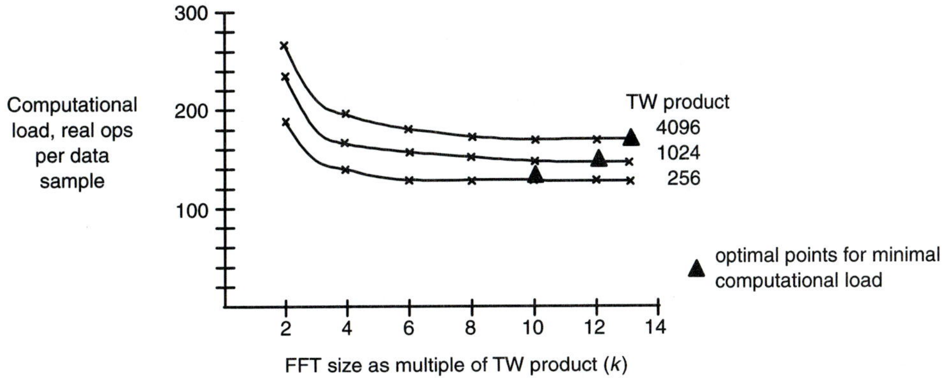

Figure 6–31 Effect of section size on computational load of sectioned fast convolution.

Table 6–11 Range processing load.

Operation	Real Operations per Complex Sample
fix-to-float	7
digital I/Q	8
fast convolution	$11 \log_2 M_r + 45$
TOTAL	$11 \log_2 M_r + 60$

I/Q conversion are combined with the range processing, and the simpler motion compensation and magnitude detection are assigned to the azimuth processors.

The total computation load is given by:

$$\text{total load} = \text{load/sample} \times \text{samples/pulse} \times \text{prf}$$

where:

$$\text{load/sample} = \text{range processing load} + \text{azimuth processing load} = 11 \log_2 M_r + 20 \log_2 M_{az} + 118;$$

$$\text{samples/pulse} = N_r = \Delta R/\delta x;$$

$$\begin{aligned} prf = B &= \frac{2V\theta_B}{\lambda} \text{ (from Eq. (4.39))} \\ &= \frac{2V\theta_B}{\lambda} \cdot \frac{R}{R} \text{ (identity)} \\ &= \frac{2VL_{eff}}{\lambda R} \quad (\text{from } L_{eff} = R\theta_B) \\ &= \frac{2V}{\delta x} \quad \left(\text{from } \delta x = \frac{\lambda R}{2L_{eff}}\right). \end{aligned} \tag{6.52}$$

So the total load L is given by

$$L_p = (11 \log_2 M_r + 20 \log_2 M_{az} + 118)\frac{V\Delta R}{\delta x} \text{ real operations/sec.} \tag{6.53}$$

Fig. 6–32 plots the total load L in MFLOPS versus image resolution δx for two values of TW for both airborne and satellite SAR. This example of a satellite

Table 6–12 Azimuth processing load.

Operation	Real Operations per Complex Sample
corner turn	10 (est)
motion compensation	6
fast convolution	$20 \log_2 M_{az} + 32$
magnitude	10
TOTAL	$20 \log_2 M_{az} + 58$

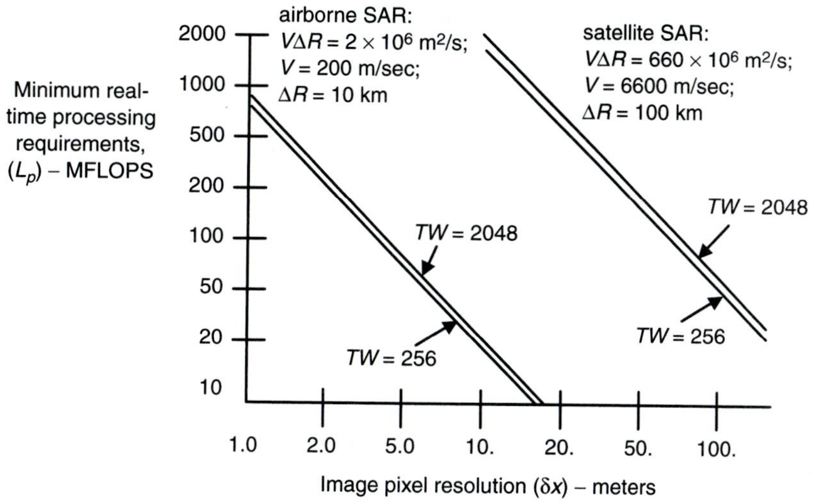

Figure 6–32 Computational load of stripmap SAR image formation versus image resolution.

SAR has a swath width ΔR that is 10 times that of airborne SAR, a velocity V that is over 30 times greater, and a scan rate $V\Delta R$ that is over 300 times greater. Thus, it has a much higher computational load for a given resolution.

The computation load L varies approximately as the inverse of the square of image resolution, but because the time-bandwidth product TW must be increased with resolution, L varies a bit faster than the inverse square of resolution.

The previous analysis considers the overall MFLOPS needed. Any particular architecture will perform some types of operations (e.g., convolution) at a faster rate than others (e.g., complex multiply). Section 6.6.6, on mapping to a particular processor, will consider the fact that all MFLOPS are not equal in the amount of load they place on an architecture.

Turning from processing load to memory requirements, a separate computation can be made for range and azimuth processing. For range processing, a double buffer is needed for both input and output for each of the P_r range processors, for a capacity of $4N_r$ complex samples. The total for P_r processors is thus $4P_rN_r$ complex samples. Each complex sample can be stored as two 16-bit integers (4 bytes per sample), although block floating point is necessary to assure that numbers are kept within proper ranges. Samples must be reconverted to two 32-bit floating point numbers before azimuth compression begins. Thus, range processing storage is given by $L^r = 16\,P_rN_r$ bytes.

Each azimuth processor requires at least the memory shown in Table 6–13. The total azimuth memory requirement summed over P_{az} processors is thus 2.5 N_rK.

Table 6–13 Memory requirement for azimuth processing.

Operation	Size
corner turning double buffer	$2(N_r/P_{az})K$ complex samples
output image buffer	$(N_r/P_{az})\,K$ real samples
TOTAL	$2.5\,(N_r/P_{az})\,K$ complex samples

Since $N_r = \Delta R/\delta x$ and $K = TW = R\lambda/2(\delta x)^2$ [see (6.44)], the total memory load L_m^{az} is given by

$$L_m^{az} = 2.5\,\frac{R\lambda\Delta R}{2(\delta x)^3}\text{ complex samples.} \tag{6.54}$$

Again reserving four bytes per complex sample, the storage is given by:

$$L_m^{az} = 5\,\frac{R\lambda\Delta R}{(\delta x)^3}\text{ bytes.} \tag{6.55}$$

Fig. 6–33 shows the dependence of memory versus resolution for airborne and satellite conditions, using two 16-bit integer storage locations for each complex sample. The plot also reflects the practice of overlapping the azimuth frames by 50 percent, such that each sample contributes to two frames. This increases the azimuth memory requirement by another factor of two. Therefore, the total azimuth memory is given by $L_m^{az} = 2.5N_rK = 5N_rTW$ complex samples, or for 4-byte storage for each complex sample, $10\,N_rTW/P_{az}$ per azimuth processor.

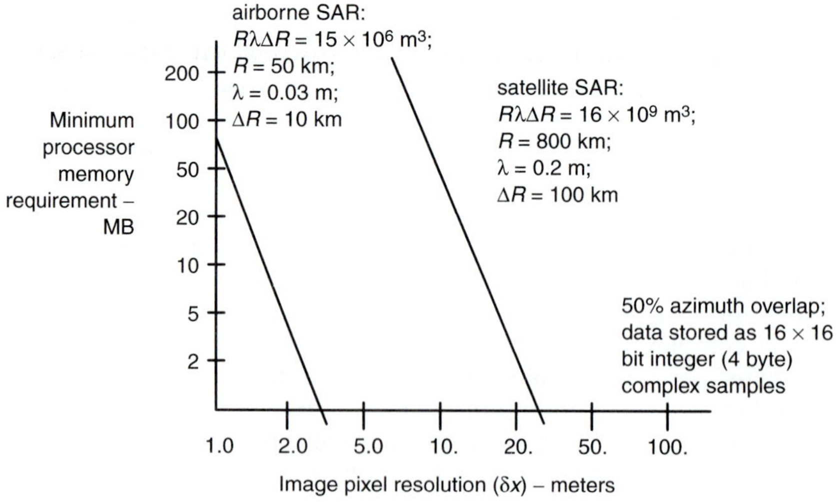

Figure 6–33 SAR stripmap processing memory requirements versus resolution.

The memory load varies as the inverse cube of image resolution. The ratio of memory to processing load, $(\delta x)^3/(\delta x)^2 = \delta x$, grows with SAR resolution: Highest-resolution SAR processing is more likely to be memory-limited than processing-limited.

The analysis of resource requirements for SAR-stripmap-image formation are summarized as follows:

- SAR processing load varies in proportion to the input sampling rate and as the inverse of the square of pixel resolution, ranging from 10^1 to 5×10^5 MFLOPS;
- SAR memory requirements vary as the inverse cube of resolution and as $R\lambda\Delta R$ product of application, ranging from $10^1 - 2 \times 10^4$ Mbytes;
- Image resolution is the principle driver of processing and memory requirements, and memory requirements increase faster with resolution than processing requirements;
- Processing is easily cast into parallel form: The data is distributed by pulse return for range processors and by range cell for azimuth processing.

6.6.6 Mapping to Specific Architecture

Mapping SAR stripmap image processing to a specific processor requires including the processor efficiency for each instruction rather than using a cross-instruction average value of computation throughput. The processing load is converted from a total number of MFLOPS to a rate for each type of operation performed on the target processor. As seen above, the memory requirement for range processing is 16 N_r bytes/processor and for azimuth processing is 10 N_rTW/P_{az} bytes/processor. The maximum bandwidth is equal to two times the input data rate divided by the number of parallel processors per stage.

The steps for determining the processing load and resulting number of processors required are:

1. Optimize equivalent MFLOP throughput of the selected processor type for each processing step, which yields the MFLOPS throughput for each processing step;
2. Determine the total MFLOP requirement of each processing step (number of operations/sample × input data rate)
3. Divide #2 by #1 to obtain the equivalent number of processors needed for each step;
4. Sum the results of #3 over all processing steps to obtain total number of processors required;
5. Add a design margin for error and growth (about 30 percent).

Table 6–14 Effective throughput of SHARC processor for SAR image formation operations.

Operation	Throughput
fix-to-float conversion	80 MFLOPS
FIR filter	45 MFLOPS
complex multiply	27 MFLOPS
fast convolution	94 MFLOPS
corner turn	80 MFLOPS
magnitude (square)	38 MFLOPS

The equivalent throughput for the SHARC digital signal processor on each type of operation used in SAR image formation can either be measured on a processor or emulator with the operation programmed in a loop or by counting cycles from each instruction in the segment of code that implements the operation. In the case of direct measurement, the time to perform the jump and any time to empty and refill the pipeline must be accounted for in determining the throughput. The results of such analysis on the SHARC yield the effective throughputs shown in Table 6–14.

A representative set of parameters for an airborne SAR system for this computation are shown in Table 6–15. From these system parameters, several derived parameters are established (Table 6–16). For these parameters, the total number of range processors is computed as shown in Table 6–17. In like manner, the required number of azimuth processors may be computed as shown in Table 6–18.

The grand total of processors needed is the sum of range and azimuth processors:

Table 6–15 Typical parameter values for airborne SAR application.

Parameter (units)	Symbol and value
Resolution (m)	$\delta x = 1$
Swath width (m)	$\Delta R = 20{,}000$
Platform velocity (m/sec)	$V = 200$
Maximum range (m)	$R = 100{,}000$
Pulse width (μsec)	$\tau = 10$
Radar wavelength (m)	$\lambda = 0.03$

Table 6–16 **Values of derived parameters for typical airborne SAR application.**

Derived Parameter (units)	Expression	Value
Data rate (samples/sec)	$Q = (V\Delta R)(\delta x)^2$	4×10^6
Azimuth processing memory (bytes)	$L_{az} = 10(R\lambda\Delta R)/(\delta x)3$	600×10^6
Range processing memory (bytes)	$16P_r(\Delta R/\delta x)$	$320P_r \times 10^6$
Range samples (count)	$N_r = \Delta R/\delta x$	20×10^3
Pulse repetition frequency (Hz)	$prf = V/\delta x$	200
Range time-bandwidth product	$TW_r = (c\tau)/2(\delta x_r)$	1500
Azimuth time-bandwidth product	$TW_{az} = (R\lambda)/2(\delta x)^2$	1500

Table 6–17 **Computation of number of range processors.**

Step	Effective Processor Throughput (MFLOPS)	Real Time Requirements (MFLOPS = FLOP/sample * sample/sec) [Q = 4 x 10^6]	Number of Processors
Fix to float conversion	80	$7 \times Q = 28$	0.35
FIR filter for digital I/Q	45	$8 \times Q = 32$	0.71
Fast convolve (10% overlap)	94	$(11 \log_2 TW + 45)Q = 116Q = 664$	7.06
Float to fix conversion	80	$7 \times Q = 28$	0.35
Sum		752	8.47
+ 30% margin			2.54
TOTAL			11

Table 6–18 **Calculation of number of azimuth processors.**

Step	Effective Processor Throughput (MFLOPS)	Real Time Requirements (MFLOPS = FLOP/sample * sample/sec) [Q = 4 x 10^6]	Number of Processors
Corner turn	80	$10 \times Q = 40$	0.50
Fix-to-float conversion	80	$7 \times Q = 28$	0.35
Complex multiply (motion compensation)	27	$6 \times Q = 24$	0.89
Fast convolution (50% overlap)	94	$252 \times Q = 1008$	10.72
Magnitude	26	$10 \times Q = 40$	1.52
Float-to-fix convert	80	$7 \times Q = 28$	0.35
Sum		1168	14.35
+ 30% margin			4.30
TOTAL			19

Range processing	11
Azimuth processing	19
Grand total	30

The processor count has been determined by computational throughput requirements—the processors must be assessed for memory requirements. As discussed previously, complex samples will be stored as pair of 16-bit integers, saving a factor of two in memory from the original 2 × 32-bit floating point and providing a 90-dB dynamic range. Total range memory required has been shown to be $16\ P_r N_r = 16 \times 11 \times 20 \times 10^3 = 3.5$ Mbyte, or spread among the eleven range processors, 0.32 Mbyte/processor. Azimuth processing memory, given by $10\ N_r * TW = 10\ (20 \times 10^3)\ (2048) = 410.2$ Mbyte, amounts to 21 Mbyte for each of the nineteen processors. As expected, the memory required for each azimuth processor is nearly 100 times that for each range processor.

Finally, the processor interconnection bandwidth must be compared to the required interprocessor communication rate. The total throughput, Q, is the same through the entire processing chain and equals the input sample rate. It is spread across $P_r = 11$ processors in the range processing section and $P_{az} = 19$ processors in the azimuth processing sections, and so is divided by P_r or P_{az}. Each processor

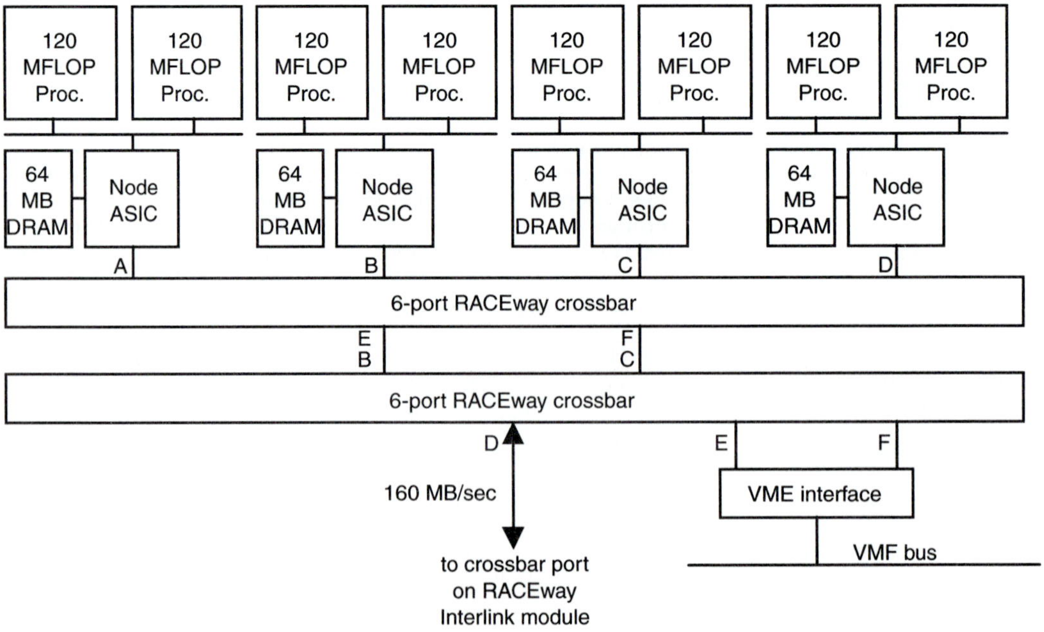

Figure 6–34 Eight-processor module from which thirty-six processor SAR image formation processor is constructed.

must handle both input and output, requiring Q to be multiplied by two, and each sample consists of two 32-bit floating point numbers, requiring multiplication of Q by another factor of 8 bytes/complex sample. Therefore, the total bidirectional data throughput rate is $16\ Q = 64$ Mbyte/sec, well below the 160 Mbyte/sec offered by the RACEway single path capability. The total I/O rate at each range processor is 6 Mbyte/sec, and at each azimuth processor is 3.4 Mbyte/sec. Thus all data rates for this airborne SAR example are well within the capabilities of the system, and intercommunication requirements are easily met.

To conclude this example, the physical partitioning of processors and memory onto boards or modules is examined. The system is comprised of modules, each with multiple processors, that are combined to achieve the required number of processors. The basic node consists of two SHARC processors, 64 MB of DRAM shared by the two processors, and a bus interface integrated circuit. Four nodes (eight processors) plus the two RACE crossbar and a bus interface are combined on one VME 6U form factor board (about 6″ × 6″). This is shown in Fig. 6–34.

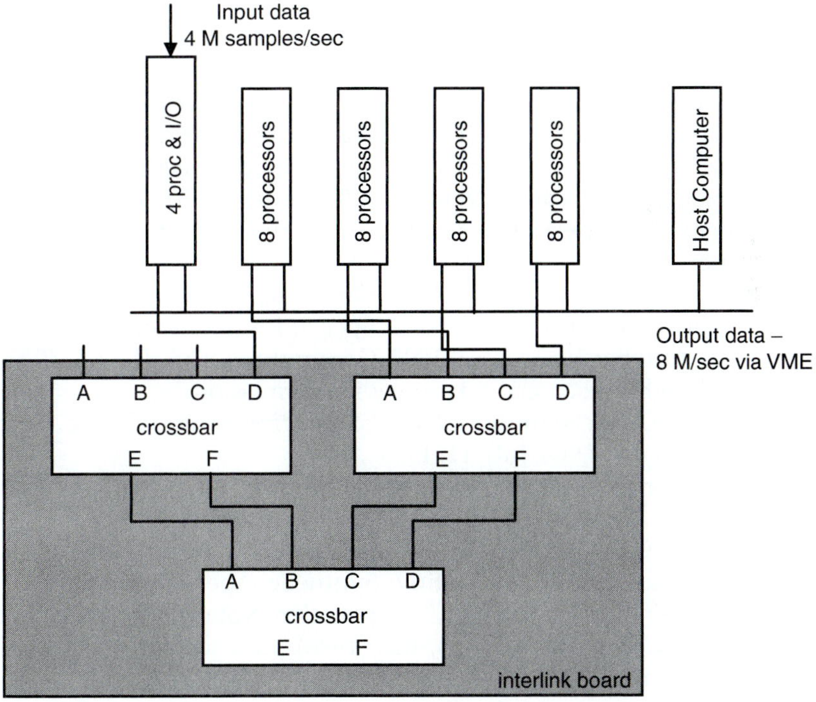

Figure 6–35 SAR stripmap processing system combines four eight-processor and one four-processor input module for a total of thirty-six processors.

When the modules of Fig. 6–34 are combined into a system, two of the nodes (four processors) are removed on the first module and replaced with a data interface that accepts the 4 Msample/sec input. This four-SHARC board is combined with four eight-SHARC boards for a total of thirty-six processors (Fig. 6–35).

The RACE interconnection is used both within one module and across modules, achieving multiple concurrent communication. A concurrent direct memory access (DMA) capability separates the communication activity from the processing activity, allowing parallel processing and memory transfer activities which are used in corner turning. This architecture can also support a heterogeneous mix of processors, as shown by the replacement of two processing nodes on one module by a high-speed input processor. Because the 64 Mbyte of memory at each two-processor node is much greater than the 0.32 Mbyte needed per range processor, it could be reduced for range processors only. As newer and faster processors emerge, the architecture supports their insertion as replacements for the present processor, as long as compatibility with the RACE interconnection is maintained at the inputs, outputs, and interfaces.

REFERENCES

1. Pierpaolo Baglietto and others, "Image Processing on High Performance RISC Systems," *Proc. IEEE*, 84 (June, 1996), 917–930.
2. Brian Bouzas, "Mapping an Application to RACE® SHARC™ Computer Node," Application Note 207.1 (Chelmsford, MA : Mercury Computer Systems, Inc., 1995). Courtesy of Mercury Computer Systems, Inc., used with permission.
3. J. G. Ackenhusen and Y. H. Oh, "Single-Chip Implementation of Feature Measurement for LPC-Based Speech Recognition," *AT&T Technical Journal*, 14 (October 1985), 1787–1805. Copyright © 1985, *The AT&T Technical Journal*. All rights reserved. Adapted with permission.
4. John G. Ackenhusen, *Linear Predictive Speech Coding Arrangement*, U.S. Patent 4,847,906, July 11, 1989.
5. Winser E. Alexander, Douglas S. Reeves, and Clay S. Gloster, Jr., "Parallel Image Processing with the Block Data Parallel Architecture," *Proc. IEEE*, 84 (July 1996), 947–968.
6. Thomas Einstein, "Realtime Synthetic Aperture Radar Processing on the RACE® Multicomputer," Application Note 203.0 (Chelmsford, MA : Mercury Computer Systems, Inc., 1996). Courtesy of Mercury Computer Systems, Inc., used with permission.

7

Basic Hardware and Software Computing Elements

Devising a real time implementation of a sophisticated signal processing algorithm involves decomposing the end-to-end processing into a set of modules, identifying and comparing alternate choices for implementing these modules, and combining the chosen set of modules into an integrated whole. The following four chapters provide a selection of hardware and software modules that can be applied to elements of a signal processing application that result from this hierarchical decomposition.

7.1 INTRODUCTION

Chapters 7 through 10 divide these computing elements by their types of function. In this chapter, some general components are examined. These components may occur in a significant portion of all signal processing applications. Tools for signal analysis and modeling, by which a signal is characterized in terms of such parameters as frequency or by its fit to a parametric model, are discussed in Chapter 8. Signal compression may use signal characterization techniques to extract information from a signal, transmit the information at a reduced bandwidth, and then reconstruct the signal. It is the topic of Chapter 9. Extracting some representation of information content from a signal allows the comparison of one signal to another or to a library of patterns extracted from a diversity of signals. Thus, tools for signal comparison are discussed in Chapter 10.

Several elements of this toolkit have already been introduced within the examples of Chapter 6. These topics are mentioned again here and appropriate reference is made.

7.2 BASIC COMPONENTS

Basic components are the components whose use transcends the specialty of an application. They include the fundamentals of computing a complicated function, such as the transcendental functions (logarithm and the trigonometric functions). Next, basic concepts of interconnecting components are discussed which range from the ways to connect low-level logic elements, through methods of interconnecting processing modules, to ways to communicate among entire processing systems. Higher level interconnection, i.e., those at the processor and system level, may also be required to work over long distances. Special-purpose buffers form another fundamental class of signal processing components. These elements capture the repetitive, predictable, but complex addressing schemes typical of signal processing and implement them in hardware and software structures that speed their execution. Finally, some fundamental structures for data-dependent processing are introduced. Data-dependent computing structures recognize that the nature of signal processing algorithms is to convert vast amounts of data into compact pieces of information. As the transformation to information occurs, the processing becomes dependent upon that information.

7.2.1 Transcendental Functions

The basic arithmetic logic unit (ALU) is capable of performing addition, subtraction, logical, and (often) multiplication operations. However, more complex operations, including division and such operations as exponentiation, logarithm, or trigonometric functions must be performed by decomposition into the more basic operations that can be handled by the ALU.

7.2.1.1 Table Lookup. One method of computing complex functions is by table lookup. This method is most effective for functions of at most two input arguments, spanning over a limited range of argument values. In the table lookup method, the argument is used as an address into a large table in memory, and the table value stored at the address is the value of the function at that argument value. Table lookup usually requires limiting the range of the input to limit the size of the table; it may also result in limiting the precision of the input. For example, an angle x can be limited to the range $(0, \pi/4)$, and represented in the 256 values provided by 8 bits. A table of 256 entries stores the sequential values of $\sin(x)$ for $x = 0, \pi/4 \cdot 1/256, \pi/4 \cdot 2/256 \cdot, \ldots, \pi/4 \cdot 255/256$. When $\sin(x)$ is required, x is first placed in the range of $x' \epsilon [0, \pi/4]$ by simple trigonometric identities, the value of $\sin(x')$ is looked up in the table, and any changes required by trigonometric identities are then performed to compute sin(x). The trigonometric identities are used to exploit the fact that sin(x) can be computed for all values of x by combinations of translating and negating the value of the first eighth cycle of the sine wave.

The precision of the output value is governed by the width of the memory, although if the input precision is limited to a smaller number of discrete values (smaller number of bits), the precision gained by a wide memory width is lost. If one moves to a function of two variables, the need for memory goes up quadratically, since the number of bits that determine the address has doubled. Table lookup is fast if argument normalization is not required, but significant manipulation may be needed for normalization in the most general case.

The use of simple bit manipulation on the argument provides an alternative method that can exploit an ALU's ability to perform rapid bit manipulation. This technique also requires normalization of the argument and requires more operations than the table lookup method, but it uses less memory and can cover a larger range of input values. A bit manipulation technique usually fits a portion of the desired function with a $\log_2$ function of the argument, then performs linear interpolation to refine the result. For example taking the following steps, illustrated in Figure 7–1, may approximate the log2(x) function:

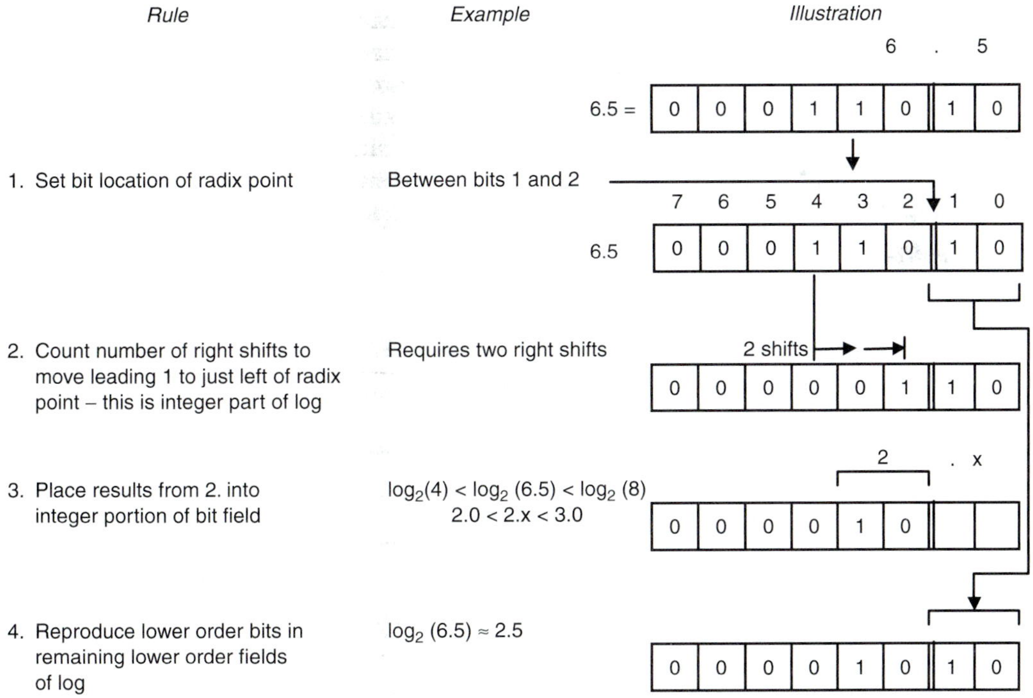

Figure 7–1 Example calculation of $\log_2(x)$ for $x = 6.5$; technique yields $\log_2(6.5) = 2.5$, while actual value of $\log_2(6.5) = 2.7$.

1. Set a fixed bit boundary to serve as the binary radix (the binary "decimal point" that sets the boundary between integer and fractional fields; in the example, it is the boundary between bits 1 and 2).
2. Count the number of shifts required to normalize the argument so that its most significant bit is just to the left of this radix point. The sign of the count is used to reflect whether the shifts were to the right or to the left (in the example two right shifts are required). This number is used as the integer portion of the value of $\log_2(x)$.
3. Copy the lower-order bits (those to the right of the leading one) to the right of the binary log value, just next to the integer (these are the values 1, 0 in the example).

The result is an approximation to the logarithm. Precision may be improved by increasing the number of bits kept in the fraction (this example uses 2 bits) and by scaling up the argument prior to computation, then reducing the log to reflect such scaling afterward.

7.2.1.2 Taylor Series Expansion. A more accurate, more computationally intense method of generating the value of a transcendental function is through the use of its Taylor series expansion. A function $f(x)$ may be computed by the Taylor series expansion about its known value at a point x_0 if the value of all its derivatives are also known at x_0:

$$f(x) = f(x_0) + \frac{f^{(1)}(x_0)}{1!}(x - x_0) + \frac{f^{(2)}(x_0)}{2!}(x - x_0)^2 + \ldots + \frac{f^{(n)}(x_0)}{n!}(x - x_0)^n + \ldots \quad (7.1)$$

Here, the notation $f^{(n)}(x_0)$ means the nth derivative of the function, evaluated a point x_0. In practice, the series is truncated in practice to the first few terms, keeping sufficient terms to obtain the desired accuracy.

For example, to find $\sin(\pi/6)$ by expanding about 0, where the values $\sin(0) = 0$ and all derivatives, $f^{(1)}(x = 0) = \cos(0) = 1$, $f^{(2)}(x = 0) = -\sin(0) = 0, \ldots$, are known, the computation proceeds by evaluating the first few terms of:

$$\sin\left(\frac{\pi}{6}\right) = \sum_{n=0}^{\infty} \frac{\sin^{(n)}(0)\left(\frac{\pi}{6}\right)^n}{n!}. \quad (7.2)$$

Table 7–1 shows the result of the evaluation of each term and the cumulative value of the sum up to that term.

The lower curve in Fig. 7–2 compares the term-by-term convergence of the Taylor series expansion about 0 of $\sin(\pi/6)$ to the Taylor series of $\sin(2\pi)$, which should equal 0. As shown, larger deviations from the expansion point require more terms to reach a specified degree of error. Thus, it is often effective to expand the function about several points and use fewer terms about each point, as

Table 7–1 Taylor expansion of sin (π/6) about 0 converges to within 0.02% of correct value (0.5) with terms through the fifth of the series.

n	Term	Value	Current Sum
0	$\sin(0)(\pi/6)^0$	0.0	0.0
1	$\cos(0)(\pi/6)^1$	0.523198	0.523198
2	$-\sin(0)(\pi/6)^2 \cdot 1/2$	0.0	0.523198
3	$-\cos(0)(\pi/6)^3 \cdot 1/6$	-0.023925	0.499273
4	$\sin(0)(\pi/6)^4 \cdot 1/24$	0.0	0.499273
5	$\cos(0)(\pi/6)^5 \cdot 1/120$	0.0003279	0.500009

was demonstrated with the raised cosine (Hamming) window described in Section 6.4.2.

Transcendental functions can be computed in hardware, either using combinations of table lookup, bit manipulation, or Taylor series as described above, or instead with logic that performs interpolation of a function about several anchor values. The use of interpolation is similar to Taylor series, but it is simpler and can be used to generate several different functions within the same logic cell.[1]

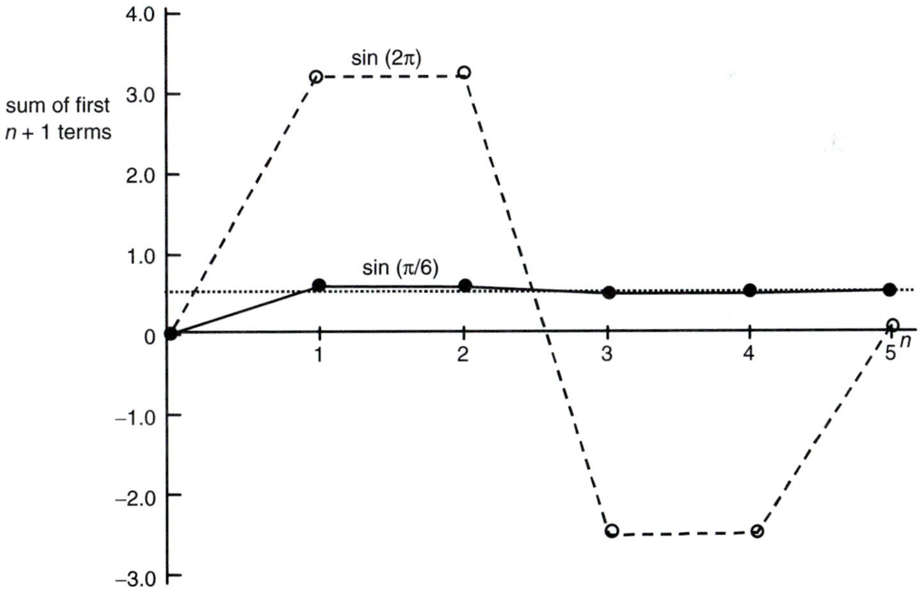

Figure 7–2 Taylor series expansion of sin(x) about 0 converges more quickly for argument values that are closer to the expansion point.

In this approach, the argument x is represented as a series of bits, $x = \sum_{b=0}^{B-1} 2^b x_b$, or $[x_{B-1}, \ldots, x_{B-M}, x_{B-M-1}, \ldots, x_0] = [\mathbf{x}_m \mathbf{x}_l]$. The M most significant bits of x, given by the vector x_m, are used to define a uniform grid that samples the function, and exact values are stored at those points. The sample points are represented by x_i, x_{i-1}, ..., and the function values at those points are represented by $f(x_i) \equiv f_i, f(x_{i-1}) \equiv f_{i-1,\ldots}$. The function value of any x between x_i and x_{i-1} is computed in logic by polynomial interpolation:

$$P_n(x) = f_i + \alpha(f_{i+1} - f_i) + (x - x_{i+1})(x - x_i)c_2 + \ldots + (x - x_{i+1})(x - x_i)^{n-1}c_{n-1}. \quad (7.3)$$

Here, $\alpha = ((x - x_i)/(x_{i+1} - x))$ and provides simple linear interpolation; values of c_n are chosen to match left-hand derivatives for added precision beyond simple linear interpolation.

Fig. 7–3 shows how an input value of x shown as 24 bits is divided into two sets—the eight most significant bits and the sixteen less significant bits. Formulas that express the coefficients c_2 and c_3 in terms of differences of values of f_i are executed on the right hand side to refine the values of f_i computed on the left.

7.2.2 Selected Buses and Interconnections

Buses and other interconnections provide the communication paths among the modules that make up a signal processor. The modules to be connected range from the very simple, such as transistors interconnected with single wires, to very complex, such as workstations connected with a wide area network. This discussion begins at the high level where the signals have a correspondence with data or control blocks in a block diagram. At this level, physical connection is usually established by means of a *bus*, or parallel group of wires. A *multi-drop bus* has several modules connected along a bus and provides means to resolve the bus contention that can arise when two modules seek to use the bus at the same time.

7.2.2.1 ISO Model and Types of Interconnection. The International Standards Organization (ISO) Reference Model captures the hierarchy of systems communications for Open Systems Interconnection, or ISO/OSI. This model provides an interconnection approach in which separate layers provide certain services and call upon the services of other layers. As shown in Table 7–2, the layers range from the Application Layer (Layer 7) at the top, which accomplishes communications between entire applications, such as sharing data across spreadsheets, to the Physical Layer (Layer 1), which establishes the electrical conduction provided by wiring.

Several characteristics distinguish one type of interconnection from another. The complexity of modules being interconnected has already been mentioned as one characteristic. The distance over which the signals are sent, from within a board through across boards to across boxes or even buildings governs the nature of the bus. The width of the bus, measured in number of parallel channels, can range from a single serial line, into which individual data words are multiplexed

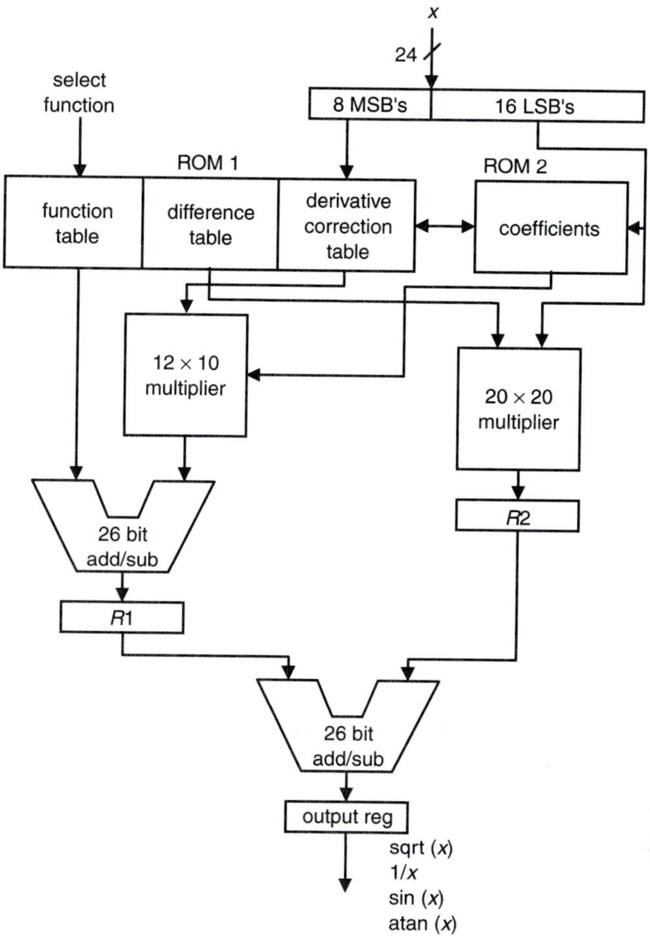

Figure 7–3 Circuit that computes multiple transcendental functions of 24-bit input x based on 16-bit polynomial interpolation across values determined by the most significant 8 bits.[1] © 1993 IEEE.

bit by bit, to multi-bit parallel buses, in which each bit has its own path. Data throughput distinguishes one bus type from another and is determined by the number of transfers per second times the number of bits per transfer. A parallel bus will transfer more data per second than a serial bus of the same type. However, certain serial buses are generally faster than electrical-based parallel buses, such as those that use optical fibers and convert the electrical signal to and from the optical domain before and after transmission. The interconnection topology distinguishes bus type—some buses are intended for point-to-point connection,

Table 7–2 Seven layers of ISO Open Systems Interconnection Model.

Layer	Name	Description
7	Application	Manages communication between applications
6	Presentation	Defines common syntax for exchange of data between applications
5	Session	Control structure for managing communications (e.g., setup, control, and terminate communication session)
4	Transport	Movement of data from one point to the next, including provision for error protection
3	Network	Determines routing of the data
2	Link	Provides data transfer across physical links, including any definition of blocks, error detection, and flow control
1	Physical	Actual connection (e.g., wire, optical fiber) over which transmission occurs

while others allow for multiple sources and receivers to share the same bus. The communication protocol of a bus determines the nature of its use—some protocols provide synchronous communication in which a separate clock signal is used to time each data transfer. Other protocols provide asynchronous communication that is not restricted to specific time slots. Asynchronous communication uses a start and stop bit to identify the beginning and end of each transfer block. Bus protocol can range from simple transfers at the low end of the ISO/OSI reference model to elaborate, high-end protocols with complicated access, validation, and acknowledgement operations.

The following discussion describes selected buses according to the following categories:

- *control:* allows multiple processor to interoperate in a system with exchange of commands and limited amounts of data;
- *data:* augments the control bus with a dedicated path for data transfer;
- *test/maintenance:* low-bandwidth interconnection, small and minimally intrusive, connected to major hardware modules of a system for test and maintenance;
- *input/output:* similar to data bus, but emphasizes raw speed over error checking and data interpretation;
- *peripheral:* combines the features of control, data, and input/output into a single bus used to communicate with computer peripherals;
- *area network:* emphasizes longer-distance transmission to support box-to-box (or building-to-building) communication.

Features of the first five types of buses, intended for communication within one machine, are discussed below and summarized in Table 7–3. Table 7–4 and its subsequent discussion summarizes the area network connection, intended for computer-to-computer communication.

7.2.2.2 Control. The VME (Versa Module Europa) bus has become a standard in Europe and United States. It includes variants for memory extension (VMX), serial operation (VMS), high-speed I/O (VSB), and high throughput (VME-64). The VME is a shared multidrop bus, using tri-state logic so that only one node drives the bus while other nodes monitor. The bus signals are grouped into functional clusters of data transfer, arbitration, priority interrupt, and utility. VME is a parallel asynchronous bus, with capability of data widths of 8, 16, 24, 32, and 64 bits. Its maximum transfer rate is 10 MHz (10 million transfers/sec), which for the maximum data width of 64 bits supports an 80 Mbyte/sec data rate. VME serves as the parent from which several more recent buses have been developed, including the Futurebus+, RACEway, and Scalable Coherent Interface (SCI).

The Futurebus+ is the VME bus with enhancements to achieve a speed limited only by the physics of the backplane. It started as part of the U.S. Navy Next Generation Computer Resources (NGCR) program, which advocated it as a single common bus to which the wide variety of Navy computing devices could connect. It supports a backplane of up to 32 nodes, and its base form provides 64 parallel lines, with variants that accommodate 32, 128, and 256 lines. The Futurebus+ increases throughput by the use of Backplane Transceiver Logic (BTL), which provides incident-wave switching that eliminates setup and hold times. By requiring low capacitance, high current drive, and low (1 volt) swings, it achieves a transfer rate of 25 MHz and a throughput of 200 Mbit/sec (for 64 bits) to 3.2 Gbit/sec (for 256 bits). It is specified by the IEEE Standard 896. While the VME bus provides greater on-board integration and is more mature and widely used, the Futurebus+ provides over an order of magnitude greater data throughput.

The S bus is used by the widely available Sun workstation family. The S bus operates in either a 32-bit burst mode, in which it achieves a throughput of 88 Mbyte/sec, or 64-bit extended transfer mode, attaining 160 Mbyte/sec.

7.2.2.3 Data. The Scalable Coherent Interface (SCI), a descendant of the VME bus, interconnects nodes with a set of unidirectional point-to-point paths. It is *scalable*, since the connection is point-to-point—as processing nodes are added, the number of paths grow, thereby avoiding the bottleneck of shared buses that occurs as an increased number of nodes competes for a fixed amount of bandwidth. The SCI bus is *coherent* in that it uses a cache-coherent protocol to assure data consistency, so that even data that is local to one node is the same across all processors. The SCI achieves a data transfer rate of 1,000 Mbyte/sec using a 16-bit parallel data path.

Table 7–3 Selected internal buses and interconnections.

Usage	Name	Characteristics	Description
Control	VME (Versa Module Eurocard)	• 8-, 16-, 24-, 32-, and 64-bit data • Up to 80 Mbyte/sec, asynchronous	• Evolving to 160 Mbyte/sec • Greater integration than Futurebus • Basis for Futurebus+ and SCI
	Futurebus+	• 64-bit address and data path • 32-bit, 128-bit, and 256-bit variants • 200 Mbyte/sec (at 64-bit data path)	• Speed limited by backplane physics • Will achieve 3.2 Gbyte/sec @ 256-bit data path • VME-to-Futurebus+ bridge available
	S-bus	• 32-bit burst or 64-bit extended • 88 Mbyte/sec (32-bit burst) • 160 Mbyte/sec (64-bit extended)	• S-bus computer system includes controller (for arbitration), master (to initiate transfer), and slave • Used in SUN computers
Data	SCI (Scalable Coherent Interface)	• 16-bit data path • 1000 Mbyte/sec	• Point-to-point • Bandwidth grows as processing nodes are added
	Mercury RACEway® Interlink	• Crossbar switch • 32-bit channels • 160 Mbyte/sec per port	• Allows dynamic bus configuration to obtain ring, crossbar, bus, data-flow, and mesh interconnects • Seeks to decouple software from topology
Test/ Maintenance	JTAG (Joint Test Action Group)	• 4 bits • Small and slow • Serial scan	• Connected to all major modules in circuit • Tests modules by placing in series, inserting known data, and decoding serial response

Input/Output	RS-232/RS-422	• Serial • 20 Kbit/sec (RS-232) to 100 Kbit/sec (RS-422)	• Widely used • Works over long distance
	SCSI (Small Computer System Interface)	• 8-bit data path • 4 MHz transfer rate • 4 Mbyte/sec throughput	• SCSI-2—32 bits, 40 Mbyte/sec throughput becoming available
	Fibre Channel (FC)	• Serial • 12.5 Mbyte/sec – 530 Mbyte/sec • 10m to 10,000m range	• Intended for large block transfers • Serial bidirectional • Point-to-point
Peripheral	PCI (Peripheral Component Interconnect)	• 132 Mbyte/sec for 32-bit transfer • 264 Mbyte/sec for 64-bit transfer	• Displaces older 5 Mbyte/sec ISA personal computer bus • Motivated by PC requirements for multimedia, CD-ROM
	USB (Universal Serial Bus)	• 12 Mbit/sec for medium speed devices; 1.5 Mbit/sec for lower-speed devices • 4-wire cable • 5 meter maximum distance	• Motivated by PC user need for simple bus for multimedia peripherals, joystick, audio,... • Auto-senses new devices
	Firewire	• Serial • 200 Mbit/sec • Governed by IEEE Standard 1394	• Higher-bandwidth complement to USB • Serves devices like digital video camcorders

Table 7–4 Selected external buses and interconnections.

Name	Characteristics	Description
Telecommunications (T1, T2, T3, & T4)	• T1—1.544 Mbit/sec; T2—6.312 Mbit/sec; T3—44.7 Mbit/sec; T4—274.2 Mbit/sec • Serial transmission	• T1—2 wire pairs, one for send, one for receive • T1—24 channels at 8-kHz sampling rate (8 bit samples)
Myrinet	• 8-bit data path • 128 Gbit/sec • 10 m length at full transfer rate; 25 m length at ½ transfer rate	• Point-to-point bidirectional link provides bandwidth that scales with number of nodes
Ethernet	• 10 Mbit/sec (traditional) • 100 Mbit/sec (100 Base-T Fast Ethernet) • 1000 Mbit/sec (Gigabit Ethernet) • 2.5 km length, 0.5 km node spacing	• Uses bus topology – nodes attached to trunk segment • One break can bring down entire bus • Operates over copper wire (twisted pair) • IEEE Standard 802.3
ATM – Asynchronous Transfer Mode	• 155.52 Mbit/sec • 627.08 Mbit/sec and 2.488 Gbit/sec coming	• Fiber – chosen for speed and ability to transmit multimedia (variable-length transmissions of voice, data, imagery) on same line
SONET – Synchronous Optical Network	• 51.84 Mbit/sec to 2.488 Gbit/sec • Supports basic T3 channel	• Optical fiber with electrical/optical conversion at each end
FDDI (Fiber Distributed Data Interface)	• 100 Mbit/sec • LAN circumference of 100 km & connects up to 500 nodes @ 2 km spacing	• Ring topology (2 opposing rings) – reconfigure on break • Optical fiber (multi- or single-mode) • Proposed ANSI standard
TCP/IP – Transmission Control Protocol / Internet Protocol	• Suite of protocols developed for use on Internet • Used for most UNIX implementations	• Protocols include SMTP (Simple Mail Transfer Protocol), FTP (File Transfer Protocol), Telnet (Terminal Emulation Capability), TCP, IP, and others

The RACEway® Interlink bus, also derived from the VME bus and using one of its standard connector types, has been developed by Mercury Computer Systems and is being proposed as an open IEEE standard for connecting multiple processors. It provides six input/output channels of 32 data bits and 8 control bits each, operating at a 40-MHz transfer rate to achieve 160 Mbyte/sec for each data path. The six data paths are arranged as a crossbar, and up to three can operate concurrently. A fat-tree interconnection topology is typically used. Data is transferred in packets, to which higher or lower priority can be assigned and honored at the various crosspoints. Packets can traverse as many as nine crossbars, which allow scaling a system of up to 1,000 processors. The manufacturer provides a performance analyzer that allows tuning application code to hide communication activities behind processing intervals.

A standard software library known as MPI[2] (Message Passing Interface) supports interprocessor communication for several types of parallel multiprocessing supercomputers, including the IBM SP-1 and SP-2, TMC CM-5, Intel Paragon, Touchstone Delta, IPSC 860, Ncube 2, and Kendall Square KSR-1 and KSR–2. It is particularly suited for parallel machines with distributed memory and it provides a set of platform-independent library calls that support such operations as point-to-point communication, collective communication, process groups, communication contexts, and process topologies. It provides bindings for both Fortran 77 and C, and it includes a communication profiling interface. MPI was described in Section 4.2.2.

7.2.2.4 Test and Maintenance. One common test and maintenance bus is JTAG, or Joint Test Action Group. This bus is intended to communication with each significant module within a chip-level or board-level system with a minimum of overhead. It emphasizes smallness at the expense of speed. Thus, it is minimally sized to consist of four lines: 1) test clock, which provides a synchronizing signal for bit-serial data, 2) test mode, which indicates that the bus is in test rather than passive mode, 3) test data in, providing test data patterns into modules, 4) test data out, showing the resultant module response. Its use is discussed more thoroughly in Section 11.6.3.

To use JTAG, a circuit must be designed with the JTAG bus serially connecting all logic modules to be tested. Asserting the test mode signal places the circuit in test mode, neutralizing normal data communication and placing the modules to be tested in series, connected in daisy chain style via JTAG. Known stimulus data is then scanned into the chain, and the series of response data passes through the chain. Test processing decodes the responses of the various modules in the chain and compares the result with an expected value. Differences can be traced back to the failure of specific modules in the chain. The JTAG interface is governed by IEEE Standard 1149.1-1990.

7.2.2.5 Input/Output. The most familiar input/output interconnection, which is also suited for long distance networking, is the RS-232 serial link and its subsequent improved version, the RS-422. The RS-232 can operate at up to 20 Kbyte/sec, with the RS-422 extending that to 100 Kbyte/sec.

The Small Computer System Interface (SCSI) is also used for longer-distance connection. It operates at a 4-MHz transfer rate and uses an 8-bit data bus to attain a throughput of 4 Mbyte/sec. Improvements include the SCSI-2, which is 32 bits wide and can achieve 40 Mbyte/sec data throughput.

The Fibre Channel (FC) interface, originally intended for optical fiber but also used on coaxial cable and twisted pairs, is used for the transfer of large blocks of data such as to and from a disk. It establishes a serial bidirectional point-to-point connection and performs asynchronous communication with variable-length frames. It achieves a scalable data throughput of 12.5 Mbyte/sec to 530 Mbyte/sec, operating at distances from 10 m to over 10,000 m.

7.2.2.6 Peripheral. The PCI, or Peripheral Component Interconnect, arises from the ISA bus common in the IBM-PC-compatible personal computer. The 5 Mbyte/sec (40 Mbit/sec) data rate of the ISA bus was quickly overwhelmed by the communication demands of multimedia CD-ROM data rates. The PCI operates at a 33-MHz rate and provides both 32-bit transfers, achieving a data throughput of 132 Mbit/sec, and 64-bit transfers, for a 264-Mbit/sec rate.

By using reflected wave techniques, the PCI has to drive only half of the required high or low level of the receiver, providing faster turn-on and turn-off times. The PCI was developed by Intel Corporation and has since become an open standard under the coordination of the PCI Special Interest Group.

As multimedia uses of the personal computer increase, so does the need for a simple peripheral bus. The Universal Serial Bus (USB) addresses these needs by providing a serial bus that can be configured in a hierarchical manner. Only one connection is needed to the PC—USB devices can themselves serve as connection ports for other USB peripherals. The bus has both a low-speed version (1.5 Mbit/sec) for slower peripherals like digital tablets and joysticks, and a higher-speed version (12 Mbit/sec) for audio and multimedia output devices.

Extending both the bandwidth and the simplicity of the USB is the Firewire serial bus. Capable of bandwidths up to 200 Mbit/sec, the Firewire provides the speed necessary for such devices as digital video camcorders and professional musical equipment. It is governed by IEEE standard 1394.

7.2.2.7 Network. The most ubiquitous network interconnection is the digital telecommunication carrier used in North America, Australia, and Japan. The telecommunication channel provides two wire pairs, one for transmit and one for receive, to attain full duplex operations. It is based on a collection of voice channels each operating at an 8-KHz transfer rate of 8-bit samples, for a serial bit-stream of 64 Kbit/sec. Twenty-four such channels are multiplexed into one 192-bit frame that contains one sample for each channel (Table 7–4). An additional frame is added for synchronization. Thus, a total bit rate of 1.544 Mbit/sec is provided. Several of these basic channels, known as T1, may be multiplexed into a T2 channel (4 T1s, or 6.312 Mbit/sec), a T3 channel (28 T1s, or 44.736 Mbit per sec), or T4 channel (168 T1s, or 274. 176 Mbit/sec).

The Myrinet[3] provides an open standard interconnection that achieves a low-cost, high-speed local area network (LAN). Myrinet is a point-to-point bidirectional link operating at a transfer speed of up to 1.28 Gbit/sec, over lengths of 10 m. Data is transmitted in variable-length packets. An S bus-to-Myrinet converter and an interface chip, known as LANai, are commercially available.

Ethernet is a common local area network that operates at Level-1 and Level-2 (Physical and Datalink) levels of the ISO OSI Reference Model. Ethernet uses a topology that attaches nodes to a single bus contained in a single coaxial cable, making it easy to route. To transmit, the node determines if the bus is busy, transmits if it is not, and then retransmits if its own receipt of its broadcast transmission is corrupted due to data collision from another source. An Ethernet link may extend to 25 km with a node spacing of less than 0.5 km. Because it is limited to a single line, one break can bring the Ethernet down. Three versions of Ethernet are available, with faster ones being upward compatible with slower ones. Traditional Ethernet offers a data rate of 10 Mbits/sec. The faster 100 Base-T Fast Ethernet provides a tenfold increase in throughput, reaching 100 Mbit/sec, and the newest Gigabit Ethernet provides another order of magnitude improvement, achieving 1000 Mbit/sec.

Displacing the Ethernet for higher-speed applications is ATM (Asynchronous Transfer Mode). ATM is a packet-switched broadband network architecture that forms the core of the ISDN (Integrated Subscriber Digital Network) that underlies modern telecommunications systems. It uses a fixed-length packet of 53 bytes (48 bytes of data preceded by a 5-byte header). The use of a fixed-length packet eases the mixing of multiple sources of variable-length transmissions such as imagery and voice. In its various versions, ATM achieves 155.52 Mbit/sec, growing to 627.08 Mbit/sec and then 2.488 Gbit/sec.

SONET (Synchronous Optical Network) is a fiber-optic-based interconnection with electrical/optical conversion at its input and output. It supports the T3 channel rate of 51.84 Mbit/sec (T3 rate of 44.7 Mbit/sec plus overhead) and can extend up to 2.488 Gbit/sec.

The FDDI (Fiber Distributed Data Interface) serves as a local area network that can extend over a circumference of 100 km and connect as many as 500 nodes at 2-km spacing. It may be implemented by a single or multimode optical fiber and can operate at a throughput of up to 100 Mbit/sec. To counter the single-point failure vulnerability of the Ethernet, the FDDI consists of dual counter-rotating rings—a single break of the fiber does not prevent transmission for FDDI.

Moving from interconnection to protocols, the TCP/IP (Transmission Control Protocol/Internet Protocol) is a suite of protocols developed for use on the Internet and used in most UNIX® implementations. The TCP/IP suite includes the following: SMTP, simple mail transfer protocol; FTP, file transfer protocol; Telnet, terminal emulator capability; SNMP, simple network management protocol; TCP, transmission control protocol; and IP, Internet protocol.

7.2.3 Special Buffers and Structures

Digital signal processing algorithms are characterized by repetitive accesses to data, especially toward the input end of a processing chain as was described in Section 3.2.8. Section 3.2.7 provided examples of typical repeating address patterns. The need for speed and compactness imposed by real-time signal processing applications often requires the use of special structures that exploit this regularity to accelerate the generation of addresses. In software, this may involve placing coefficients in memory in order of use, so that simple address increment operations can access them. An example of placing coefficients in memory and padding an array with zeros to obtain regular addressing was given in Section 6.4.3. Alternatively, simple hardware logic can be developed to implement address generation.

7.2.3.1 Double Buffer. The double buffer, or ping-pong buffer, allows the filling of one half of a memory array for data associated with one analysis block, while the other half, containing data for the previous analysis block, is emptied. This multiplexing is useful for two processors connected in series. While the upstream processor produces results $x_n(m)$, $m = 0,1,\ldots, L-1$ for analysis frame n, the downstream processor begins the next stage of processing by ingesting $x_{n-1}(m)$, which are the results that were produced by the upstream processor during the previous analysis frame. Fig. 7–4 outlines these concepts. While this approach is generally used for block processing, it can be modified for stream processing by reducing the block size L to 1. This technique is used between range and azimuth processing blocks in the example of SAR processing in Section 6.6.3. The input and output data rates into the buffer are the same. The upstream processor can be performing a data-rate reduction calculation as shown in the example of LPC processing in Section 6.4, and this buffer can again be used to keep a downstream processor working on the data-reduced representation. It can also be used as an elastic buffer between processor with different input and output data rates, as shown in Section 4.2.3.

One way of achieving regular, repetitive addressing is through the use of a *circular buffer* (Fig. 7–5). The circular buffer is used for cases in which limited amounts of data are accessed in sequence over and over. For example, filter coefficients are accessed in sequence to multiply a consecutive set of samples, and then the samples are shifted by one and the same set of filter coefficients are again accessed. The address cycle may proceed in increments of one and reset to zero after reaching a maximum value of $N-1$, thereby implementing a modulo N counter. Alternatively, the address cycle may increment by a number $m > 1$, implementing a striding memory that accesses every mth location until exceeding $N-1$, then resetting either back to zero or to the remainder of the quotient N/m. If N is an integer power of two, $N = 2^i$, a simple hardware solution can pick the least significant i bits of a counter to provide modulo N cycling. In software, simple bit masking by means of a logical AND can be used for modulo address generation.

7.2.4 Structures for Data-Dependent Processing

As a signal moves from the input toward the output of an information extraction algorithm, the processing migrates from highly regular operations repeated identically on all data to general, diverse operations performed on information-rich portions of the data. The processing steps carried out on the data toward the output of such algorithms depend upon the data itself.

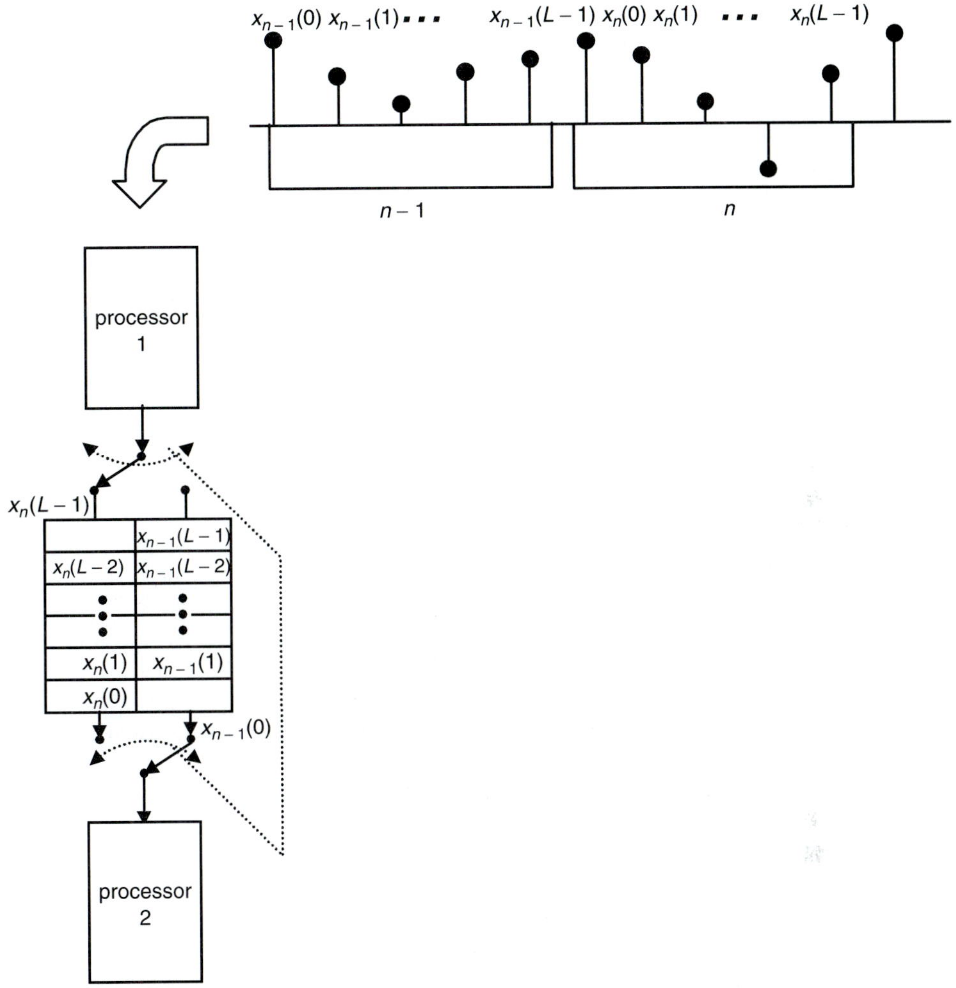

Figure 7–4 Double buffer is filled on one side by Processor 1 while being emptied on the other by Processor 2—after a block of L samples, input and output sides of buffer are exchanged.

7.2.4.1 Sorting. An example of data-dependent processing is sorting a list of numbers by magnitude while keeping track of which position in the original list each number in the sorted list came from. Furthermore, the smallest number is picked from the sorted list and its separation from the second smallest is checked—if the separation is great, this smallest number is returned as well as its position in the original list. This sequence of steps is typical of a signal recognition operation in which a signal is decomposed into multiple adjacent (or sequential) pieces. Each piece is compared to a reference set of known signal pieces, and

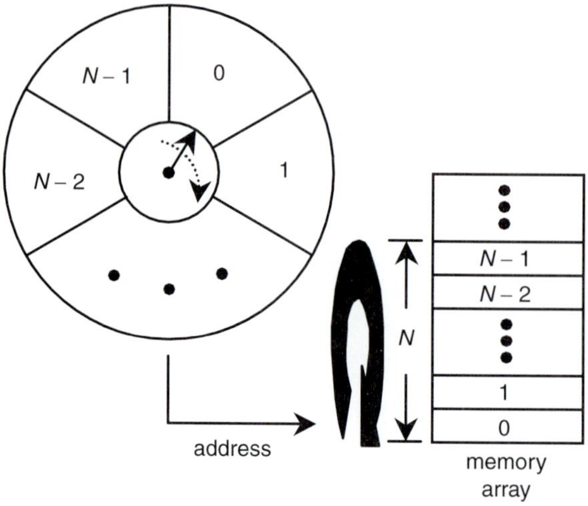

Figure 7–5 Cyclic buffer provides modulo N circular addressing.

the identity of the least-distant reference piece, as determined as the piece with the minimum distance or dissimilarity measure, is used to represent the corresponding section of the unknown signal. Recognition of the entire signal is achieved by combining the results of recognition of these recognized subunits.

7.2.4.2 Minimum Selection. An important set of operations to accomplish hierarchical recognition is the sequence of choosing a minimum from N numbers, identifying the argmin, or index of the minimum number, adding the minimum to an accumulating distance score, and passing back the argmin and updated distance score. The structure of Fig. 7–6 performs these operations. It receives N numbers that correspond to distance values, as shown on the left, and outputs the minimum value s' and the index of that minimum value, t'. Alternatively, the dis-

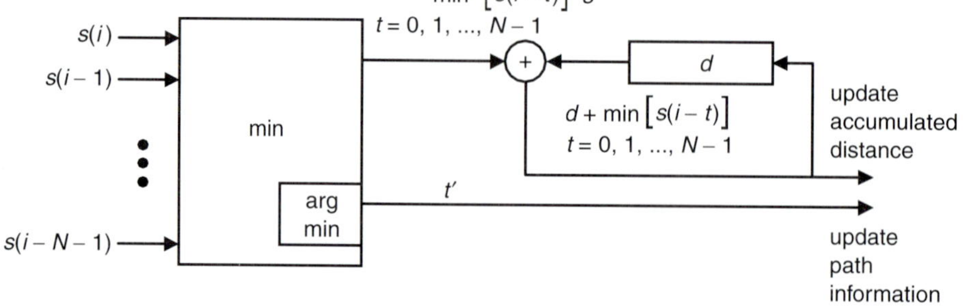

Figure 7–6 Structure for hierarchical recognition processing performs a local minimization over the range given by N and updates a global distance d and path information with t'.

tance measure, which is small for similar patterns, could be replaced by a similarity measure, which increases as two patterns become identical, and the min function would be replaced by a max function. The minimum distance s' is added to a global distance that is the accumulation of many such operations in the past from the scoring of other recognition subunits and the global distance is thus updated. The argmin is sent to a linked list structure that concatenates appropriate recognition subunits that have been identified as local minima in past such computations.

A second structure that supports linked lists receives t' and concatenates the appropriate recognition subunit to a global pattern that is being constructed. This construct, called a traceback structure, consists of two memories (Fig. 7–7). The first is a linked list, and the second is an assembly buffer. The linked list is itself a sequence of paired memory locations in which the first location stores an output object O_i and the second stores a pointer to a previous location in the list that identifies the object to which O_i is adjacent in the global pattern. In the figure, the object associated with an index i is placed at the head of the list, and a pointer $(i-2)$ indicates that this object is preceded in the global pattern by the object stored in location $i-2$. In this manner, an entire pattern can be constructed as the result of recognizing and combining several recognition subunits. This type of processing is used in Viterbi decoding, which finds the most likely state transitions given a sequence of signals corrupted by noise. It is also used in dynamic programming, which finds an overall minimum as a series of local minima, and in speech or image recognition, where phrases or objects are constructed from subword units or object primitives.

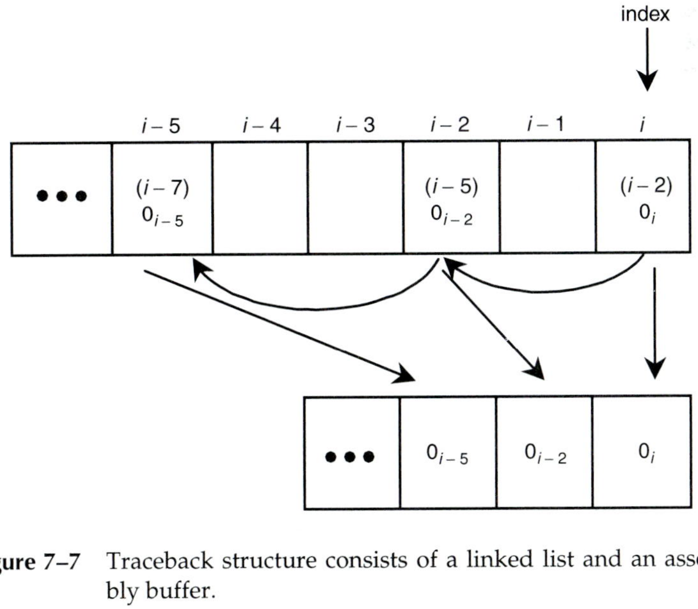

Figure 7–7 Traceback structure consists of a linked list and an assembly buffer.

REFERENCES

1. Vijay K. Jain and Lei Lin, "Square-Root, Reciprocal, Sin/Cosine/Arctangent Cell for Signal and Image Processing," *Proc. IEEE Inter. Conf. Acoustics, Speech, Signal Processing,* Minneapolis, MN, April, 1993, pp. II-521–II-524.
2. Argonne National Laboratory, "MPI: A Message Passing Interface Standard" http://www.msc.anl.gov/mpi/mpi-report/mpi-report.html, 1994.
3. "Myrinet: A Brief Technical Overview," http://www.myri.com/myrinet/overview.html, 1998.

8 Components for Signal Analysis

Signal analysis transforms a signal from one domain to another, for example from the time domain to the frequency domain. By transforming the signal, the intent is to emphasize information in the signal and cast it into a form that is easier to extract. Closely allied to signal analysis is signal modeling. A signal model is a mathematical expression, often based in physical phenomenology, of how a signal is produced. For example, a model of the vocal tract excited by the vocal cords may be used to represent speech signals, and a set of connected regions that form objects can represent an image. Both analysis and signal modeling are used in signal compression, which seeks to reduce the number of bits in a signal without reducing its information content. Signal compression is treated in the next chapter. This chapter explores real-time signal analysis via the Fourier transform and the wavelet transform and examines signal modeling via linear predictive coding, revisiting a discussion presented in Section 6.4 on detailed mapping of LPC to a signal processor.

8.1 BACKGROUND

Two properties underlie the analysis of signals:

1. The inner product of two functions may be used as a measure of the similarity between them.

2. A function can be decomposed into the weighted sum of another set of functions.

Signal modeling uses the first property to seek a function that best matches a model of the data, and signal analysis, using such means as the Fourier or wavelet transform, uses the second property.

8.1.1 Inner Product

To examine the first property, the inner product of two *vectors*, $(\mathbf{g}_1, \mathbf{g}_2)$, is given by $(\mathbf{g}_1, \mathbf{g}_2) = |\mathbf{g}_1||\mathbf{g}_2| \cos \theta_{g1 \text{ and } g2}$, where θ is the angle between them. This inner product is maximum when θ is zero, which corresponds to both vectors pointing in the same direction. The inner product of two vectors may be expressed in terms of the coordinates of the end of each vector in an n-dimensional space, or $(\mathbf{g}_1, \mathbf{g}_2) = \Sigma_{n=1}^{N}(\mathbf{g}_1^n \cdot \mathbf{g}_2^n)$.

In the case of the inner product of two functions as used for signal analysis, one function φ_n is a member of a class of functions that share the same basic properties (e.g., shape) but that are distinguished by one or more parameters, n. The second function is the signal to be modeled, shown here in the time domain as $x(t)$. The inner product of these two functions, where n defines the particular φ_n that is used, is:

$$c_n = \int_a^b \varphi_n(t)x(t)dt. \tag{8.1}$$

Like the vector inner products, the function inner product consists of a sum (or actually, an integral) of products of function components. Also like the vector inner product, the function inner product will be maximum for the value of n that renders $\varphi_n(t)$ most similar to $x(t)$. For example, if $\varphi_n(t) = \sin n\omega_0 t$ and $x(t) = \sin \omega t$, over the interval $[0, 2\pi]$, then

$$c_n = \int_0^{2\pi} \sin(n\omega_0)t)\sin(\omega t)dt. \tag{8.2}$$

This function achieves its maximum when the integrand equals $\sin^2 \omega t$, which occurs when $n\omega_0 = \omega t$, providing $c_n = 2\pi$.

8.1.2 Decomposition into Basis Functions

The second property states that a function $x(t)$ can be *decomposed* into a sum of functions. Again, $\varphi_n(t)$ represents the function set into which $x(t)$ will be decomposed. Specifically, for $x(t)$ defined on the interval $a \le t \le b$, let $\varphi_1(t), \varphi_2(t), \ldots, \varphi_n(t)$ be functions that are all piecewise continuous on this interval. To verify this, we develop a formula for computing the weights c_n such that:

$$x(t) = \sum_{n=1}^{\infty} c_n \varphi_n(t). \tag{8.3}$$

We multiply both sides of (8.3) by $\varphi_m(t)$ and integrate term by term:

$$\int_a^b x(t)\varphi_m(t)dt = \sum_{n=1}^{\infty}\varphi_m(t)\varphi_n(t)dt. \tag{8.4}$$

If φ_n is cleverly chosen so that

$$\int_a^b \varphi_m(t)\varphi_n(t)dt = 0 \text{ for } m \neq n \tag{8.5}$$

then all integrals on the right side of (8.4) become zero except for $n = m$:

$$\int_a^b x(t)\varphi_n(t)dt = c_m\int_a^b [\varphi_m(t)]^2 dt \tag{8.6}$$

so

$$c_m = \frac{\int_a^b x(t)\varphi_m(t)dt}{\int_a^b [\varphi_m(t)]^2 dt}. \tag{8.7}$$

The denominator is a simple gain factor that can be represented as B_m, so:

$$c_m = \frac{1}{B_m}\int_a^b x(t)\varphi_m(t)dt. \tag{8.8}$$

A system of functions $\{\varphi_n(t)\}$, $n = 1, 2, \ldots$, is termed an *orthogonal system* in the interval $a \leq t \leq b$ if φ_m and φ_n are orthogonal for each pair of indices m and n, i.e., if (8.5) is true. If also $\int_b^a [\varphi_n(t)]^2\, dt = 1$, $n = 1,2, \ldots$, then $\{\varphi_n\}$ is also an orthonormal system. Orthonormal functions can be used to provide an invertible transformation of a signal that allows an exact reconstruction of the signal as the sum of basis functions, $\{\varphi_n\}$. However, a signal can be decomposed onto a set $\{\varphi_n\}$ even if $\{\varphi_n\}$ is not orthonormal, as will be shown.

It is often useful to project a signal onto a set of basis functions. Invertible transformations unambiguously represent the signal, and more involved operations such as parameter estimation, coding, and pattern recognition can be performed in the transform domain, where relevant properties may become more apparent.

The choice of function used for $\{\varphi_n\}$ can be made to emphasize signal information content. One goal may be to obtain the broadband spectrum of a signal. The frequency content of a signal can provide a more compact representation of signal information; while a sine wave requires many points to describe in the time domain, it is described by only three numbers—amplitude, frequency, and phase—in the frequency domain. The response of a linear shift-invariant system to a sinusoid is another sinusoid, with phase and amplitude determined by the system. Since any input can be represented in terms of sinusoids (or complex exponentials), the system response may be computed by decomposing a complicated input into sinusoids, computing the system response to each, and then summing the responses to sinusoids to compute the response to the complicated input. Thus, representing signals in terms of sinusoids or complex exponentials is very useful in system analysis.

Therefore, φ_n becomes a complex exponential:

$$\varphi_n \rightarrow e^{-j2\pi ft} \tag{8.9}$$

then

$$c_n \rightarrow X(f) = \int_{-\infty}^{\infty} x(t)e^{-j2\pi ft}dt. \tag{8.10}$$

$X(f)$ is the global frequency content of a signal of frequency f, which is computed as the inner product of the signal and an infinite exponential at frequency f.

8.1.3 Time-Frequency Decomposition—Short Time Fourier Transform

To get the best representation of frequency content, a maximum number of samples must be used. However, if a signal suddenly changes frequency in (8.10) using all the signal samples will cause the frequency change to be smeared over time. A *stationary* signal is one whose properties (e.g., frequency content) do not change over time. To model a time-varying signal, the signal can be broken into sequential analysis blocks, each of which is separately analyzed. A trade-off arises—long analysis blocks provide finer frequency information, but cannot as quickly follow underlying parameter changes. Short blocks more accurately follow the underlying parameter changes, but provide less frequency resolution. Thus, resolution in both time and frequency cannot become arbitrarily small. In fact the product of frequency resolution Δf and time resolution Δt is bounded by $\Delta f \Delta t \geq 1/4\pi$.

The concept of frequency changing in time leads to the idea of using a two-dimensional time-frequency representation of a signal, $s(t, f)$. Like a musical score, this representation places time on the horizontal axis and represents frequency by vertical displacement (Fig. 8–1). Unlike a musical score, Fig. 8–1 has higher frequencies at lower positions on the vertical axis. This is to preserve the vertical axis correspondence with the concept of spatial *scale*, with higher scale (longer dimensions) being placed higher on the vertical axis.

The Short Time Fourier Transform (STFT) is a Fourier transform centered at time τ with an analysis window $g(t - \tau)$, which has some finite width:

$$\text{STFT}\ (\tau, f) = \int x(t)g^*(t - \tau)e^{-j(2\pi ft)}dt. \tag{8.11}$$

The plot of Figure 8–1 shows the STFT and can be viewed as either:

1. All frequencies of STFT of a version of the signal windowed at time τ (column); or
2. Filtering the signal at all times with a bandpass filter that has as its impulse response the window function modulated to frequency f (row).

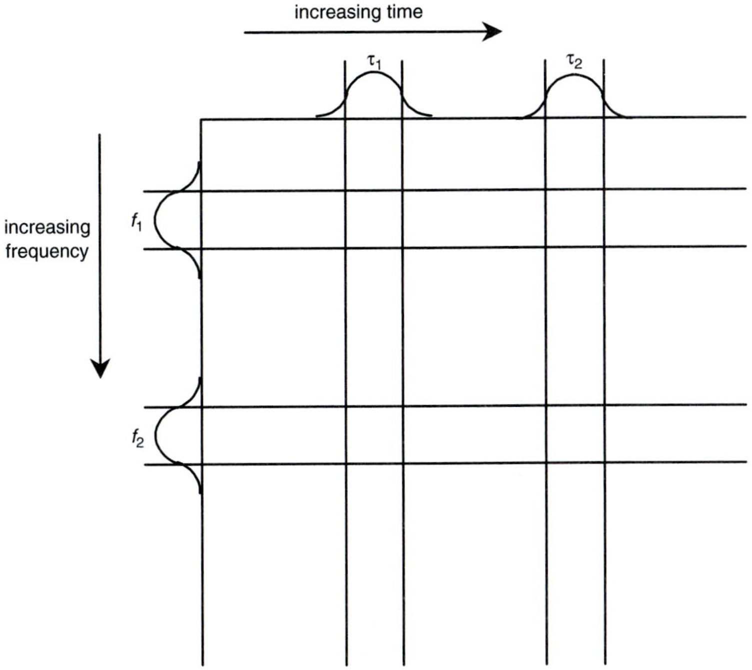

Figure 8–1 Two-dimensional musical score representation of time and frequency, $s(t, f)$.[2] © 1991 IEEE.

8.1.4 Time-Frequency Decomposition—Wavelet Decomposition

The shape and duration of the STFT window is fixed, regardless of frequency. As an alternative to the fixed-window STFT, the wavelet transform uses short windows at high frequencies and long windows at low frequencies. This provides better time resolution but poorer frequency resolution at high frequencies, and better frequency resolution with poorer time resolution at low frequencies.

Like the Fourier transform, the wavelet transform may be regarded as a signal decomposition onto a set of basis functions. For the wavelet transform, these basis functions are obtained from a single prototype wavelet $h(t)$ by means of dilations and contractions, known as *scalings*, as well as translational shifts along the time axis. The scale is modified to achieve a constant ratio of the filter frequency to its bandwidth, $f/\Delta f = Q$. In this way, the prototype wavelet behaves like a constant-Q bandpass filter.

The scaling of the prototype wavelet is governed by the scale parameter a as follows:

$$h_a = \frac{1}{\sqrt{a}} h\left(\frac{t}{a}\right). \tag{8.12}$$

As a increases, $h_a(t)$ is stretched out in time, and as a decreases, $h_a(t)$ is contracted. Since the local frequency is no longer linked to that of the window, the term "scale" is used instead of "frequency." Translation corresponds to moving along the time axis by an amount b*:

$$h_b(t) = h(t - b). \tag{8.13}$$

Scaling and translation are combined into:

$$h_{a,b}(t) = \frac{1}{\sqrt{|a|}} h\left(\frac{t - b}{a}\right). \tag{8.14}$$

The factor $1/\sqrt{a}$ is introduced to assure that the total energy of $h_{a,b}(t)$ remains independent of a. As a result, the wavelet transform $\mathbf{W}(a, b)$ of a function $x(t)$ is given by

$$W(a,b) = \frac{1}{\sqrt{a}} \int x(t) h^*\left(\frac{t - b}{a}\right) dt. \tag{8.15}$$

Recall that the inner product of two functions can be used as a measure of correlation, or similarity, between two signals. Thus, another view of the wavelet transform is to regard $W(a, b)$ as finding which basis function of shape $h(t)$ is most similar to the signal, as indicated by the a and b that provide the maximum $W(a, b)$. The original waveform may be synthesized by adding elementary signals of the same shape as $h(t)$ but of various scales and translations. For an orthonormal set of $H_{a,b}(t)$, an analysis/synthesis operation allows exact reconstruction of $x(t)$:

$$x(t)c = c \iint_{a>0} W(a,b) h_{a,b}(t) \frac{da\,db}{a^2}. \tag{8.16}$$

Even if the $h_{a,b}(t)$ functions are not orthogonal, they behave orthogonally in that the reconstruction formula (8.16) works as long as $h(t)$ is of finite energy and is bandpass (i.e., oscillates over a limited extent, like a short wave or wavelet). If both properties are true, then $\int h(t)dt = 0$, which provides an easy test on the validity of possible $h(t)$ functions.

8.2 FOURIER TRANSFORM

The Discrete Fourier Transform (DFT) is first examined to understand the basic operations involved; the more computationally efficient Fast Fourier Transform (FFT) is then examined in detail and hardware structures are described to speed its execution.

---

*Having introduced the concept of time translation by an amount τ, the variable τ is now replaced by b for consistency with the customary notation for time in the wavelet literature.

8.2.1 Discrete Fourier Transform

The discrete Fourier transform (DFT) sets the basis function φ_n to become

$$\varphi_n \rightarrow e^{-j(2\pi nk/N)} = W^{nk}, \text{ where } W = e^{-j(2\pi/N)}. \tag{8.17}$$

The DFT of a function $x(n)$ is then given by

$$X(k) = \sum_{n=0}^{N-1} x(n)e^{-j(2\pi nk/N)} = \sum_{n=0}^{N-1} x(n)W^{nk};\ 0 \leqslant k \leqslant N. \tag{8.18}$$

The DFT requires the order of N^2 (henceforth written $O(N^2)$) operations to compute.

8.2.2 Fast Fourier Transform

Another version of the discrete Fourier transform known as the fast Fourier transform (FFT) requires $O(N \log_2 N)$ calculations, which for large N is significantly less than the computation required by the DFT. However, the FFT has two requirements that can render it less efficient than the DFT in some situations:

1. The FFT must be computed at all N frequencies at once, while the DFT can be calculated at just a few frequencies of interest. Thus, answering such questions as "How much 60 Hz component is there in this signal?" can be computed for that one frequency far more simply with the DFT than with the FFT.
2. The FFT operates only with *complex* data. Therefore, if one is working with real data only, half of the FFT results must be computed and never used. Although these unused values are easily discarded, computational effort is expended to produce them and further effort is expended to remove them.

The key computation kernel to the FFT is the butterfly (Fig. 8–2) which consists of:

1. Take two complex words at a time from memory;
2. Perform a complex multiply and two complex adds;
3. Put answers back into memory.

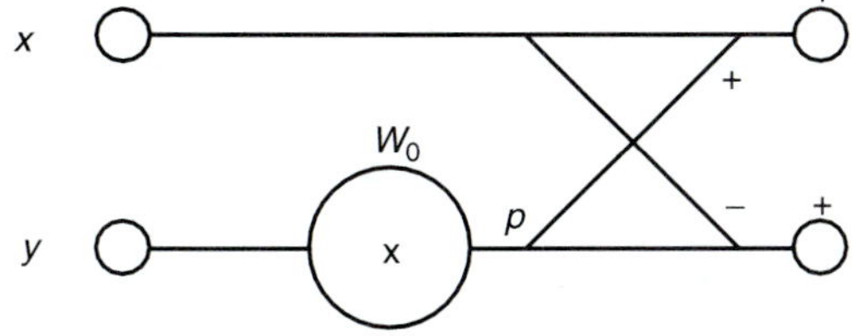

Figure 8–2 FFT butterfly takes complex words x and y from memory, multiplies by W_0 to obtain p, then adds p to x for one output and subtracts it from x for the other output.

The complex factor W_n is defined as

$$W_n = e^{j\frac{2\pi n}{N}} = \cos\left(\frac{2\pi n}{N}\right) + j\sin\left(\frac{2\pi n}{N}\right). \tag{8.19}$$

It is used in the butterfly to form the product p with one of the inputs, y, as shown in Fig. 8–2.

Two arrangements of the FFT are commonly used:

1. *In place:* places results back into same memory from which input was taken; requires single memory array;
2. *Write answers in location other than input array,* thereby doubling memory requirements but simplifying addressing and control.

With the definition of a butterfly in mind, the structure of an 8-point FFT can be drawn using an in-place structure (Fig. 8–3).[1] The concept of *bit reversal* is defined as reversing the order of individual bits of a number, $x = b_{N-1}b_{N-2}\ldots b_1b_0$, to obtain $x' = b_0b_1\ldots b_{N-2}b_{N-1}$. Bit reversal is used in Fig. 8–3 to reorder the storage of the input vector $x_0, x_1, \ldots, x_7$ as follows. The left column expresses the addresses in consecutive order using decimal notation, with the corresponding binary notation shown in the next column. The third column performs the bit reversal, resulting in the bit-reversed address shown in decimal form in the fourth column, next to the inputs of the butterflies. For an N-point FFT, $\log_2 N$ butterfly stages are used, with the cross

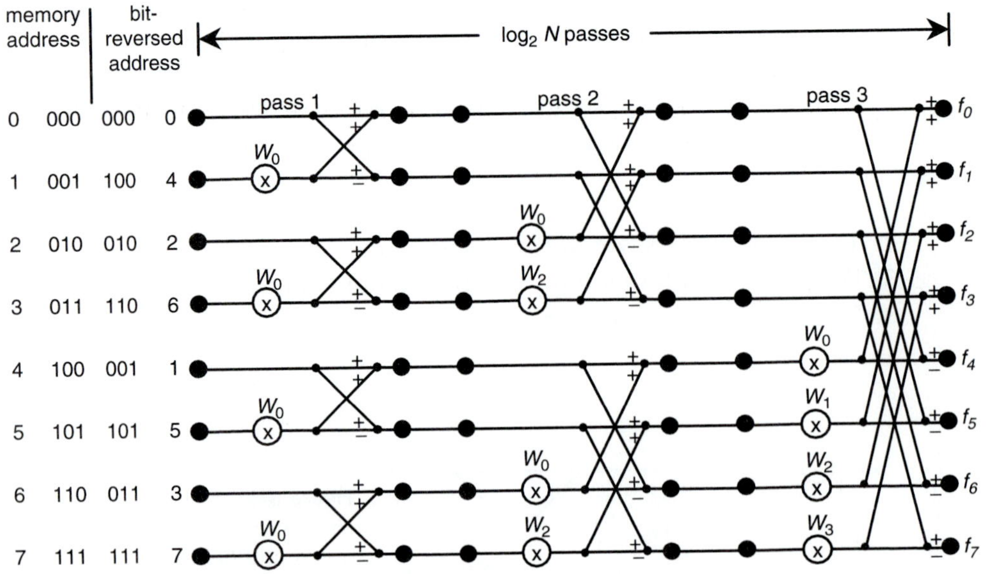

Figure 8–3 In-place FFT algorithm for $N = 8$ includes $\log_2 N = 3$ stages of butterflies which reach across 2^n, $n = 1, 2, \ldots, \log_2 N$.

branches reaching across one row in the first stage, two rows in the second, and four rows in the third stage. The output presents the values $X(0), X(1), \ldots, X(7)$.

A basic FFT circuit is shown in Fig. 8–4. The controller counts g groups and stimulates the execution of b butterflies per group. The controller generates addresses for W_n and for the RAM containing input data and output results. The butterfly circuit includes a multiplier and adder and is devoted to computing the butterflies. The sin/cos generator is either a table lookup or logic to generate the necessary values of sine and cosine.

Two variations on the in-place FFT are now considered. In the first (Fig. 8–5), the inputs, outputs, and W_n are kept in natural order, with no bit reversal needed. This arrangement requires twice the memory as the in-place algorithm. The memory is divided into two N-complex-word blocks, with one serving as sender memory and the other acting as receiver. After the completion of one pass of N points, the memories are switched in ping-pong style, reversing the roles of sender and receiver. As shown in Fig. 8–5, the x and y butterfly inputs are always applied to branches that are four rows apart—the x address is always 0, 1, 2, or 3, and the y address is always $x + 4$.

In a second variation, the addressing may be kept constant for each pass, as shown in Fig. 8–6. This constant geometry architecture requires bit reversal of address values, but because the address sequence is the same for each pass, it may be stored as a sequence in memory rather than computed.

Yet other variations on the FFT are possible. For example, a four-point butterfly ("Radix-4") may be used, combining the first and second butterflies of the first pass with the first and second butterflies of the second pass. In the FFTs described so far, the input stream has been divided into smaller subsequences. Instead, the output sequence may be similarly subdivided, working backward from the output. This technique is referred to as decimation in frequency, in contrast to the decimation in time approach discussed above.

To apply the FFT to a two-dimensional data set such as an image, the notation $W_N^{nk} = e^{-j(2\pi/N)}$ is used to write:

$$\begin{aligned} X(k,l) &= \sum_{m=0}^{M-1} \sum_{n=0}^{N-1} x(m,n) e^{-j(2\pi km/M)} e^{-j(2\pi ln/N)} \\ &= \sum_{m=0}^{M-1} \sum_{n=0}^{N-1} x(m,n)\, W_M^{km}\, W_N^{lm}; \end{aligned} \tag{8.20}$$

where $k = 0, 1, \ldots, M-1$ and $l = 0, 1, \ldots, N-1$.

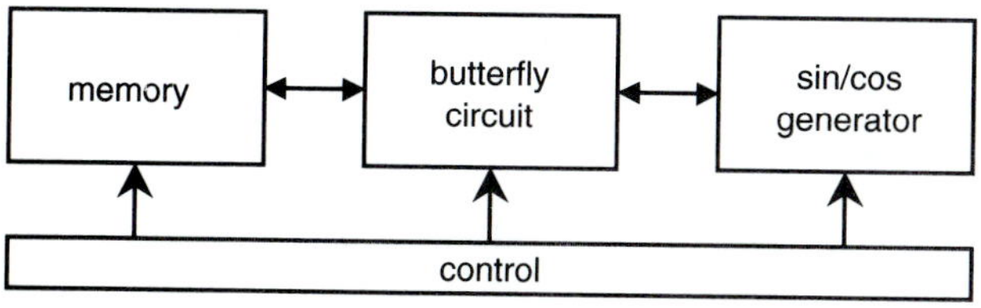

Figure 8–4 Basic circuit for FFT.

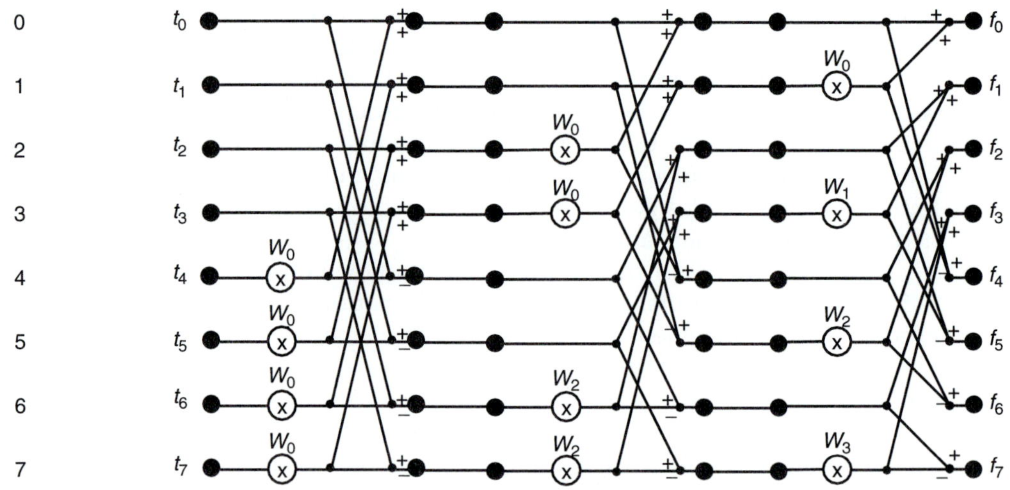

Figure 8–5 Alternative structure for FFT simplifies addressing by removing need for bit reversal at the expense of double the memory requirement of the in-place technique.

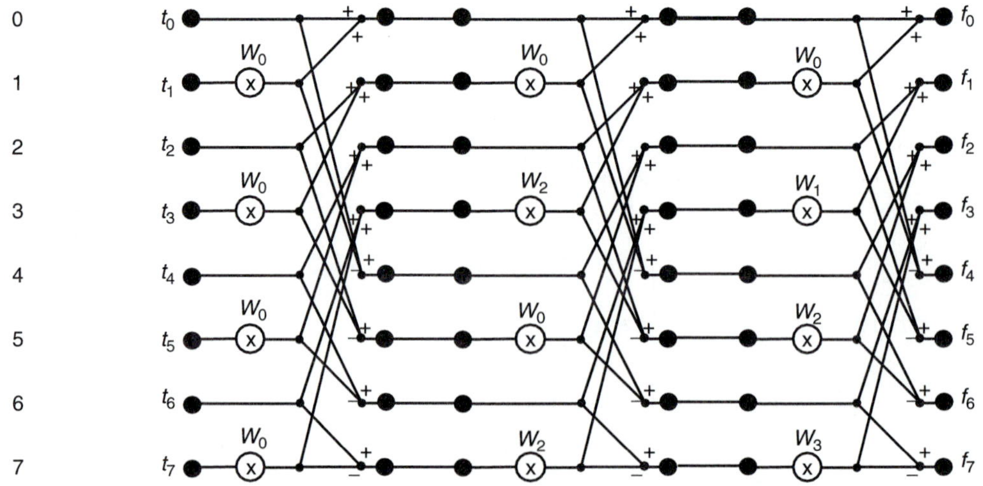

Figure 8–6 Constant geometry FFT architecture keeps address calculation constant for each pass, but requires bit-reversal addressing.

The expression (8.20) may be viewed as M discrete Fourier transforms (DFTs) of form $A(m, l) = \Sigma_{n=0}^{N-1} x(m, n) W_N^{ln}$ followed by N DFTs of form $x(k, m) = \Sigma_{m=0}^{M-1} A(m, l) W_M^{km}$. An FFT rendering of (8.20) requires $NM/2 \log_2(NM)$ complex multiplications, instead of the N^2M^2 multiplications required by a direct FFT approach.

8.3 WAVELET TRANSFORM[2,a]

The wavelet transform, $W(a, b)$, of a signal $\tilde{x}(t)$is given by

$$W(a,b) = \frac{1}{\sqrt{a}} \int \tilde{x}(t) h\left(\frac{t-b}{a}\right) dt \tag{8.21}$$

where $h(t–b/a)$ is the wavelet corresponding to the prototype wavelet $h(t)$ at scale a and time b.

8.3.1 Continuous versus Discrete Wavelet Transform

In this most general case, $\tilde{x}(t)$ is a continuous signal, and scale and time parameters a and b are continuous. This is referred to as the continuous wavelet transform (CWT). For $x(t)$, which is a sampled version of $\tilde{x}(t)$, the wavelet transform can be discretized on a grid whose samples are arbitrarily spaced in both time b and scale a:

$$W(a,b) = \frac{1}{\sqrt{a}} \sum_{t=b}^{t=aL+b-1} x(t) h\left(\frac{t-b}{a}\right); \tag{8.22}$$

where $h\left(\frac{t-b}{a}\right)$ is obtained by sampling the prototype wavelet;

L = size of support of h (extent of h);
N = number of input samples;
J = number of scales, $1 \le a \le J$.

To motivate a particular discrete sampling of a and b, two analysis cases are examined.[2] The first case builds a multiresolution pyramid.[3] The input signal $x(t)$ has a bandwidth B (Fig. 8–7). From it, a lower-resolution signal is derived by low pass filtering with bandpass filter of bandwidth $B/2$ having impulse response $g(t)$. Since the bandwidth of this signal has been halved, its sampling rate may also be halved, which is performed by dropping every other sample.

This low pass signal $y(t)$ is now used to construct an approximation of the original signal. It is upsampled by a factor of two, which is accomplished by placing a zero sample between each sample. The resulting signal is interpolated with

---

[a]Portions adapted, with permission, from Olivier Rioul and Martin Vetterli, "Wavelets and Signal Processing," *IEEE Signal Processing*, 8 (October, 1991), 14–38.

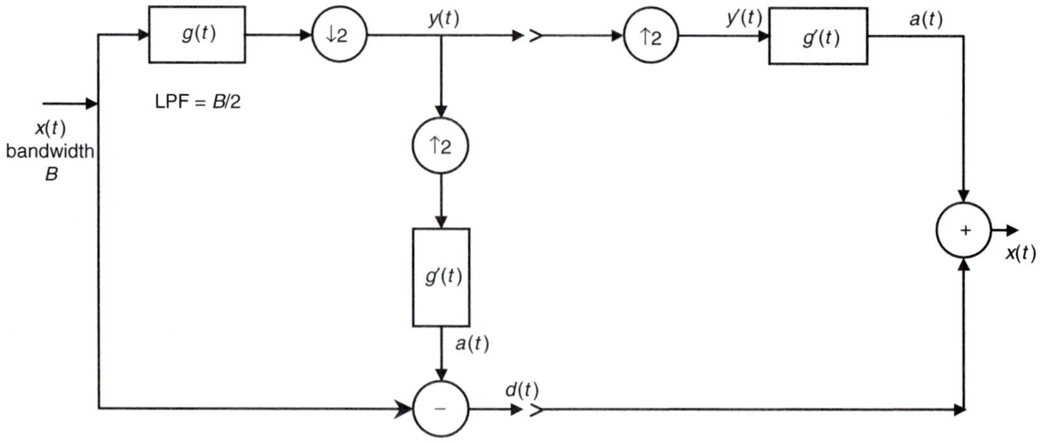

Figure 8–7 Multiresolution pyramid decomposition of $x(t)$ creates subsampled lowpass signal $y(t)$ that is then reconstructed and differenced with the original signal.[2] © 1991 IEEE.

a filter $g'(t)$ to obtain the approximation $a(t)$. The sample-by-sample difference $d(t)$ is computed between $a(t)$ and the original signal $x(t)$. The original signal can then be reconstructed exactly by forming $x(t) = a(t) + d(t)$. If the filters are perfect half-band low pass filters, then the frequency of $d(t)$ is restricted to lie between $B/2$ and B, meaning that its sampling rate can be reduced by a factor of two. By this manner, the signal $x(t)$ is separated into a coarse approximation $a(t)$ plus some added detail contained in $d(t)$. A hierarchy of lower resolution signals at lower scales can be created by repeating the process on $y(t)$.

The second analysis approach is that of the subband filter bank.[4] Like multiresolution analysis, this approach begins with the calculation of a low pass subsampled version of the signal. Instead of computing a difference signal, the subband case computes a high pass version of $x(t)$ and subsamples it by a factor of two. Thus, unlike the case of the multi-resolution pyramid, the total sampling rate is maintained, although it is now distributed across two channels. The signal is reconstructed by upsampling each branch, passing them through an interpolation filter, and adding them back together (Fig. 8–8). If the filters have the property that $h(L - 1 - t) = (-1)^t g(t)$ and L is even, then perfect reconstruction is achieved.

The low pass filtering and downsampling can be continued on the low-frequency branch. This recursion improves frequency resolution as the bands become narrower, but decreases the temporal resolution as the sampling rate becomes lower (Fig. 8–9).

In the Discrete Wavelet Transform (DWT), the role of the wavelet is played by the high pass filter h(t) and the cascade of low pass filters. These filters correspond to octave band filters (a factor of two change in frequency is termed an oc-

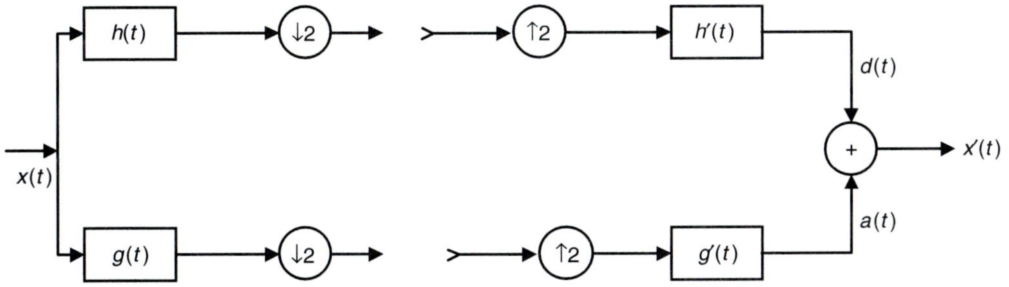

Figure 8–8 Subband analysis approach splits signal $x(t)$ into upper halfband with filter $h(t)$ and lower halfband with filter $g(t)$, subsamples, then reverses the process to reconstruct $x(t)$.[2] © 1991 IEEE.

tave). For a sample sequence of length N, the DWT produces $N/2$ samples at the output of the highest resolution ($y_1(t)$), $N/4$ at the next ($y_2(t)$), and so forth. Thus, the scale variable a is a power of two: $a = 2^k$. The scale index k ranges from 1 to J, where J is the total number of scales. For a total number of samples N, in which N is a power of two, $J \leq \log_2 N$.

The original definition of the continuous wavelet transform (CWT) was a transformation with no decimation in time or in scale, and with arbitrary tiling of the scale space, given by a, and time translation, given by b. More recently,[5] the CWT has been defined in manner similar to the DWT, but without the twofold subsampling of each channel, so that for an input of N samples, *each* scale provides N samples. The CWT has also been used to provide sampling with finer

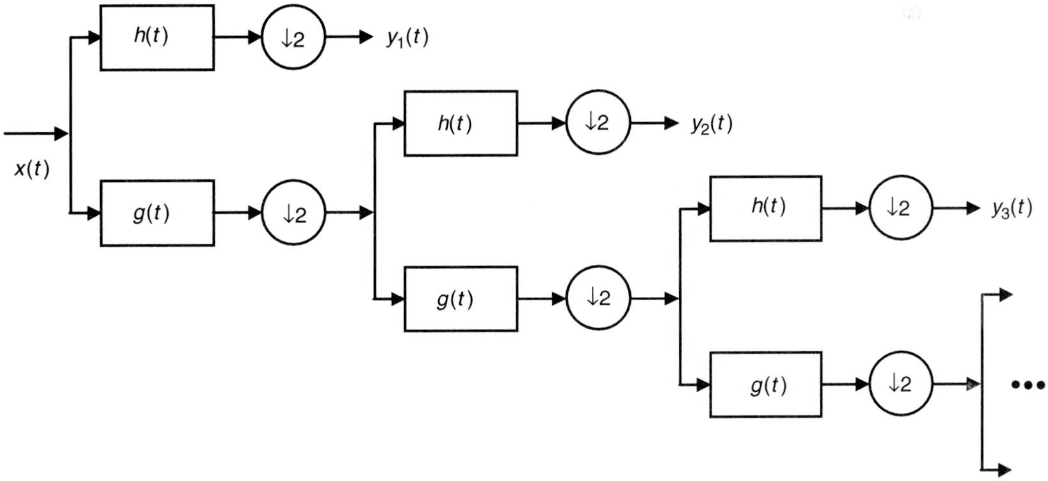

Figure 8–9 Recursive application of two-channel subband frequency decomposition: $h(t)$ is highpass filter; $g(t)$ is lowpass filter.[2] © 1991 IEEE.

scale resolution than the octave: Instead of scale $a = 2^k$, a scale $a = a_0^{\ k}$, $1 \le a_0 \le 2$, is used, where if $a_0 = 2$, the octave filtering is recovered.

8.3.2 Structures for Computing the DWT

Structures for computing the DWT must address both the application of the filter and the subsampled recursion through the scales. To implement the mother wavelet, a form of the regular polyphase filter structure known as the *quadrature mirror filter* (QMF) has been used.[6] The two-band QMF structure takes advantage of the fact that coefficients of the upper and lower bands of the filter are identical except for the signs of the odd-numbered coefficients, and they exploit the fact that subbands are reduced in sampling rate by a factor of two. Fig. 8–10 shows a block diagram of a two-band QMF filter bank implemented on a single-chip programmable digital signal processor. Because of the sampling rate reduction, the computation can be alternated over two time slots, Cycle 0 and Cycle 1. In Cycle 0, the odd taps for the filter are computed for the low band, and on Cycle 1, the even taps are computed for the high band. In the figure, most operations of the upper filter are computed on Cycle 0, while the operations of the lower filter occur on Cycle 1. The results of the two branches are summed with cross-branch connectors to obtain two bands.

Another technique[8] to improve the efficiency of filter computation is particularly effective for long filters ($L > 64$). This method applies the FFT to compute the filter in a manner similar to computing long convolution by performing the FFT, performing multiplications, and then performing the inverse FFT as discussed in Section 4.4.5. The use of the FFT changes the complexity of the DWT for a execution length of L from $2L$ to $4 \log_2 L$. In practice, however, most DWT filters are too short to gain from applying the FFT.

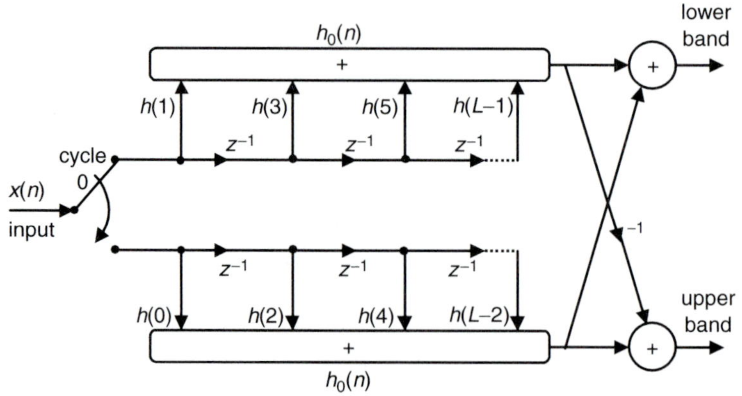

Figure 8–10 Efficient quadrature mirror filter bank structure.[7] © 1982 IEEE.

The recursive subbanding and subsampling of the DWT may be used to populate a plot of time versus scale processing and apply a processing cell to each row (Fig. 8–11).[9] Each input processing cell contains a high pass filter $h(t)$ and a low pass filter $g(t)$, whose output is passed to the next scale. At each scale, the number of processing intervals decreases by a factor of two.

The complexity of the DWT remains bounded as J, the number of scales, increases. This may be seen as follows. Each filter $h(t)$ and $g(t)$ of length L requires L multiplications and $L - 1$ additions for every two inputs. At octave J, $1 \le j \le J$, the input is subsampled by 2^{J-1}, so the total complexity for J octaves is $(1 + 1/2 + 1/4 + \ldots + 1/2^{J-1}) = 2(1 - 2^{-J})$ times the complexity of $h(t)$, or $2\,L\,(1 - 2^{-J})$ multiplications per point and $2\,(L - 1)\,(1 - 2^{-J})$ additions per point. The complexity remains similar to that of a filter of length $2L$, regardless of J. If each filter is executed directly and without exploiting the dilation property of $h(t)$, the computation becomes proportional to JL. As a result, the DWT algorithm may be considered a fast algorithm as a result of the complexity reduction from JL to L. These techniques do not make any assumptions about the wavelet itself, except for its length. If the wavelets are orthogonal, the analysis and synthesis filters are the same and computation drops by 25 to 50 percent.

It is instructive to examine the timing of computing DWT stages on a single computation unit (like a programmable signal processor) with a computation latency T of one time unit and with $L\,(\log_2 N - 1)$ cells of storage.[9] The Recursive Pyramid Algorithm (RPA) is used to rearrange the order of N outputs such that an output is scheduled as soon as possible. A precedence relation is used to schedule outputs, as follows. If the earliest instance of octave i clashes with that of octave $i + 1$, then octave i is scheduled first. Thus, the first octave produces outputs at every other cycle, and the higher octaves are scheduled between the first

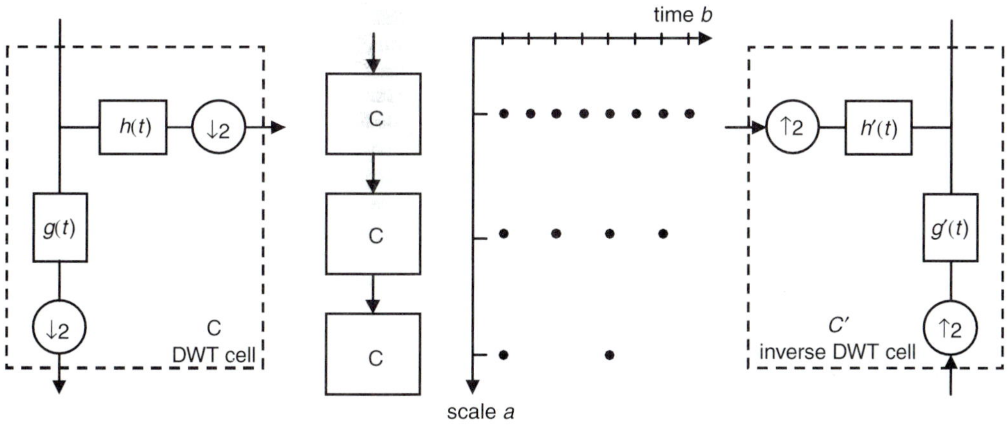

Figure 8–11 DWT processor cell C and inverse DWT processor cell C′ are each used at each DWT scale to perform computation.[8] © 1992 IEEE.

octave. For $y(t)$ as the nth output of octave i, the schedule for $N = 8, J = 3$ is $y_1(1)$, —, $y_2(1)$, —, $y_1(2)$, —, $y_3(1)$, —, $y_1(3)$, ...

For structures with a computational latency of greater than one cycle ($T > 1$), the RPA algorithm is modified to allow scheduling within computational latency times (MRPA, for modified RPA). A quantity off(t) is introduced as representing the additional delay in the scheduling of the first output at the kth octave that is caused by a scheduling conflict with octaves of indices less than k. Inputs for the ith octave are fed in at times given by:

$$(i - 1)T + \sum_{k=1}^{i} \text{off}(k) + 2^i m; \text{ where } 0 \leqslant m_i \leqslant \frac{n}{2^i}. \tag{8.23}$$

A recursive definition of off(k) is given by

$$\text{off}(k) = 1, \text{ for } k = 1; \text{ else}$$

off(k) is the smallest integer such that

$$\text{off}(k) \neq 2^j m_j - (k - j)T - \sum_{p=j+1}^{k-1} \text{off}(p); 0 \leqslant m_j \leqslant \frac{N}{2^j}. \tag{8.24}$$

Fig. 8–12 shows the timing diagram, indicating a latency of 4 and slower data rates at higher octaves.

A custom architecture for executing the DWT uses a pair of parallel filters (one for low pass and one for high pass) and uses the MRPA scheduling algorithm (Fig. 8–13). The architecture contains LJ storage locations to store J subbands of L inputs each, implemented as J serial in/parallel out registers of length L. Each parallel filter uses L fixed multipliers and a tree of $L - 1$ adders. For a multiply time of t_m and an addition time of t_a, a latency of $t_m + t_a \log_2 L$ applies. Every $2i - 1$ cycles, an output is written into the ith shift register, $1 \leq i \leq J$. All L elements of the shift register are passed to the parallel filters every $2i$ cycles. Control signals c_i and a_i, $1 \leq i \leq J$, enable the AND-gating in of signals that flow to or from the filters at the times specified by the MRPA schedule. The circuit consists of $2L$

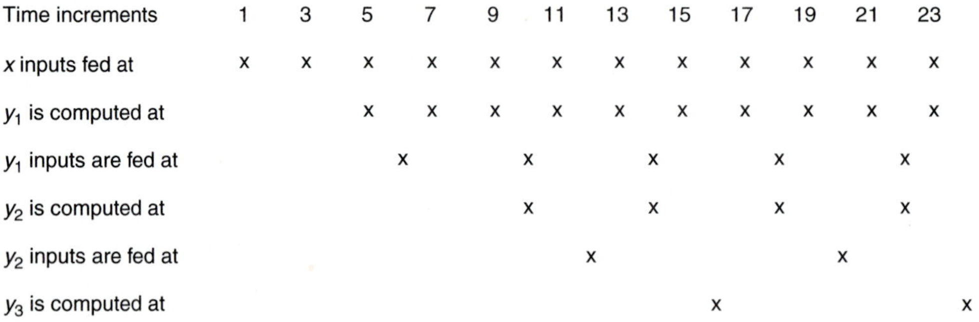

Time increments
1 3 5 7 9 11 13 15 17 19 21 23
x inputs fed at
y1 is computed at
y1 inputs are fed at
y2 is computed at
y2 inputs are fed at
y3 is computed at

Figure 8–12 Timing for DWT architecture that has latency $T > 1$ ($T = 4$ in this example).[9] © 1995 IEEE.

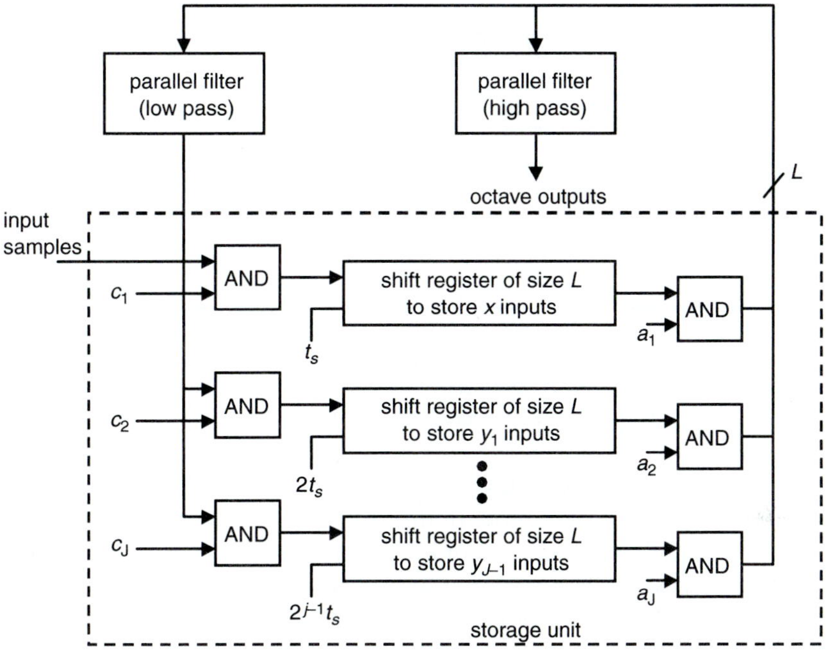

Figure 8–13 Parallel filter implementation of one-dimensional DWT.[9] © 1995 IEEE.

multipliers, $2(L - 1)$ adders, and JL storage locations. Its computation period is approximately equal to the number of samples, N.

A single instruction multiple data (SIMD) processor array may be used for faster execution of the DWT. A key element of a SIMD array implementation is a processor with a reconfigurable bypass switch. If the switch is set to 1, data passes through (or around) the processor with no delay and no processing, and if the switch is set to 0, the processor performs one cycle of processing on the input and produces an output. In Fig. 8–14, for octave i, $1 \le i \le J$, an N-processor array is configured with the bypass switches to become an $N/2^{i-1}$ array, with processors $p(2^{m-1}j)$ active. In the SIMD array, all outputs of any particular octave are computed at once. The high pass and low pass coefficients are broadcast to each processor. Each processor computes the products of the data elements and the filter coefficients and updates the partial results of the low pass and high pass outputs. The data element is then sent to the right-hand neighbor. In the linear array, the time to compute an octave is L; the time to compute J octaves is LJ.

The DWT computation period can be reduced to L by using a multigrid architecture of size $2N - 1$ processors ($N/2^i$ processors in each level, $0 \le i \le \log_2 N$). For a DWT on N points, a multigrid of size N is used (Fig. 8–15). The computation at octave i outputs high and low bands, sending the low band to level $i + 1$.

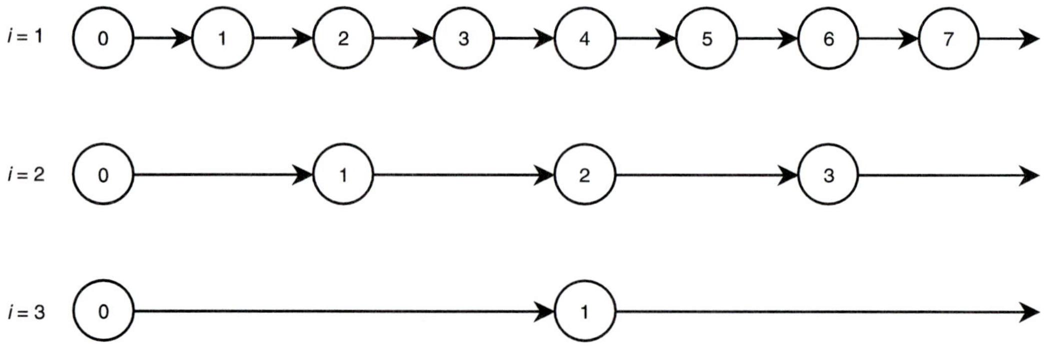

Figure 8–14 SIMD array bypasses intermediate processors for higher octaves ($N = 8, J = 3$).[9] © 1995 IEEE.

The above circuits for computing the one-dimensional DWT can be compared on the basis of number of multipliers, area, and computation period. In this comparison, B is the number of bits of precision used in the implementation. One commercial DWT chip, known as Accupress from Aware, Inc,[10] uses a four-tap filter and external memory and relies on software from computing the DWT. Table 8–1 shows the available tradeoffs of area and space.

The two-dimensional DWT is used for two-dimensional signals such as occur in image processing. As with the one-dimensional case, both a parallel filter and SIMD array are discussed.

In the parallel filter implementation, a two-dimensional $L \times L$ filter is centered on various rows in the array, with the row spacing a function of scale and

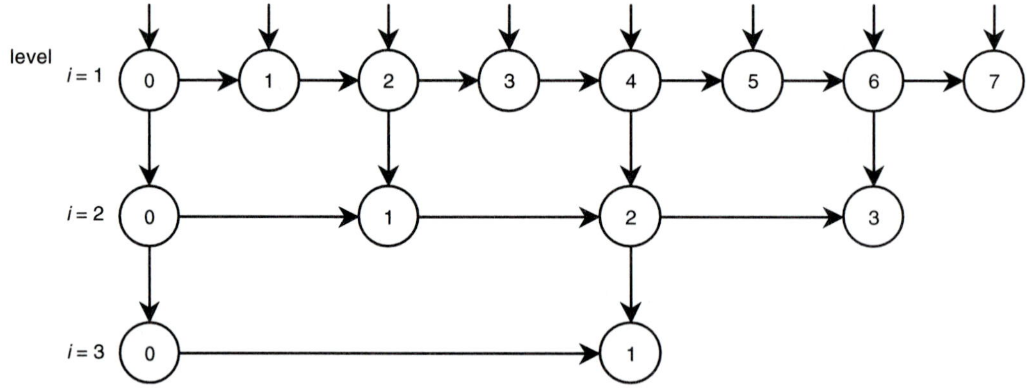

Figure 8–15 SIMD grid applies N processors to a DWT on N points, with $N/2$ processors on the first row and each subsequent row with $1/2$ the number of processors of the preceding row ($N = 16$ in this example).[9] © 1995 IEEE.

Table 8–1 Comparison of commercial and conceptual architectures for computing one-dimensional DWT (*N* = number of points in DWT, *J* = number of octaves, *L* = length of filter, *B* = number of bits).[9] © 1995 IEEE.

Architecture	# Multipliers	Area	Period
Aware WTP	4	$O(NB)$	$O(N \log N)$
Parallel filter	$2L$	$O(LJB)$	$\approx N$
SIMD linear array	$2N$	$O(NB)$	LJ
SIMD multigrid	$2NJ$	$O(NJB)$	L

higher-octave computations interspersed between lower-octave processing. A specific technique of interspersing rows is:

- place lowest octave on even rows, higher octaves on odd rows;
- center the first octave computation on rows 0, 2, 4, 6, 8, 10, 12, and 14;
- center second octave on 1, 5, 9, 13;
- center third octave on 1, 9, . . .

The filters consist of L^2 programmable multipliers and a tree of $L^2 - 1$ adders. Each filter computes the output of two bands, and the output of the lower band is stored in a storage unit for input to higher octave computations. The filter latency is $t_m + 2\, t_a \log_2 L$, where t_m is the multiplier latency and t_a is the adder latency.

Fig. 8–16 shows the parallel filter architecture for the two-dimensional DWT structure. A total of J shift registers provide serial in, parallel out operation. Each shift register has a two-dimensional array of storage cells of size $L \times N/2^{i-1}$, where $1 \le i \le J$. New data is shifted into the last row of the unit, and when the last row is filled, the data is shifted up by one row in the array. Each shift register has a row clock, with a period that is 2^{1-i} times the sample period, and a column clock, whose period is N times the row period, with a duty cycle of 2^{1-i}. Data is written into a shift register unit when both row and column clocks are high. For $i = 2$, L data values are read out when every second row clock and every second column clock are high. As before, the timing of the clock signals is set by the MRPA algorithm. The parallel filter architecture for the two-dimensional DWT consists of $2L^2$ multipliers, $2(L^2 - 1)$ adders, and NL storage cells, as well as a control unit. Both the computation time and the period are proportional to N^2.

The two-dimensional DWT can also be mapped onto an SIMD array of $N \times N$ processors. This mapping is achieved by following the row computations of each octave by column computations for that octave. The SIMD array is then reconfigured for octave i, using bypass switches as discussed for the one-dimensional DWT, to form a $N/2^{i-1}$—row array of size $N/2^{i-1}$ for the row computations, which is then reconfigured into $N/2^i$ elements of size $N/2^{i-1}$ for the col-

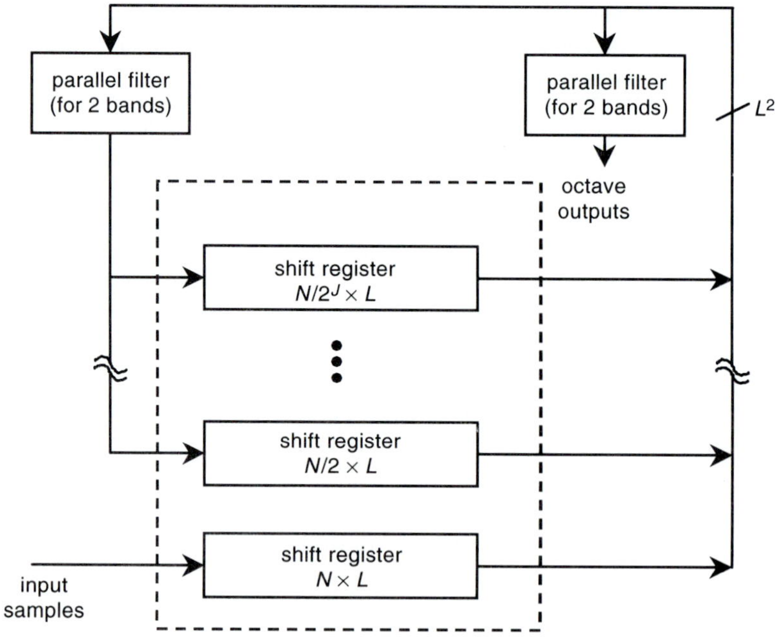

Figure 8–16 Parallel filter architecture for two-dimensional DWT.[9] © 1995 IEEE.

umn octaves. Because all outputs of an octave are computed at once, both the computation time and the period are $2LJ$.

Fig. 8–17 shows the use of bypass switches to reconfigure the original $N \times N$ array (used for $i = 1$), to $N/2^{i-1} \times N/2^{i-1}$ for computation of the subsequent octaves.

The above discussions of the two-dimensional DWT apply to the most general case of nonseparable wavelets. Separable wavelets allow a filter $h(x, y)$ to be separated into the product $h(x, y) = h_x(x)h_y(y)$. Therefore, for a separable two-dimensional wavelet filter, the two-dimensional DWT can be obtained by the separate application of the one-dimensional DWT in the horizontal and vertical directions. Fig. 8–18 shows a structure that performs two separate dimensional DWTs to accomplish a two-dimensional DWT.

Table 8–2 provides a comparison of circuit complexity, area, and period of the parallel filter and SIMD two-dimensional array for calculating the two-dimensional DWT.

8.3.3 Structures for Computing the CWT

As shown by the discussion above, the DWT is in fact an octave filter bank. Octave filter banks have regular structures and can be implemented by repeated use of identical operations (in software) or cells (in hardware). As a result, to the ex-

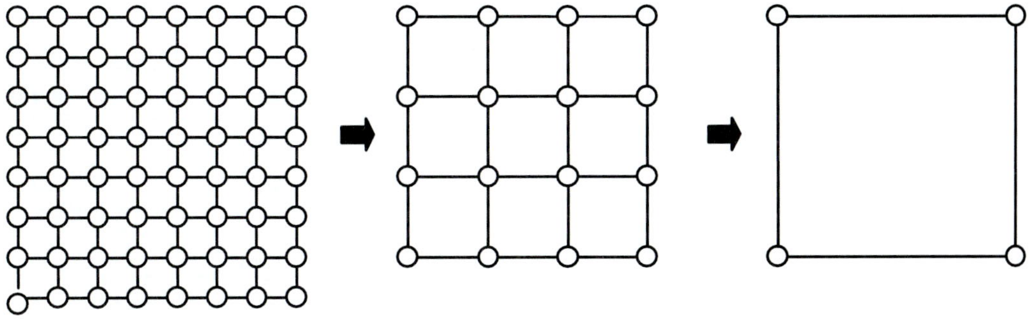

Figure 8–17 Reconfiguration of SIMD array for two-dimensional DWT by bypassing alternate processors for higher octaves.[9] © 1995 IEEE.

tent that a CWT can be made to look like a DWT, it can be placed in a regular structure.

Fig. 8–19[8] contrasts a structure for computing the CWT versus a structure for the DWT. The downsampling by a factor of two is removed for the CWT and the low pass filter outputs [from $g(t)$] alternate between the two output paths. One path provides the even-sampled outputs (like the DWT), but instead of discarding every other sample, these samples are output through the other lead.

The CWT cell, with one input shown vertically from the top, one high pass output exiting from the right, and two low pass outputs exiting from the bottom, can be cascaded to form a multi-cell CWT structure (Fig. 8–20). Each cell is indicated with an identifying shape. The shape is used in the two-dimensional dis-

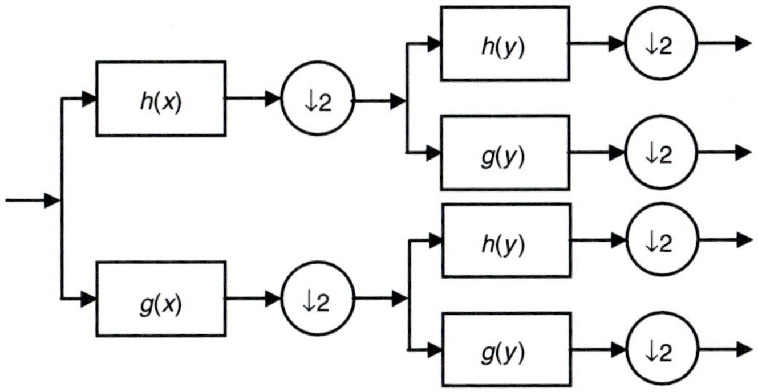

Figure 8–18 Separation of two-dimensional DWT into horizontal and vertical one-dimensional DWTs.[2] © 1991 IEEE.

Table 8–2 Comparison of architecture alternatives for two-dimensional DWT.[9] © 1995 IEEE.

Architecture	# Multipliers	Area	Period
Parallel filter	$2L^2$	$O(NLb)$	$\approx N$
SIMD 2-D array	$2N^2$	$O(N^2b)$	$\approx L^2J$

play of time versus scale. It is seen that only one in every 2^{i-1} coefficients in the filter is non-zero at octave i. The term "a'trous" ("with holes") has been used to describe this effect.[11] The a'trous structure for the one-dimensional CWT has 2^{i-1} cells at level i, with each cell working at a rate of $1/2^{i-1}$ of that of the first level.

The CWT, like the DWT, can be accelerated by the use of the FFT. Since the CWT filters are effectively twice as long as the DWT, the efficiency of the FFT is gained on shorter values of L than for the DWT.

In a multi-cell implementation of J stages for filters of length L, $2L$ multiplications and $2(L-1)$ additions are required per input point per cell. There are 2^{i-1} elementary cells at octave i in Fig. 8–20. All are identical but they work at different rates. A cell at octave i is fed by an input stream which has been subsampled by a factor of 2^{i-1} as compared to the original. Thus, the total complexity for J stages is $\Sigma_{j=1}^{J} 2^{j-1}/2^{i-1} = J$ times the complexity of one cell. As a result, the complexity grows linearly with the number of octaves. In a more naïve direct implementation that does not exploit the dilation relationship among the wavelets, the computation would grow exponentially with J.

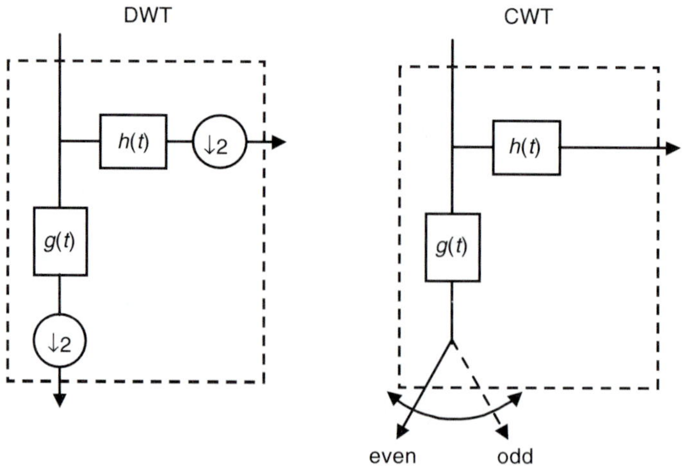

Figure 8–19 CWT cell omits factor of two downsampling of filter outputs and places the alternate outputs of $g(t)$ onto two output lines.[8] © 1995 IEEE.

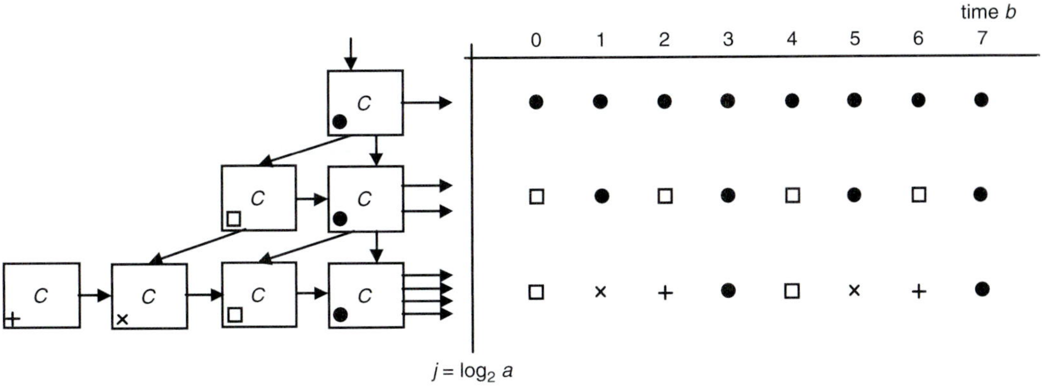

Figure 8–20 Cascade of CWT cells from Fig. 8–19 and corresponding outputs of CWT in time-scale plane.[8] © 1992 IEEE.

The On-line CWT Algorithm (OCA) may be implemented with a parallel filter structure that is reminiscent of the structure used for the DWT.[9] The case of $a_0 = 2$, corresponding to octave filters, is considered. The OCA has N outputs per octave. If the processor latency T is one clock cycle, the outputs of the various octaves are scheduled one after the other. Thus, for $J = 4$, the outputs are $y_1(1)$, $y_2(1)$, $y_3(1)$, $y_4(1)$, $y_1(2)$, $y_2(2)$, $y_3(2)$, $y_4(2)$, $y_1(3)$, ... However, if $T > 1$, then as with the MRPA scheduling algorithm, the lower octaves have precedence. For $J = 4$ and $T = 4$, the octave outputs are scheduled as: $y_1(1)$, –, –, –, $y_1(2)$, $y_2(1)$, –, –, $y_1(3)$, $y_2(2)$, $y_3(1)$, -, $y_1(4)$, $y_2(3)$, $y_3(2)$, $y_4(1)$, $y_1(5)$, $y_2(4)$, $y_3(3)$, $y_4(2)$, $y_1(6)$, ...

A parallel filter implementation of the CWT is shown in Fig. 8–21. The cells of Fig. 8–20 have been collapsed into a single parallel filter unit in Fig. 8–21. The circuit consists of J computation units, where J is the number of octaves. The inputs to the ith computation unit are subsampled by a factor of 2^{i-1} by storing data in a delay line of length $2^{i-1}L$ and tapping the delay line every 2^{i-1} units. J outputs are computed every cycle.

The complexity of the parallel filter CWT circuit at level i consists of $2L$ multipliers, $2(L-1)$ adders, and $2^{i-1}(L-1)$ storage cells. Summing across J levels, the complexity is $2JL$ multipliers, $2J(L-1)$ adders, and $(2^J-1)(L-1)$ storage units. The period is proportional to N, the number of samples.

The parallel filter architecture may also be used to implement a parallel CWT (Fig. 8–22). In the two-dimensional case, each parallel filter has L^2 multipliers, arranged as L subunits with L multipliers each. The output of each subunit is summed by a tree of $L-1$ adders. The inputs to the ith filter unit have to be subsampled by a factor of 2^{i-1} along both rows and columns. To accomplish the subsampling, the data is stored in a delay line of length $2^{i-1}L$ which is tapped every 2^{i-1} units. Two-dimensional inputs are provided in a raster scan mode, one line at

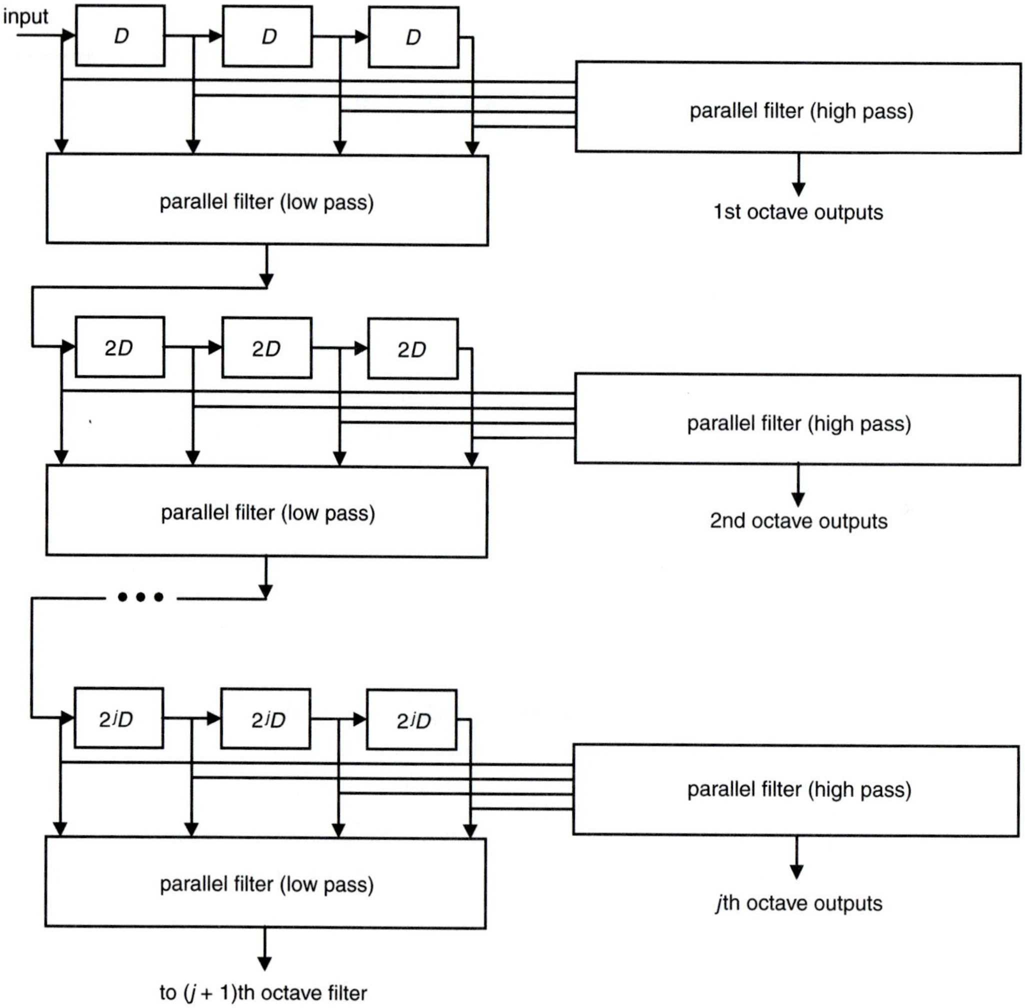

Figure 8–21 Parallel filter architecture for one-dimensional CWT (filter size $L = 4$; ND indicates a delay of N units).[9] © 1995 IEEE, adapted with permission.

a time. Thus, subsampling along columns is obtained by storing the data in a delay line of length $2^{i-1}LN$ which is tapped every $2^{i-1}N$ units.

The total hardware content for this parallel filter implementation of the two-dimensional CWT is $2JL^2$ multipliers, $2J(L^2 - 1)$ adders, and $(2J - 1)(N + L)(L - 1)$ storage units. For a bit width of B the circuit area is proportional to N^2LB, and the computational period is proportional to N^2.

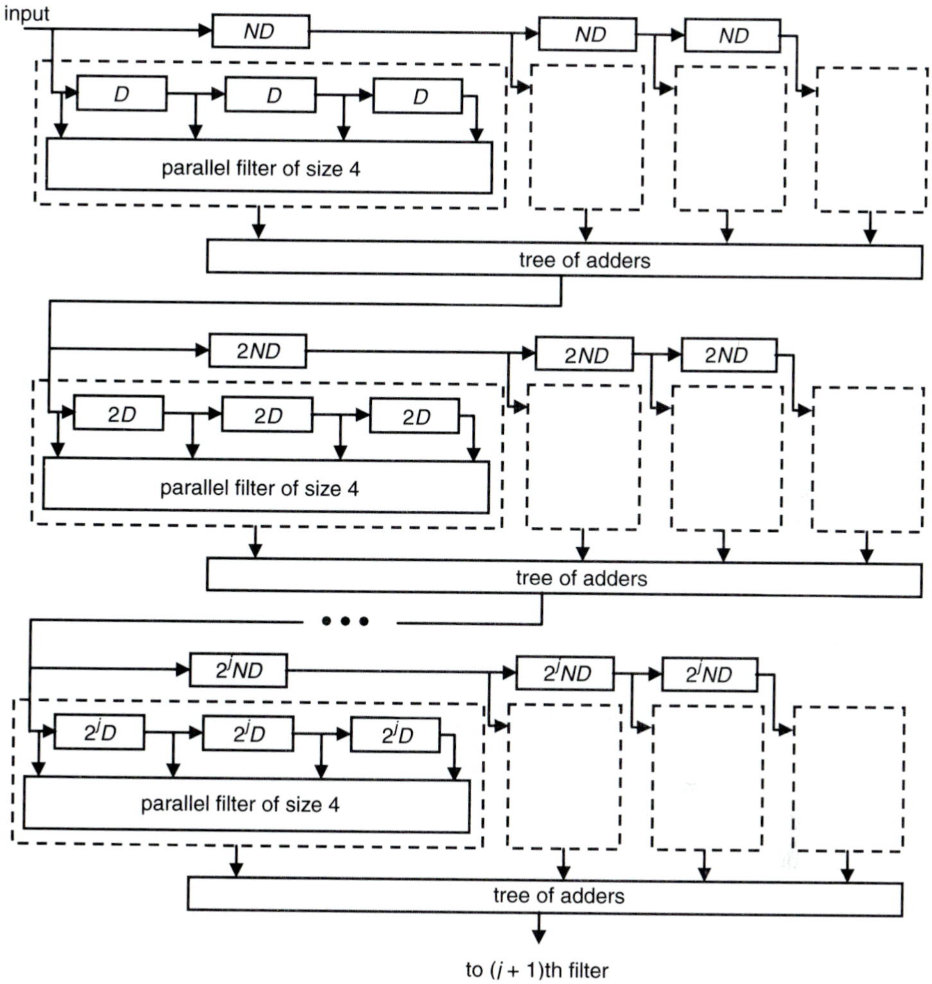

Figure 8–22 Parallel filter architecture for two-dimensional CWT for $L = 4$ (4×4 filter). *ND* indicates a delay of *N* units.[9] © 1995 IEEE.

8.4 LINEAR PREDICTIVE CODING

The analysis of signals as discussed above is supplemented by the modeling of signals, in which a small number of parameters are used to characterize a large number of signal sample values. Indeed, the Fourier and wavelet analyses represent a signal as a few parameters that provide the nature and weighting of a set of basis functions that may be summed to generate an approximation to the signal.

8.4.1 Equations

Linear predictive coding[12] is a modeling technique that seeks to predict the next sample of a signal, $x(n)$, by a weighted sum of the p previous samples, where p is the model order. Thus, the estimate $x'(n)$ is performed by:

$$x'(n) = \sum_{k=1}^{p} a_k x(n-k). \tag{8.25}$$

The difference between the true signal and the estimate, known as the prediction error $\delta(n)$, is given by

$$\delta(n) = x(n) - x'(n). \tag{8.26}$$

A physical basis for linear prediction applied to speech signals is provided by an all-pole model of speech production, which models the vocal tract as an acoustic tube with poles. The transfer function is given by

$$H(z) = \frac{A}{1 - \sum_{k=1}^{p} a_k z^{-k}}. \tag{8.27}$$

The numerator A represents the energy of excitation and is proportional to the energy of the prediction error $\delta(n)$ summed over samples—the $\delta(n)$ values are modeled as signals that excite the acoustic tube. Short time analysis is used by performing a modeling of the analysis blocks of the speech signal over time windows that are short compared to the time of vocal tract changes. These frames, typically 10–50 msec in duration, are indexed by l, which is suppressed in the following discussion to focus on one frame.

To find the values of a_k, the energy of the prediction error is minimized over an analysis block of N samples:

$$E = \sum_{n=0}^{N-1} \left[x(n) - \sum_{k=1}^{p} a_k x(n-k) \right]^2. \tag{8.28}$$

The minimum is found by setting $\partial E / \partial a_k = 0$, $k = 1, 2, \ldots, p$ to obtain the following system of equations:

$$\sum_{n=0}^{N-1} x(n-j)x(n) = \sum_{k=1}^{p} a_k \sum_{n=0}^{N-1} x(n-j)x(n-k);\ 1 \leqslant j \leqslant p. \tag{8.29}$$

This system of p equations and p unknowns may be simplified by defining

$$\varphi(j,k) = \sum_{n=0}^{N-1} x(n-j)x(n-k) \tag{8.30}$$

to get

$$\sum_{k=1}^{p} a_k \varphi(j,k) = \varphi(j,0);\ j = 1,2,\ldots,p. \tag{8.31}$$

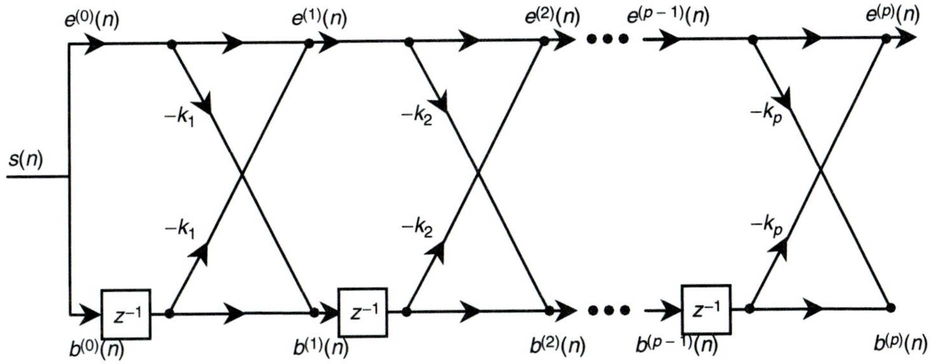

Figure 8–23 Lattice method of linear prediction.[13] © 1978 Prentice-Hall, Inc.

This system may be represented in matrix notation as

$$\Phi \cdot \mathbf{a} = \Psi. \tag{8.32}$$

To solve for **a**, the matrix ϕ is inverted.

One method for performing this inversion is the autocorrelation method. A mapping of the autocorrelation method to a programmable digital signal processor was discussed in Section 6.4.

8.4.2 Lattice Method[a]

Another method of computing the a_k is the lattice method.[13,14] Unlike the autocorrelation method, the lattice method obtains the predictor coefficients directly from the speech samples, bypassing the intermediate calculation of the autocorrelation function. The lattice method is more compatible with the stream processing that is typical of sample-by-sample signal processing. It iterates through successive optimum linear prediction, beginning with one-sample prediction ($i = 1$) and recursively progressing to the p-sample prediction.

Fig. 8–23 shows the lattice structure that gives this method its name. Predictions proceed from left to right, iteratively computing both forward prediction errors $e^{(i)}$ and backward prediction errors $b^{(i)}$, $i = 1, 2, \ldots, p$. Reflection coefficients k_i are computed to connect the forward and backward branches, and the final prediction error $e^{(p)}$ emerges from the right.

The ith reflection coefficient is computed as a function of the forward and backward prediction errors from iteration $i - 1$:

[a]Material in this section based on *Digital Processing of Speech Signals* by Rabiner/Schafer, © 1978. Adapted by permission of Prentice-Hall, Inc., Upper Saddle River, NJ.

$$k_i = \frac{2\sum_{n=0}^{N-1}[e^{(i-1)}(m)b^{(i-1)}(m-1)]}{\sum_{m=0}^{N-1}[e^{(i-1)}(m)]^2 + \sum_{m=0}^{N-1}[b^{(i-1)}(m-1)]^2}. \tag{8.33}$$

Forward and backward prediction errors for order i are computed from order i-1 by:

$$e^{(i)}(m) = e^{(i-1)}(m) - k_i b^{(i-1)}(m-1) \tag{8.34}$$

$$b^{(i)}(m) = b^{(i-1)}(m-1) - k_i e^{(i-1)}(m). \tag{8.35}$$

Finally, just as when Durbin's recursion was applied to autocorrelation coefficients in the LPC analysis of Section 6.4, a matrix of intermediate values $\alpha_j^{(i)}$ is developed and used at the end to extract the predictor coefficients, a_j:

$$\alpha_1^{(i)} = k_i \tag{8.36}$$

$$\alpha_j^{(i)} = \alpha_j^{(i-1)} - k_i\alpha_{i-j}^{(i-1)};\ 1 \leq j \leq i-1 \tag{8.37}$$

$$a_j = \alpha_j^{(p)} \tag{8.38}$$

The lattice method is then executed by the following steps:

1. Initialize $e^{(0)}(m) = s(m) = b^{(0)}(m);\ m = 0, 1, \ldots, N-1$.
2. $k_1 = \alpha_1^{(1)}$—compute from (8.33).
3. Determine forward prediction error $e^{(1)}(m)$ from (8.34) and backward prediction error $b^{(1)}$ from (8.35).
4. Set $i = 2$.
5. Determine $k_i = \alpha_i^{(i)}$ from (8.33).
6. Determine $\alpha_j^{(i)}$ for $j = 1, 2, \ldots, i-1$ from (8.37).
7. Determine $e^{(i)}(m)$ and $b^{(i)}(m)$ from (8.34) and (8.35).
8. Set $i = i + 1$.
9. If $i \leq p$ go to (5).
10. Compute predictor coefficients a_k from (8.38) and stop.

The lattice method of linear prediction computation provides an alternative to the autocorrelation method that is more suited for the stream processing that is typical of real time implementation.

REFERENCES

1. Louis Schirm IV, "FFTs for Non-FFT People," (Redondo Beach, CA : TRW, 1979).
2. Olivier Rioul and Martin Vetterli, "Wavelets and Signal Processing," *IEEE Signal Processing Magazine*, 8 (October, 1991), 14–38.

3. P. J. Burt and E. H. Adelson, "The LaPlacian Pyramid as a Compact Image Code," *IEEE Trans. Comm.*, 31 (April, 1983) 532–549.
4. R. E. Crochiere, S. A. Weber, and J. L. Flanagan, "Digital Coding of Speech in Subbands," *Bell Syst. Tech. J.*, 55 (Oct. 1976), 1069–1085.
5. Rioul and Vetterli, "Wavelets and Signal Processing."
6. R. E. Crochiere and others, "Real-Time Implementation of Sub-Band Coding on a Programmable Integrated Circuit," *Proc. IEEE Inter. Conf. Acoustics, Speech, Signal Processing,* Atlanta, GA, March–April, 1981, pp. 455–458.
7. Ronald E. Crochiere, Richard V. Cox, and James P. Johnston, "Real-Time Speech Coding," *IEEE Trans. Communications* COM-30 (April 1982) 621–634.
8. Olivier Rioul and Pierre Duhamel, "Fast Algorithms for Discrete and Continuous Wavelet Transforms," *IEEE Trans. Information Theory,* 38 (Nov. 1992), 569–586.
9. Chaitali Chakrabarti and Mohan Vishwanath, "Efficient Realizations of the Discrete and Continuous Wavelet Transforms: From Single Chip Implementations to the Mapping on SIMD Array Computer," *IEEE Trans. Signal Processing,* 43 (March, 1995), 759–771.
10. Aware, Inc., 40 Middlesex Turnpike, Bedford, MA, 01730, (617) 276–4000.
11. M. Holschneider and others, "A Real-Time Algorithm for Signal Analysis with the Help of Wavelet Transforms," in *Wavelets, Time-Frequency Methods, and Phase Space*, J. M. Combes, A. Grossman, and Ph. Tchamitchian, Eds. (Berlin : Springer, IPTI, 1989), 286–297.
12. Ronald W. Schafer and Lawrence R. Rabiner, "Parameteric Representations of Speech," in *Speech Recognition*, D. Raj Reddy, Ed. (New York : Academic, 1975), 99–150.
13. J. Makhoul, "Stable and Efficient Lattice Methods for Linear Prediction," *Trans. Acoust., Speech, Signal Processing,* ASSP-25 (October, 1977), 423–428.
14. L. R. Rabiner and R. W. Schafer, *Digital Processing of Speech Signals* (Englewood Cliffs, NJ : Prentice-Hall, 1978).

9

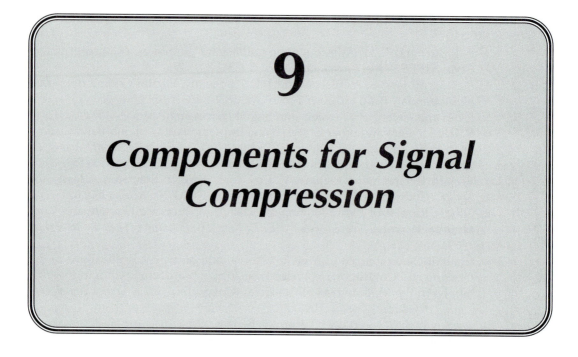

Components for Signal Compression

The process of signal analysis and modeling described in the previous chapter results in a compact formulation of the information-bearing portions of the signal. This compact representation can be used to compress the signal to allow its transmission over limited-bandwidth channels or its storage within limited space.

This section discusses the real time implementation of signal compression, also known as coding. Various coding schemes are distinguished by several properties:

- *compression ratio*: the amount of compression achieved, determined by the ratio of the size of the original signal to the size of the compressed signal;
- *reconstruction quality*: some compression schemes are lossless, providing a reconstruction waveform that exactly matches, sample for sample, the original signal. Other methods achieve a higher compression ratio through lossy compression, which does not allow exact reconstruction of the waveform, but instead seeks to preserve its information-bearing portions;
- *fixed versus variable transmission bit rate*: bit rate can vary in some schemes that are based on encoding the rate of change of the properties of the signal—a signal that is relatively stable, such as a sustained single-frequency sine wave, will require fewer bits per second than a speech signal;

- *delay (latency) of coding*: a greater compression ratio can be achieved if a large sequence of samples are collected, statistically analyzed, and the results of the statistical analysis are sent; this aggregation of samples introduces a delay which may be unacceptable;
- *computational complexity*: this is generally higher for high compression ratios.

This chapter examines speech compression, vector quantization (as applied to speech or image coding), and image compression. Compression, like other transmission-related activities, is aided by the use of standard formats that ensure interconnectivity. For each application, this chapter discusses algorithms and describes the related activity of relevant standards organization. The section examines computational structures, including both programmable digital signal processors and custom processors, to implement coding in real time.

9.1 SPEECH CODING

Speech coders fall into one of two classes. *Waveform coders* generate a reconstructed waveform (after coding, transmission, and decoding) that approximates the original waveform, thereby approximating the original speech sounds that carry the message. *Voice coders*, or vocoders, do not attempt to reproduce the waveform, but instead seek to approximate the speech-related parameters that characterize the individual segments of the waveform. Speech coding systems usually operate either within telephone bandwidth (200 Hz to 3.2 KHz) or wideband (up to 7 KHz, used in AM radio-commentary audio, multimedia, etc.).

9.1.1 Waveform Coders

The simplest waveform coder is pulse code modulation, or PCM. As shown in Fig. 9–1, the waveform is passed through a low pass filter to remove high-frequency components and then is passed through a sampler. The sampler per-

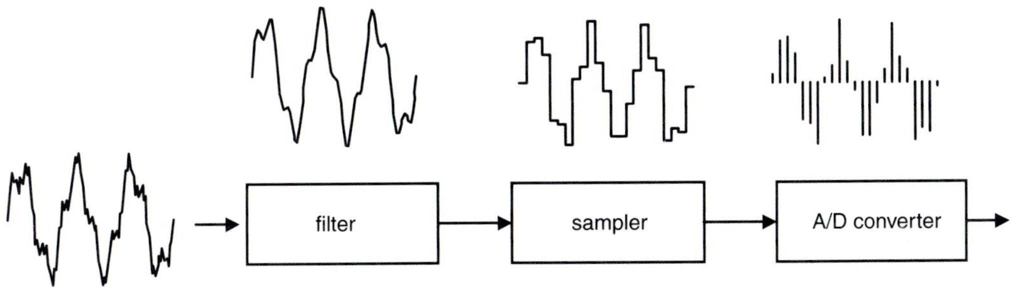

Figure 9–1 Pulse code modulation represents analog signal by low pass filtering, sample-and-hold, and analog-to-digital conversion.

forms a sample-and-hold operation by capturing the instantaneous value of the waveform at each sampling instant and holding it at that value, resulting in a stair step pattern. During the hold interval, an analog-to-digital converter computes and outputs a digital representation of the current analog value.

The sampling rate is at a frequency f_s, and each sample is represented by a B-bit word. The sampling frequency is set by the bandwidth of the low pass-filtered signal, W. This relationship is set by the Nyquist criterion, which requires a minimum of two samples to determine a frequency, so $f_s = 2W$.

Successive samples may be similar in a PCM system, especially when the bandwidth is well below $f_s/2$. This sample-to-sample correlation may be exploited to reduce bit rate by predicting each sample based previous sample values, comparing the predicted sample value to the actual sample value, and encoding the difference between the two. This difference is usually smaller than either sample, so fewer bits are needed to encode it accurately. Extending the predictor to include the weighted sum of the previous p samples can extend this method, known as differential PCM or DPCM, to improve the prediction of a sample:

$$\tilde{x}(n) = \sum_{i=1}^{p} a_i \hat{x}(n - 1). \tag{9.1}$$

The difference $e(n)$ between the predicted sample $\tilde{x}(n)$ and the actual sample $x(n)$ is given by

$$e(n) = x(n) - \tilde{x}(n), \tag{9.2}$$

where $\hat{x}(n - i)$ is the encoded and decoded $(n\text{–}i)$th sample. The quantizer produces $\hat{e}(n)$, which is the quantized version of $e(n)$. Fig. 9–2 shows the structure of a DPCM coder and the quantized prediction error, $\hat{e}(n)$, that is output.[1]

The quantizer and predictor may be adapted over time to follow the time-varying properties of the signal and adapt the use of bits to the signal. Adaptive DPCM (ADPCM) performs such adaptation. An adaptive predictor changes the number of bits, based on one or more of the following signal properties:

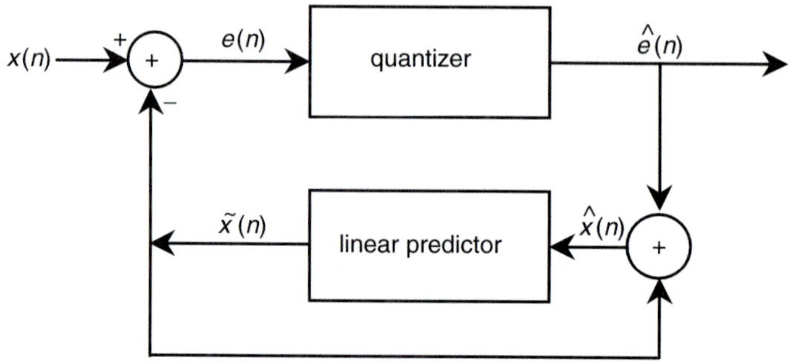

Figure 9–2 DPCM coder.

- probability density function (histogram of values) of the signal;
- mean value of input signals;
- dynamic range (variance from mean).

The ADPCM predictor is thus based on a shorter-term average of signal statistics than the long-term average. Adaptation may be performed in the forward direction, backward direction, or both directions. In forward adaptation, a set of N values is accumulated in a buffer and a set of p predictor coefficients is computed. This buffering induces a delay that corresponds to the acquisition time of M samples. For speech signals, this is typically 10 msec. The delay is compounded for telephone-routing paths that perform multiple stages of encode/decode. Backward adaptation uses both quantized and transmitted data to perform prediction, thereby reducing the delay.

An example of ADPCM implementation[2] uses a backward adaptive algorithm to set the quantizer size. It uses a fixed first-order predictor and a robust adapting step size. It is implemented on a low-cost fixed point digital signal processor with a B-bit fixed uniform quantizer [shown at point (1) in Fig. 9–3]. A pair of tables (2) stores the step size and its inverse for adaptive scaling of the signal before and after quantization. A step-size adaptation loop (3) generates table addresses.

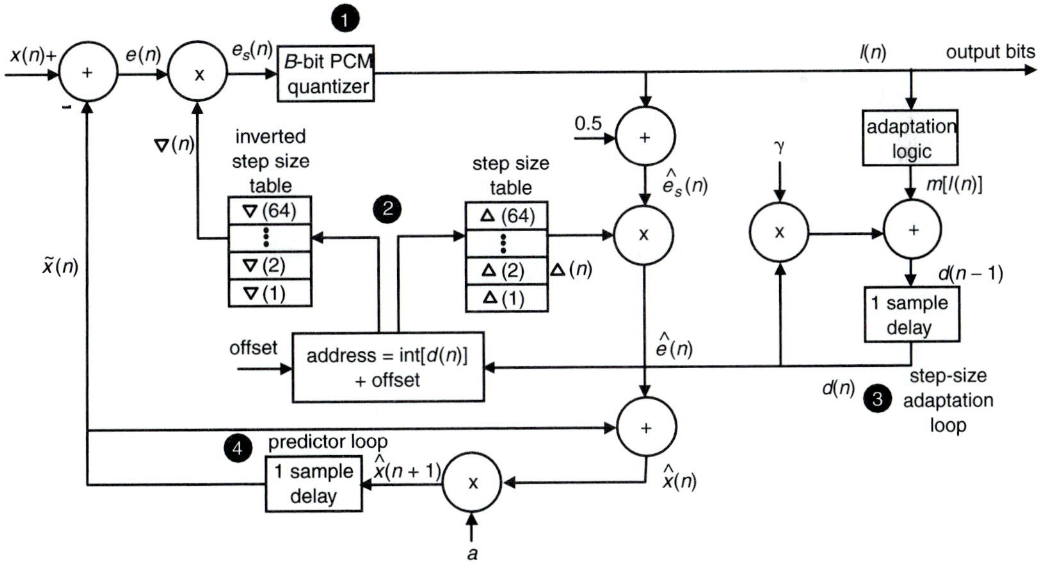

Figure 9–3 ADPCM coder implemented on fixed-point digital signal processing chip[2] consists of (1) fixed uniform quantizer, (2) tables to store step size and inverted step size, (3) step-size adaptation loop, and (4) fixed-predictor loop. © 1982 IEEE, adapted with permission.

The ADPCM fixed predictor loop (4) generates a predictor signal $\tilde{x}(n)$ by multiplying the previously encoded-and-decoded sample $\hat{x}(n-1)$ by the predictor coefficient a and subtracting it from the current sample $x(n)$, forming the difference signal $e(n)$:

$$e(n) = x(n) - \tilde{x}(n). \tag{9.3}$$

The difference signal $e(n)$ is then adaptively quantized.

The step size used to quantize $e(n)$ is a function of the amplitude of $e(n)$. Using a prediction loop applied to $e(n)$ sets the step size. The step size and inverse step size are chosen using locked pointers, the placement of which is set by the weighted prediction error of $e(n)$.

To reconstitute the signal, the b-bit integer $I(n)$ resulting from the quantizer (1) is scaled by the step size and the predicted sample $\hat{x}(n)$ is added back.

Table 9–1 compares the features of waveform coding methods used for speech signals. In the table, "toll quality" refers to a quality level consistent with the best-quality 3.2-KHz bandwidth telephone speech, transmitted at 64 Kbit/sec and encoded with μlaw-companded PCM. Communication quality is less than toll quality, but preserves the characteristics that allow identifying the talker.

The above coders use a single B-bit quantizer, forming 2^B quantization levels, for digitizing all amplitude samples. Another approach decomposes the signal into a set of components, each of which is separately digitized with possibly different sampling rates. The signal is then reconstructed during decoding by summing these components. One method to separate a signal is by frequency. A multifrequency decomposition has the advantage of allowing control of the quantization error for each frequency band, based on the number of bits allocated to each band. As a result, the overall spectrum of the error signal can be shaped to allow placing most of the error spectrum within frequency areas that are less easily perceived. The error spectrum is thus shaped to complement the human perception of error.

An example of multifrequency decomposition is subband coding. Like the discrete wavelet transform discussed in the previous section, subband coding provides successive approximations to the waveform by recursively decomposing it into a low-frequency and high-frequency portion. Subband coding divides the input channel into multiple frequency bands and codes each band with ADPCM. The quadrature mirror filter described in connection with the wavelet transform

Table 9–1 Comparison of waveform coding methods for speech.

Method	Bit Rate	Quality	Relative Complexity
PCM	64 kb/sec	toll	low
DPCM	24 kb/sec–32 kb/sec	communications	med
ADPCM	32 kb/sec	toll	high

was originally applied to subband coding, and it can be implemented on a programmable signal processor as was shown in Chapter 8. The ADPCM implementation just described may be used for the ADPCM portion of a subband coder.

9.1.2 Voice Coders

In contrast to the waveform coder, the voice coder seeks to preserve the information and speech content without trying to match the waveform. A voice coder uses a model of speech production (Fig. 9–4).

One model consists of a linear predictive coding (LPC) representation of the vocal tract. The input speech signal is then filtered with the inverse of the vocal tract filter. Because the filter is not exact, a residual signal is obtained at the output of the inverse filter. This residual is regarded as the excitation signal for the filter. A characterization of its periodicity is made, and if strongly periodic, the section of speech is declared as *voiced* with the measured pitch period. If not, the speech sound is declared *unvoiced* and the excitation is modeled with random noise. In either case, the overall amplitude of the excitation is also measured to preserve amplitude characteristics.

The LPC analysis and its mapping to real time processing were discussed in Sections 6.4 and 8.4.2. To characterize the excitation, a pitch-period estimator, based on the autocorrelation analysis of the signal, can be implemented. In this pitch-period estimator, the speech signal is first low-pass filtered to remove energy above the highest likely pitch period (> 800 Hz). Next, the unnormalized autocorrelation is computed for lag values m, $m_{min} \le m \le m_{max}$. Here m_{min} and m_{max} are sample lags based on the minimum and maximum expected pitch period. The computation is performed on a windowed version of the speech, where $w(n)$ is the window:

$$r_n(m) = \sum_{l=-\infty}^{\infty} w(n-l)x(l-m);\; m_{min} \le m \le m_{max}. \tag{9.4}$$

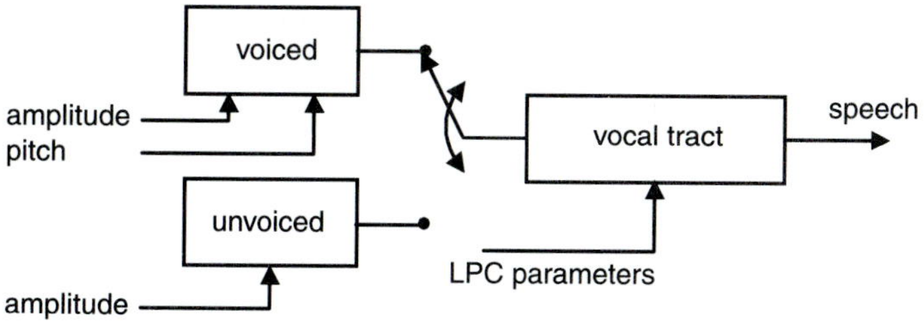

Figure 9–4 Voice coder model of speech production uses a vocal tract representation that is excited by either a voiced or unvoiced signal.

To place the pitch period estimation in a form suited for real-time implementation, the algorithm is cast into a stream-processing form. An exponential function is used for windowing:

$$w(n) = \begin{cases} \gamma^n; & n \geq 0 \\ 0; & n < 0. \end{cases} \tag{9.5}$$

The autocorrelations are computed on a sample-by sample basis for each m:

$$r_n(m) = \gamma\, r_{n-1}(m) + x(n)\cdot x(n - m). \tag{9.6}$$

The computation load is reduced by updating the autocorrelation only every jth sample (e.g., $j = 4$):

$$r_n(m) = \gamma'\, r_{n-j}(m) + x(m)\cdot x(n - m). \tag{9.7}$$

A value of $\gamma' = 0.95$ is typical. The range of lags $m_{min} \leq m \leq m_{max}$ is distributed over j samples, performing $1/j$th of the $m_{max} - m_{min} + 1$ values. The pitch is quantized to 6 bits, or 64 values, which are distributed over a range from 66.7 Hz to 320 Hz. Fig. 9–5 shows a structure for real time implementation of the pitch period estimation algorithm. A lowpass filter (LPF) is followed by a j-fold downsampling, and every jth sample is entered into a shift register of length $m_{max} - m_{min} + 1$ locations. The shift register is used in the autocorrelation update of $r_n(m)$, and the results are written in an autocorrelation buffer. The buffer is scanned for a peak value, and the pitch period associated with the lag of the peak is output.

A standardized version of an LPC vocoder has been used in the implementation of the Secure Telephone Unit Version 3, or STU-III. The STU-III is combined with an encryption system and error protection processing to implement a secure

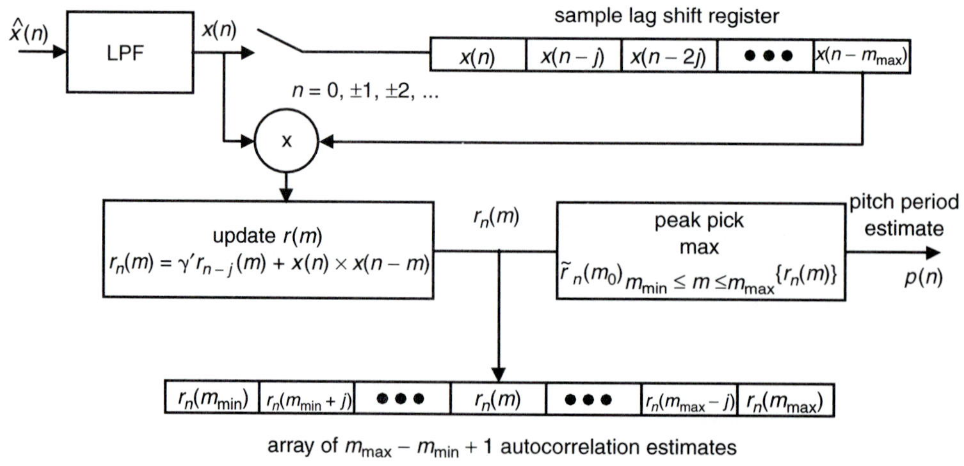

Figure 9–5 Computational structure for computing pitch period estimate by autocorrelation analysis.[2] © 1982 IEEE, adapted with permission.

voice communication link. In its original version,[3] an enhanced 10th-order LPC analysis known as LPC-10e was used. In LPC-10e, the excitation signal is categorized as voiced or unvoiced, with pitch period transmitted for voiced and gain (energy) encoded for both voiced and unvoiced speech. The analysis properties of Table 9–2 are used for LPC-10e, and its coding and low bit rate render a buzzy quality to the speech that masks most characteristics of talker identification.

Computational blocks for LPC-10e include the LPC analysis, pitch and voiced/unvoiced decision, encoding, and communication processing. Pitch detection is performed by the Average Magnitude Difference Method (AMDF), which avoids the multiplications needed by the autocorrelation method. The AMDF subtracts a delayed version of the waveform from the incoming waveform at various lags and averages the differences across samples. The lag that produces the smallest difference is selected as the pitch period estimate. An 800-Hz low pass filter is applied, and 60 possible pitch values are accommodated. These values are not uniformly spaced, but follow a sequence of lags given by {20, 21, . . ., 39, 40, 42, . . ., 78, 80, 84, . . ., 156}, corresponding to pitches at the 8-KHz sampling rate that range from 51.3 Hz to 400 Hz.

In addition to LPC analysis, and pitch-period estimation, the LPC-10e algorithm requires parameter encoding and "communication" processing. Parameter encoding, described by an example below (Table 9–3), assigns particular bit locations for the various LPC, pitch, and other parameters that are transmitted. In the transmit mode, communication processing includes parallel to serial conversion and forward error correction. In the receive mode, it includes initial acquisition of synchronization, frame-to-frame maintenance of synchronization, de-interleaving of frame information, serial-to-parallel conversion, error correction, and parameter decoding.

The LPC-10e algorithm was developed to run on a bit-sliced 16-bit computer with a dedicated multiplier, at a time when single-chip digital signal processors were a rarity. On a programmable signal processor, the block form of

Table 9–2 Analysis parameters used in LPC-10e coding standard.

Parameter	Value
Sampling rate	8 KHz
Frame period	22.5 msec
Speech samples/frame	180
Output bits/frame	54
Bit rate	2.4 Kb/sec
Bits per sample (average)	0.3
Compression factor (average)	30
LPC analysis method	Covariance
Transmission format	Serial

covariance-based LPC analysis requires a relatively large RAM. One implementation[4] uses a standard microprocessor and three fixed-point digital signal processors to implement an LPC-10e encoder/decoder pair. The partitioning is shown in Fig. 9–6 and assigns non-repetitive operations such as voiced/unvoiced decision, pitch tracking, coefficient encoding, and synchronization to the microprocessor. The signal processors perform such repetitive tasks as LPC analysis, pitch period estimation, and LPC synthesis.

In an alternative implementation,[5] custom-integrated circuits implement the LPC analysis, synthesis, and AMDF pitch analysis, and three microprocessors complete the pitch-period estimation, control the gain, perform error correcting, and format the coefficients (Fig. 9–7).

The efficient encoding of LPC and pitch parameters for transmission is exemplified by the method used in the speech synthesizer of the commercial learning aid known as "Speak and Spell," developed by Texas Instruments.[6] The encoded parameters consist of the frame energy, pitch (which is set to 0 to indicate unvoiced speech), and 10 LPC-derived reflection coefficients (Table 9–3). A frame period of 25 msec provides a rate of 40 frames/sec, and a bit rate of 1200 bits/sec is achieved. For the first voiced frames, 49 bits are used as shown in Table 9–3; a separate code for "repeat" transmits subsequent frames with identical LPC parameters (pitch and energy of repeated frames can vary) at 10 bits each. Unvoiced frames are transmitted at 28 bits each.

The quality of speech coding can be improved by allowing more flexibility in modeling the prediction residual than permitted by the binary choice of voiced/unvoiced. These two states can be blurred into a continuum for each analysis frame by exciting the computed LPC synthesis filter with a variety of candidate excitation functions, comparing the synthesized speech to the original

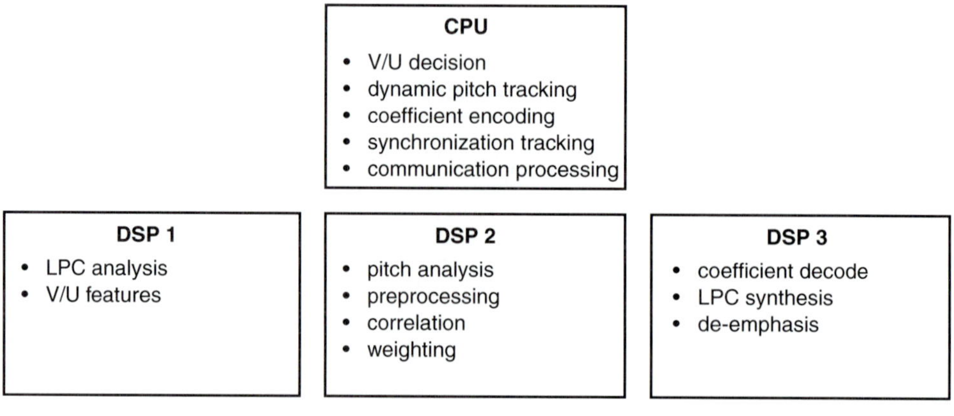

Figure 9–6 Implementation of real-time LPC-10 encoder/decoder uses three first-generation digital signal processing chips (NEC μPD 7720) and a microprocessor CPU.

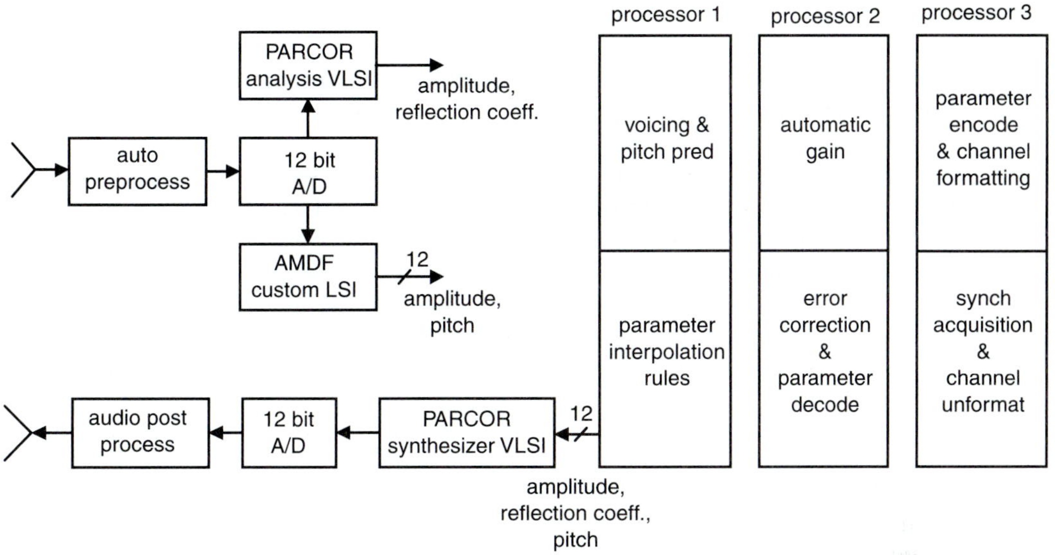

Figure 9–7 Custom-integrated circuit implementation of LPC analysis and synthesis, augmented by three microprocessors, for real time implementation of LPC-10 speech encoder/decoder.

voiced/unvoiced. These two states can be blurred into a continuum for each analysis frame by exciting the computed LPC synthesis filter with a variety of candidate excitation functions, comparing the synthesized speech to the original waveform within the encoder, and picking the excitation function that minimizes the difference between the re-synthesized and original speech, using a distance measure based on human perception. This selection of an excitation function replaces both the voiced/unvoiced decision and the pitch period excitation.

For example, *code-excited linear prediction* (CELP)[7] uses vector quantization (VQ), by which a predetermined set of excitation signals is stored in a codebook.

Table 9–3 Encoding method to achieve 1200 bit/sec average rate for LPC10 parameters E (energy), P (pitch), K(*n*) (*n*th LPC-based reflection coefficient), R (repeat flag).[6] © 1982 IEEE, adapted with permission.

Frame Type	How Determined	# Bits/Frame	Parameters Sent (# bits)
Voiced	$E \neq$ or 15; $P \neq 0; R = 0$	49	$E(4)$, $P(5)$, $R(1)$, $K1(5)$, $K2(5)$, $K3(4)$, $K4(4)$, $K5(4)$, $K6(4)$, $K7(4)$, $K8(3)$, $K9(3)$, $K10(3)$
Unvoiced	$E \neq 0$ or 15; $P = 0$; $R = 0$	28	$E(4)$, $P(5)$, $R(1)$, $K1(5)$, $K2(5)$
Repeated	$E \neq 0$ or 15; $R = 1$	10	$E(4)$, $P(5)$, $R(1)$
Zero Energy	$E = 0$	4	$E(4)$
End of Word	$E = 15$	4	$E(4)$

For each frame, the codebook is searched for the particular excitation sequence that, upon recreating a speech waveform through the synthesis filter, minimizes a perceptually weighted distance.

A CELP coder consists of an LPC filter and a VQ excitation section, which in turn includes a computation of the distance metric and a codebook search mechanism. Fig. 9–8 shows the reconstruction of the speech waveform, its comparison with the original, and the creation of a perceptually weighted filter controlled by both the LPC synthesis parameters a_k and a frequency-weighted perceptual weight α that depends on the sampling frequency f_s.

Two techniques are used for compiling the codebook of possible excitation waveforms for CELP. The first, *stochastic excitation*, assumes that the best excitation sequence cannot be predicted on the basis of such simplifications as pitch or voiced/unvoiced categories. Instead, each entry is a different sequence of random numbers. However, a stochastic codebook has no intrinsic order and is thus difficult to search. Several alternatives that ease the search include sparse-excited codebooks, which contain a large number of zeros; lattice-based codebooks, which have regularly spaced arrays of points; trained codebooks, which are built up from clustering a large number of previously-gathered excitation sequences (as will be described below); and multiple codebooks, which consist of both a stochastic and an adaptive codebook.

An adaptive codebook uses the set of excitation samples from the previous frame and performs a search for the optimal time lag at which to present them in the current frame. After this excitation, a stochastic codebook is then searched to minimize the perpetual difference between original and resynthesized waveforms, and this stochastic entry is added to the lag-adjusted excitation and the sum is used to represent frame excitation.

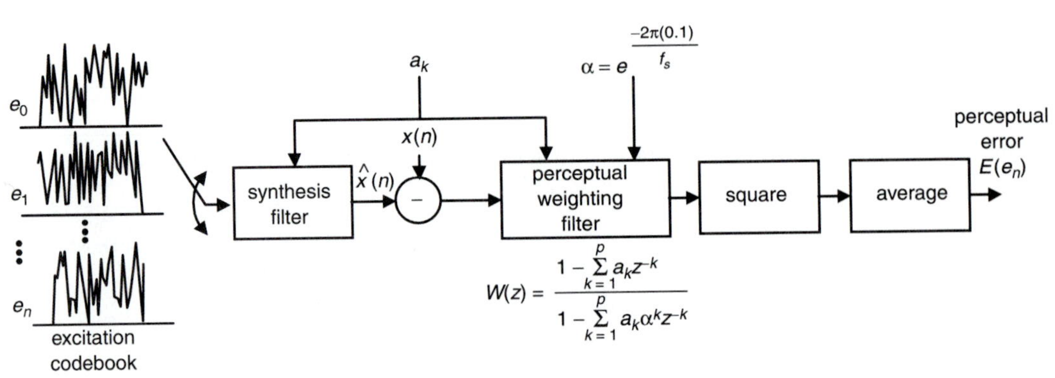

Figure 9–8 Computation of perceptual distance metric versus excitation source for code-excited linear prediction (CELP).

9.2 VECTOR QUANTIZATION

Vector quantization has been mentioned as a method used to generate and select possible excitation waveforms for code-excited linear prediction. It is more widely used, extending to both speech and image compression. To understand vector quantization in its more general application, one can envision a signal that generates samples of B bits each at a rate of R samples/sec. Because the number of possible values from B bits is 2^B, each sample may be regarded as a *symbol* taken from a dictionary (or "alphabet") of 2^B elements, and for R such samples/sec, a bit rate of BR bits/sec results. Not all symbols occur with equal probability—for example, samples of maximum amplitude are usually less likely to occur than samples near zero magnitude. A method of lossless compression known as *entropy coding* assigns short indices to the highest-probability symbols and longer indices to lower-probability symbols. To further increase compression, a lossy method may be introduced that reduces the number of alphabet, or codebook, entries to a number less than 2^B by concentrating the smaller number of available symbols on values that are likely to occur. A large amount of collected data, used for training, is placed in the feature space, allowing a distance between two samples to be defined. The distance measure is used to define clusters among the data. As shown in Fig. 9–9, a codebook entry is placed at the centroid of each cluster. Each actual data value is replaced by its nearest codebook entry, introducing some distortion. The codebook-generation algorithm, described below, minimizes the total amount of distortion.

A codebook of J codewords requires $\log_2 J$ bits to transmit each codebook index. If the number of entries in the code book is less than the number of values available with B bits ($J < 2^B$), then a compression factor of $B/\log_2 J$ results. Sending the index of the nearest of J codebook entries instead of the exact value reduces the data rate, but at the expense of increased distortion.

9.2.1 Vector Encoding/Decoding

Vector quantization can be applied to time-domain or image signals directly, but more recently and effectively, it has been applied to the residual after the signal passes through a matched inverse filter. As with CELP, the signal is encoded by sending the filter parameters and codebook index of the model.

More precisely, for a feature vector $\mathbf{v}$, a codebook consisting of J codewords, $\{\mathbf{w}(i), 1 \le i \le J\}$ and a distance function $d[\mathbf{v}, \mathbf{w}(i)]$ defined between feature vector $\mathbf{v}$ and codeword $\mathbf{w}(i)$, vector quantization finds the particular codeword index i^* that is associated with the codeword $\mathbf{w}(i)$ that is the minimum distance from $\mathbf{v}$:

$$i^* = \underset{1 \le i \le J}{\arg\min}\, d[\mathbf{v}, \mathbf{w}(i)]. \tag{9.8}$$

Then instead of transmitting feature vector $\mathbf{v}$, the index i^* of the best-matching codeword is sent. Fig. 9–10 shows a flowchart of the vector-quantization coding operations for each input vector $\mathbf{v}$.

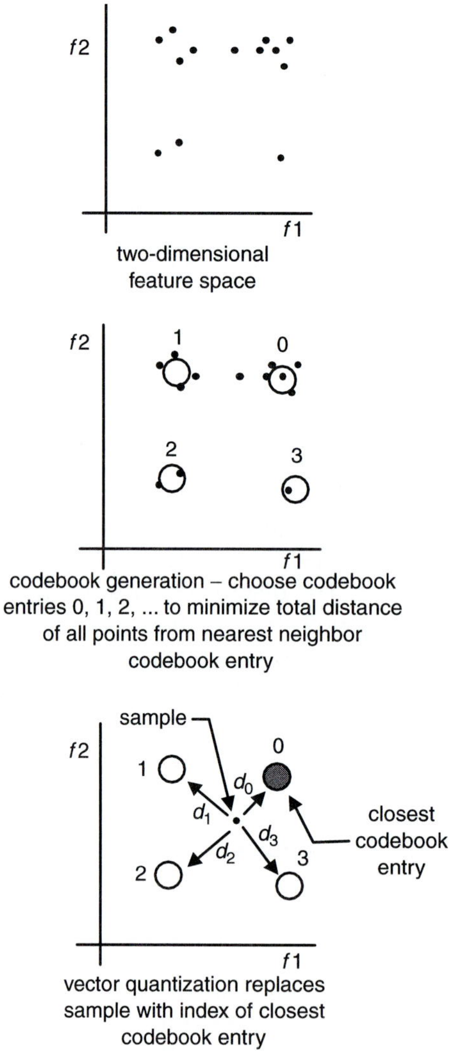

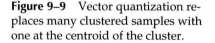

Figure 9–9 Vector quantization replaces many clustered samples with one at the centroid of the cluster.

To meet the requirements of real time encoding, the feature vectors must be encoded as fast as they arrive. For a full-search encoding algorithm, J comparisons, one against each codebook entry, are required every feature vector period. Each comparison must access and examine each of the values in each feature vector. The computation is regular, but it requires high throughput.

Parameters that define a particular instance of vector quantization and impact real time requirements are:

J = number of codebook vectors;
d = type of local distance selected (e.g., sum of products, sum of difference, ratio,...);

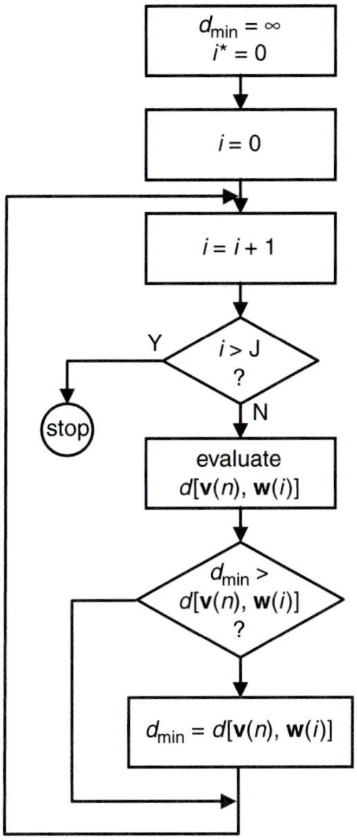

Figure 9–10 Flowchart for vector quantization encoding of feature vector $\mathbf{v}(n)$ by J-element codebook $\mathbf{w}(i)$, $0 \le i \le J$.[8] © IEEE 1995, adapted with permission.

P = number of features per distance comparison;
τ = frame period; $1/\tau$ = number of frames/sec;
l = number of codebook indices submitted per frame (for images, image is divided into l multiple subblocks with one index submitted per subblock; for speech, l =1);
Method of codebook search (full, tree, trellis, . . .).

For an on-line, dynamically adapted codebook as described below, additional parameters influence the throughput:

k = size of training set;
F = frequency of adaptation.

Table 9–4 provides an estimate of codebook throughput for a codebook size J of $1024 = 2^{10}$.

The computation of vector quantization encoding can be partitioned to a linear array.[8] Each processor is assigned one codebook entry; J processors cover the entire codebook. The vector to be encoded, $\mathbf{v}(n)$, enters each codebook processor as shown in Fig. 9–11. An initial value of $d_{min} = \infty$ is inserted at the left-hand entry point into the array. Each processor computes $d[\mathbf{v}(n),\mathbf{w}(n)]$ and outputs

Table 9–4 Computational speed requirement for real time vector quantization for speech and image coding.

Type	Frame period τ	Samples/ frame $f_s\tau$	Features/ frame P	Throughput	Computation time per coefficient
Speech	10 msec	80 (f_s = 8 KHz)	10	100 frames/sec $\times 2^{10}$ compares/frame $\times$ 10 ops/compare = 2^{13} coeff/sec	~ 10 μsec
Image	33 msec	512 x 512 pixels	16 x 16 block = 256	$(512/16)^2$ blocks/image $\times$ 256 pixels/block $\times$ 30 frames/sec = 10^{14} pixels /sec	0.01 picosec/ pixel

min(d_{min}, d) to its right-hand neighboring processors as well as its corresponding value of i. From the right side of the array emerges $\mathbf{v}(n)$, d_{min}, and i^*, which is the index of the best-matching code word. Multiple processors allow pipeline computation—while $\mathbf{v}(n)$ is being compared to $\mathbf{w}(n)$ in processor i, $\mathbf{v}(n + 1)$ is being compared to w(i – 1) in processor i – 1. This pipeline provides a J-fold speedup.

9.2.2 Codebook Generation

The codebook itself may either be produced offline or may be adapted or regenerated in real time. If the codebook is adapted in real time, its changes must be communicated to the receiving end (in addition to the message encoded with the current codebook), thereby introducing a tradeoff of total bit rate and adaptation rate. A commonly-used codebook generation algorithm is the one proposed by Y. Linde, A. Buzo, and R. M. Gray known as the LBG algorithm.[9] The algorithm begins with an initial codebook of J vectors, which may be generated in several ways:

- random set of J training vectors;
- J vectors that uniformly sample the feature space;
- previous codebook (especially for an adaptive system).

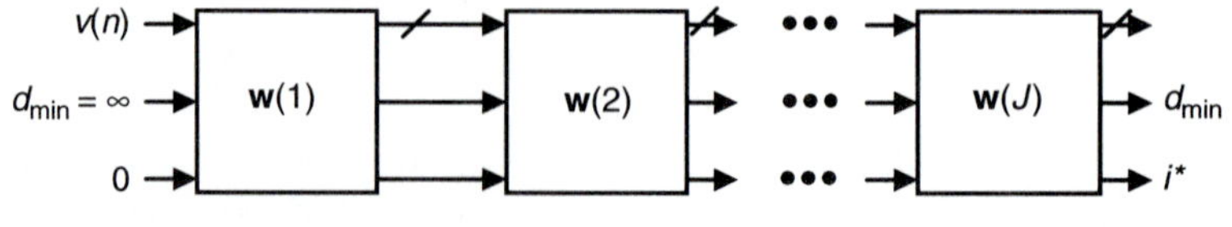

Figure 9–11 Array of J processors, each assigned to one of J codebook entries, for real-time vector quantization.[8] © 1995 IEEE.

Codebook generation proceeds according to the following steps:

1. Initialize.
2. For each of the K training vectors, find the closest codebook vector (this requires computing the distance of the training vector from each codebook vector and selecting the minimum).
3. Add the distance between the training vector and its closest codebook neighbor to an accumulating sum of overall distance.
4. After assigning each training vector to a codebook value, replace that codebook value with a vector computed as the centroid of the set of training vectors that are closest to it.
5. Compare the total distance (across all pairs of training vectors and the nearest codebook entry of each) with the total distance from the previous iteration—if the change is less than a present convergence criterion, stop; else go to 2.

Fig. 9–12 represents the process graphically by showing a set of training values as ●, a set of codebook vectors (o), and the association of a set of training

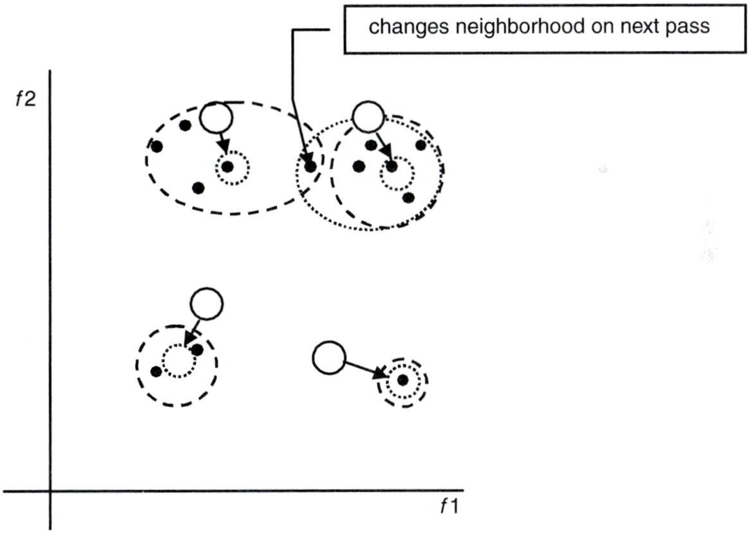

Figure 9–12 Vector quantization codebook generation via the LBG algorithm: Each training vector (●) is associated with its nearest codebook vector (o), where the large dashed circle shows association; the next training iteration is begun by moving each codebook vector to the centroid of the training vectors that were mapped to it, shown by the dotted small circle. This movement may change the association of a borderline training vector to another codebook vector.

vectors with a codebook vector by a dashed circle around the training vectors associated with a codebook vector. The arrow shows the movement of the codebook vector upon the next iteration to the centroid of its training set neighbors that were mapped to it. This movement may bring new training vectors into the neighborhood of the newly placed codebook entry, indicated by the dotted hollow circle.

The LBG training algorithm may be described in pseudocode,[8] tailored for high-speed implementation on parallel processors (Fig. 9–13). Specifically, a stream processing adaptation of the updated of centroid location is implemented in which each relevant training vector is added into **wnew**(i), and after all training vectors have been assigned, a division by the number of training vectors assigned to the codebook vector is performed.

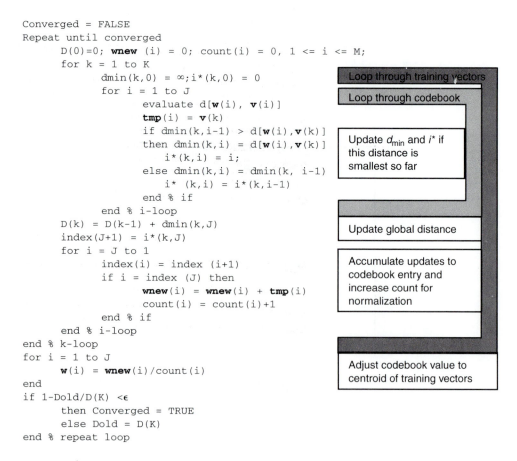

```
Converged = FALSE
Repeat until converged
      D(0)=0; wnew (i) = 0; count(i) = 0, 1 <= i <= M;
      for k = 1 to K
             dmin(k,0) = ∞;i*(k,0) = 0
             for i = 1 to J
                    evaluate d[w(i), v(i)]
                    tmp(i) = v(k)
                    if dmin(k,i-1) > d[w(i),v(k)]
                    then dmin(k,i) = d[w(i),v(k)]
                       i*(k,i) = i;
                    else dmin(k,i) = dmin(k, i-1)
                       i* (k,i) = i*(k,i-1)
                    end % if
             end % i-loop
      D(k) = D(k-1) + dmin(k,J)
      index(J+1) = i*(k,J)
      for i = J to 1
             index(i) = index (i+1)
             if i = index (J) then
                    wnew(i) = wnew(i) + tmp(i)
                    count(i) = count(i)+1
             end % if
      end % i-loop
end % k-loop
for i = 1 to J
      w(i) = wnew(i)/count(i)
end
if 1-Dold/D(K) <ε
      then Converged = TRUE
      else Dold = D(K)
end % repeat loop
```

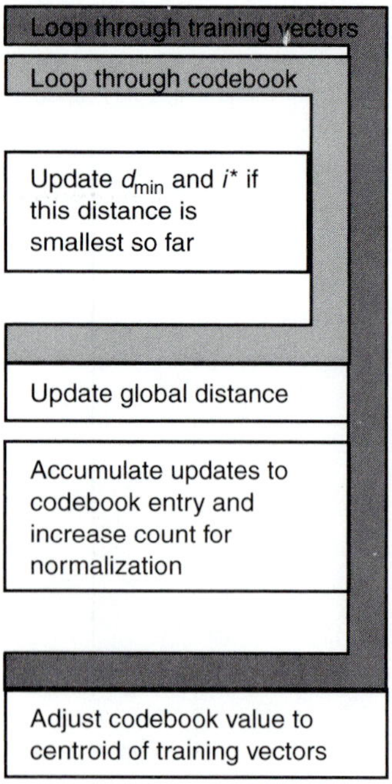
Loop through training vectors
Loop through codebook
Update d_{min} and i^* if this distance is smallest so far
Update global distance
Accumulate updates to codebook entry and increase count for normalization
Adjust codebook value to centroid of training vectors

Figure 9–13 Pseudocode listing of LBG algorithm for realtime implementation.[8] © 1995 IEEE, adapted with permission.

For a high-speed implementation of training, the parallel VQ encoding array of Fig. 9–11 can be augmented to allow execution of the LBG algorithm by adding a second processor array that receives the training vectors closest to each codebook entry and computes a new codebook vector. In Fig. 9–14, a dotted arrow between **w**(i) and **wnew**(i) indicates a transfer for each training iteration of K samples.

9.3 IMAGE COMPRESSION

Two types of image compression, or coding, are discussed here. Single-image frame coding compresses a still picture, while video coding is built up from single-frame coding by adding interframe coding techniques to compress the image sequence that makes up a video stream. Methods for single-frame coding include the discrete cosine transform (DCT), described below, and subband (or wavelet) encoding described in Section 8.3. Interframe coding supplements single-frame coding techniques with motion estimation, using search methods from frame to frame to predict and encode object motion.

9.3.1 Single-Frame Coding Methods

The discrete cosine transform (DCT) is an important element in image coding. It is performed on a two-dimensional image by acting upon subblocks of adjacent pixels within the image. For example, a 512 × 512 pixel image may be broken up into an array of 8 × 8 pixel blocks and a DCT may be performed on each block. The DCT packs most of the energy of image data into a few coefficients. It is approximately equal to a $2N$-point FFT of a reflected version of the signal sequence

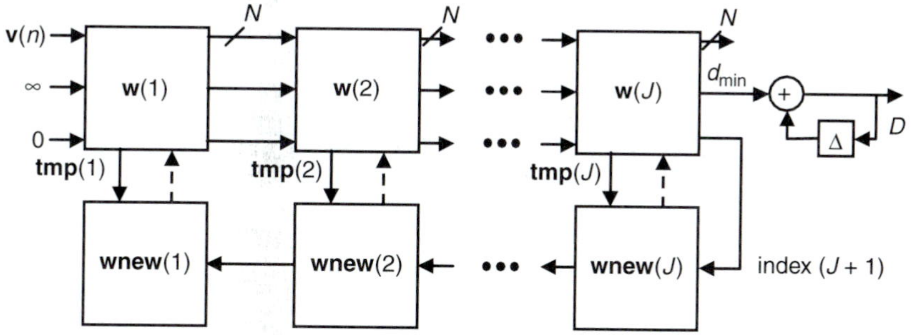

Figure 9–14 Processor array for VQ training is created from the linear array for VQ encoding (top) by adding a second processor below each coding processor which computes the new location of each codebook entry.[8] © 1995 IEEE, adapted with permission.

concatenated with the N-point sequence itself, and it exploits even symmetry and the restriction of image data values to the real domain.

The DCT may be compared to the FFT on a one-dimensional sequence (Fig. 9–15).[10] The FFT operates on a waveform segment formed by a finite-duration analysis window, and the waveform behaves as if it were periodically extended beyond the analysis frame. At the point of extension, the signal experiences a discontinuity ("glitch") as the low-amplitude windowed tail is abutted to the full-amplitude center of the next analysis window. The glitch at this joint introduces high-frequency components into the Fourier spectrum. Alternatively, the DCT causes the waveform to behave as if it were first reflected and then periodically extended, such that the low-amplitude tail of one analysis frame is abutted to the low-amplitude head of the next in a smoother transition. The DCT spectrum does not contain the high-frequency components introduced by the periodic extension of the FFT.

The one-dimensional DCT of a function $x(n)$ is given by:

$$\text{DCT: } Y(k) = \sum_{n=0}^{N-1} x(n)\cos\left[\frac{\pi}{N}k\left(n+\frac{1}{2}\right)\right] \tag{9.9}$$

$$\text{Inverse DCT: } x(n) = \frac{2}{N}\sum_{k=0}^{N-1} Y(k)\cos\left[\frac{\pi}{N}k\left(n+\frac{1}{2}\right)\right]. \tag{9.10}$$

Similarly, the two-dimensional DCT is given by:

$$\text{DCT: } Y(k,l) = \sum_{n=0}^{N-1}\sum_{m=0}^{M-1} x(n,m)\cos\left[\frac{\pi}{N}k\left(n+\frac{1}{2}\right)\right]\cos\left[\frac{\pi}{M}l\left(m+\frac{1}{2}\right)\right] \tag{9.11}$$

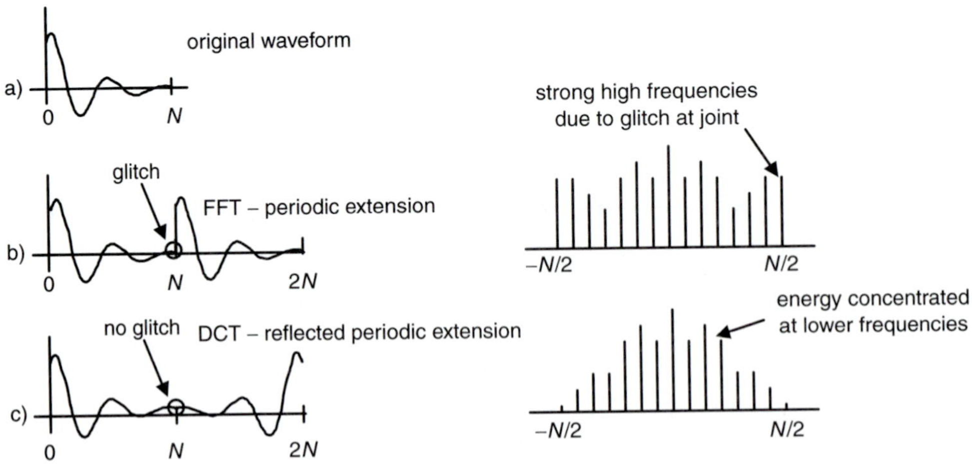

Figure 9–15 For an original waveform spectrum (a), as contrasted with the FFT (b), the DCT periodically extends a *reflected* version of the signal (c), reducing the high-frequency component resulting from the glitch when the extended signal meets the original in the FFT.[10]

$$\text{Inverse DCT: } x(n,m) = \frac{4}{NM}\sum_{k=0}^{N-1}\sum_{l=0}^{M-1} Y(k,l)\cos\left[\frac{\pi}{N}k\left(n+\frac{1}{2}\right)\right]\cos\left[\frac{\pi}{M}l\left(m+\frac{1}{2}\right)\right]. \quad (9.12)$$

The two-dimensional DCT can be computed by row/column decomposition into one-dimensional DCTs:

$$Y(k,l) = \underbrace{\sum_{m=0}^{M-1}\underbrace{\left\{\sum_{n=0}^{N-1} x(n,m)\cos\left[\frac{\pi}{N}k\left(n+\frac{1}{2}\right)\right]\right\}}_{N\text{-point DCT of columns}}\cos\left[\frac{\pi}{M}l\left(m+\frac{1}{2}\right)\right]}_{M\text{-point DCT of rows}} \quad (9.13)$$

A thorough review and comparison of real-time implementations of the DCT has identified four types of architectural approaches[11]: Direct, separate rows and columns, fast transform, and distributed arithmetic. Each will now be discussed in turn.

The direct method of DCT implementation uses the formula of (9.11) directly. For an image block size of $L \times L$ pixels, it requires L^4 multiplications and L^4 adds (alternatively, L^2 multiplies and L^2 adds per pixel). The direct implementation of processing DCT blocks can be mapped to an array of processing elements (PEs). An image of $N \times M$ pixels is divided into $L \times L$ blocks. There are no data dependencies across blocks, so a separate DCT processor may be devoted to each block.

A separable implementation of the DCT performs L one-dimensional DCTs on the rows and L one-dimensional DCTs on the resulting columns. This requires $2L^3$ multiplications and $2L^3$ additions ($2L$ of each per pixel). To avoid the need to rearrange the coefficient memory and processors between the row and column

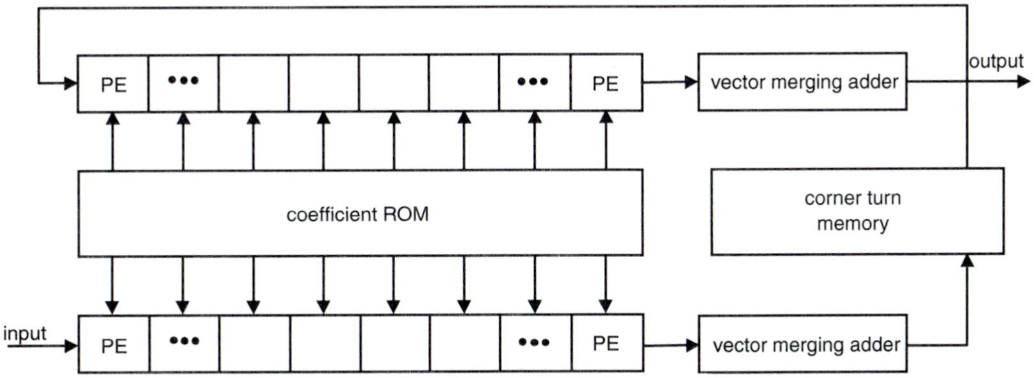

Figure 9–16 Two arrays of processing elements (PE), for rows and columns, interspersed with a corner turning memory for real-time DCT.[11] © 1995 IEEE.

computations, a corner-turn memory is interposed between the row and column processors that switches rows with columns. The corner turn memory was described in Section 6.6.4 in the example of synthetic aperture radar image formation. Fig. 9–16 shows two linear arrays of processor elements with the corner turn memory in between. It pipelines successive frames into rows and columns.

The DCT, like the Fourier transform, can be cast into a fast form by decomposing it into smaller DCTs. Such transformation reduces the computation on an $L \times L$ block from $2L^4$ operations to $2L^2\log_2 L$. This proceeds as follows. The one-dimensional DCT can be written in matrix form:

$$[\mathbf{Y}] = [\mathbf{C}][\mathbf{X}] \tag{9.14}$$

where $[\mathbf{C}]$ is an $L \times L$ matrix of coefficients based on the cosine function and $[\mathbf{X}]$ and $[\mathbf{Y}]$ are L-point input or output vectors.

For a specific instance, the matrices and vectors are written out for $L = 8$:

$$\begin{bmatrix} Y_0 \\ Y_1 \\ Y_2 \\ Y_3 \\ Y_4 \\ Y_5 \\ Y_6 \\ Y_7 \end{bmatrix} = \begin{bmatrix} c_4 & c_4 & c_4 & c_4 & c_4 & c_4 & c_4 & c_4 \\ c_1 & c_3 & c_5 & c_7 & -c_7 & -c_5 & -c_3 & -c_1 \\ c_2 & c_6 & -c_6 & -c_2 & -c_2 & -c_6 & c_6 & c_2 \\ c_3 & -c_7 & -c_1 & -c_5 & c_5 & c_1 & c_7 & -c_3 \\ c_4 & -c_4 & -c_4 & c_4 & c_4 & -c_4 & -c_4 & c_4 \\ c_5 & -c_1 & c_7 & c_3 & -c_3 & -c_7 & c_1 & -c_5 \\ c_6 & -c_2 & c_2 & -c_6 & -c_6 & c_2 & -c_2 & c_6 \\ c_7 & -c_5 & c_3 & -c_1 & c_1 & -c_3 & c_5 & -c_7 \end{bmatrix} \begin{bmatrix} x_0 \\ x_1 \\ x_2 \\ x_3 \\ x_4 \\ x_5 \\ x_6 \\ x_7 \end{bmatrix} \tag{9.15}$$

where $c_i = \cos(i^*\theta)$; $i = L/16$.

This requires $L^2 = 64$ multiplications and 64 additions. The matrices and vectors can be decomposed into two $L/2 \times L/2$ matrix and vector products to save computation:

$$\begin{bmatrix} Y_0 \\ Y_2 \\ Y_4 \\ Y_6 \end{bmatrix} = \begin{bmatrix} c_4 & c_4 & c_4 & c_4 \\ c_2 & c_6 & -c_6 & -c_2 \\ c_4 & -c_4 & -c_4 & c_4 \\ c_6 & -c_2 & c_2 & -c_6 \end{bmatrix} \begin{bmatrix} x_0 + x_7 \\ x_1 + x_6 \\ x_2 + x_5 \\ x_3 + x_4 \end{bmatrix}$$
$$\begin{bmatrix} Y_1 \\ Y_3 \\ Y_5 \\ Y_7 \end{bmatrix} = \begin{bmatrix} c_1 & c_3 & c_5 & c_7 \\ c_3 & c_7 & -c_1 & -c_5 \\ c_5 & -c_1 & -c_7 & c_3 \\ c_7 & -c_5 & c_3 & -c_1 \end{bmatrix} \begin{bmatrix} x_0 - x_7 \\ x_1 - x_6 \\ x_2 - x_5 \\ x_3 - x_4 \end{bmatrix}. \tag{9.16}$$

This requires $2(L/2)^2 = 32$ multiplications and $2^*2(L/2)^2 = 64$ additions for computation. A flowgraph that results from successive decompositions provides a fast version of this algorithm (Fig. 9–17), as proposed by B. G. Lee.[12]

The final of the four DCT implementation approaches uses distributed arithmetic to avoid multiplications. Given that the DCT can be performed as a set of scalar products as shown above, all multiplications can be replaced with addi-

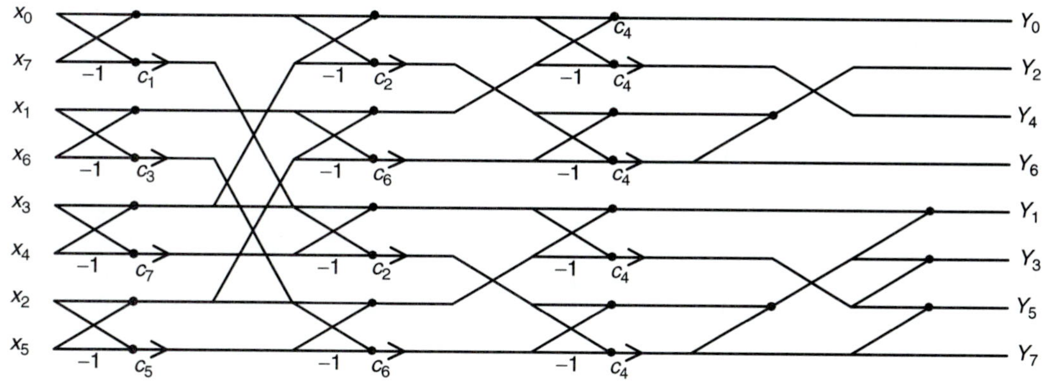

Figure 9–17 Flowgraph representation of fast one-dimensional DCT for $L = 8$.[11] © 1995 IEEE.

tions. Distributed arithmetic converts the scalar product of the matrix multiply into an efficient form by explicitly treating both the data value x and coefficient c as bit-weighted powers of two:

$$x_i = -x_{i,0} + \sum_{b=0}^{B-1} x_{i,b} \cdot 2^{-b} \tag{9.17}$$

$$C_i = -c_{i,0} + \sum_{b=0}^{B_c-1} c_{i,b} \cdot 2^{-b}. \tag{9.18}$$

The product $[\mathbf{Y}] = [\mathbf{C}] \cdot [\mathbf{X}]$ becomes:

$$\begin{aligned} Y &= \sum_{i=0}^{B_c-1} c_i x_i \\ &= \sum_{i=0}^{B_c-1} c_i \left[-x_{i,0} + \sum_{b=0}^{B-1} x_{i,b} \cdot 2^{-b} \right]' \end{aligned} \tag{9.19}$$

where $\boldsymbol{B}$ is the number of bits in the binary representation of x and B_c is the number of bits in the binary representation of C_i.

Next, each sum that multiplies c_i is expanded and grouped as shown [example for term (0)]:

$$c_0[x_{0,0} \cdot 2^{-0} + x_{0,1} \cdot 2^{-1} + \ \ldots \ + x_{0,B_c-1} \cdot 2^{-(B-1)}] \tag{9.20}$$

to obtain:

$$Y = \sum_{j=0}^{b-1} C_j \cdot 2^{-j}, \text{ where}$$

$$C_b = \sum_{i=0}^{B_c-1} c_i \cdot x_{i,b}. \tag{9.21}$$

Since C_j has only 2_c^B possible values, which are determined by x_{ib}, a read-only memory (ROM) is used to store its value, and B_c $\{x_{0,b}, x_{1,b}, \ldots, x_{b_c,b}\}$ bits are used to access and retrieve C_i values. Fig. 9–18 shows a circuit to implement distributed arithmetic, in which a ROM stores the appropriate values.

These four structures for real-time implementation of the DCT are compared in Table 9–5.

Having examined various structures for efficient implementation of the DCT, image compression algorithms that use the DCT are now discussed. The DCT is an important component of both the still-frame JPEG (Joint Photographic Experts Group) and MPEG (Motion Pictures Experts Group) standards for image coding.

In the JPEG standard for still pictures, the image is broken into 8 × 8 pixel blocks, and the DCT is applied to each block (Fig. 9–19). The DCT results are quantized by spatial-frequency-dependent quantization, which divides each of the 64 (8 × 8) DCT coefficients by a corresponding value in the quantization table. The array of 64 coefficients are then stored in order of increasing spatial frequency, using a zig-zag search that starts with DC (0 frequency), then moves to next-lowest horizontal spatial frequency, then the next lowest vertical, etc. As a result, the coefficients with the highest spatial frequencies are positioned sequentially toward the end of the 64 coefficients. If these values are zero, as they often are as a result of the low-frequency energy packing property of the DCT, they are easily encoded as a string of consecutive zeros using run-length encoding.

For the coding of coefficients across neighboring blocks, the DC coefficient does not vary much, so differential coding is used to transmit the differences between the DC values of successive blocks. The higher-frequency coefficients are encoded using run-length encoding as mentioned above. Huffman coding at the output uses predefined codes based on the statistics of the image, assigning shorter codes to frequently occurring values and longer codes for less frequently occurring values.

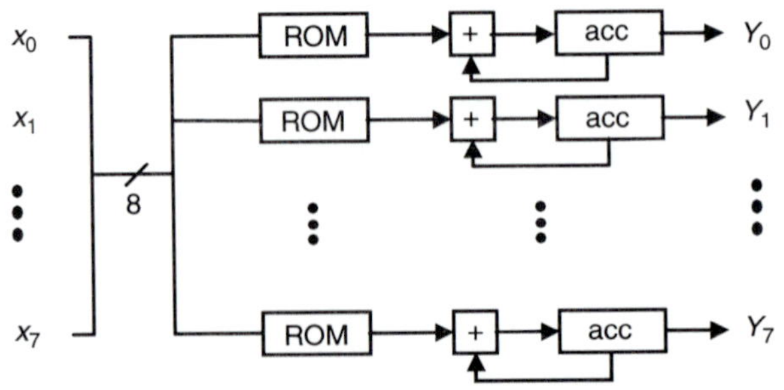

Figure 9–18 Distributed arithmetic implementation of eight-point DCT replaces multiplication with ROM lookups.[11] © 1995 IEEE.

Table 9–5 Comparison of alternative structures for real time implementation of discrete cosine transform for block size of *L* x *L* pixels.[11]

Method	# Multiplies/Pixel	# Adds/Pixel
Direct	L^2	L^2
Separable	$2L$	$2L$
Fast algorithm	$\log_2 L$	$2\log_2 L$
Distributed arithmetic	0	$32/L$

9.3.2 Moving Picture Encoding

The coding of image sequences for moving picture transmission includes both JPEG-based still-frame encoding and frame-to-frame coding of object movement. Motion estimation for interframe coding is accomplished by block matching. A frame is divided into blocks. Based on the frame rate, a postulate is made of the maximum number of pixels that an object can move between frames, which is labeled as w pixels in each direction. Block matching is based on a search of $\pm w$ blocks in all directions on one frame to find the best matching region with the previous frame. A mean absolute distance is used:

$$D(m,n) = \sum_{k=0}^{N-1} \sum_{l=0}^{M-1} \left| x_{i+1}(k,l) - x_i(k+m,l+n) \right| \tag{9.22}$$

$$\mathbf{v} = \arg \min_{m,n} [D(m,n)]$$

The search proceeds as shown in Fig. 9–20.

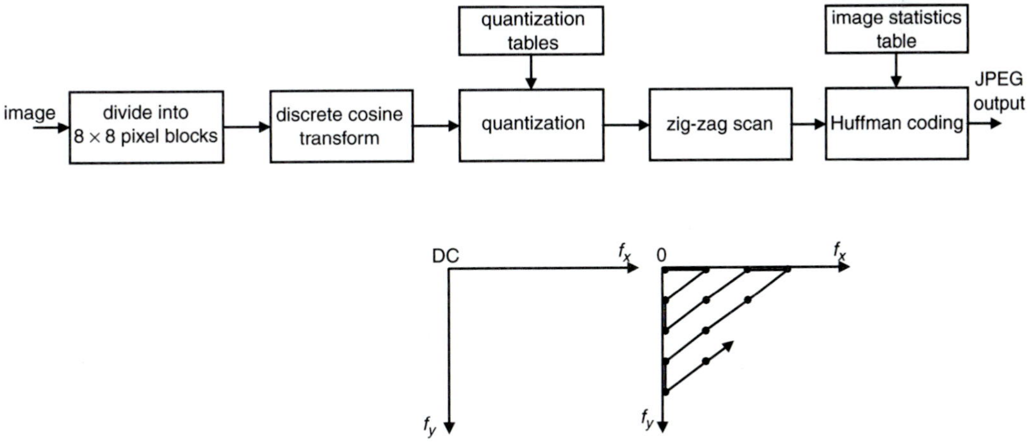

Figure 9–19 JPEG encoding algorithm performs DCT on 8 × 8 subblocks of image, then quantizes and orders the DCT coefficients in order of increasing frequency for final Huffman coding.

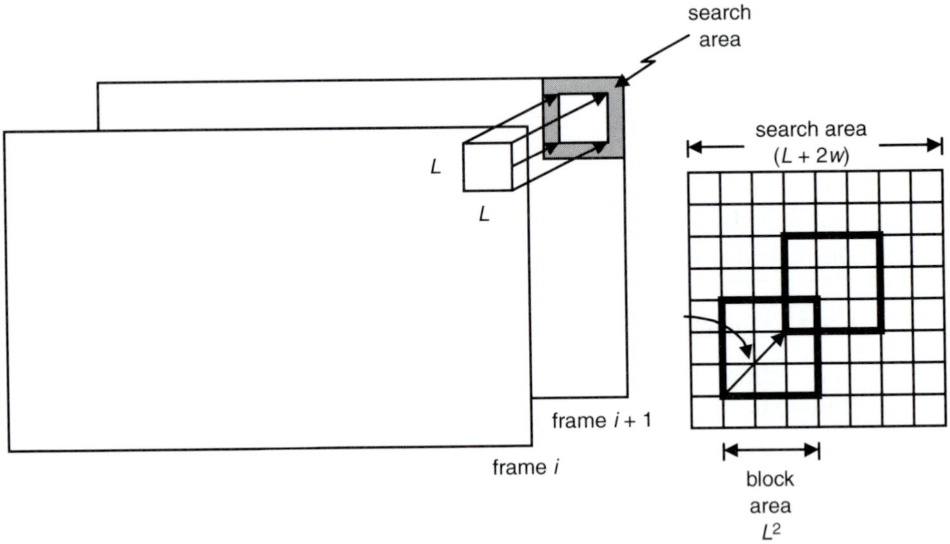

Figure 9–20 For block matching motion estimation, a block of size $L \times L$ in frame i is translated by an amount of w pixels in all directions in frame $i + 1$, and the position of best match is found and indicated by the motion vector, **v**.[11] © 1995 IEEE.

To develop an implementation of block matching, a dependency graph of the search operation is generated (Fig. 9–21). In this example, the block size L is 4 pixels, and the window excursion is ±1 pixel in each direction. Each minimum selection processor P_M computes at one search position of the window, and each absolute difference processor P_{AD} computes the absolute difference for a pair of pixels chosen from frame i and frame $i + 1$. The sum over the 4×4 block is generated by accumulating the individual values across all processors P_{AD}.[11]

This processor arrangement can be made more efficient by exploiting the locality of reference, that is, by taking advantage of the fact that data for adjacent pixel positions has already been used and is in the processor. To do this, a two-dimensional shift register is used, which stores the search window of size $L(2w + L)$ and can shift the coefficients up/down and right to execute the search. Each processor P_M checks whether the current distortion $D(m, n)$ is smaller than the previous distortion value and if so, it updates the $D_{\min}$ register (Fig. 9–22, right processor).

Processors of type P_{AD} store $x_i(k, l)$ and receive the value of $x_{i+1}(m + k, n + l)$ that corresponds to the current position of the reference block within the search window. P_{AD} then performs 1) subtraction, 2) computation of absolute value, and 3) addition to partial result coming from the upper processing element. Each P_{AD} has one register that, when combined with other processors in an array, provides

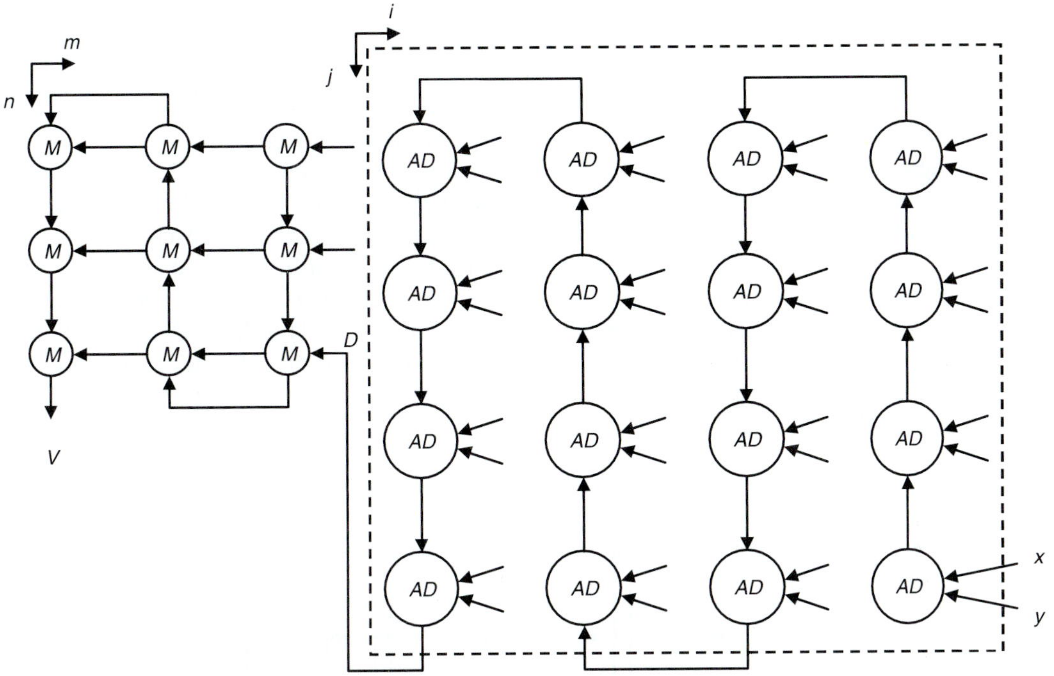

Figure 9–21 Dependence graph of block matching algorithm includes an absolute difference processor P_{AD} for each pixel in the block (shown as circle labeled AD), and a minimum select processor P_M for each search position of the block (circle M).[11] © 1995 IEEE.

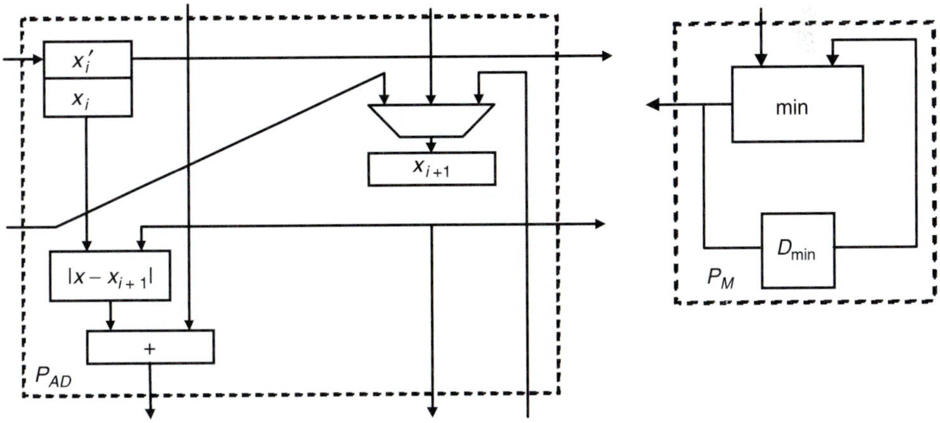

Figure 9–22 Absolute difference processor P_{AD} includes one element of shift register for storing neighborhood values, double buffer x_i and x_i', and absolute value and adder circuits; minimum processor P_M selects and identifies minimum value.[11] © 1995 IEEE.

a shift register for elements of the pixel neighborhood, and each processor obtains its particular needed value of $x_{i+1}(m + k, n + l)$. The processors P_{AD} and shift registers R are arranged as shown in Fig. 9–23. This array consists of $L(2w + L)$ processing and storage elements. Data enters serially as a new column of $2w + L$ pixels of the search area, which is stored in the shift registers R. The minimum processor P_M at the lower left selects the minimum value of D across the search area.

An MPEG coder (Fig. 9–24a) includes both the still image coder, a decoder in a feedback loop, and the motion estimator described above. In the still image coder, a variable length coder (VLC) follows the DCT. The decoder consists of an inverse quantizer and DCT (Q^{-1} and DCT^{-1}). The motion estimation section provides motion compensated prediction and provides the prediction errors to the

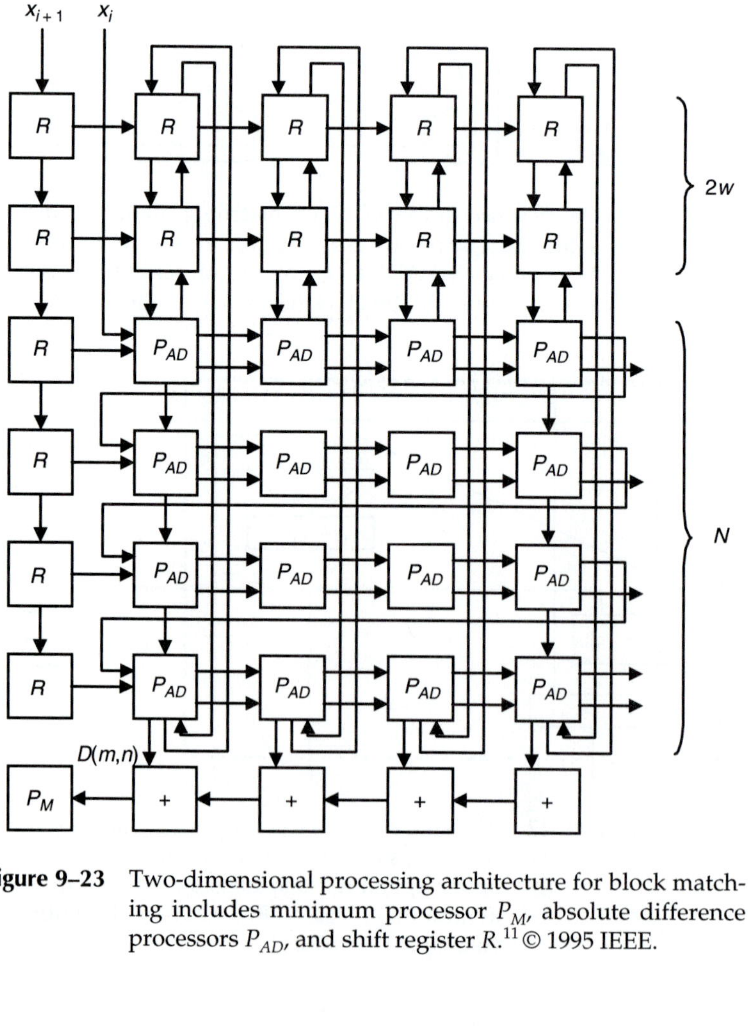

Figure 9–23 Two-dimensional processing architecture for block matching includes minimum processor P_M, absolute difference processors P_{AD}, and shift register R.[11] © 1995 IEEE.

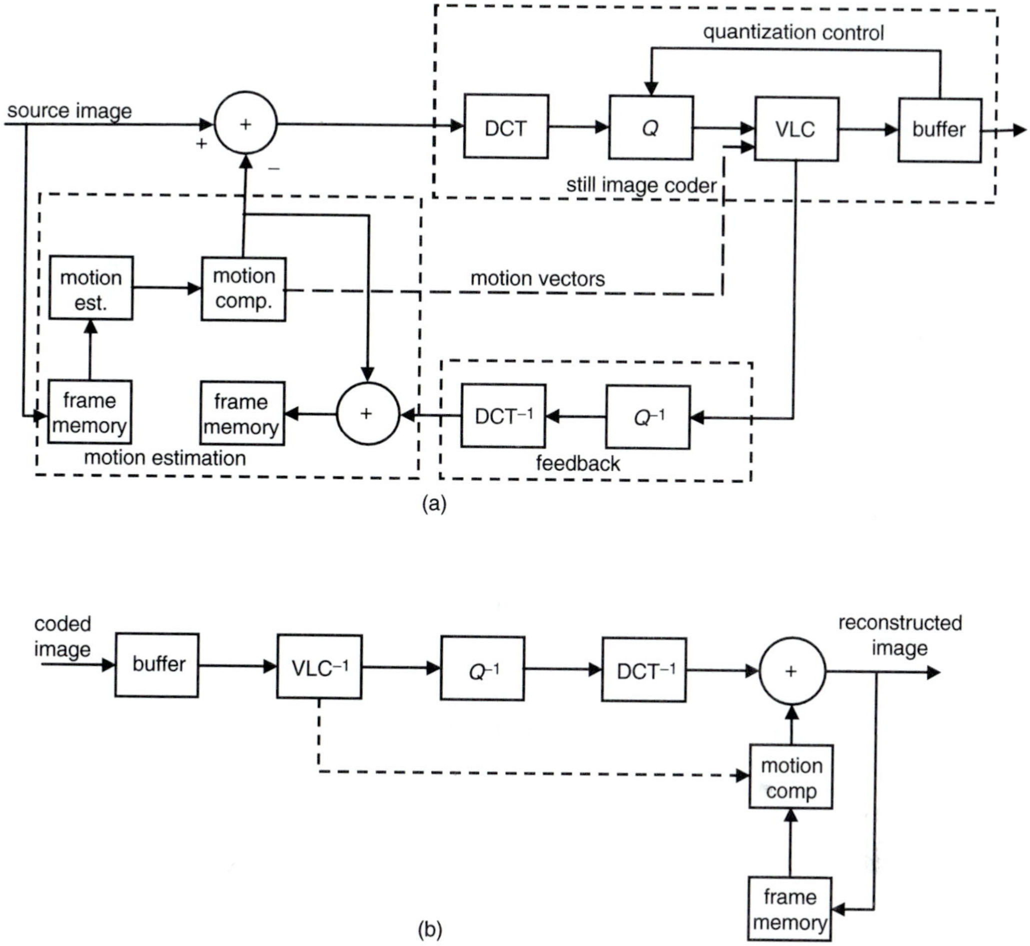

Figure 9–24 Image sequence encoder (a) and decoder (b) as used in MPEG standard.[11] © 1995 IEEE.

DCT coder. The image sequence decoder (Fig. 9–24b) includes the decoding path and motion reconstruction components and includes an inverse variable length coder (VLC^{-1}).

Both still-image and moving-image coders are subject to standards such as JPEG and MPEG. Table 9–6 shows the common still- and moving-image coding standards in use.

In addition to encoding for transmission purposes, video sequence coding for storage purposes introduces the additional requirements of random access, high-speed search, and reverse playback. These functions may be accomplished using a special predictive coding technique (Fig. 9–25). This technique, known as temporal

Table 9–6 Standards for still image and video coding.[11]

Name	Application	Typical Image Format	Coded Bit Data
JPEG	Still, photo-CD, photo videotext	Any size, 8 bits/pixel	0.25–2.25 bits/pixel
H.261	Video, videophone, video conference	QCIF, CIF; 10-Hz–30-Hz frame rate	$p \times 64$ Kbit/sec; $1 \leq p \leq 30$
MPEG1	Video CD-ROM, CD-1, computer	SIF, 25-Hz–30-Hz, CCIR	1.2 Mbit/sec
MPEG2	Video distribution & contribution	CCIR	4–9 Mbit/sec
MPEG4	Multimedia video coding—video & audio, object-based coding	CCIR CIF, TV	up to 4 Mbit/sec

predictive coding, encodes one frame of every *J* frames and interpolates for the interspersed frames. Frame are encoded according to type *I* (coded, not predicted), *P* (predicted frame, every *J*th frame), and *B* (bidirectionally interpolated frames), which are used to reconstruct the skipped frames between every *J*th frame.

It has been shown here that the structures for signal compression arise from processing to preserve the information content of the signal, which is extracted using the signal analysis computing structures discussed in the previous chapter. Once information is extracted, it may be compressed and then compared to compressed information obtained from other signals. Comparing signals based on their information content forms the subject of the next chapter.

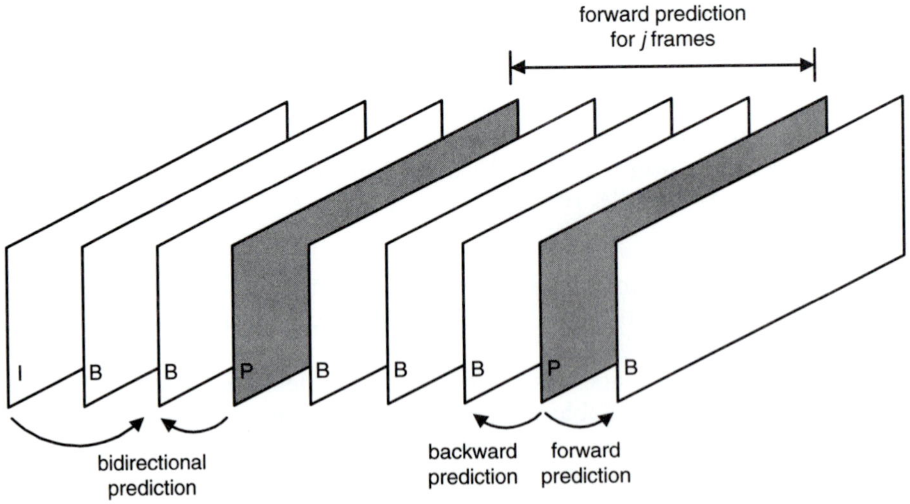

Figure 9–25 Sequential frame encoding for video storage to support rapid search and playback.[11] © 1995 IEEE.

REFERENCES

1. N. S. Jayant and Peter Noll, *Digital Coding of Waveforms* (Englewood Cliffs, NJ : Prentice Hall, 1984).
2. Ronald E. Crochiere, Richard V. Cox, and James P. Johnston, "Real-Time Speech Coding," *IEEE Trans. Communications,* COM-30 (April 1982), 621–634.
3. United States Federal Standard 1015 and United States Military Standard MIL-STD-188-133.
4. Alfred Kaltenmeier, "Implementation of Various LPC Algorithms Using Commercial Digital Signal Processors," *Proc. Inter. Conf. Acoustics, Speech, Signal Processing,* Boston, MA, April, 1983, pp. 487–490.
5. Bruce Fette, Chaz Rimpo, and Joseph Kish, "A Manpack Portable LPC10 Vocoder," *Proc. Inter. Conf. Acoustics, Speech, Signal Processing,* San Diego, CA, March, 1984, pp. 34B.1.1–34B.1.4.
6. Gene A. Frantz and Richard H. Wiggins, "Design Case History: Speak & Spell Learns to Talk," *IEEE Spectrum* (February 1982), 45–49.
7. Manfred R. Schroeder and Bishnu S. Atal, "Code-Excited Linear Prediction (CELP): High Quality Speech at Very Low Bit Rates," *Proc. Inter. Conf. Acoustics, Speech, Signal Processing,* Tampa, FL, March, 1985, pp. 937–940.
8. Yu Hen Hu, "Optimal VLSI Architecture for Vector Quantization," *Proc. Inter. Conf. Acoustics, Speech, Signal Processing,* Detroit, MI, May, 1995, pp. 2853–2856.
9. Y. Linde, A. Buzo, and R. M. Gray, "An Algorithm for Vector Quantization Design," *IEEE Trans. Communications,* 28 (January, 1980), 84–95.
10. Bruce Musicus, "Architectures for Digital Signal Processing," course notes, Massachusetts Institute of Technology, Cambridge, MA, 1985.
11. Peter Pirsch, Nicholas Demassieux, and Winfried Gehrke, "VLSI Architectures for Video Compression—A Survey" *Proc. IEEE,* 83 (February 1995), 220–246.
12. Byeng Gi Lee, "A New Algorithm to Compute the Discrete Cosine Transform," *IEEE Trans. Acoustics, Speech, Signal Processing,* ASSP-23 (December 1984), 1243–1245.

10

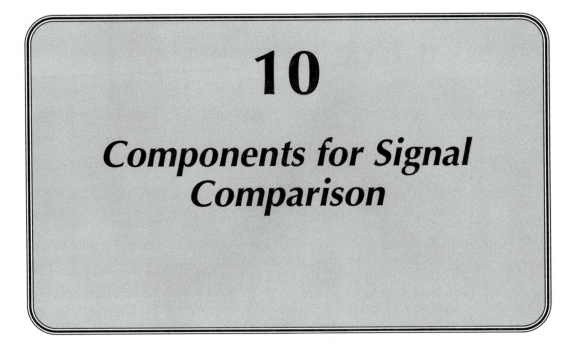

Components for Signal Comparison

Algorithms for signal comparison are distinguished by several features that affect their real-time implementation. First, these algorithms consist of not only a matching mode that compares an unknown input to a collection of previously stored data or models, but also a training mode by which patterns are built from models or data and compared to signals to effect the step of recognition. The training mode itself uses the matching mode to guide the training. Second, the pattern comparison operation consists of a hierarchy of levels. At the lowest level, a pattern's smallest units (analysis frames for time-domain signals, pixels or features for images) are compared to those of another pattern. At the next higher level, these local distances are summed along consecutive units according to various mapping paths to form global or accumulated distances. This accumulation allows the formation of words from phones or frames, or objects from pixels or features. In turn, these patterns are combined in various orders according to transition rules set by grammars to provide higher level scores. Third, during this hierarchical combination, several paths that map local distances of one pattern to those of another are analyzed, seeking to compensate for distortion by the choice of path used in the accumulation of the final global distance. Mapping-path distortions are restricted by constraints that keep pattern events in order, such as the preserving the sequence of sounds in a word while accommodating variations in duration of these sounds.

10.1 REAL TIME PATTERN COMPARISON

To cast signal comparison algorithms into real time, both algorithmic innovation and architectural structural design are important to speeding computation. Such innovations include limiting the search space and using sequential decision-making or coarse-to-fine search strategies. These algorithm innovations may be applied across a variety of recognition domains, as exemplified by the use of hidden Markov modeling for both speech and image recognition.

This section first examines the matching of patterns that consist of amplitude versus time, or time-domain waveforms. It then turns to the discussion of the recognition of two-dimensional objects occurring in images. The combining of two-dimensional objects with the time domain, as represented by video sequences, was discussed with regard to video compression (Section 9.3.2).

10.2 SPEECH RECOGNITION

Fig. 10–1 represents the hierarchical representation of a speech signal. This hierarchy underlies the structural approach that allows the construction of millions of phrases and sentences from thousands of words, in which the words are com-

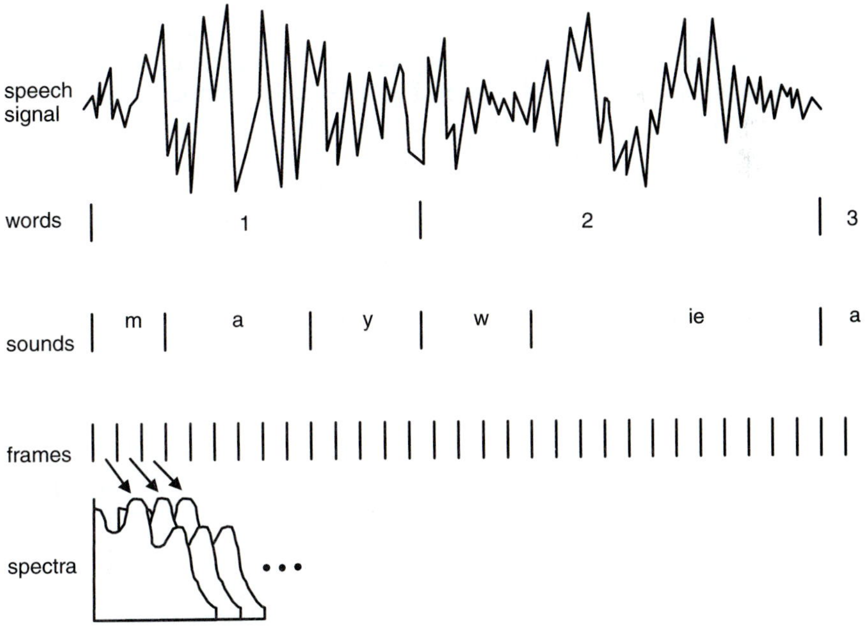

Figure 10–1 Hierarchical decomposition of speech from waveform into words, sounds, frames, and spectra.

posed of a few dozen basic sounds that are represented by sequential frames of spectral information.

An end-to-end speech recognition system (Fig. 10–2) begins by applying high-frequency preemphasis to flatten the overall spectral tilt of the signal, then applying endpoint detection to identify the start and stop times of the word or phrase. Computation of spectral-based feature vectors is used to segment the speech into a sequence of frames. The order of endpoint detection and spectral feature generation can be interchanged if the feature generator operates continuously for both silence and speech frames; or else the endpoint detector instead may turn on the feature extractor when speech is detected. The recognition of acoustic subword units, phones, phonemes, or words all occur in the stage of acoustic matching, which compares the incoming feature vector sequences with a previously stored set of units and builds up sequences of smaller units into larger units such as words. Finally, words may be combined into larger patterns such as phrases or sentences according to a language model to complete the recognition processing.

The speech recognition problem is formulated as a maximum likelihood path searching optimization process. It is efficiently solved by processing the incoming speech on a frame-by-frame basis and applying dynamic programming techniques. As an example, Fig. 10–3a shows two examples of the energy contour of a two-syllable word. Two peaks occur in either energy contour, but even though the overall lengths of the words have been normalized by linear magnification to match, the peaks do not align. At each time instance of either word, the energy value is accompanied by a spectral-based feature vector, not shown. The peaks (and features) are time-aligned by placing one pattern on the x-axis and the other on the y-axis, as shown in Fig. 10–3b. The mapping process, called *dynamic time warping*, proceeds from left to right, mapping the first frame of one word to the first frame of the other. The second frame of the x-axis pattern is next mapped to all reasonable variations of y-axis frames, which in this example are those that lie within a mapping slope of 1/2 to 2 (frames 1, 2 and 3).

The mapping computation consists of computations of local distances, by which the spectral feature vector of frame i of one speech pattern is compared to frame j of the other. Global distances, representing the sum of all local distances along a mapping path, are computed for each possible mapping path.

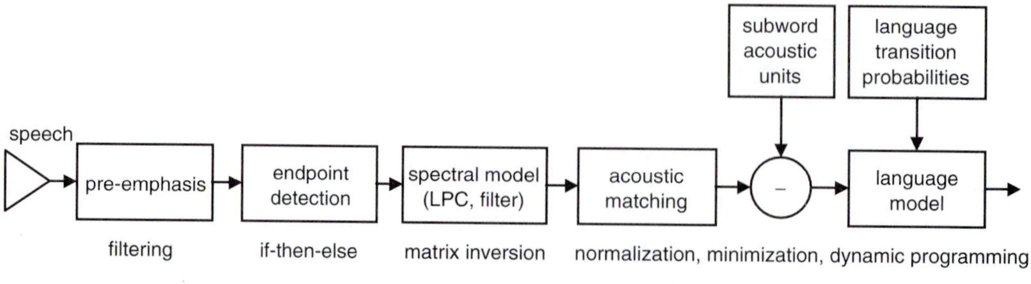

Figure 10–2 Processing sequence for automatic speech recognition.

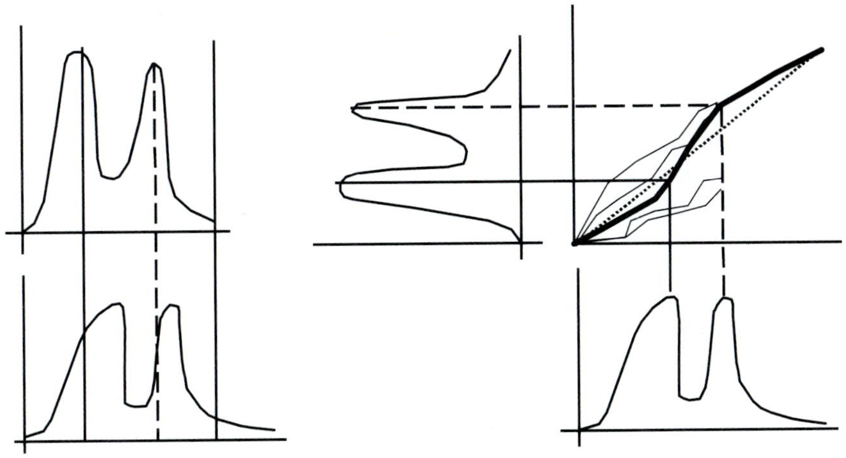

Figure 10–3 Warping of the time axis of one time-domain signal to that of another involves searching many potential mappings of time axes until one is found that best aligns the frames.

If each pattern were of length N frames, a total of approximately N^2 ways to map one pattern to another are possible. Application of global constraints, such as requiring that the first and final frame of each pattern be mapped to the corresponding frame of the other, and local constraints, which limit mapping paths to slopes of between 1/2 and 2, reduce the search space from N^2 to about $1/3\ N^2$.

To further speed the search, Bellman's principle of optimality[1] is used. This principle specifies that the best path to any node i at time t can be determined from the best paths to all nodes j at time $t - 1$, plus the best transition from node j to node i at time t. The principle of optimality suggests the use of the Viterbi[2] algorithm, developed for decoding the communication of symbols over a noisy channel. The Viterbi algorithm is discussed in the next subsection and is then applied in the form of dynamic programming to isolated and connected word recognition.

10.2.1 Viterbi Algorithm[3,a]

To describe the Viterbi algorithm, we set up the problem of recognition as a finite state machine similar to that introduced in our discussion of sequential logic in Section 2.1.3. We recall from that section the motivation for using the state machine as a way to simplify the representation of a system in which the current

---

[a]Portions adapted, with permission, from Hui-Ling Lou, "Implementing the Viterbi Algorithm," *IEEE Signal Processing Magazine* (September 1995) 42–51. © 1995 IEEE.

output is a function of the history of past inputs. Instead of keeping track of the entire past sequence of samples to determine what the next sample should be, we capture the results of the past inputs as placing the system in a particular state. We then use the current input and the current state to determine both an emitted symbol which is the basis for what we observe, and the next state. The hierarchical view of recognition is used, in which the recognition of an overall entity (say a word in optical character recognition) is decomposed into a sequence of recognition operations on smaller components (such as the characters that make up the word).

At any instant of time, the system is in one of K possible different states, and at each instant of time, the system transitions into another (or possibly the same) state. Associated with each transition is the emission of a symbol C_{xy}, denoting the symbol emitted upon transition from state x to state y. This symbol is subjected to corruption from various sources of noise, such as imperfect performance by the recognizer and the applications of "nuisance" transformations such as scaling and rotation that do not change the information content. The resulting observation at time j is derived from this symbol passing through the noisy channel and is represented as O_j. Because we do not know the state transitions that led to observation O_j, O_j does not include the indices of state transitions x,y. Fig. 10–4 represents the process of a state transition, an emitted symbol, the corruption of the symbol due to a noisy channel, and the resulting observation at time j.

The problem that the Viterbi algorithm addresses is: given a sequence of noisy observations $\mathbf{O} = \{O_1, O_2, \ldots, O_J\}$, find the underlying state sequence $\mathbf{q} = \{q_1, q_2, \ldots, q_J\}$ that best explains $\mathbf{O}$. Put in more formal probabilistic terms, given a sequence $\mathbf{O}$ of observations of a discrete time finite state Markov process in memoryless noise, find the state sequence $\mathbf{q}$ for which the *a posteriori* probability $P(\mathbf{q} \mid \mathbf{O})$ is maximum.

In our notation, time j progresses from $j = 0$ to $j = J$, and at time j, the system is in a state q_j. There are a total of K possible states, numbered by index k, $k = 1, 2,$

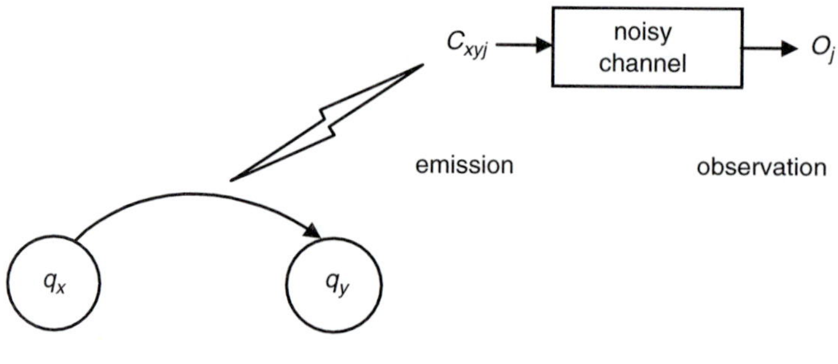

Figure 10–4 Transition from state q_x to state q_y at time j results in emission of symbol C_{xyj}, which is transmitted over a noisy channel to result in the observation O_j.

…, K. As time progresses from 0 to J, the system moves through a sequence of states given by $\mathbf{q} = (q_0, q_1, \ldots, q_J)$. An initial state q_0 may be explicitly defined. (Alternatively, q_0 may be governed by a probability density function π_m that indicates the probability of each state q_m serving as the initial state).

To define the finite state system, we must define:

1. The symbol emitted from state q at time j, and
2. The next state that the system transitions to from its current state, $q_j \rightarrow q_{j+1}$.

The output and the next state are determined by separate functions that depend on both the current state and the current input. The state diagram may be used to specify these relationships. In Fig. 10–5a, a five-state process is diagrammed. Each node (circle) represents a state, and each arc (arrow) represents a transition. Each arc is labeled with a pair of numbers, $(C/i)_{xy}$. The input (i) indicates that this transition out of state a to state b results from receipt of input i, and the transition is accompanied by the emission of symbol C. Thus, the emission of symbol 2 and transition from state 3 to state 4 upon reciept of input 1 would be shown as $(2/1)_{34}$.

Fig. 10–5(b) shows the allowed transitions in the form of a lattice. Each state k, $k = 1, 2, \ldots, 5$, is listed vertically, and time runs horizontally. Arrows show

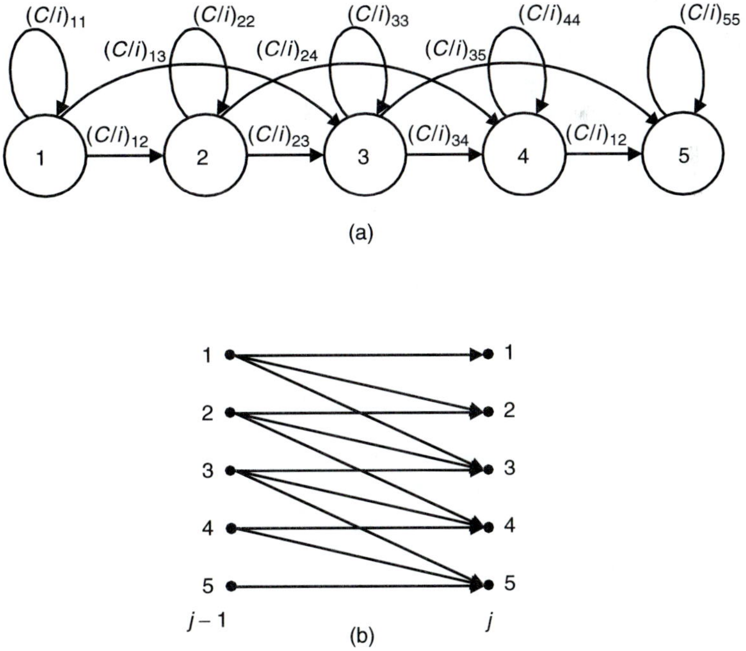

Figure 10–5 State diagram representing a five-state process (a), and the corresponding lattice diagram (b).

transitions out of each state at time $j-1$ on the left to a new state at time j on the right. Inspection of Fig. 10–5(b) shows that each state on the right has a subset of the set of states on the left that contains its allowable predecessors. The set of allowed predecessor states that lead into state k is given by the range $\mathbf{a}(k)$. In this example, $\mathbf{a}(2) = \{1, 2\}$.

One of the parameters of a Viterbi process is the particular state diagram that is used. Although we have here chosen a model that progresses from lower-numbered to higher-numbered states (known as a left-to-right model), we could have used any other combination of state interconnection, including one that allows transition from any state to any other state (known as an *ergodic* model). However, left-to-right models are typically used for modeling time-sequenced processes such as speech, so we have chosen this type.

The lattice may be used as the basis for a trellis, which indicates the evolution of the system as it makes transitions over time. In Fig. 10–6, at each time step k, the set of states is shown vertically, indexed by variable k. Time proceeds from left to right as variable j. The transition from each state at time $j-1$ to each state at time j is shown by an arrow—a sequence of these arrows defines a path representing the state sequence $\mathbf{q} = \{q_1, q_2, \ldots, q_j\}$. We have added the initial condition $q_0 = 1$. The trellis provides a $J \times K$ array and is shown for K states from time 1 to time J. A particular path, indicated by the heavy line, has been indicated as corresponding to the state sequence {1, 2, 2, 4, 5, 5, 5}. For any possible state sequence $\mathbf{q}$, there corresponds a unique path through the trellis, and vice versa.

A distance may be defined between the (ideal, noiseless) emitted symbol C_{xy} and a noisy, corrupted received observation O_j. This distance, known as a *local distance*, is a function of C_{xy} and O_j that increases as C_{xy} and O_j become less similar. (In an alternative implementation, the distance is replaced by a probability or similarity, in which lower values indicate less probability that O_j derived from C_{xy} and higher values indicate more similarity between O_j and C_{xy}, but we will

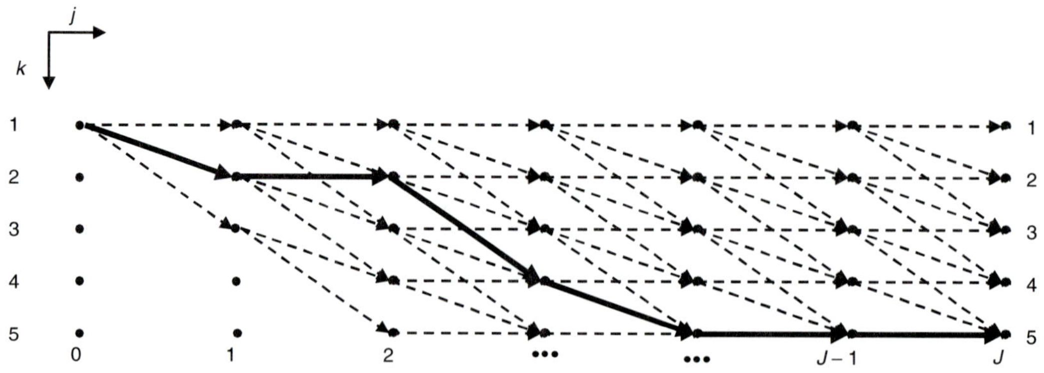

Figure 10–6 Trellis of five states, $k = 1, 2, \ldots, 5$, proceeds over time from left to right, $j = 0, 1, \ldots, J$.

only mention this equivalent case here and not treat it further.) We represent this distance as $d(x, y, j)$. If $C_{xy} = O_j$, then $d(x, y, j) = 0$, and if $C_{ab} \neq O_j$, then $d(x, y, j) > 0$. More specifically, if we are looking at the local distance of each allowed transition into k, given by $\mathbf{a}(k)$, then $x \rightarrow \mathbf{a}(k)$ in state k and $y \rightarrow k$, so $d(x,y,j) \rightarrow d[\mathbf{a}(k), k, j]$.

For every path in the trellis, an accumulated distance may be computed as the sum of all local distances along the path. The size of the accumulated distance is indicative of the difference between the particular state sequence defined by the path and the observed symbol sequence. If the observed symbols match the emitted symbols associated with the path exactly, then the accumulated distance is zero for that path, as it is the sum of local distances of value zero.

For any particular node k at time j, there will in general be several paths that reach it, one from each allowed predecessor of that node's predecessor set, $\mathbf{a}(k)$. The key concept to the Viterbi algorithm is the fact that Bellman's optimality principle may be applied: Rather than keeping track of all possible paths that transition into a state, it is sufficient to merely keep track of the one path entering the state that has the lowest accumulated distance. This lowest-distance path for each node is called the *survivor path* for that node. The best-scoring (lowest-scoring) path to any node k at time j can be determined from the survivor path to every allowed predecessor node at time $j - 1$, plus the lowest-distance transition into node k. Therefore, to proceed with the computation of all paths through the lattice at time j, given all paths up to time $j - 1$, we proceed as follows for each node at time j:

1. Compute the local distance scores for all transitions into node k;
2. For each allowed predecessor for node k, $a(k)$, add the corresponding local distance to the accumulated distance from the predecessor node to produce an updated accumulated distance and updated path.
3. Find the path into node k with the lowest accumulated distance score and select it as the one transition path into node k for time j.
4. Make note of which predecessor had the winning low score and store it for future path traceback.

These steps are repeated for all nodes at time j, and then time is advanced to $j + 1$ and the process is again repeated for all nodes. When time $j = J$ is reached, the process is repeated and then the list of predecessors recorded at step 2 above is traced from $j = J$ to $j = 1$ to reconstruct the state sequence.

More formally, executing the Viterbi algorithm consists of three steps:

1. For each state k at time j, compute the local distance between the received symbol O_j and each ideal noiseless symbol transitioning into k, C_{xy}, to produce the set of local distances $d(x, y, j)$.
2. For each path entering a state, add the local distance for that path to the accumulated distance (sum of local distances along the path) and then com-

pare the accumulated distance to the accumulated distance of all other paths entering that state and select the minimum:

$$s(k,j) = \min_{a(k)}[s(a(k),j - 1) + d(x,y,j)] \tag{10.1}$$

where $a(k)$ defines the search range. The value of a that produces the minimum over $s(a(k), j - 1) + d(x, y, j)$, given by $a' = \arg\min_{a(k)} [s(a(k), j - 1) + d(x, y, j)]$ or simply a', is stored in a memory array at a location associated with time t;

3. Trace the sequence of a' values that is stored, one for each input symbol, backward from the end to the beginning to recover the most likely sequence of states that led to the observations.

In real time implementation, several other considerations arise. Because the input may be continuous, the computation 3 of the most likely sequence must be conducted at an arbitrary time, rather than waiting until the end of the sequence. Specifically, at any time interval, the best-matching path reaching each state at that time can be traced back to the point at which all paths converge on the same state. Then, all best paths will have the same state sequence from the beginning of the sequence to the last frame of common best-matching path. Also in the case of continuous input, the accumulated distance can become arbitrarily large and, if not renormalized, can overflow available registers. Thus every time that a traceback is conducted, a renormalization is performed by subtracting the minimum accumulated distance at that frame from all accumulated distances. A basic kernel of add-min-argmin-update, as described in Section 7.2.4.2, is used repeatedly in performing the search over accumulated distance and their pattern.

We proceed with a specific example to illuminate the Viterbi algorithm. We use a two-state process $k = \{1, 2\}$ and examine $j = 0, 1, 2, \ldots, 5$. At each instant j, the system is in either state 0 or state 1. Also at each instant j, the system makes a transition, either again entering its current state or instead changing to the other state. These transitions form an input signal that may take on values of 0 or 1. From its current state, the system will emit a symbol, either –2, 0, or 2 in this example. The choice of symbol emitted depends upon both the state at time $j(q_j)$ and the input received at time j. The system will also transition to a new state at time $j + 1$, and the transition at time j depends upon both the current state and the current input. The system can be represented by a state transition diagram as shown in Fig. 10–7a or by a lattice as shown in Fig. 10–7b. The two states are shown as nodes labeled 0 and 1. From each node are two arcs, one leading to the other state and the other looping to remain in the same state. Each arc is labeled with the input that determines which transition is made and with the output that is emitted for that transition—the format of the pair is output/input. Thus, if an input of 1 is received while in state 1, the symbol 2 is output and the system stays in state 1, as indicated by the arrow looping from state 1 back to state 1 and labeled 2/1.

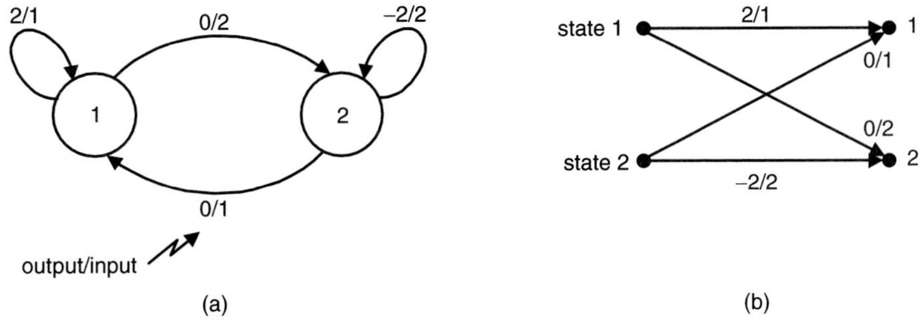

Figure 10–7 State diagram (a) and lattice (b) with two states, −1 and 1.[3] © 1995 IEEE.

If an input of 2 is received while in state 1, a 0 is output and transition is made to state 2.

Figures 10–8 and 10–9 show an example of Viterbi decoding applied to the system of Fig. 10–7. A sequence of noisy observations **O** = {0.05, 2.05, −1.05, −2.00, −0.05} arrives at time slots 1, . . . , 5, as shown at the top of Fig. 10–9. Points in the trellis are indexed by state index k and time index j as (k, j). A local distance measure $d(x, y, j) = (O_j - C_{xy})^2$ is used, computed between arriving observation O_j and ideal noiseless output C_{xy} for that transition. As discussed in our description of the local distance, this function meets the requirement of low values when C_{xy} and O_j are similar. For either state at each time stage of the trellis, the Viterbi algorithm computes the most-likely path coming into each state by computing the accumulated distances $s(k, j)$ of all paths coming into state j, then selecting the path with the minimum distance.

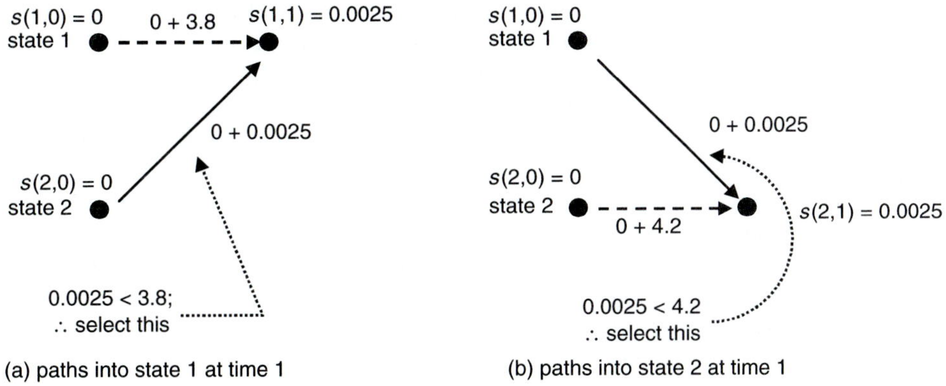

Figure 10–8 Computation of survivor path accumulated distance for time $j = 1$ for state 1 (a) and state 2 (b).[3] © 1995 IEEE.

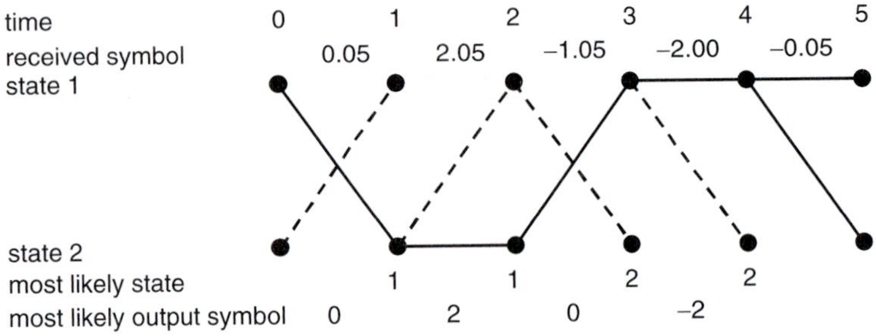

Figure 10–9 Illustration of five stages of Viterbi decoding as discussed in text.[3] © 1995 IEEE.

We begin by computing the accumulated distance for point (1,1) (Fig. 10–8a). The computation is initialized at $j = 0$ with $s(1, 0) = 0$ and $s(2, 0) = 0$. At time $j = 1$, the first symbol arrives ($O_1 = 0.05$). For state 1, time 1, the two paths that arrive are the one arriving from state 1 and the one arriving from state 2. According to the state diagram (Fig. 10–7), a branch from state 1 to state 1 produces an output of 2; the local distance between the ideal output 2 and the received output $O_1 = 0.05$ is

$$d(1,1,1) = (0.05 - 2)^2 = 3.8.$$

For the path that originates from state 2, the output for a transition from state 2 to state 1 is 0, and the local distance is

$$d(2,1,1) = (0.05–0)^2 = .0025.$$

The accumulated distance of the best path leading into state 1 at time 1 is given by min $\{0 + 3.8, 0 + 0.0025\}$, so $s(1, 1) = 0.0025$, corresponding to the transition from state 2. Similarly, the accumulated distance for the best path leading into state 2 at time 1 (Fig. 10–8b) is computed from local distances $d(1, 2, 1) = (0.05–0)^2 = 0.0025$ and $d(2, 2\ 1) = (0.05 + 2)^2 = 4.2025$, and an accumulated distance $s(2, 1) = \min\{0 + 0.0025, 0 + 4.2025\} = 0.0025$ is produced corresponding to the path from state 1.

Fig. 10–9 shows the entire sequence of computations for times $j = 0, 1, \ldots, 5$. Along the top are listed the observations **O** that are received. The best-scoring path is shown as the solid line; the associated most-likely state sequence and the corresponding noiseless transition outputs are shown at the bottom.

The Viterbi algorithm is quite powerful and can be adapted to perform in a number of hierarchical recognition applications. For the following discussions of speech recognition, we make these modifications:

1. Each state corresponds to a frame of speech, characterized by a set of spectrally based features and described by a feature vector consisting of several numbers.

2. A word becomes described as a certain sequence of feature vectors, different words are each stored during a training session as a sequence of feature vectors, and the task of Viterbi decoding becomes finding which sequence of feature vectors stored in memory as words best fits an unknown sequence of feature vectors received as the observations from an unknown utterance.
3. Indexing through the states, shown above by the index k, is changed to indexing through each frame i of every word t.
4. The observable is changed from a symbol that is emitted as the result of a transition between two states to a feature vector that is emitted as a result of being in a state.
5. Because symbols are now associated with vectors rather than with transitions, the local distance, kept inside the minimization operation of (10.1), is moved outside,

$$\text{so } s(k,j) = \min_{a(k)}[s(a(k),j-1) + d(x,y,j)] \text{ becomes}$$

$$s(k,j) = \min_{a(k)}[s(a(k),j-1)] + d(x,y,j).$$

6. The local distance is changed from the Euclidean used in the example above to a spectral measurement that compares two spectral-derived feature vectors.
7. The range of allowed search **a** corresponds to a few adjacent preceding states, thereby realizing a left-to-right model typical of speech (Fig. 10–5), rather than a more arbitrary model that allows backward transitions or transitions between any two states.

10.2.2 Dynamic Time Warping for Speech Recognition

This subsection applies the principles of dynamic programming and the Viterbi algorithm to the stretching and compressing of the time axis of one speech pattern to achieve alignment with another. The observations are spectral feature vectors and are referred to as the *test pattern. Reference patterns,* or sequences of feature vectors corresponding to words or subword units, have been previously stored during a training session. As explained above, the recognition task is to find the reference pattern that provides a sequence of feature vectors that best matches the test pattern. This approach of time-aligning and matching the test utterance pattern against each reference pattern is known as *template matching.*

The process of stretching and compressing the time axes, known as dynamic time warping, places the test pattern on the horizontal axis and the reference pattern on the vertical axis and applies the principle of optimality. This discussion first examines the case of isolated word recognition, in which the segmentation process is aided by the fact that each word is preceded and followed by a silence. The more difficult case of connected word recognition is then described, by which isolated words are concatenated without a break and the problem of coarticulation, in which the pronunciation of the beginning of one word is altered by the end of the previous word, is addressed. In connected word recognition, the sequence of isolated word patterns that best explains a connected sequence of observations is sought. Similarly, connected word recognition techniques are used to combine subword acoustic units, such as phonemes or phones, into words and phrases.

Dynamic time warping (DTW) proceeds from left to right, frame by frame, mapping each test frame to a range of reference frames for every word. Fig. 10–10 shows dynamic time warping, placing the unknown test pattern, with its feature vectors indexed by j, on the horizontal axis and the reference pattern, corresponding to a word, with its sequence of feature vectors running up the vertical axis and indexed by i. For isolated word recognition, local constraints limit the slope from 1/2 to 2. The slope constraint of 1/2 is implemented by preventing no more than two test frames in a row to be mapped to the same reference frame, and the slope constraint of 2 is implemented by allowing the mapping of test frame to reference frame to step no more than one reference frame for each step in test frame. As shown in the inset of Fig. 10–10, this amounts to enforcing the set of allowed predecessors for a frame as described by condition 7 of the previous section. Global constraints are a natural result of the local constraints and prevent any mapping path from occurring outside a parallelogram made of sides of slopes

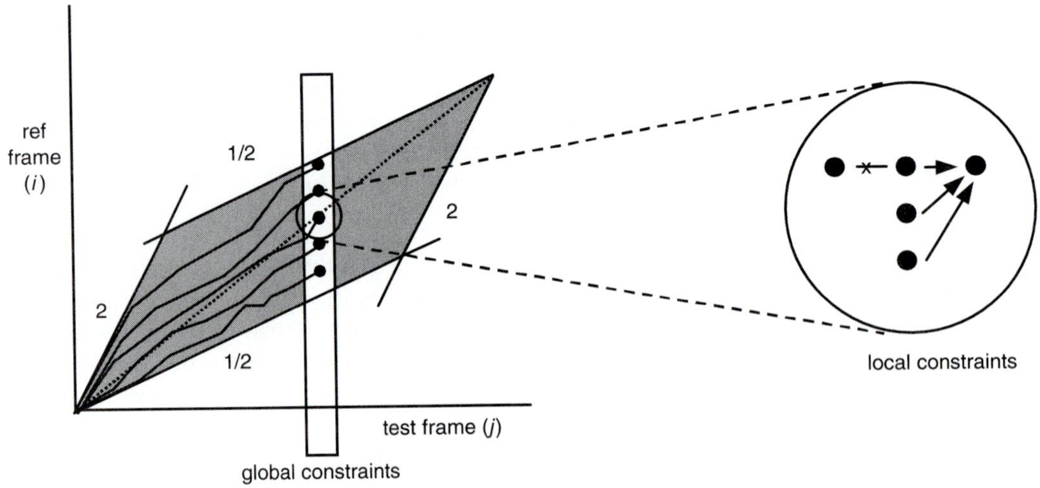

Figure 10–10 Mapping of test time axis to reference occurs by dynamic programming, with both global and local constraints limiting the deviation from the linear mapping shown by the dotted line of slope 1. Local constraints enforce a mapping slope of at least 1/2 by preventing doubly flat paths, as shown by the x.

1/2 and 2 extending from the first and last frame of the pattern. If nonlinear rate variation did not occur between test and reference utterances, the best mapping path would be the dotted line of slope 1. In practice, one of the other paths occurs. The initial and final frames of test and reference are constrained to map to each other, as shown by the points on the lower left and upper right corners of the parallelogram.

The DTW algorithm is used to match the test word pattern against each reference pattern, compute a time-aligned distance score for each, and then choose the word pattern with the smallest distance as the match choice. The confidence of this choice can be estimated by comparing the lowest distance to an absolute threshold, since even the best-matching out-of-vocabulary word will have a relatively large-distance score. Confidence can be further assured by examining the ratio of best-matching to next-best matching pattern that corresponds to a different word.

A custom single-chip architecture for performing dynamic time warping has been developed.[4] For this implementation, the local distance $d(i, j)$ between a test and a reference frame consisted of a log likelihood ratio, implemented as the logarithm of a scalar product of two feature vectors. A dedicated multiplier and log circuit for local distance computation is shown in the circuit (Fig. 10–11). The

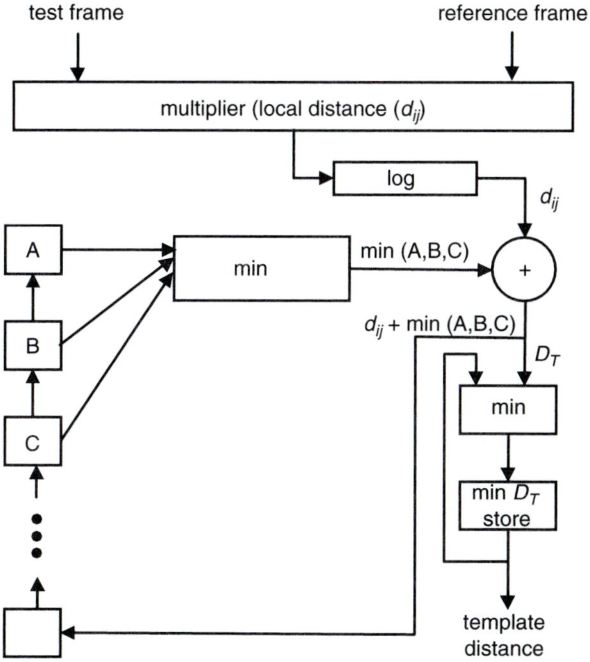

Figure 10–11 Architecture of pattern matching integrated circuit.[4]

accumulated distance scores for each path leading to the currently-used test frame, corresponding to the data ending each potential mapping path midway along the test axis in the vertical box of Fig. 10–10, are stored in a shift register. The three paths of slope 0, 1, and 2 at time $t - 1$ that lead to a specific reference frame at time t are presented to a three-way minimum selector by shift register locations A, B, and C. The minimum accumulated distance is added to the local distance and placed at the end of the shift register.

The specifics of the dynamic time warping calculation are most generally discussed in the context of connected word recognition. The three steps of connected dynamic time warping correspond to the three stages of the Viterbi algorithm described previously:[5]

1. Calculation of local (spectral) distance between a feature vector of the test and a feature vector of the reference (known as local distance processing, or for the Viterbi algorithm, the computation of the distance between the noisy received symbol and the ideal noiseless reference symbol);
2. Combining of these local distances along an optimum time alignment path to produce an accumulated distance for each of several mapping paths (accumulated distance processing, corresponding to finding the best path to each state in the Viterbi algorithm);
3. Identifying the best-matching concatenation of reference patterns from the beginning of the unknown input to the current frame (traceback processing, which in Viterbi decoding recovers the most likely sequence of states that led to the observed sequence).

As with the isolated word DTW, the recognition vocabulary consists of a set of word patterns (or, for connected DTW, subword units) which are both stretched and compressed to align them in time. These patterns are concatenated in various orders in a search to best match the incoming sequence. For computational reasons that simplify memory addressing, we choose to invert the direction of the reference index to an orientation that increases downward with time and with template number. As shown in Fig. 10–12, the test pattern axis i proceeds from left to right. Multiple reference patterns appear on the vertical axis, each with its own length and its own DTW parallelogram of paths. Here the DTW continues the sequential matching process by proceeding from the end of one word to the beginning of the next. Word reference patterns are indexed by word index t and, within a word, by frame index i. The length of word t is $n(t)$ frames.

Each word (or subword unit) consists of a sequence of frames, with each frame being described by a p-element feature vector. Thus, frame j of the test input is represented by feature vector $\mathbf{V}(j) = [v_0(j), v_1(j), \ldots, v_{p-1}(j)]$ and reference frame i of word (or subword unit) t, $t = 0, 1, \ldots, T$, is represented by feature vector $\mathbf{R}(t, i) = [n_0(t, i), n_1(t, i), \ldots, r_{p-1}(t, i)]$.

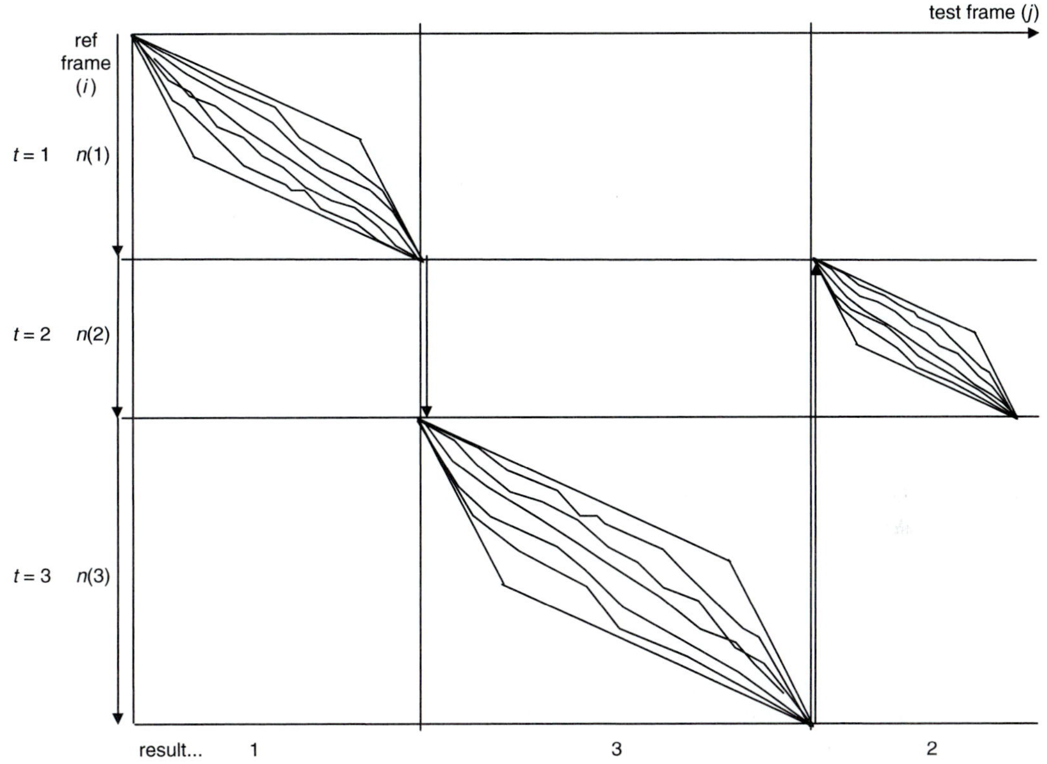

Figure 10–12 Connected word DTW proceeds from left to right along the time axis of the test pattern, indexed by *j*. Individual reference patterns are arranged in the sequence that best explains the test pattern. In this example, the sequence {1, 3, 2} is shown.

The local distance between test frame *j* and reference frame *i* of template *t* is given by:

$$d(t,i,j) = \log\left[\sum_{m=0}^{p-1} v_m(j)\cdot r_m(t,i)\right]. \tag{10.2}$$

This scalar product form of the distance is suitable for use when test and reference vectors are computed from an LPC analysis of the waveform. Alternative local distance measures, such as the Euclidean distance $d^E(t, i, j) = \sqrt{\Sigma_{m=0}^{p-1}[v_m(j) - r_m(t,i)]^2}$, may also be used for such features as rectified, low-pass-filtered outputs of a bank of bandpass filters. The computation proceeds from beginning to end of the test pattern, one frame at a time.

To update the accumulated distance from frame $j - 1$ to the current frame j, local distances are computed between the current test frame and each frame of every template. The accumulated distance $s(t, i, j)$ between test frame j and reference frame i of template t is calculated from the previous frame accumulated distance and the local distance as follows:

$$s(t,i,j) = \min_{a=0,1,2} [s(t,i - a,j - 1)] + d(t,i,j) + p(a); i > 1. \tag{10.3}$$

The term $p(a)$ is an additive distance penalty to compensate for distortion of the time axis as a function of the slope of the path connecting segment, a. For the first frame of a reference template ($i = 1$), the accumulated distance for the previous input frame to be used in updating the accumulated distance is $s'(j)$, which is the best (smallest) accumulated distance score of the final frames of all reference templates. In this manner, the best matching reference template is concatenated with the first frame of the current reference template to build up the connected template string.

$$\begin{aligned} s(t,1,j) &= s'(j) + d(t,1,j);\ i = 1, j \neq 0; \\ s(t,1,0) &= 0. \end{aligned} \tag{10.4}$$

The distance $s'(j)$ is obtained by examining the accumulated distance scores that correspond to the ends of the reference templates, $n(t)$:

$$s'(j) = \min_t [s(t,n(t),j - 1)])]. \tag{10.5}$$

This section of computation also records path information for each reference frame to allow subsequent computation of the sequence of reference templates that generated that path. The incremental path leading to cell (t, i, j) from the cell for test frame $j - 1$ is determined by the choice of a that minimizes (10.3), labeled a'. These choices of a are used to update an array $l(t, i, j)$, which records the incremental path yielded by each minimum selection for later reconstruction of the optimum pattern alignment path. The array element $l(t, i, j)$, used to keep track of test input frames that correspond to word boundaries, records the input frame index at which the best path to cell (t, i, j) ended the template previous to template t, as follows:

$$l(t,i,j) = j - 1; i = 1 \tag{10.6}$$

$$l(t,i,j) = l(t,i - a', j - 1); i > 1. \tag{10.7}$$

Traceback operations are of two types, those performed on every frame of test input, and those performed at some traceback interval of g input frames. The per-frame traceback operations are required to record path information, while the operations performed at the traceback interval constitute the identification of reference templates that match the unknown input.

At every input frame, the scores of those paths that end at the final frame of a reference pattern are examined and the lowest distance path is identified by template index $t'(j)$ and l-array index $l'(j)$ and the associated distance $s'(j)$ is found as was shown in (10.5):

$$s'(j) = \min_t [t,n(t),j)] \tag{10.8}$$

$$t'(j) = \underset{t}{\text{argmin}}\, [s(t,n(t),j)] \tag{10.9}$$

$$l'(j) = l(t',n(t'),j) \tag{10.10}$$

where argmin[$f(x)$] means the value of x that minimizes $f(x)$.

The path distance s' is sent back to the accumulated distance calculation. The values of $t'(j)$ and $l'(j)$ are stored in an array indexed by j which is known as the word link record. After processing the final frame J of the test utterance, the value $t'(J)$ contains the last best-fitting word reference pattern, and $l'(J)$ provides a pointer value j to move back in the t' array $[t'(l'(J))]$ to get the preceding best-matching word pattern. The t' array and l' array may be traced back in linked list style, until $l' = 0$ is reached, indicating that the beginning of the test utterance has been reached.

At the traceback interval g (typically a number of frames which corresponds to an average word length), the word link records are examined to determine whether an unambiguous recognition choice can be made and if so, that choice is then output from the processing.

The traceback operation proceeds as follows. First, the best path up to the current input frame is identified by finding $\tilde{i}(j)$ and $\tilde{t}(j)$ and the associated path distance, $\tilde{s}(j)$:

$$\tilde{s}(j) = \min_{t,i}\, [s(t,i,j)] \tag{10.11}$$

$$\tilde{t}(j) = \underset{t}{\text{argmin}}\, [s(t,i,j)];\; t = 1,2,\ldots,v \tag{10.12}$$

$$\tilde{i}(j) = \underset{t}{\text{argmin}}\, [s(t,i,j)];\; i = 1,2,\ldots,n(t). \tag{10.13}$$

Next, t^*, the identity of the immediately previous template that this path most recently completed, is recovered by using the proper cell of the l arrays as a pointer into the word link record:

$$\tilde{l}(j) = l[\tilde{t}(j),\tilde{i}(j),j] \tag{10.14}$$

$$t^* = t'[\tilde{l}(j)]. \tag{10.15}$$

If $\tilde{i}(j) = n(\tilde{t})$ the best path coincides with the end of a template and $\tilde{l}(j) = l[\tilde{t},n(\tilde{t})]$.

The choice t^* must next be examined to determine whether it had been put out at the last traceback, which occurred at input frame $j - g$. If $\tilde{l}(j) > j - g$, the t^* was not previously put out and is now sent out as a recognition choice. If $l(j) \leq j - g$, then the choice was put out at the last traceback and is not put out again.

There may have been more than one template choice resolved since the last traceback occurred. The entire sequence of templates that has been resolved by the input since the last traceback may be recovered in reverse order by tracing back through the word link records until the word link record of index $j - g$ has been passed. The most recent template is $t^* = t'[\tilde{l}(j)]$, the one preceding it is $t'[l'[\tilde{l}(j)]]$, and so forth, back to the beginning of the traceback interval at frame $j - g$.

We construct a simple example to further describe this process. For computational ease, we choose a simplified feature vector that consists of just one value ($p = 1$), and we use the simple absolute value of the difference of two one-valued feature vector values as the local difference between feature vectors. Table 10–1 shows a sequence of test frame features running from left to right (shaded row) and a set of two reference patterns of three frames each, running from top to bottom down the left (shaded column). Within the matrix formed by the shaded row and column, the local distance between each test and reference frame of every template is shown, obtained by taking the absolute value of the difference of the values in accordance with our definition. For example, for $t = 1, j = 1$, and $j = 1$, the local distance is $|7.0–7.2| = 0.2$.

The matrix of local distances is used as input to the calculation of accumulated distances, their corresponding mapping paths, and their traceback information (Table 10–2). Again, the column headings correspond to successive values of j, and the row headings correspond to each frame of every reference pattern. At each cell (t, i, j), the value of $s(t, i, j)$ and $l(t, i, j)$ are written. Also added to the end of the table are rows that provide the word link record.

The computation begins by initializing all accumulated distances $s(t, i, 0)$ and all traceback pointers $l(t, i, 0)$ to zero. For reference frames $(t, i, 1)$ in the example, all previous values $s(t, i, 0)$ and $l(t, i, 0)$ are 0, and so the values of $s(t, i, 1)$ are simply the local distances $d(t, i, 1)$. The traceback pointers $l(t, i, 1)$ are all 0.

For the next test frame ($j = 2$), as on all subsequent test frames, computation of $s(t, 1, j)$ proceeds in a manner to allow concatenation to the end frame of the best-matching reference pattern that has a path exiting at $j - 1$. In this case, $d(1, 1, 2) = 0.3$ is added to the minimum of those accumulated distances for paths that end a reference pattern in the previous frame, in accordance with (10.4) and (10.5). The minimum is selected from $s(1, 3, 1) = 2.5$ and $s(2, 3, 1) = 2.6$, and so 2.5 is selected and added to the local distance $d(1, 1, 2) = 0.3$ to obtain 2.8. The traceback pointer value for the first frame of a reference pattern is specified by (10.6) to

Table 10–1 Feature vector values for test pattern (shaded across top) and two reference patterns (shaded column) and associated local distance matrix for example of connected DTW.

		$j =$	1	2	3	4	5	6
t	i		7.0	7.5	9.1	4.1	8.2	9.9
1	1	7.2	0.2	0.3	1.9	3.1	1.0	2.7
	2	6.5	0.5	1.0	2.6	2.4	1.7	3.4
	3	9.5	2.5	2.0	0.4	5.4	1.3	0.4
2	1	4.2	2.8	3.3	4.9	0.1	4.0	5.7
	2	8.0	1.0	0.5	1.1	3.9	0.2	1.9
	3	9.6	2.6	2.1	0.5	5.5	1.4	0.3

Table 10–2 Corresponding array of accumulated distances, traceback indices, and per-frame traceback information—first frame of each reference pattern (shaded) selects minimum-distance path from frames that end patterns (bold).

	$j =$	**0**	**1**	**2**	**3**	**4**	**5**	**6**
t	i	s, l for $(t, i, 0)$	s, l for $(t, i, 1)$	s, l for $(t, i, 2)$	s, l for $(t, i, 3)$	s, l for $(t, i, 4)$	s, l for $(t, i, 5)$	s, l for $(t, i, 6)$
1	1	0.0, 0	0.2, 0	2.8, 1	4.1, 2	4.7, 3	8.0, 4	5.8, 5
	2	0.0, 0	0.5, 0	1.2, 0	3.8, 0	6.2, 2	6.4, 3	9.8, 3
	3	**0.0, 0**	**2.5, 0**	**2.2, 0**	**1.6, 0**	**7.0, 0**	**6.0, 3**	**6.4, 3**
2	1	0.0, 0	2.8, 0	5.8, 1	7.1, 2	1.7, 3	11.0, 4	8.8, 5
	2	0.0, 0	1.0, 0	1.5, 0	2.6, 0	6.5, 0	1.9, 3	3.8, 3
	3	**0.0, 0**	**2.6, 0**	**3.1, 0**	**2.0, 0**	**7.5, 0**	**3.1, 3**	**2.2, 3**
traceback	$s'(j)$	0	2.5	2.2	1.6	7.0	3.1	2.2
record	$t'(j)$		1	1	1	1	2	2
	$l'(j)$		0	0	0	0	3	3

be set to $j - 1$, which is 1 in this case. Thus, the cell at (1, 1, 2) contains $s\,(1, 1, 2) = 2.8$ and $l\,(1, 1, 2) = 1$.

For the next reference frame ($i = 2$), the accumulated distance is selected from $s(1, 1, 1)$ and $s(1, 2, 1)$ in accordance with (10.3). (Since $a = 2$ would cross a template boundary, it is ignored. Alternatively and more accurately, $a = 2$ could be fed by the minimum of all accumulated distances of paths that have ended templates on frame $j - 1$, as above.) Thus, $s(1, 2, 2) = \min\{0.2, 0.5\} + d(1, 2, 2) = 1.2$. The traceback pointer $l(1, 2, 2)$ is created by pulling forward the same traceback pointer associated with the path that provided the minimum accumulated distance, in accordance with (10.7). Because the minimum selected was from $s(1, 1, 1)$, the value of $l(1, 1, 1)$ is copied into $l(1, 2, 2)$. The resulting pair is $s(1, 2, 2) = 1.2$; $l(1, 2, 2) = 0$. The computation proceeds in similar manner for (1, 3, 2).

Cell (2, 1, 2), computed next, is again the first frame of a reference pattern, and thus requires special treatment. Such first frames are shaded in Table 10–2. For this frame, the accumulated distances of all paths that end reference patterns at the previous time slot are examined and the minimum is selected [(10.4) and (10.5)]. Also, again in accordance with our special treatment of first frames, the trace back pointer is given the value $j - 1$ [(10.6)]. The computation then concludes for the remaining two cells (2, 2, 2) and (2, 3, 2).

Upon ending all reference patterns for the current j, the lowest-scoring path that ends a frame is selected in accordance with (10.8), its value is recorded as $s'(2)$, the associated template index is recorded as $t'(2)$, and the corresponding traceback pointer is recorded as $l'(2)$. Thus, $s'(2) = \min[2.2, 3.1] = 2.2$;

$t' = \text{argmin}[2.2, 3.1] = 1$, and $l' = l(1,3,2) = 0$. A similar set of computations was conducted at the end of test frame 1 and recorded in $s'(1)$, $t'(1)$, and $l'(1)$.

The computation of $s(t, i, j)$, $l(t, i, j)$, $s'(t)$, $t'(j)$, and $l'(j)$ continues in the above manner until it is computed for the final test frame $J = 6$. At this point, the final pattern of the best matching string is held in $t(J) = t(6) = 2$, the accumulated distance for that matching path is held in $s'(6) = 3.7$, and a pointer to the array element of array t' that contains information on the previous pattern is stored in $l'(6) = 3$. Then stepping back to $j = 3$, the identity of the preceding pattern in the connected string is found ($t'(3) = 1$) and this traceback pointer $l'(3) = 0$ points to the frame that precedes the first frame of test input, indicating that the traceback is completed and no further templates remain in the string. A heavy box at the right lower corner of Table 10–2 shows the location that holds the identity of the final template in the unknown string, as well as the identity of the preceding template in $t'(3)$, as pointed to by $l'(6)$.

An example of a processor for performing connected dynamic time warping provides a description of the structures that exploit the regularity of the computations. The processor, known as the Connected Dynamic Time Warp Processor, or CDTWP,[6] consists of three processors, one for each of the three computations. These are the local distance processor (LDP), accumulated distance processor (ADP), and traceback processor (TBP). A double buffer pipeline (DBP) between the ADP and TBP allows the output from the ADP of results from test frame j to fill one side simultaneous with the input to the TBP of results from frame $j - 1$. (Fig. 10–13).

The local distance processor (LDP) compares a single test frame $t(j)$ to all reference frames $r(t,i)$, calculating the dot products of the coefficients and storing them in memory. In Fig. 10–14, the locations of the test and reference coefficients are displayed. A reference memory contains the reference coefficients $r_a(t,i)$ for all templates, ordered first by feature order a ($a = 0, 1, \ldots, 8$), then frame i, and then template t. The test memory contains one frame of test coefficients, $t_a(j)$. Test

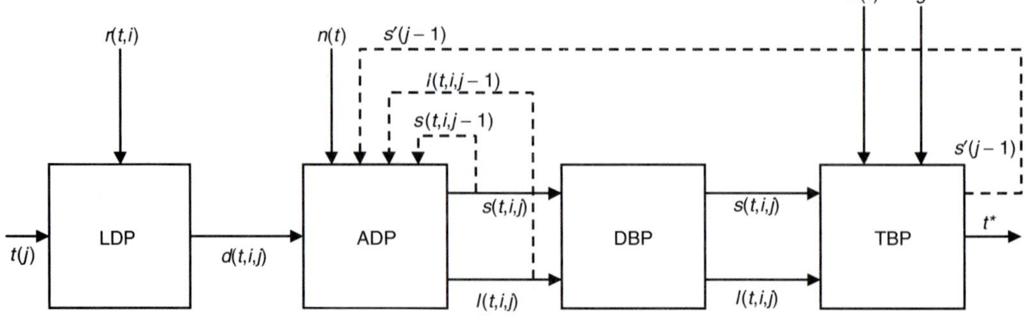

Figure 10–13 Processing sections of connected dynamic time warp processor (CDTWP).[6] © IEEE 1984.

memory is cyclically addressed and reference memory is sequentially addressed to simultaneously present corresponding orders of t_a and r_a to the multiplier-accumulator. After nine multiply-add operations, the accumulated result $\tilde{d}(t,i,j)$ is written to a third memory array and the accumulator is zeroed. This third memory array is also addressed sequentially.

The computation of the local distance dot products may be represented as the multiplier with data inputs and outputs moving vertically through the reference and distance memories and cyclically through the test memory as shown by the arrows in Fig. 10–14.

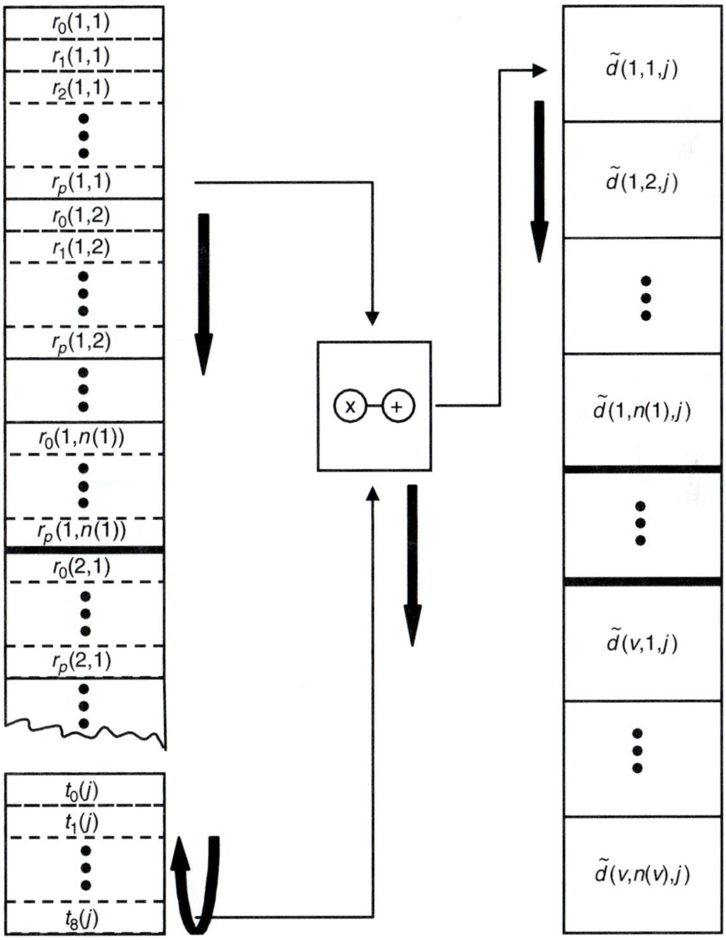

Figure 10–14 Processing sequence and memory for local distance processor (LDP).[6] © IEEE 1984.

The accumulated distance processor (ADP) receives the antilog of the local spectral distances $\tilde{d}(t,i,j)$, calculates the logarithm (10.2), combines each local distance with an accumulated distance allowed by local path constraints, and updates the accumulated distance with local distances for that frame [(10.3) and (10.4)].

In Fig. 10–15, the local distances $\tilde{d}(t,i,j)$ are shown at the left, arranged sequentially in the $\tilde{d}$ memory, having been placed there by the LDP (Fig. 10–14). Before the ADP begins computing the current test frame j, the values of the accumulated distances through frame $j - 1$, $s(t, i, j - 1)$, are stored sequentially in one array and the l array values for each cell are stored in a second array $l(t, i, j - 1)$.

The ADP also has access to the values of the end frame indices of each reference template, $n(v)$, which are used to mark the boundaries of templates.

For values of i that are not at the ends of reference templates, the updating of $s(t, i, j)$ begins at the highest address, corresponding to $s(v, n(v), j - 1)$, and moves to the lowest address, $s(1, 1, j - 1)$ in a single pass. As shown in Fig. 10–15, to update an accumulated distance $s(t, i, j - 1)$ to $s(t, i, j)$, the values of

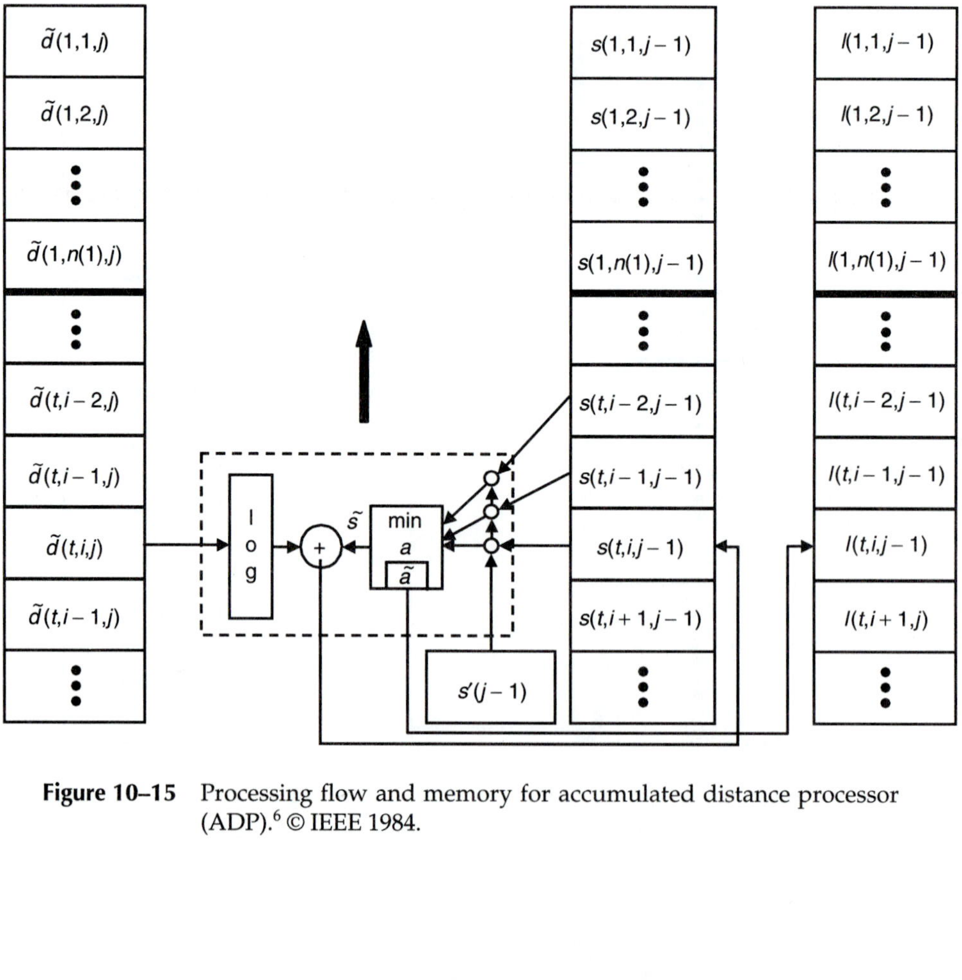

Figure 10–15 Processing flow and memory for accumulated distance processor (ADP).[6] © IEEE 1984.

$$\tilde{s}(t,i,j) = \min_{a=0,1,2} [s(t,i - a,j - 1)] \tag{10.16}$$

and

$$\tilde{a}(t,i,j - 1) = \arg\min_{a=0,1,2}[s(t,i - a,j - 1)] \tag{10.17}$$

are calculated. The logarithm is calculated,

$$d(t,i,j) = \log \tilde{d}(t,i,j), \tag{10.18}$$

and the sum

$$s(t,i,j) = d(t,i,j) + \tilde{s}(t,i,j - 1) \tag{10.19}$$

is formed and written to the memory location that held $s(t, i, j-1)$. The value of $\tilde{a}(t,i,j - 1)$ is used to update $l(t, i, j - 1)$ to $l(t, i, j)$ by transferring the value of $l(t,i - \tilde{a},j - 1)$ to the memory location that held $l(t,i,j - 1)$ in accordance with (10.7). After the update, the index i is decremented, the computational unit (in dotted box in Fig. 10–15) moves up one unit, and the minimum selector receives a new value for $s(t,i - 2, j - 1)$, while previous values of $s(t, i - 2, j - 1)$ and $s(t,i - 1, j - 1)$ become $s(t, i - 1, j - 1)$ and $s(t,i,j - 1)$ respectively. By performing the update from last frame to first in the reference template, only one storage array is needed for both the arrays $s(t,i,j - 1)$ and $s(t,i,j)$.

For values of i that correspond to the first frame of a reference pattern ($i = 1$), $s'(j-1)$, the minimum accumulated distance found at the end of all reference templates for test frame $j-1$ is used instead of the array value of $s(t, i, j)$ in accordance with (10.4). In Fig. 10–15, this is shown as the alternative distance $s'(j-1)$ which may be input to the minimum selector instead of $s(t,i,j-1)$, $s(t,i-1, j-1)$, or $s(t,i-2, j-1)$.

The traceback processor (TBP) operations that are performed at every input frame are described in Fig. 10–16; those performed at the traceback interval are shown in Fig. 10–17. At every frame, the TBP calculates $s'(j)$ and sends it to the ADP. It also calculates a word link record entry for the current input frame, which is then stored in the word link record (WLR) as described in (10.9) and (10.10) (Fig. 10–16).

At the traceback interval, the best path to the current frame is identified [(10.11) through (10.13)] and the corresponding $\tilde{l}(j)$ is used to index the WLR and identify the template that this best path last completed. If this template was completed after the last traceback, the template identity, stored in the WLR, is output and the WLR may then be cleared of dead paths.

A single-chip processor for performing the ADP and TBP operations has been developed.[7] This processor is suitable for both isolated word and connected word recognition, as well as for the implementation of finite-state grammars and hidden Markov modeling

10.2.3 Hidden Markov Modeling

The use of dynamic time warping on templates collected from training examples represents a nonparametric pattern recognition technique—actual data is used,

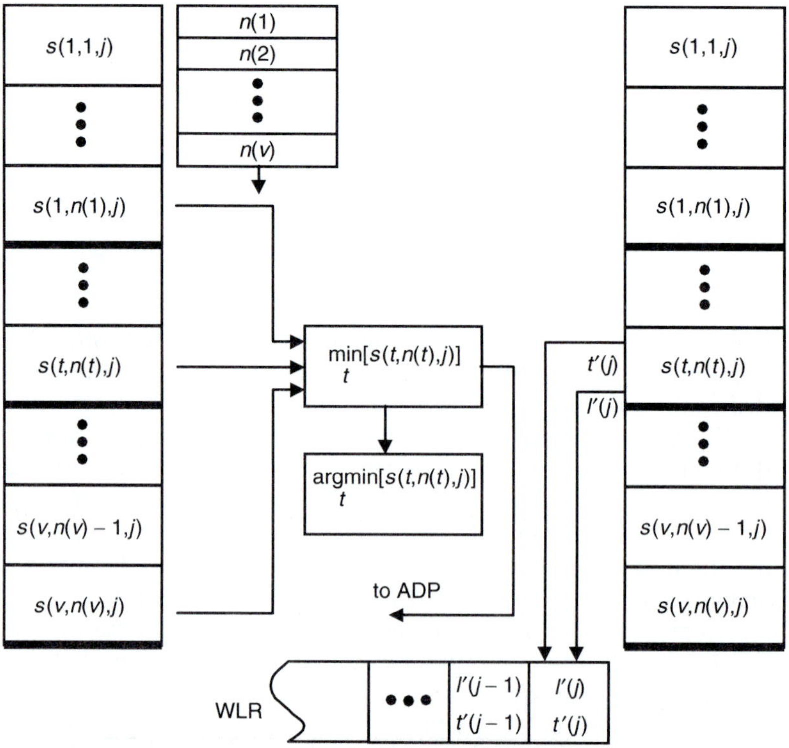

Figure 10–16 Processing flow and memory for trace back processor (TBP) per-frame operations.[6] © IEEE 1984.

and attempts are made to collect over the range of possible variations that occur across utterances, speakers, and (for connected word recognition) preceding and following words. The computation for training is relatively simple, but recognition requires examining large numbers of patterns over all possible variation, presenting a formidable computational burden.

Parametric techniques, which use a probability distribution function whose shape is governed by a few parameters, seek to describe pattern differences as differences in distribution parameters and use more extensive training and modeling to set parameters with sufficient precision to discriminate words. Parametric techniques are particularly relevant for real time implementation because the increase in the computational complexity of training is accompanied by a significant decrease in the computational complexity of pattern identification. In operational systems, pattern identification must usually be performed in real time, while training may occur more slowly as an off-line operation. Thus, speeding identification at the expense of slowing training speed is often an acceptable trade.

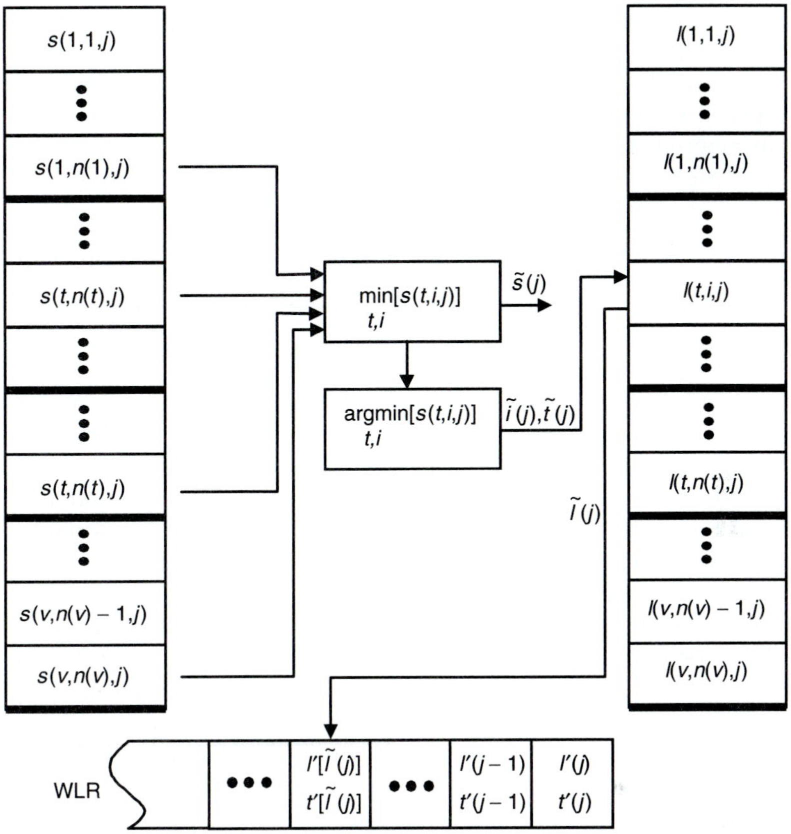

Figure 10–17 Processing flow and memory for traceback processor operations performed at each traceback interval. [6] © IEEE 1984.

As before, a hierarchical approach to recognition is used, in which longer multi-word utterances are built up from sequences of words or subword acoustic units. The speech utterance is described as a series of frames taken at a uniform frame rate. A frame at time t is regarded as an observation O_t, and the total utterance of length T frames is given by $O_1O_2, \ldots, O_T$.

In a hidden Markov model, or HMM, each observation is treated as having been emitted as the result of a model being in a particular state, represented as q_j. The HMM defines the probability of emitting observation O_t at time t while in state q_j. A probability density function applies to the emission of a specific symbol from a vector quantization codebook or from a parameterized distribution like a Gaussian mixture density. The HMM also includes a set of transition probabilities a_{ij}, indicating the probability of transitioning from state i to state j. Finally, an initial probability function π_i indicates the probability of each state i being the initial state of the system at time $t = 1$.

As an example, Fig. 10–18 shows a left-to-right HMM typically used for speech recognition in which a lower-numbered state always precedes a higher-numbered state. The probability for each allowed transition, a_{ij}, and the observation probability for each state, b_i, are shown.

Therefore, an HMM is defined by:

$$\mathbf{A} = a_{ij} = \mathrm{P}\{q_j \text{ at time } t + 1 | q_i \text{ at time } t\} - \text{state transition probability} \quad (10.20)$$

$$\mathbf{B} = b_j(k) = \mathrm{P}\{O_k \text{ at } t | \, q_j \text{ at } t\} - \text{observation probability in state } j \quad (10.21)$$

$$\Pi = \pi_i = \mathrm{P}[q_i \text{ at } t = 1\} - \text{initial state distribution.} \quad (10.22)$$

More recent work has allowed HMMs to explicitly model the duration of time spent in a state. Instead of the ever-decreasing probability of staying in the same state over multiple time intervals that is imposed by $a_{ii} \times a_{ii} \times \ldots$, where $a_{ii} < 1$, *explicit* durational modeling allows the ability to model a most likely duration for each state that is a peaked function of time. The quantity $d_j(\tau)$ is used as the probability that the process will emit τ observations when in state j, $\tau = 1, 2, \ldots, D$. Computational complexity is increased by $O(D^2)$ for including explicit durational modeling for durations that extend up to length D frames. With explicit durational modeling, a specific HMM is defined by $\lambda = \{(a_{ij}), d_j(\tau), b_j(k), \pi_i\}$.

To apply HMMs to isolated word recognition, the following steps are used:

1. In an off-line training session, build the HMM for each word in the vocabulary, λ^v, $v = 1, 2, \ldots, V$.
2. For each unknown word in the incoming test pattern $O = O_1O_2, \ldots, O_T$ and for each word model λ^v, calculate $P_v = P(O|\lambda^v)$.
3. Choose the word for which the model probability is the highest, i.e., $v^* = \underset{1 \leq v \leq V}{\mathrm{argmax}} \, [P_v]$.

Fig. 10–19 summarizes these steps for HMM-based isolated word recognition.

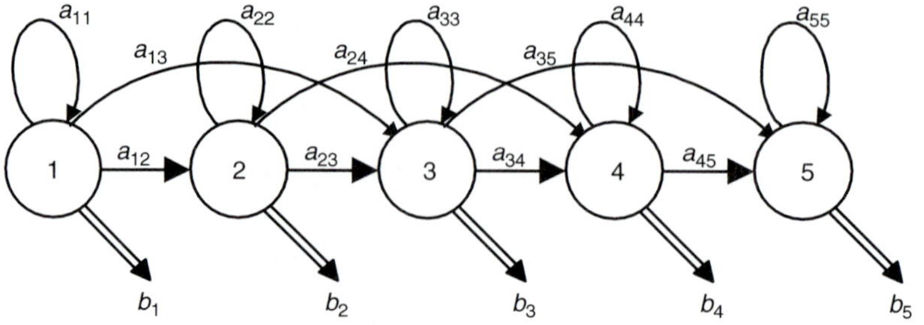

Figure 10–18 Five-state left-to-right hidden Markov model with states 1 to 5, state transitions a_{ij}, and state-dependent observation probabilities b_1 through b_5.

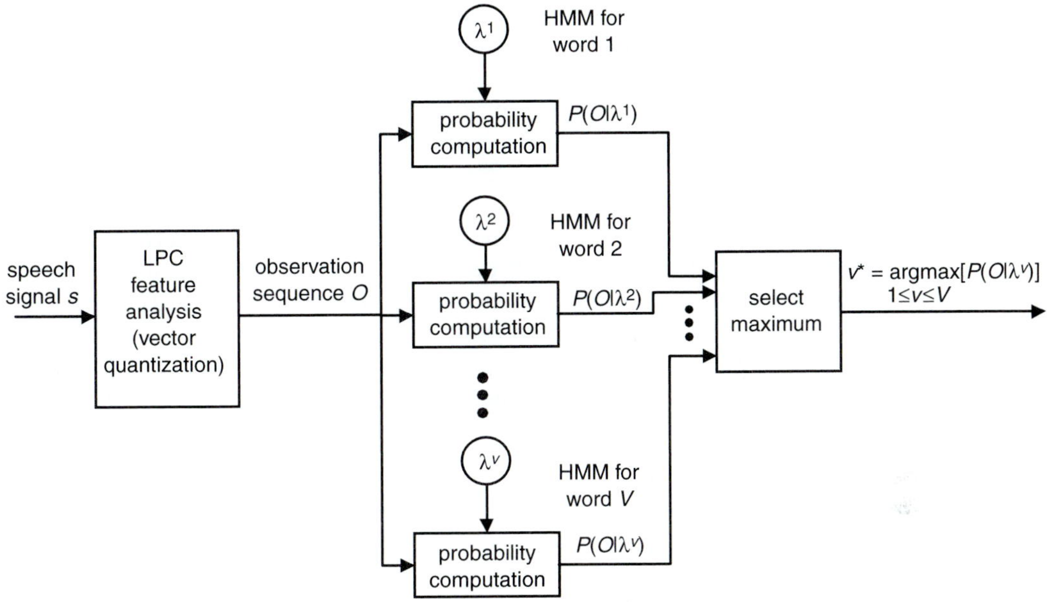

Figure 10–19 HMM-based isolated word recognition computes probability that an observation sequence **O** was produced by each of V hidden Markov models λ^v.[8] © 1989 IEEE.

The calculation load to perform recognition of V isolated words using an N-state HMM with an input duration of T frames is approximately $V \times N^2 \times T$, which is fairly modest (about 10^5 for $V = 100$, $N = 5$, and $T = 40$). The Viterbi algorithm is used for performing Step 2 above, and the same computational structures discussed for the dynamic time warp calculation can be used.

Training the HMM is quite computationally demanding, although in most cases, training may be performed off line, free from real-time constraints. The following steps are used to train an HMM:[8]

1. Find the probabilities of partial paths beginning at $t = 1$ and ending on the current frame by using the *forward algorithm*, described below, on the training sequence $\mathbf{O}_T = \{O_1, O_2, \ldots, O_T\}$;
2. Find the probabilities of partial sequences from the current time to ending time T by using the *backward algorithm*.
3. Combine forward and backward probabilities to form a new estimate of the model parameters: λ: $P(\mathbf{O}_T|\hat{\lambda}) > P(\mathbf{O}_T|\lambda)$.

The forward algorithm of Step 1 used in training is similar to the Viterbi algorithm, but it keeps track of the sum of *all* partial paths (not just the most likely, as does the Viterbi). The result of the forward algorithm is an array of NT partial path probabilities (for N states and T observations). The backward algorithm (Step 2) is like the forward, but propagates paths that end at the T to earlier times back into the sequence. Step 3 is accomplished using the Baum-Welch reestimation algorithm, which counts each event (i.e., transition) as the sum of probabilities of all paths for which the event occurs and divides by the sum of all path probabilities. Fig. 10–20[9] includes the flow of both training and recognition on a VQ-based HMM recognizer.

To build an HMM from training data, we must compute several intermediate probabilities and produce a formula with which we can translate count of events from training data to a set of model parameters $\lambda = \{(a_{ij}), d_j(\tau), b_j(k), \pi_i\}$. Thus, training the HMM amounts to estimating the probability of events (state transitions, observations) from the frequency of their occurrence in the training data set. The forward recursion computes a parameter $\alpha_t(j)$. The parameter $\alpha_t(j)$ is known as the *forward variable* and provides the probability of the partial observation sequence $\{O_1, O_2, \ldots, O_t\}$ being in state S_j at time t, given the model λ:

$$\alpha_t(j) = P(O_1 O_2 \ldots O_t, q_t = S_j \mid \lambda). \tag{10.23}$$

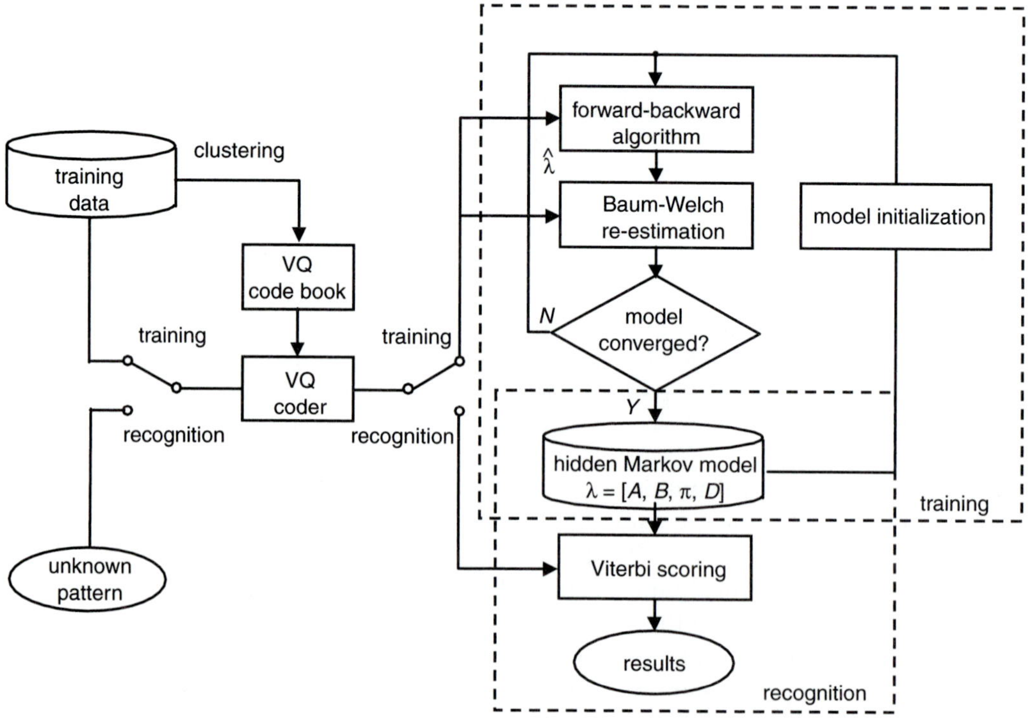

Figure 10–20 Training and recognition modes for hidden Markov modeling using vector quantization (VQ).[9]

We assume that state 1 is the initial state and state N is the final state:

$$\alpha_1(1) = 1 \tag{10.24}$$

$$\alpha_1(j) = \sum_{i=1}^{N} \alpha_1(i) a_{ij} d_j(0) \text{ for } j = 2,3, \ldots, N. \tag{10.25}$$

For $t = 1$ to T and $j = 1$ to J:

$$\alpha_t(j) = \sum_{i=1}^{N} \sum_{\tau=1}^{D} \alpha_{t-\tau} a_{ij}(i) d_j(\tau) \prod_{s=1}^{\tau} b_j(O_{t-s+1}). \tag{10.26}$$

Inspection of (10.26) shows that $\alpha_t(j)$ is the sum of all partial sequences arriving at state s_j that lead to observation O_t. The sum of all partial sequences ending at $q_t = s_j$ is $\Sigma_{i=1}^{N} \alpha_t(i) a_{ij}$; the probability of emitting observation O_t from state s_j (ignoring duration for the moment) is $b_j(O_{t+1})$.

In similar manner, the backward algorithm computes the parameter $\beta_t(i)$ as follows:

$$\beta_T(N) = 1 \text{ (assume path ends in state } N) \tag{10.27}$$

$$\beta_T(i) = \sum_{j=1}^{N} a_{ij} d_j(0) \beta_T(j);\ i = 1,2, \ldots, N-1. \tag{10.28}$$

Then for $t = T - 1$ to 1:

$$\beta_t(i) = \sum_{j=1}^{N} \sum_{\tau=1}^{D} a_{ij} d_j(\tau) \beta_{t+\tau}(j) \prod_{s=1}^{\tau} b_j(O_{t+s}). \tag{10.29}$$

In manner similar to $\alpha_t(j)$, the backward parameter $\beta_t(i)$ provides the probability of the partial observation sequence $\{O_{t+1}, O_{t+2} \ldots O_T\}$ from time $t + 1$ to the end, given state s_i at time t and the model:

$$\beta_t(i) = P(O_{t+1} O_{t+2} \ldots O_T \mid q_t = s_i, \lambda). \tag{10.30}$$

For convenience in the upcoming discussion on real time implementation, the quantity

$$w_{tij}(\tau) = a_{ij} d_j(\tau) \beta_{t+\tau}(j) \prod_{s=1}^{\tau} b_j(O_{t+s})$$

is defined to obtain:

$$\beta_t(i) = \sum_{j=1}^{N} \sum_{\tau=1}^{D} w_{tij}(\tau) \tag{10.31}$$

We now compute the probability of being in state S_i at time t and S_j at time $t + 1$. The probability of being in state S_i at time t is the sum over i of all partial observation sequences that end at state S_i, time t. This is given by $\alpha_t(i)$. The probability of being in state S_j at time $t + 1$ is computed from the backward probability as $\beta_{t+1}(j)$. Finally, the probability of transitioning from S_i to S_j is a_{ij}. Thus, the forward and backward probabilities are combined to form a new estimate of model parameters $\hat{\lambda}$ defining the parameter

$$\xi_t(i,j) = P(q_t = S_i, q_{t+1} = S_j \mid O, \lambda) \tag{10.32}$$

where $\xi_t(i,j)$ gives the probability of being in state S_i at time t and S_j at time $t+1$:

$$\xi_t(i,j) = \frac{\alpha_t(i) a_{ij} b_j(O_{t+1}) \beta_{t+1}(j)}{P(O \mid \lambda)} \tag{10.33}$$

where

$$P(O \mid \lambda) = \sum_{i=1}^{N} \sum_{j=1}^{N} \alpha_{ij} b_j(O_{t+1}) \beta_{t+1}(j). \tag{10.34}$$

Next, $\gamma_t(i)$ is defined as the probability of being in state S_i at time t, given the observation sequence and the model, so

$$\gamma_t(i) = \sum_{j=1}^{N} \xi_t(i,j). \tag{10.35}$$

Then $\gamma_t(i)$ can be summed over all time to estimate the total number of transitions out of state S_i:

$$\sum_{t=1}^{T-1} \gamma_t(i) = \text{expected number of transitions out of state } S_i. \tag{10.36}$$

To obtain the total number of transitions from S_i to S_j, $\xi_t(i, j)$ is summed over t:

$$\sum_{t=1}^{T-1} \xi_t(i,j) = \text{expected number of transitions from } S_i \text{ to } S_j. \tag{10.37}$$

These quantities are used to form the ratio of expected counts to reestimate the model parameters and obtain $\hat{\lambda}$:

$$\hat{\pi}_i \text{ at times } t = 1, \text{ expected frequency (number of times) in state } S_i = \gamma_1(i) \tag{10.38}$$

$$\hat{a}_{ij} = \frac{\text{expected \# transitions from } S_i \text{ to } S_j}{\text{expected \# transitions from } S_i} = \frac{\sum_{t=1}^{T-1} \xi_t(i,j)}{\sum_{t=1}^{T-1} \gamma_t(i)} \tag{10.39}$$

$$\hat{b}_j(k) = \frac{\text{expected number of times in state } j \text{ and observing } v_k}{\text{expected number of times in state } j}$$

$$= \frac{\sum_{t=1 \text{ s.t. } O_t = v_k}^{T} \gamma_t(j)}{\sum_{t=1}^{T} \gamma_t(j)}. \tag{10.40}$$

With the equations now defined for recognizing and training a hidden Markov model, the real time implementation of these equations is now considered. For recognition, the equations and structures are the same as for the dynamic time warping algorithm and the computing structures are the same. Because recognition and Viterbi scoring depend upon the relative magnitude of distances for minimum selection rather than their absolute values, the scores can be cast into the logarithm domain and combined via addition, thereby removing the need for table lookups or multiplication.

Training is much more challenging to implement in real time, particularly when the explicit modeling of state durations is included. An implementation that uses a massively parallel fine-grained architecture has been developed—its examination is instructive in exploring the issues of real time HMM training.[10] The computer is the MasPar MP-1, which consists of a total of 16,384 processing elements arranged in a 128 × 128 array. Each processor consists of a 4-bit arithmetic logic unit and 16 KB of memory. Two modes of interprocessor communication are provided (Fig. 10–21):

1. *Fast (single cycle) communication*, restricted to operating between processors that are situated as nearest neighbors (left/right, up/down, or diagonal);
2. *Global communication*, which has slower transfer, but allows communication between any pair of processors in the array.

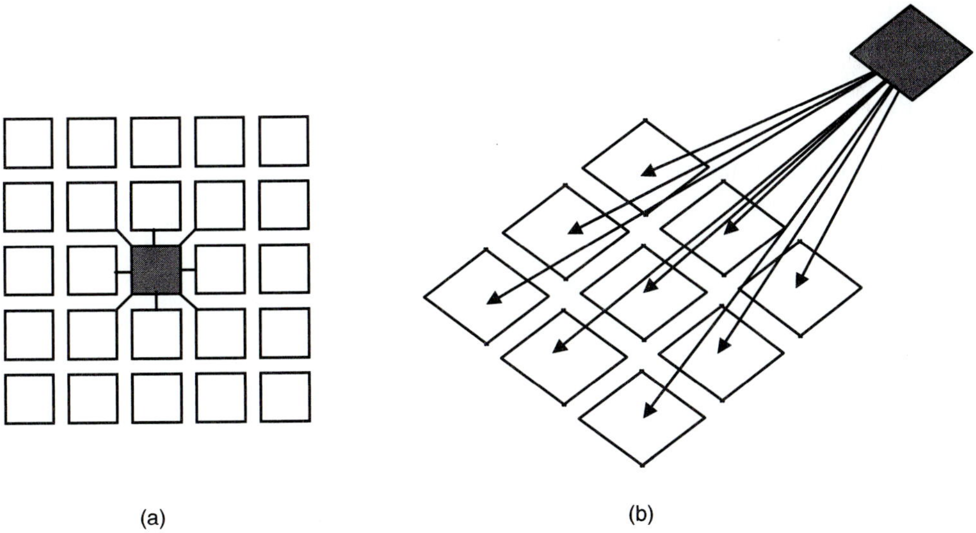

Figure 10–21 Two modes of interprocessor communication: a) high-speed communication to immediate neighbors; b) lower-speed broadcast to all processsors.

For this implementation, forty-five subword units, corresponding to phones, are used as basic building blocks, and they each are modeled with a left-to-right model with a common start and end state (Fig. 10–22).

For the probability distribution of $b_j(O_t)$, this implementation selects a Gaussian distribution with mean μ_j and variance σ_j:

$$b_j(O_t) = \frac{(2\pi)^{-1/2}}{\prod_{m=1}^{M}\sigma_{j,m}} \exp\left\{-\frac{1}{2}\sum_{m=1}^{M}(O_{t,m} - \mu_{j,m})^2/\sigma_{j,m}^2\right\} \tag{10.41}$$

where M = dimension of observation (feature) vector. Here, the duration is explicitly modeled.

Mapping the forward algorithm consists of allocating the expression for $\alpha_t(i)$ to multiple processing elements (PEs):

$$\alpha_t(i) = \sum_{i=1}^{N}\sum_{\tau=1}^{D}\alpha_{t-\tau}(i)a_{ij}d_j(\tau)\prod_{s=1}^{\tau}b_j(O_{t-s+1}). \tag{10.42}$$

In the implementation, each PE computes the summand inside the τ loop for a particular value of τ and particular state i. Thus D processors, one for each possible duration, are assigned to one state, and N states are computed, using a total of ND processors for this step. The D PEs that are allocated to each state are chosen to be adjacent neighbors in the same row so that the faster communication mode may be used (Fig. 10–23).

In the computation of the product of Gaussians of $b_j(O_{t-s+1})$ at time t [by (10.41)] all terms but one have been computed in previous times. When first computed, these terms are stored in a rotating buffer that is implemented by shifting through the storage for adjacent PEs to align the $b_j(O_{t-s+1})$ value with the proper PE.

The computation of output probabilities (10.41) updates the forward probability for each state as follows. For each time t:

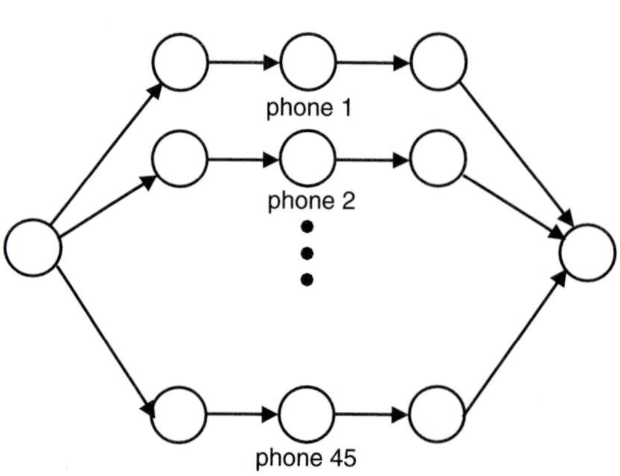

Figure 10–22 HMM used for phone recognition in real time parallel implementation.[10] © 1995 Academic Press.

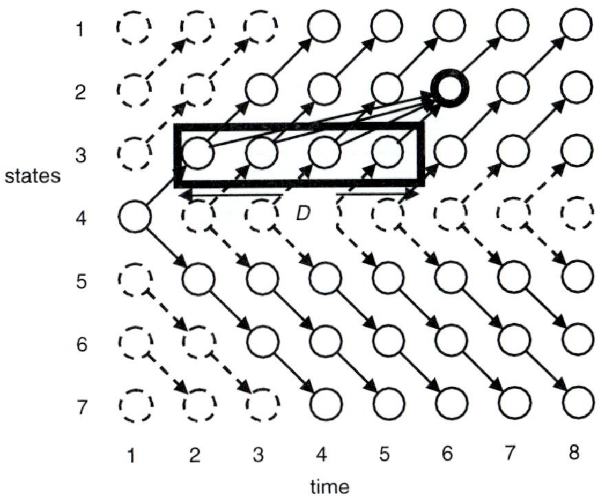

Figure 10–23 The calculation of $\alpha_t(i)$ for two branches of the phone model of Fig. 10–22 is assigned to a row of processors for each state and D adjacent processors for each row (here, $D = 4$).[10] © 1995 Academic Press.

1. Divide the partial product at time $t - 1$ by the probability of O_{t-D} in each of the D PEs.
2. Shift the product of probabilities in alignment with the processors for time t.
3. Multiply the first PE by the probability of O_t.

To compute the probability of O_t, most of the D PEs assigned to an HMM state are used. The mean and covariance diagonal for the Gaussian expression of $b_j(O_t)$ are stored and shared across a horizontal group of D PEs. Also, M PEs are used to compute $(O_{t,m} - \mu_{j,m})^2/\sigma_{j,m}^2$, one for each value of m, and the results are summed. This partitioning requires that $M < D$, which is usually the case.

The sequence of computation of the output probabilities is shown in Fig. 10–24, numbered in accordance with the following steps. The figure captures the processing allocation for one state at time t, and $D = 4$, $M = 3$ is used in the sketch. For each time t;

1. At time $t - 1$, the probabilities $O_{(t-1)-\tau}$, $\tau = 1,2,\ldots, D$ have been computed and are arranged across a group of D horizontal processors.
2. These probabilities are shifted right by one processor to compute each term of the sum for $b_j(O_t)$.
3. M processors are used to compute each term of the sum for $b_j(O_t)$.
4. The final processor is used to perform the summation.
5. The exponentiation is performed and result is copied to all D processors.
6. The new probabilities are multiplied by the shifted probabilities of Step 2.

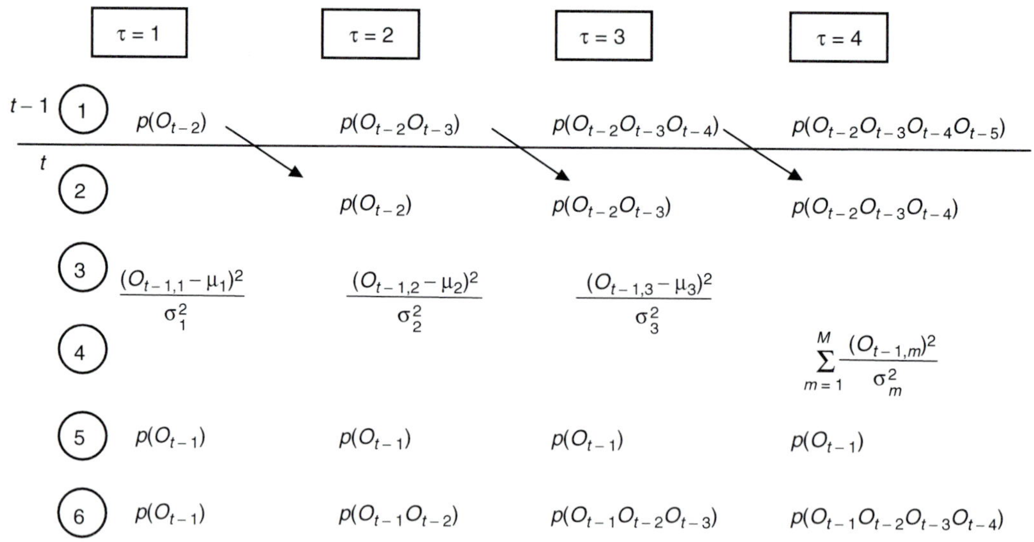

Figure 10–24 Updating the $b_j(O_t)$ used in forward probabilities $\alpha_t(i)$ for one state using the numbered sequence of steps described in the text.[10] © 1995 Academic Press.

Fig. 10–25 shows the processor array in the process of calculating $\alpha_t(i)$. An α array for $t - 1$ is passed over the boundary, multiplied by the appropriate product of observation probability, and used to compute an updated array of $\alpha_t(i)$'s, which are then combined by the lower right processor. As with the computation of output probabilities, the old values of α are stored in a rotating buffer.

The backward algorithm is combined with Baum-Welch reestimation in this implementation, partly because the term $w_{tij}(\tau)$ occurs in both. For example, updating the estimate of the mean $\mu'_{i,m}$ in the Gaussian expression for $b_j(O_j)$:

$$\beta(i) = \sum_{i=1}^{N} \sum_{\tau=1}^{D} w_{tij}(\tau) \text{ for } i = 1,2,\ldots, N, \text{ where}$$

$$w_{tij}(\tau) = a_{ij} d_j(\tau) \beta_{t+\tau}(j) \prod_{s=1}^{\tau} b_j(O_{t+s}); \tag{10.43}$$

$$\mu_{i,m}(j) = \sum_{t=0}^{T-1} \alpha_t(i) \sum_{i=1}^{N} \sum_{\tau=1}^{D} O_{t+\delta} W_{tij}(\delta); \text{ where } W_{tij}(\delta) = \sum_{\tau=\delta}^{D} w_{tij}(\tau). \tag{10.44}$$

Fig. 10–26 shows the mapping of the reestimation of $\mu'_{i,m}$. The portion of computation shared between (10.43) and (10.44) is the calculation of $w_{tij}(\tau)$. The terms of $w_{tij}(\tau)$ appear for each value of τ in a column of PEs in Fig. 10–26. The

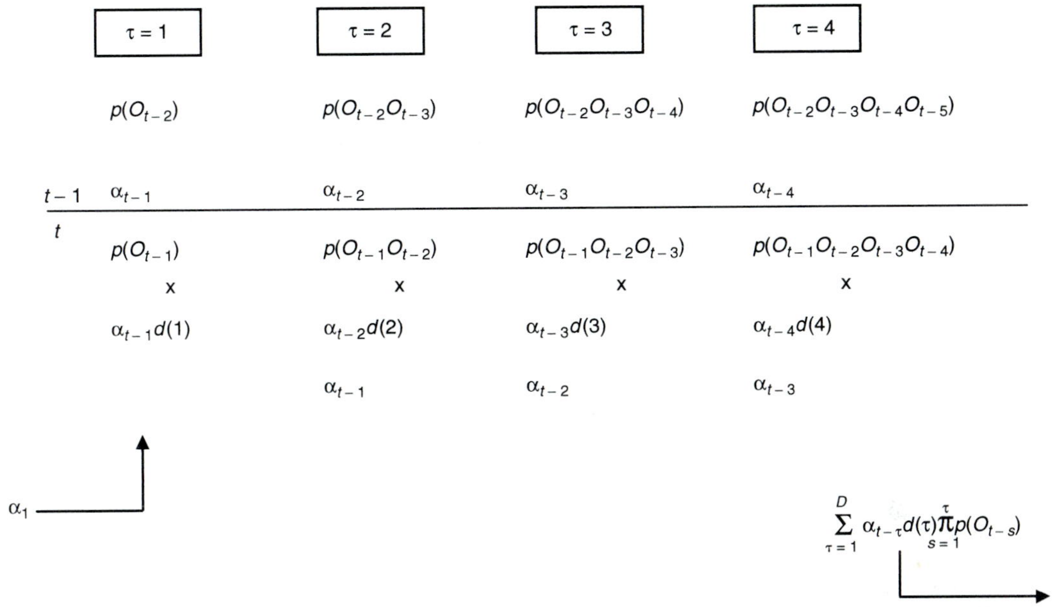

Figure 10–25 Updating forward probabilities $\alpha_t(i)$ for one state.[10] © 1995 Academic Press.

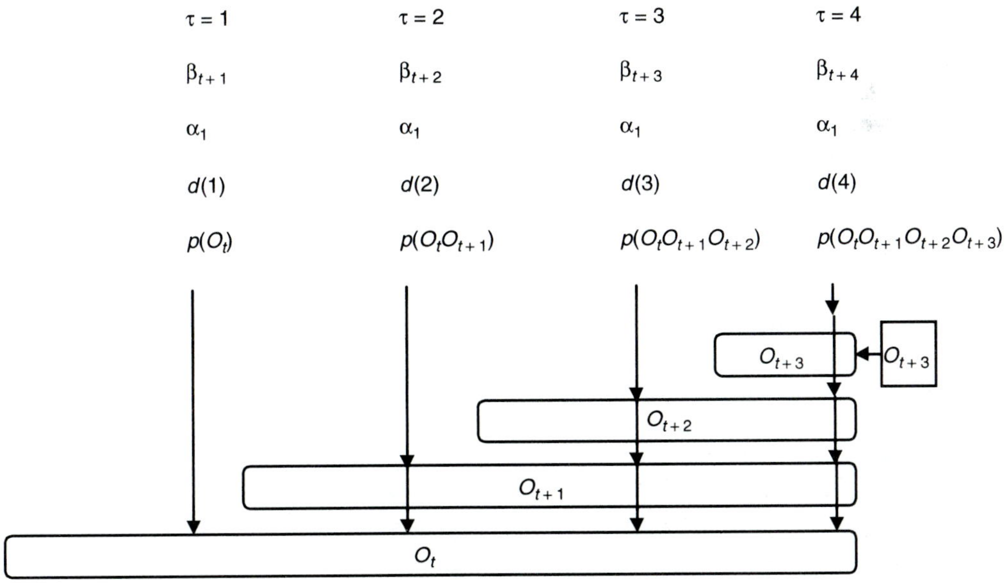

Figure 10–26 Mapping the calculation of the mean update μ at time t to processor array.[10] © 1995 Academic Press.

rectangles at the bottom indicate how the results for the $\tau = 1$ column of processors enter the O_t result, results for $\tau = 1$ and $\tau = 2$ enter O_{t+1}, and so forth, as specified by (10.43).

In implementing the training algorithm on finite-precision machines, scaling of parameters becomes an issue. The probability $\alpha_t(i)$ is the sum of many terms of the form $\Pi_{s=1}^{t-1} a_{q_s} a_{q_{s+1}} \Pi_{s=1}^{b} b_{q_s}(O_s)$. Since each a_{ij} and $b_j(k)$ is a probability of a relatively rare event, they are significantly less than one. As t becomes large and approaches 100 observations, each term of heads exponentially to zero, quickly underflowing even double-precision numeric representations.

Therefore, $\alpha_t(i)$ is scaled up by multiplying it by a scale factor that is independent of i and depends only on t.[8] One scale that may be used is $c(t) = 1/\Sigma_{i=1}^{N} \alpha_t(i)$.

The computational time of HMM training for T observations, N states, and maximum state duration D depends upon the type of architecture used. In a serial architecture, the computation time is $O(NDT)$. For a fully parallel implementation, it is $O[(\log N + \log D)T]$. For the implementation described here, the computation time is $O[(N+D)T]$.

In this MP-1 implementation, for the typical values of $D = 32$, $M = 26$, $N = 46$, operating at a rate of 100 frames/sec, 10 min of speech (600,000 frames) may be trained in 5 passes, where each pass requires 2 min. This yields a total training time of 10 min, equal to the amount of input speech and thus achieving real time training.

10.2.4 Other Considerations

To attain real time implementation, the pattern search is often speeded by limiting the search space. In all cases, limiting the search space degrades performance, although the intent and practice is generally to obtain factors of two to five speedup with no more than a 10 percent increase in error rate. The acceleration techniques include:

- limiting search of DTW to more modest deviations from linearity by cutting the obtuse-angle corners off the parallelogram of Fig. 10–10;
- limiting the variety of reference frame values by using vector quantization and pre-computing all codebook entry pairs of distances, to allow on-line use of table lookup of local distance values;
- using two-pass pattern matching—the first pass subsamples frames to screen non-contending words, and the second pass performs full matching on the reduced number of word candidates;
- aborting the comparison of a template in a left-to-right mode if the accumulated distance increases too quickly—while this saves computation, the control structure becomes irregular because each pattern takes an unpredictable amount of time to be matched;

- developing and applying a grammar that reflects the probability of one word following another, based on a language model—this limits the search to those words likely to follow a recognized word (this is the one technique that can both speed and improve recognition).

So far, the problem of finding the word or utterance within the silence or background noise has not been addressed, yet it is as important in speech recognition as finding the object of interest within a scene is to performing object identification. *Endpoint detection* distinguishes speech frames from nonspeech frames and accommodates both the silences that occur within the normal flow of speech (due, for example, to stop consonants) and from the speech-like pulses of background noise.

Three methods of endpoint detection may be used. The first is to use the signal energy contour and duration rules to discriminate speech from non-speech. In Fig. 10–27, the energy contour of an utterance is preceded and followed by noise. Two energy thresholds, k_1 and k_2, are dynamically adjusted according to the long-term average of the short-term minimum frame energy. A timer measures the duration of high-energy pulses and intervening silences, and multiple endpoint candidates are then presented for recognition.

A second method of endpoint detection builds a model of the background silence. It uses features (and energy) in an identical manner to feature use in word recognition and applies it to all incoming frames. Silence templates are included with word templates. These are either compared against all speech frames to declare non-speech if no match is found, or are compared against both speech and background noise frames and used to declare speech vs. non-speech.

Finally, utterances may be detected by extending HMMs so that background is included as a possible state within the first and last state of the model. Again, this amounts to explicitly modeling silence and including it within the recognition set.

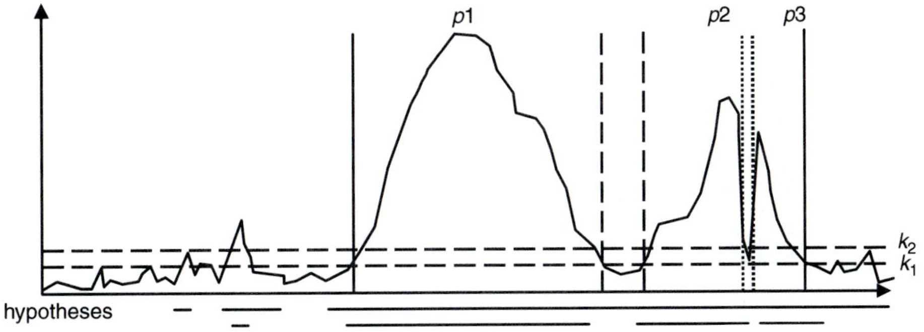

Figure 10–27 Multiple energy thresholds and pulse durations are used in endpoint detector to identify speechlike pulses, which are combined to form multiple endpoint candidates.

10.3 IMAGE RECOGNITION

Tasks of image recognition include the analysis and recognition of entire scenes, the recognition of objects within a scene, and the recognition of printed or handwritten characters from the printed page. The complexity of image recognition algorithms depends upon the range of position of objects to be recognized as well as the characteristics of the objects themselves. In simpler applications, the objects do not rotate and can be identified by their outer and inner boundaries. These applications include optical character recognition, shape recognition of closed contours, and face recognition. More difficult are the applications in which objects are rotated or translated, or in which objects are viewed from a spherical set of elevation and azimuth angles, or in which they have internal non-rigid movement such as changes in articulation (e.g., tank with turret turned to various positions). In these more difficult cases, objects may not have both inner and outer boundaries. These applications include the difficult task of automatic target recognition, which seeks to identify vehicles or other moveable objects in an image.

The concept of invariance is important to image recognition—the identity of an object is constant regardless of how its image changes due to rotation, scale distortion, or occlusion. Image recognition algorithms must be able to "normalize" the variations away in the image.

Fig. 10–28 shows an end-to-end scheme for image recognition. The processing chain starts with a scene from which areas of interest that are likely to include targets are selected. Within some of these areas, objects of interest are located and segmented, and features are then extracted from them that emphasize the invariant information in a compressed form, thereby reducing the processing load downstream. Features are transformed to symbolic representations of the object

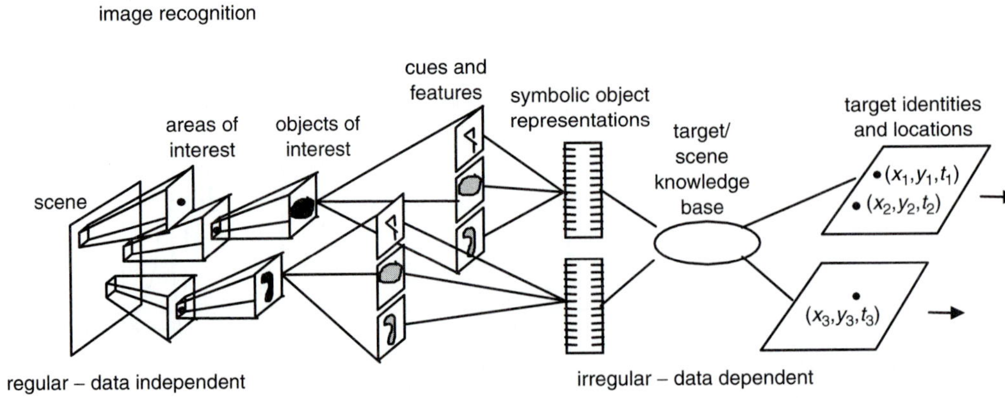

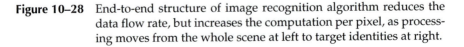

Figure 10–28 End-to-end structure of image recognition algorithm reduces the data flow rate, but increases the computation per pixel, as processing moves from the whole scene at left to target identities at right.

and compared against members of a previously trained database to arrive at an estimate of target locations and identities.

Several classes of object-recognition algorithms have arisen to accommodate variations in pose and lighting conditions. The first, based on statistical pattern recognition, applies a sequence of stages to segment objects from the image. These algorithms then form a feature representation of the object and proceed to classify it based on its features. The matched filter or template-based approach finds whether the object in the image matches any of a set of pre-stored patterns whose identities are known. Model-based approaches adjust a model to make a predicted pattern from that model match the pattern of the unknown object. Most recent approaches have emphasized the template or model-based approaches, and five methods are examined in this section.

The first method, based on image templates, collects numerous views of the object across a variety of geometric and lighting conditions, seeking to densely sample the range of geometry variation by covering the range with a large number of templates. Another approach, known as deformable templates, creates templates that are adjusted with a few parameters and then adjusts these parameters to search for a match. A simple example of a deformable template uses the position of the upper left and lower right corner to parameterize a rectangle, and then moves these corners to change the size and shape in a search to match a particular unknown pattern. Approaches based on hidden Markov modeling have been applied to image recognition, either by turning the two-dimensional problem into a one-dimensional problem by some sort of data reordering transform, or by developing a full two-dimensional approach. A recent method known as model-based vision first examines the gross properties of an unknown target signature to estimate possible target classes and geometries, and then inputs these estimates to a signature prediction simulation to obtain what would be seen if the unknown target and pose were indeed what was simulated. The unknown and predicted patterns are compared and the results are used to refine object type and pose estimates, predict new signatures, and iterate until the best match is found. Finally, some of the above approaches may be combined. For example, actual template data may be used at the geometry conditions of the unknown object that match their collection, and then model-based signature prediction can be used to supplement areas of the template that vary with geometry. This approach combines the accuracy of actual data with the ability to model any geometric variation provided by the signature predictor.

The overall real time processing throughput requirements for image recognition may be roughly estimated from the resolution required for recognition and the amount of area to be examined per unit time. The "Johnson Criterion"[11] indicates that for a person to detect an object, a resolution of one line pair is needed across the minimum dimension of the object. With four line pairs along one dimension, the object may be classified, and it may be identified if eight line pairs of resolution are provided. In two dimensions and for *automatic* target recognition, the following rules have been presented:[12]

- detection requires at least 10 pixels on target;
- classification requires 200 pixels on target;
- identification requires 1000 pixels on target.

Thus, the resolution requirements are set by object size. To classify vehicles, resolution of 0.3–1.0 m is needed, but to identify ships, 3–6 m resolution is sufficient. The area coverage rate is determined based upon the size of the area to be searched and the amount of time available to search it, which is loosely based upon the speed and number of moving objects in the scene that must be tracked.

To determine overall throughput, a sensor resolution is selected based on the target size and level of recognition function (detect, classify, identify) required. An area coverage rate is determined, and then the recognition algorithm is analyzed in terms of the number of operations per pixel required at each stage. Finding areas of interest requires 100–1000 3×3 pixel image neighborhood operations, in which a neighborhood operation comprises replacing each pixel with a linear or nonlinear combination of its adjacent neighbors (and itself). One 3×3 neighborhood operation corresponds to 30 to 150 traditional computer instructions. Detection requires approximately 10^6 operations per cue, and identification requires around 10^6 operations per detection.

To complete the computation of throughput requires an estimate of the number of targets to be searched per scene, the number of candidate target classes against which each unknown target is to be matched, and the number of false alarms expected per scene. These quantities provide the number of possible cues and targets to which the above detection and identification operations will be applied.

10.3.1 Extraction of Region of Interest

Like endpoint detection in speech recognition, the first problem in image recognition is to find the object(s) of interest in the much larger scene (or lines of text within a document to be read), segment the potential objects from the scene, and pass them on to further recognition processing. Region-of-interest extraction may be implemented in two or more stages. The first stage selects any region that has any possibility of containing a target—anything missed here cannot be recovered later, so a very forgiving detection threshold is set and many false detections are allowed. A stage of second-level detection follows and determines whether received regions correspond to targets. While the first-level extraction may be restricted to pixel brightness, second-level detection often uses features such as object size, shape, and texture, as well as context information such as proximity to roads (for vehicle detection).

One method of region-of-interest extraction is the Constant False Alarm Rate (CFAR) detector. CFAR is adaptive to the local statistics across an image. It is based on the premise that optical clutter may be modeled as a Gaussian random process with possibly rapidly varying mean and a more slowly varying covariance over the image. It is called "constant false alarm rate" because the algorithm allows its detection threshold to be set for a constant probability of false alarm, regardless of intensity.

The CFAR algorithm defines a neighborhood of dimension $a \times a$ and computes $\overline{x}$, the mean pixel intensity across the neighborhood, and σ, the standard deviation across the neighborhood. It then compares the value of the center pixel, s, to those neighborhood quantities and compares them against a threshold, T, to declare target present or target absent:

$$\frac{x - \overline{x}}{\sigma} > T\text{: target present.} \tag{10.45}$$

This test indicates how different the center pixel is from its local neighborhood, measured in multiples of the standard deviation for the neighborhood.

For an image of M pixels, a load of $2M$ multiplies, $3M$ divides, M square roots, and $4\ M^*A$ additions is needed, where $A = a \times a$ (neighborhood area). Fig. 10–29 shows the sliding window of area A and its center pixel sliding over an image of size M pixels.

The divide and square root operation are more time consuming than the other operations on a single chip digital signal processor and require a Taylor series operation. Thus, while multiplies and adds can be performed in a single cycle, these more difficult operations may require ten cycles. With this factor in mind, the CFAR is estimated to require about forty-six instruction cycles per pixel, and for 512 × 512 images arriving at a rate of 30 frames/sec, a throughput of 360 million instructions/sec is needed. This computational load outstrips the capability of a single processor, and so the algorithm must be allocated across multiple processors. For multiprocessor implementation, a portion of each image is assigned to each processor. Processors of adjacent portions of the image must share borders of each subimage, because the neighborhood strip of $a - 1$ pixel

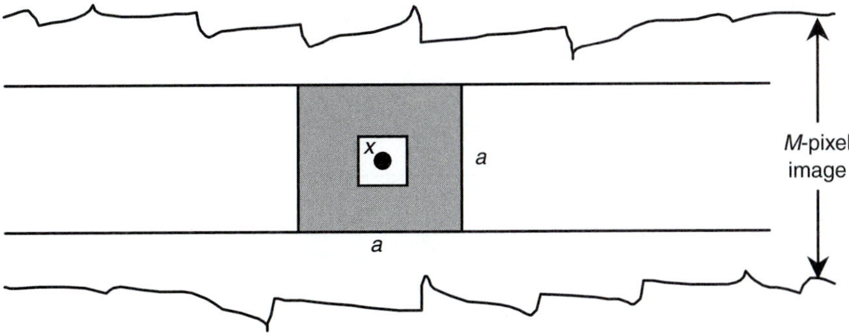

Figure 10–29 CFAR window moves over all pixels of the image, computing the mean and variance over an $a \times a$ neighborhood and comparing the center pixel to that value.

width is common to abutting subimages. A multiprocessor array implementation must also accommodate the overhead time need to partition the image and move it in and out of each processor.

A limitation of the performance of a CFAR detection results from the fact that actual image statistics are not truly Gaussian. An alternative technique uses image morphology, applying nonlinear operations such as logical OR, minimum select, etc. to the neighborhood of each pixel. Image morphology can be implemented very efficiently on a neighborhood-based processor, as will be described below. It often provides better detection performance, due to its ability to handle a wide range of statistical distributions.

Region of interest detection can be combined with image recognition techniques, providing a decision-directed context-dependent guide to restricting the analysis to interesting regions of the image. For example, region extraction may be combined with SAR image formation. As a SAR image is being formed, a sequential decision process is used to decide whether it contains clutter and therefore need not be completely computed, or target-like objects, indicating the need to finish its computation.[13] The process operates on a pulse-by-pulse basis and monitors an evolving statistic versus radar pulse $x_1, x_2, \ldots$ and applies a binary hypothesis test to choose either the target present hypothesis (H_1) or target absent (H_0). The technique determines whether the probability density function (pdf) of the incoming pulse matches a clutter-like pdf, or a target-plus clutter pdf, H_0: $(x_1, x_2, \ldots) \approx f_{n_0}$ or a target plus clutter pdf, $H_1 : (x_1, x_2, \ldots) \approx f_{n_1}$. Here, f_{n1} is a function of n samples that models the likelihood that the samples came from a target; f_{n0} is a function of n samples modeling the likelihood of originating from clutter. A log likelihood ratio is computed pulse by pulse and compared to a threshold T:

$$L_n = \frac{f_{n1}(x_1, x_2, \ldots, x_n)}{f_{n0}(x_1, x_2, \ldots x_n)} \geqslant T. \tag{10.46}$$

The specific function used is an autoregressive process model by which pulse autocorrelation is computed for various orders. Fig. 10–30 shows the evolution of L_n with pulse number, n, for the three cases of clutter only (H_0), target pre-

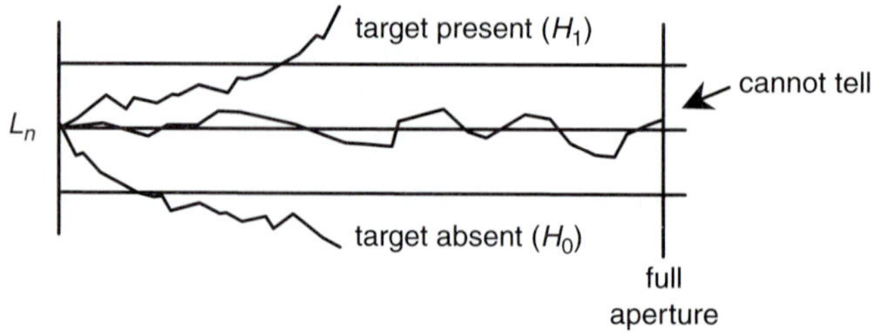

Figure 10–30 The evolution of decision statistic L_n versus pulse number provides early indication as to whether a SAR image consists of clutter only or target plus clutter.

sent (H_1), and cannot tell. Generally, an accurate decision can be made within the first 10 percent of the pulses.

This statistic may be used to implement decision-directed SAR image formation by applying a multiresolution approach to SAR image formation.[14] A coarse-resolution image is formed with a portion of the pulses, the decision statistic is compared to threshold, and the computation proceeds to finer resolution only if the clutter-only threshold is not exceeded. In an allocation of multiresolution SAR image formation to multiple processor, each processor can be allocated one patch of the total image and can form the image patch, compares its decision statistic to the threshold, and abort if a target is not likely to be present. This processor then becomes available for speeding computation upon other parts of the image.

10.3.2 Template Matching

Template matching of two image objects involves sliding a reference pattern over the unknown object in all directions (left, right, up, down), computing the amount of overlay at each portion, and determining the position and score of the best-matching position. Template matching is similar to the block matching operation described in Section 9.3.2 that was used in frame-to-frame alignment of objects in video compression.

Specifically, we define an input image I as an array of size $N_1 \times N_2$ pixels. If W is a template of size $K \times K$, then the template matching of W by I is given by:

$$TM(i,j) = \sum_{m=0}^{k-1} \sum_{n=0}^{k-1} I(i + m, j + n)W(m,n); \quad 0 \leq i \leq N_1 - K; 0 \leq j \leq N_2 - K. \quad (10.47)$$

Fig. 10–31 shows the image I, template W, and the moving of the template about the image.[15]

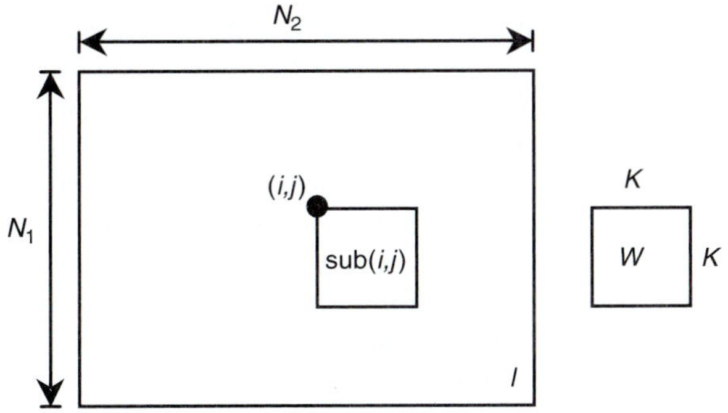

Figure 10–31 A $K \times K$ template of an object W is matched to an image I by comparing its overlap to each subimage sub(i, j) across the image.[15] © 1991 Academic Press.

The computational load of matching one template may be computed by first examining the computation required for each pixel-to-pixel distance. In (10.47), that distance corresponds to the product of two pixel values, $I(i + m, j + n) \cdot W(m,n)$. Other alternatives include a Euclidean distance, $\sqrt{[I(i + m, j + n) - W(m,n)]^2}$.

A total of K^2 distances must be computed for each template position. For an image of $N_1 \times N_2$ pixels, a total of $(N_1 - K + 1)(N_2 - K + 1)$ template positions must be used. The local distance measure requires α instruction cycles of time T_c, where α is dependent upon the particular local distance used and where the acceleration due to pipelining a large sequence of distance is reflected in α. Thus the time to compare one template to the image is

$$T_v = (N_1 - K + 1)(N_2 - K + 1)K^2 \alpha T_c. \tag{10.48}$$

To map this to a multiprocessor array, a total of P processors is assumed, each with an instruction time T_c. In the ideal case of speedup proportional to P, the number of processors needed for the multiprocessor case is $T_v^P = T_v / P$. For a frame rate of f_F frames/sec, the template compare time must be less than the frame period to keep up, or $N\ T_v^P < 1/f_F$, where N is the number of templates being matched.

Then the total number of processors needed to achieve real time operation is

$$P = (N_1 - K + 1)(N_2 - K + 1)K^2 \alpha T_c f_F. \tag{10.49}$$

If $T_c = 20$ nsec and $\alpha = 8$, for a template size of 8×8 pixels and an image typical of the Video Telephone Standard ($N_1 = 258$, $N_2 = 353$, $f_F = 10$), a total of $P = 10$ processors are needed per template. For NTSC standard imagery ($N_1 = 512$, $N_2 = 480$, $f_F = 10$), twenty-five such processors per template are needed. In a realistic application of automatic target recognition, choosing from among ten target classes, with each target represented by templates at 5° increments in both rotation and elevation, over 10^5 templates are needed, and well over 10^6 processors are needed. In addition to the large real time processing burden, a communication bottleneck arises in moving templates into and out of a million processors in real time.

A shift register with multiple tapout points may be used to simplify communication and speed template matching. One such architecture[16] uses a shift register buffer to store K rows of an image and present them to K^2 processors (Fig. 10–32).

The application of a multi-tapped shift register can also be used for inserting appropriate delays to present adjacent pixels from a raster-scanned image to a connected array of processors. One such example is the Geometric Arithmetic Parallel Processor (GAPP).[17] The GAPP is a *cellular* processor that is used for fine-grained massively parallel multiprocessors. A GAPP processor may be connected to neighbors in north, south, east, west, up, and down. The processor has four 1-bit latches (registers) with an input multiplexer, a 1-bit full adder/subtractor, and a 128-bit RAM. The GAPP is configured as 192 processors on a chip (16×12 array), operating at a rate of 25 MHz (in the most publicized version) or 40 MHz

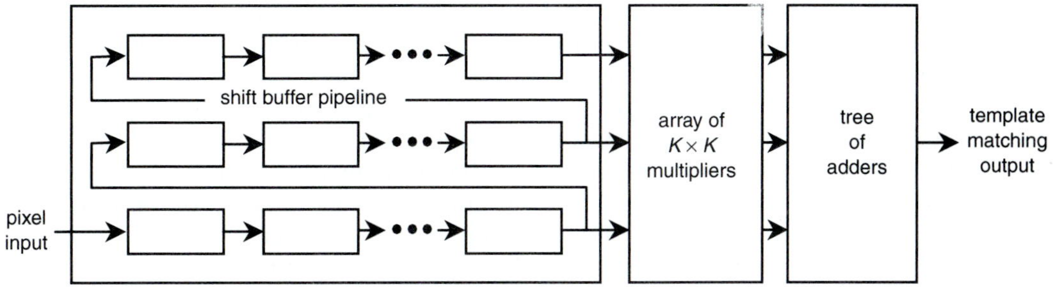

Figure 10–32 Architecture uses a shift buffer pipeline to present K rows of the image to a $K \times K$ multiplier array.[15] © 1991 Academic Press.

(more recent version). An array of 192 × 480 GAPPs (92,600) operating at 25 MHz is capable of performing automatic target recognition based on the two-dimensional fast Fourier transform via pairwise comparison of all the targets for a 10-Hz frame rate, using five target types with two aspects of each.

10.3.3 Image Morphology

Image morphology has already been mentioned as a means to find regions of interest in an image. It may also be used for pattern comparison and it lends itself to implementation on cellular processor arrays.

A basic operation of morphological-based recognition is *erosion*. For an image I and an object template image A, the erosion of I with A, denoted as $I \ominus A$, is defined as the locus of all points that A can be positioned such that I contains (or covers) A. The complementary operation of *dilation*, given by $I \oplus A$, is the locus of points outlined by A as it is slid over all positions in I. Fig. 10–33 shows these operations on an array of data.[18]

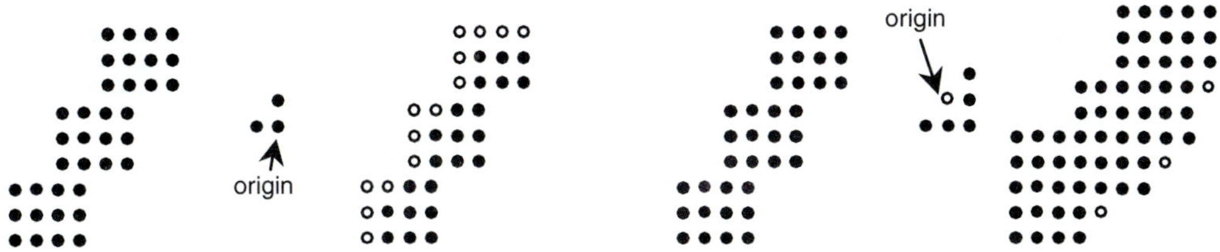

Figure 10–33 Operations of erosion and dilation of image I by structuring element A roughly correspond to the *AND*-ing and *OR*-ing of the two patterns, respectively.[18, a]

---

[a]*Fundamentals of Digital Image Processing* by A. K. Jain, © 1989. Adapted by permission of Prentice-Hall, Inc., Upper Saddle River, NJ.

The operations of erosion and image complement (changing all on-pixels to off and off-pixels to on) can be used to search for an object in a binary image that matches a template,[19] and the technique can also be extended to gray-scale images. To locate an object A in image I, I is eroded by A ($I \ominus A$) resulting in the set of all locations of I where I has a foreground that matches A (left three images of Fig. 10–34). The complement of I ($\bar{I}$) and the complement of A ($\bar{A}$) are formed and their erosion is performed ($\bar{I} \ominus \bar{A}$, right images). The AND-ing of these two erosion results provides a dot at the origin of object A in image I.

Like the template matching described above, image morphology can be implemented as a sequence of operations on a processing neighborhood region of pixels that is scanned to cover the image. Indeed, image neighborhood operations, in which each pixel is replaced by a linear or nonlinear function of the pixels around it (Fig. 10–35) are one of four basic classes of image operations.

All local transformations can be implemented as a sequence of operations on 3×3 neighborhoods with certain constraints. These form a complete and sufficient set of operations for the design of image pattern matching.

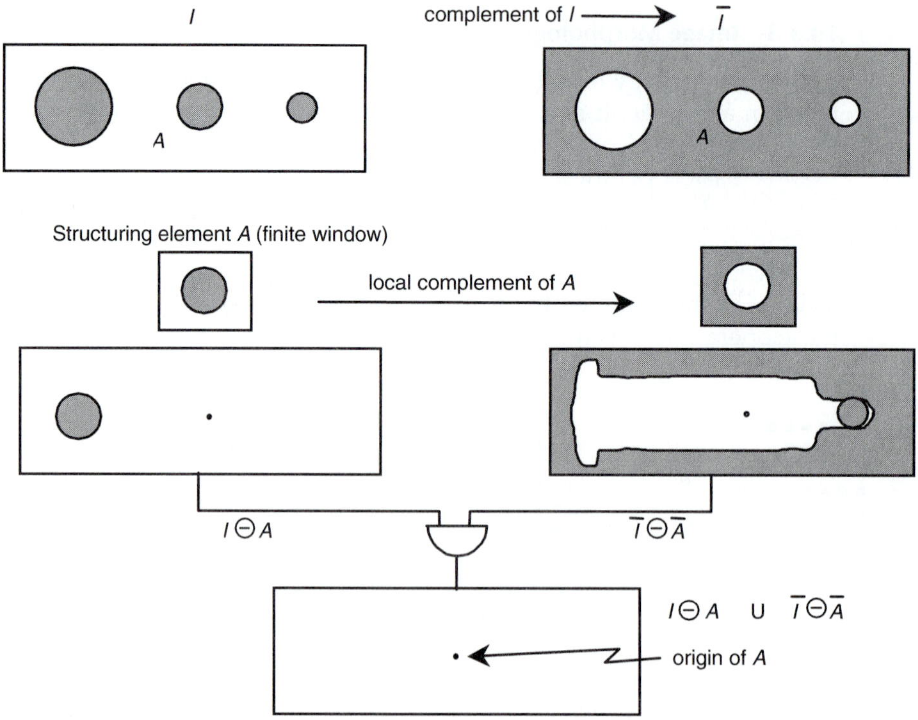

Figure 10–34 Morphology-based search of image I for structuring element A takes the AND of the erosion of I by A and $\bar{I}$ by $\bar{A}$.[19] © 1989 IEEE.

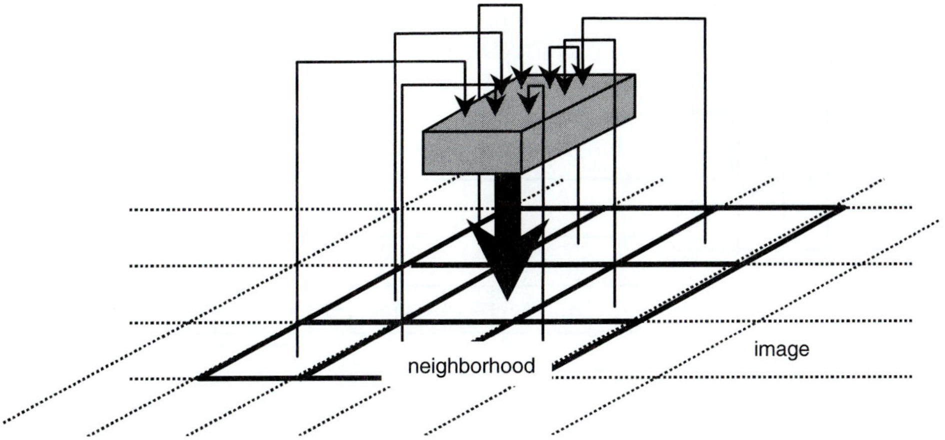

Figure 10–35 Nearest neighbor operation used in image morphology, convolution, and filtering.

A second class of image operations is multiple image combination. These operations combine multiple images using either set-theoretic operations (union, intersection, etc) or arithmetic (pixel-by-pixel add/subtracts, multiplies, etc). Geometric transformations comprise the third class of operations, which include image rotation and rubber-sheet-like warping of an image, including magnification and minification. The fourth class includes information extraction operations, receiving image as inputs and providing symbols as outputs, such as numbers describing count, location, or pass/fail of a criterion.

Neighborhood image operations can be accelerated by processors of various types:[20]

- *full array*: each pixel has its own processor, and each processor is connected to all of its neighbors (N, S, E, W, and sometimes diagonal). A full array can never be as large as the largest images, which may include over one million pixels;
- *parallel subarray*: same concept as a full array, but on an array that is less than the image size;
- *raster subarray*: this architecture presents a neighborhood of pixels to the processor at the same time. The processor enables the performance of each operation on the multiple pixels in the neighborhood in a single instruction cycle.

Fig. 10–36 shows an example of a raster subarray processor. The multitapped shift register places the correct number of delays, based on image size, to present an entire neighborhood of pixels to the processor at once.

It is instructive to compare the number of instructions needed to perform a neighborhood operation on a conventional processor to the single-instruction ca-

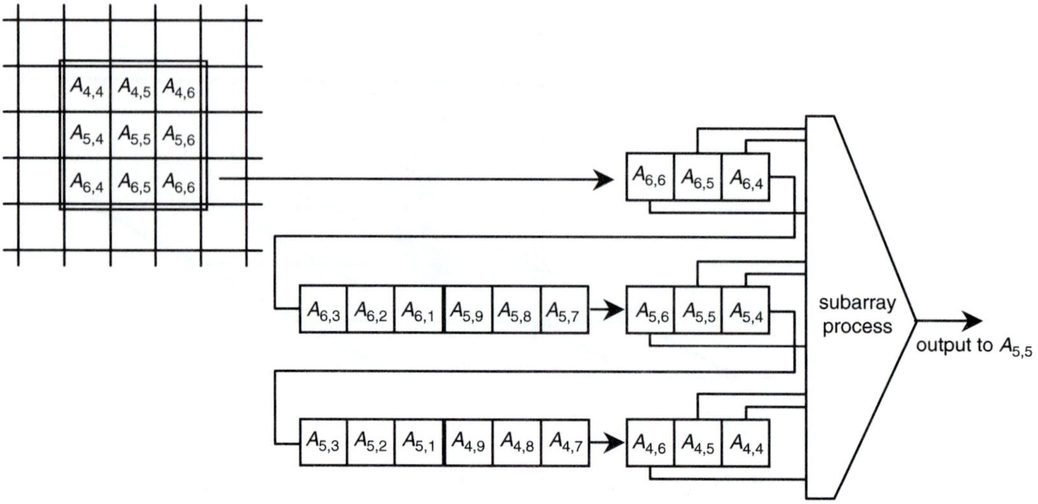

Figure 10–36 Using programmable delay registers of length $L - 3$ to synchronize presentation of adjacent pixels from image of width L to neighborhood processor.

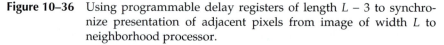

pability of a raster subarray. To perform a 3×3 convolution on a single-instruction-stream, single-data-stream general purpose processor, for each of the nine pixels (eight neighbors plus center pixel), the processor must multiply, add, and change the address pointer to the next neighborhood pixel. This requires from 3 to 15 instructions per pixel, or 30–150 instructions per neighborhood. When compared to the single instruction time of the raster subarray cellular processor, the acceleration to single-instruction operation afforded by the raster subarray is significant.

An example of decomposing a complex image recognition algorithm into sequences of 3×3 neighborhood operations is shown in Fig. 10–37. This example implements a multi-sensor algorithm to detect minefields, in which the three sensors are a 1.06 μm wavelength reflectance return, polarization return, and passive thermal image (8–12 μm).[21] Each sensor is processed by a separate channel as shown in the flowcharts and the results are then combined for further processing. For a throughput of 700 pixels/scan line and 1,024 scan lines/sec, a pipeline of cellular processors is used. The processor, known as the Pipeline Processing Element,[22] is capable of both nonlinear (morphological) and linear (convolutional) operations and is used in this example in a baseline configuration of twenty pipelined processors. The data is recirculated through the pipeline of processors multiple times to achieve the effect of an arbitrarily long pipeline.

10.3.4 From Templates to Models

This subsection begins a transition from template-based image recognition techniques of the previous subsection to full model-based vision discussed in the next

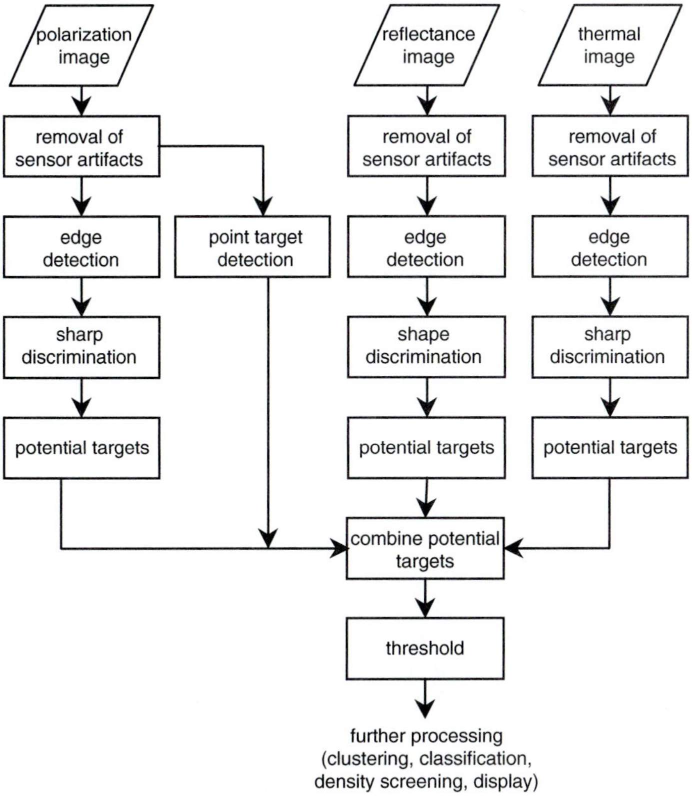

recirculations in 20-stage pipeline of neighborhood processors

operation	Pol	Ref	Ther	TOT
remove artifacts	1	1	1	3
edge/point detection	2	1	1	4
shape discrimination	4	4	4	12
develop potential targets	1	1	1	3
combine and threshold targets				1
TOTAL				23

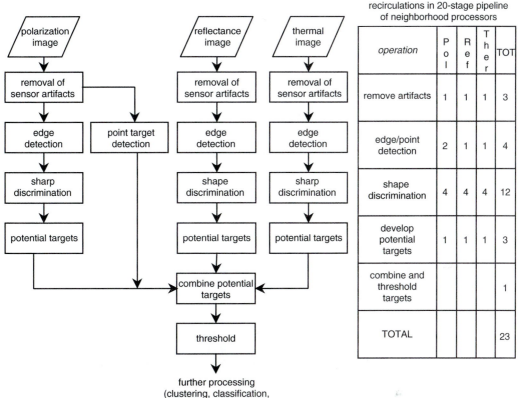

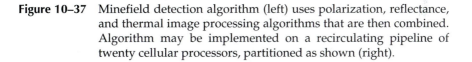

Figure 10–37 Minefield detection algorithm (left) uses polarization, reflectance, and thermal image processing algorithms that are then combined. Algorithm may be implemented on a recirculating pipeline of twenty cellular processors, partitioned as shown (right).

one and describes deformable templates, dynamic planar warping, and hidden Markov models. By describing image recognition algorithms that use the techniques for which real time implementations have already been discussed, a means to extend the applicability of the computational structures of these implementations to image recognition algorithms becomes available.

As a supplement to capturing each possible geometric or lighting condition with its own template, a smaller number of templates can be used if the templates can be adjusted, through distorting or deforming, by the variation of a small set of parameters. The *deformable template*[23] is a parameterized model of an object in which different parameter values result in different object configurations.

A deformable template consists of:

- a parameterized object model, including prior probabilities $P(s)$ for the parameter s;
- an image model, specifying $P(I \mid s)$, or the probability of obtaining image I given the parameter value s;
- an algorithm for finding the optimum parameters, given an image:

$$P(s \mid I) = \frac{P(I \mid s)P(s)}{P(I)} \quad \text{(Bayes Rule).} \tag{10.50}$$

The problem of applying deformable templates is to learn the prior model M^* and thereby its parameters from a set of training images, then adjust prior parameters to maximize $P(s \mid I, M^*)$ for a given image. As with other recognition techniques, both learning and matching modes are used. In the learning mode, the task is to find the prior parameters which maximize $P(M \mid I)$. This set of parameters provides the representation for a general class of objects. For matching, given a fixed prior parameter M^*, recognition is achieved by finding configuration parameters to maximize $P(s \mid I, M^*)$ for a given image J, thereby matching the image to the model. The training and parameter adjustments are reminiscent of using the hidden Markov model for speech recognition, and are also used in the upcoming approach of model-based object recognition.

In practice, the application of deformable templates proceeds as a set of patches applied to an image as a general model. The modeling starts with a rectangular grid and then allows the patches to move or change intensity, while maintaining connectivity of each patch with its neighbors.

Another mode of deforming templates is the matching of a reference that was taken from a different view than the test template. It is based on the techniques that underlie dynamic time warping of speech signals. The image counterpart uses planar warping rather than time-axis warping and hence is called dynamic planar warping (DPW).[24] The problem of dynamic planar warping is to align a reference pattern $r(x,y)$ with an elastically distorted test image $t(x,y)$, in which the size of the test and reference are given by XY and $X_r Y_r$, respectively. As with the dynamic time warp, the matching proceeds from one end of the test pattern to the other, applying the principle of optimality to build up a global warping of reference to test through a sequence of local mappings. As a result of the similarity of the DPW for images to the DTW for speech, the kernel structure of add-min-argmin-update can be used.

Dynamic planar warping performs the mapping of a test lattice to a reference lattice via the mapping function F:

$$\begin{bmatrix} \tilde{x} \\ \tilde{y} \end{bmatrix} = F\begin{bmatrix} x \\ y \end{bmatrix} \tag{10.51}$$

such that the global distance, as measured by the summation of pixel-to-pixel local distances $d[r(\tilde{x},\tilde{y}),t(x,y)]$ into a global distance s, given by

$$s[r(\tilde{x},\tilde{y}),t(x,y)] \equiv s = \sum_{x=1}^{X_T} \sum_{y=1}^{Y_T} d[r(\tilde{x},\tilde{y}),t(x,y)] \qquad (10.52)$$
$$\begin{bmatrix} \tilde{x} \\ \tilde{y} \end{bmatrix} = F\begin{bmatrix} x \\ y \end{bmatrix}$$

is minimized for some path $F\begin{bmatrix} x \\ y \end{bmatrix}$. Global constraints map boundaries (e.g., corners) of one object to the corresponding boundaries of the other, while local constraints assure a monotonicity of the mapping function.

The DPW proceeds by warping the reference to the test, as shown in Fig. 10–38. Formally, the mapping proceeds by defining the following quantities:

1. Define a set of N_T overlapping test subshapes $\{\theta_n\}$ such that θ_n contains θ_{n-1}, $\Delta\theta_n = \theta_n - \theta_{n-1}$ = pixels for row n, where n is a convenient one-dimensional parameterization, and θ_{N_T}= entire test pattern.
2. Define a set of m warping sequences φ_m, $1 \leq m \leq M$ that satisfy the constraints: φ_m is a sequence of X_T reference pixels, $\varphi_m = \{(\tilde{x}_1^m, \tilde{y}_1^m), (\tilde{x}_2^m, \tilde{y}_2^m), \ldots, (\tilde{x}_x^m, \tilde{y}_x^m), \ldots, (\tilde{x}_T^m, \tilde{y}_T^m)\}$.
3. Each sequence φ_m determines a subset $\Lambda_m$of sequences, $\Lambda_m = \{\varphi_k : \tilde{y}_x^k < \tilde{y}_x^m, 1 \leq x \leq X_T\}$, so that if φ_m is a candidate warping sequence for row n at the test image, the preceding row $n - 1$ can only be matched with a warping sequence from $\Lambda_m$to satisfy vertical monotonicity.
4. Define $F_{m,n}$ to be the set of submapping functions from the nth test rectangle θ_n that satisfy the constraints and match as the nth row of the test, $\Delta\theta_n$, with φ_m.

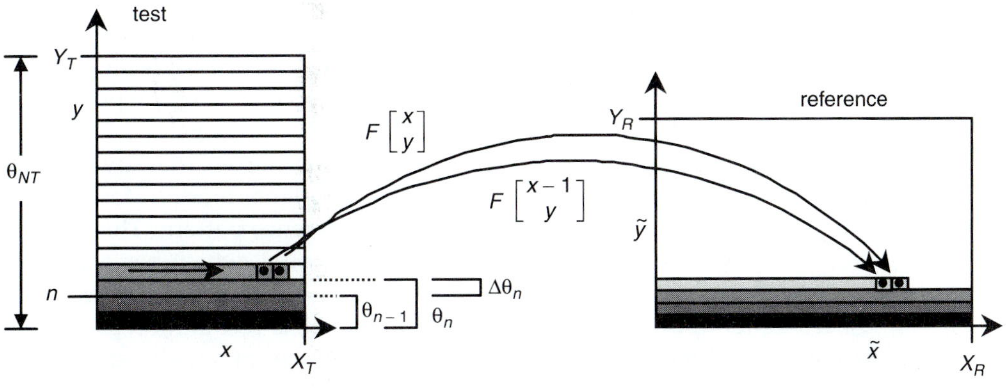

Figure 10–38 In dynamic planar warping, mapping of test to reference proceeds from left to right in quadrilaterals of ever-increasing size (θ_{n-1}, θ_n, . . . , with $\Delta\theta_n$ = next row of test).

The DPW then iterates to level n by assuming that the optimal warpings of the $(n-1)^{\text{th}}$ rectangle of the test image with warping function φ_m are known for $1 \le m \le M$. We find the optimal warping of the nth test rectangle that matches the nth test image row to the lth warping sequence. As with the DTW, an accumulated distance $s(m, n)$ is built up from a previous accumulated distance $s(m, n-1)$ and a local distance:

$$s(l,n) = \min_{\substack{m \\ \varphi_m \in \Lambda_m}} s(m,n-1) + \sum_{x=1}^{X_T} d[r(\tilde{x}^l,\tilde{y}^l),t(x,n)]. \tag{10.53}$$

For the minimum-scoring φ_m, the corresponding mapping function is:

$$F_{l,n}(x,y) = \begin{cases} F_{m,n-1}(x,y) & \text{for } (x,y) \in \theta_{n-1} \\ (\tilde{x}^l_x,\tilde{y}^l_x) & \text{for } (x,y) \in \Delta\theta_n \end{cases} \tag{10.54}$$

where m is the argmin of (10.53). The iteration begins for $n = 1$ by setting

$$s(1,m) = \sum_{x=1}^{X_T} d[r(\tilde{x}^m_x,\tilde{y}^m_x)t(x,1)\delta(\tilde{y}^n_x - 1)], \tag{10.55}$$

where $\delta(\cdot)$ is the Kronecker delta function.

Despite the use of Bellman's optimality principle, this DPW is of exponential computational complexity ($O[XYX_RY_R]^X$) and is therefore intractable for all but the most modest image sizes. However, the addition of two more constraints can reduce the complexity to $O[X_RXY_RY]$:

- Divide each image into subimages and find the (suboptimal) solution for each subimage; reconstruct the subimage mappings into a (suboptimal) solution for the full image;
- Limit the permitted set of warping sequences with further constraints consistent with the specific application; e.g., constrain possible mappings to those for which vertical distortion is independent of horizontal position.

Dynamic planar warping can also be extended to a hidden Markov model formulation, which is gained by regarding the states of a planar hidden Markov model as having states on the rectangular lattice of the reference pattern. This approach provides a true two-dimensional hidden Markov model, which has exponential complexity unless constraints such as those mentioned above are introduced.

As a simpler approach, the two-dimensional object recognition problem can be cast into a one-dimensional form to use the full power of one-dimensional hidden Markov modeling techniques (as well as their computing architectures). The key step to accomplishing this is that of "data ordering," or changing the two-dimensional image data into a one-dimensional sequence consistent with hidden Markov models.

One-dimensional hidden Markov models have been applied to images that have been converted to a sequence of one-dimensional values by data reordering for object recognition in synthetic aperture radar.[25] The process proceeds as follows:

1. Segment the image to be recognized into the four regions of background, target shadow, dim, and bright, using a thresholded connectivity-based algorithm.
2. Apply the Radon transform on each of the four segmentation images. The Radon transform computes the sum of pixels values $f(m, n)$ along each parallel line that spans the extent s of the object. Radon transforms are computed at various angles θ (Fig. 10–39) according to $F(\theta,s) = \Sigma_m \Sigma_n f(m,n)\ \delta(s - m\cos\theta - \sin\theta); 0 \le \theta < \pi$. This operation is performed for each of the four image regions. A sequence of values is built by concatenating all s values in order of s at one angle θ, then progressing to the next angle and adding its s values to the sequence. The result is four time-domain-like signals.

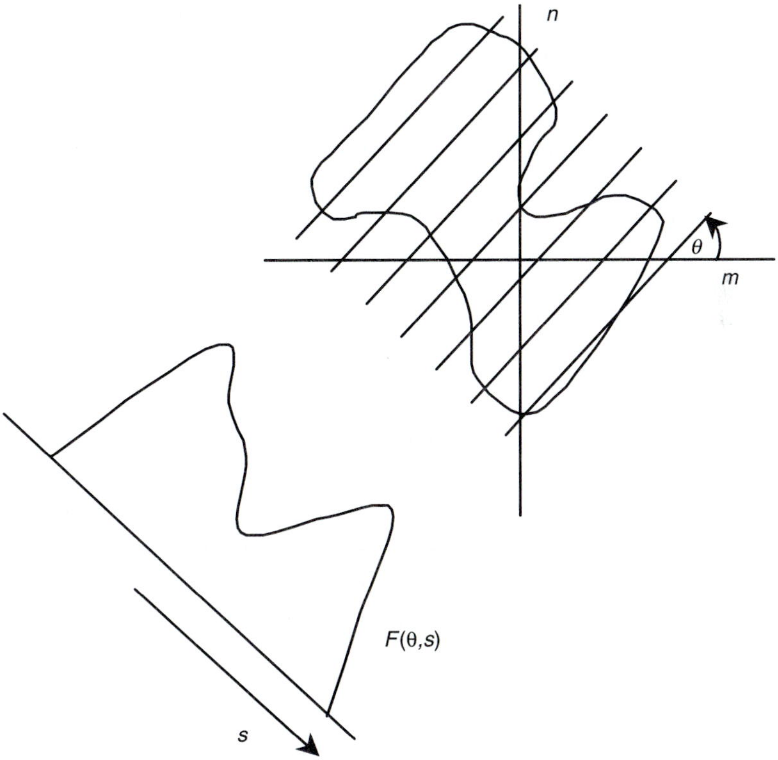

Figure 10–39 Radon transform computes the sum of pixels along each parallel line s through an object at angle θ. © 1997 IEEE.

3. Build target models by training a hierarchical hidden Markov model using the forward-backward algorithm and Baum-Welch reestimation. At the lowest level, the hidden Markov model is applied to the projection segments, shown by PS1, PS2, . . . in (Fig. 10–40). At the next level, it is a network of projection segments, where the order of the segments is specified by the network.
4. Score the unknown pattern using the Viterbi algorithm.

10.3.5 Model-Based Vision

Model-based vision[26] involves the explicit use of three-dimensional models of object, sensors, and image formation, with the use of a formal calculus (e.g., Bayes rule) for accumulating evidence. A mission hypothesis space is formed that contains the product of several spaces: (*target types*) × (*configuration*) × (*articulation*) × (*pose*) × (*deployment*) × (*acquisition geometry*).

Fig. 10–41 outlines the structure of a model-driven recognizer. It consists of two major sections: hypothesis reduction and hypothesize-and-test. These are supplemented by inputs of information about the sensor, image formation, and acquisition parameters, as well as supplementary information on context, such as map information that indicates which areas of a scene have terrain that is impassible to vehicles.

Hypothesis space reduction reduces the input image to areas of interest using region-of-interest extraction techniques as described earlier. The hypothesis space is reduced by an indexing step, which uses crude, easily-performed mea-

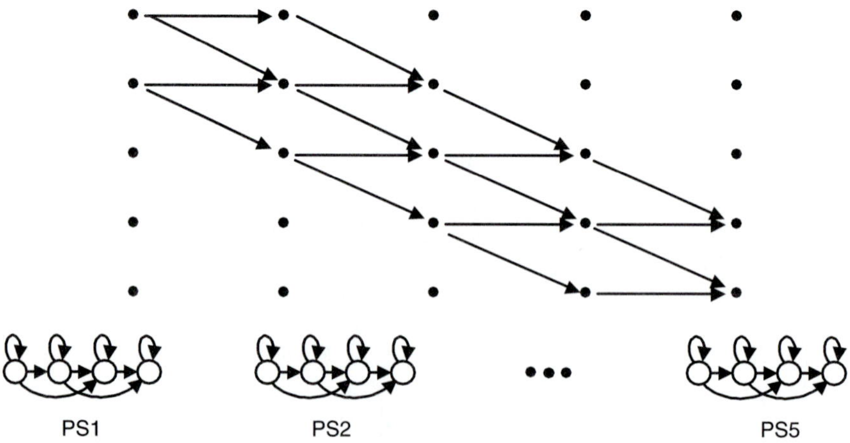

Figure 10–40 Hierarchical target grammar for hidden Markov model object recognition.[25] © IEEE 1997.

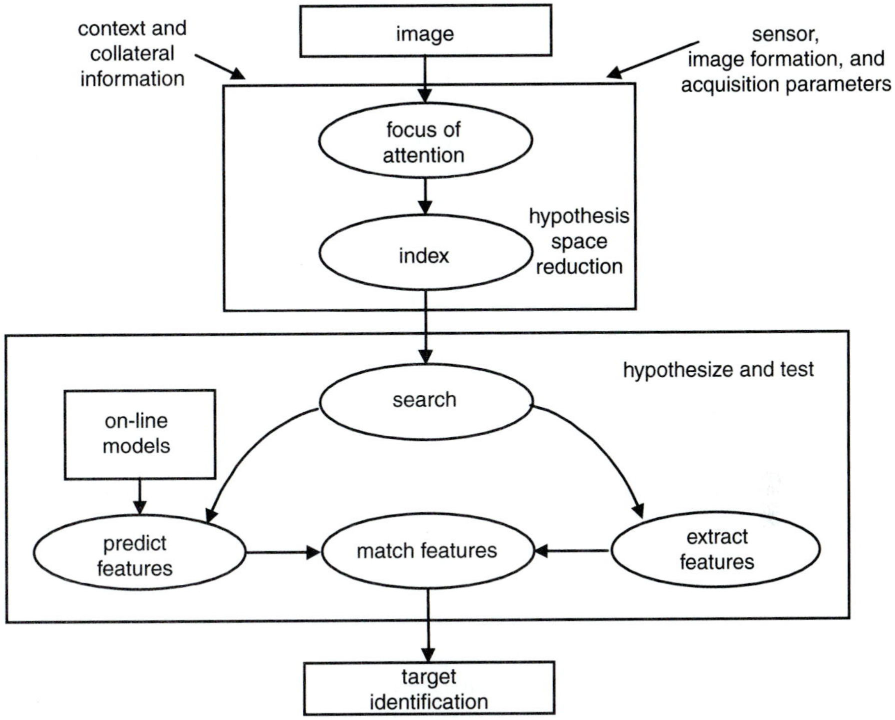

Figure 10–41 Processing flow for model-driven recognition includes both hypothesis space reduction, which refines image areas and target types for further exploration, and hypothesize-and-test, which uses a predict-extract-match-search paradigm.

surements of size, shape, and orientation to limit the number and type of target types and orientations to a subset of the total.

In the hypothesize-and-test section, the prediction module uses target and sensor models to predict which features should be seen, given a hypothesized target type and orientation. The extraction module pulls out corresponding features from the unknown image, which are compared to the predicted features by the matching module. The search module refines the hypothesis space and repeats the hypothesize-and-test section, rejects the choice, or accepts the choice as the proper identification.

The computational load of model-driven recognition is significant during the recognition operation. The processing time is dominated by feature prediction and storage. The feature prediction consists of both off-line preprocessing and on-line matching. Prior to recognition (offline), the feature predictor is used to perform a full high-resolution ray tracing of three-dimensional models of each expected target type. This step provides estimates of returns from a multiple-scattering-center decomposition of the target model. For 6-in. resolution on a $10' \times 20'$ vehicle, in which azimuth and elevation space are sampled every three degrees, over 4×10^6 ray traces are needed. For 20 target types, 120 different articulations of each target, and eight configuration alternatives for each target, over 10^{12} operations are needed to perform off-line computation at 100 operations per ray trace.

On-line processing completes the signature prediction by summing over all scatterers and computing the resulting image probability density function for a particular sensor and vehicle geometry based on combining amplitude and phase probability density functions of signature pixels.

A hybrid approach that combines the advantages of template-based recognition with those of model-driven recognition has been proposed.[27] Template-based recognition burdens the identification process with an up-front requirement for the exhaustive collection of data by sampling all possible values in geometry. It is computationally simpler in real-time matching, but generally is hindered because the full range of input variability is not represented in the template set, so matches are less accurate. Model-based techniques are fully able to respond to any combination of input parameters, but the computation is performed during the time-constrained on-line recognition process and is in the simulation domain, raising questions about the computed signature fidelity.

The hybrid approach collects a target signature at certain sampled values of the range of geometric variations. It then uses a signature predictor in conjunction with the collected signatures to extend the coverage of a signature collected in one geometric condition into new geometric conditions. This is done by decomposing the variation of a target signature vs. geometry that occurs into regions of stability in an image template, which do not change as geometry is varied, and regions of variation, which do change with geometry. A hybrid template is then constructed that mixes pixels produced from actual collected data (for regions of stability) with pixels produced by simulated data for the selected geometric conditions (for regions of variation with geometry). This hybrid approach keeps signature data anchored in the reality of collected data while using models to extend the range of geometric coverage of each templates.

In summary, real time algorithms for pattern matching consist of both a recognition and a training mode. To speed real time throughput, recognition processing may be simplified through the use of models, which increase the burden of training. Recognition proceeds from lower to higher levels of a structural hierarchy, and the searching of various ways to combine lower level elements into a higher one, across multiple potential mapping paths, introduces data-dependent processing flow into real time recognition. The kernel of add-min-argmin-update is common to many hierarchical recognition schemes and is accelerated with hardware structures. Techniques to limit search space through local constraints on matching and global context and information on grammars and map information further speed recognition, although usually with higher error rate.

REFERENCES

1. R. Bellman, *Dynamic Programming* (Princeton, NJ: Princeton University Press, 1957).
2. A. J. Viterbi, "Error Bounds for Convolutional Codes and an Asymptotically Optimal Decoding Algorithm," *IEEE Trans. on Information Theory,* IT-13 (April 1967), 260–269.
3. Hui-Ling Lou, "Implementing the Viterbi Algorithm," *IEEE Signal Processing Magazine* (September 1995), 42–51.
4. M. K. Brown, R. Thorkildsen, Y. H. Oh, and S. S. Ali, "The DTWP—An LPC-Based Dynamic Time Warping Processor for Isolated Word Recognition," *Proc. IEEE Inter. Conf. Acoustics, Speech, Signal Processing*, San Diego, CA, April, 1984, pp. 25B.5.1–25.B.5.4.
5. John S. Bridle, Michael D. Brown, and Richard M. Chamberlain, "An Algorithm for Connected Word Recognition," *Proc. IEEE Inter. Conf. Acoustics, Speech, Signal Processing*, Paris, France, May, 1982, pp. 899–902.
6. John G. Ackenhusen, "The CDTWP: A Programmable Processor for Connected Word Recognition," *Proc. IEEE Inter. Conf. Acoustics, Speech, Signal Processing,* San Diego, CA, April 1984, pp. 35.9.1–35.9.4.
7. Stephen C. Glinski et al., "The Graph Search Machine (GSM): A VLSI Architecture for Connected Speech Recognition and Other Applications," *Proc. IEEE,* 75 (September, 1987), 1172–1188.
8. L. R. Rabiner, "A Tutorial on Hidden Markov Models and Selected Applications in Speech Recognition," *Proc. IEEE* 77, (February, 1989), 257–286.
9. Jun-ichi Takahashi, "A Hardware Architecture Design Methodology for Hidden Markov Model Based Recognition Systems Using Parallel Processing," *IEICE Trans Fundamentals,* E76-A (June, 1993), 990–1000.
10. Carl D. Mitchell and others, "A Parallel Implementation of a Hidden Markov Model with Durational Modeling for Speech Recognition," *Digital Signal Processing,* 5 (1995), 43–57.
11. J. Johnson, "Analysis of Image Forming Systems," *Proc. Image Intensifier Symposium*, 1988, pp. 249–273.
12. Larry D. Hostetler, "Automatic Target Recognition Techniques for Synthetic Aperture Radar (SAR)," *Proceedings of Advanced Vision Systems Workshop,* Advanced Research Projects Agency, Washington, DC, March 22–23, 1994.
13. Nikola S. Subotic and Brian J. Thelen, "Sequential Processing of SAR Phase History Data for Rapid Detection," *Proc. IEEE Inter. Conf. Image Processing,* Washington DC, October, 1995, vol. 1, pp. 144–146.
14. R. P. Perry, R. D. DiPietro, A. Kozma, and J. J. Vaccaro, "SAR Image Formation Using Planar Subarrays," MTR 94B000003 (MITRE : Bedford, MA, Sept. 1994).
15. Chaitali Chakrabarti and Joseph Ja'Ja', "VLSI Architectures for Template Matching and Block Matching," in *Parallel Architectures and Algorithms for Image Understanding,* V. K. Prasanna Kumar, Ed. (San Diego, CA: Academic Press, 1991), 3–27.

16. Y. C. Liao, "VLSI Architecture for Generalized 2-D Convolution," *Proc. SPIE* 1001, *Visual Communications and Image Processing*, 1988, pp. 450–455.
17. Eugene L. Cloud, "Geometric Arithmetic Parallel Processor: Architecture and Implementation," in *Parallel Architectures and Algorithms for Image Understanding*, V. K. Prasanna Kumar, Ed. (San Diego, CA: Academic Press, 1991), 279–305.
18. Anil K. Jain, *Fundamentals of Digital Image Processing* (Englewood Cliffs, NJ: Prentice Hall, 1989), p. 385.
19. W. M. Brown and C. W. Swonger, "A Prospectus for Automatic Target Recognition," *IEEE Trans. Aerospace Electronic Systems,* 25 (May, 1989), 401–409.
20. Robert M. Lougheed, "A High-Speed Recirculating Neighborhood Processing Architecture," *Proc. SPIE* 534, *Architectures and Algorithms for Digital Image Processing II*, 1985, pp. 22–33.
21. "Remote Minefield Detection Systems (REMIDS), " (U.S. Army Waterways Experiment Station, 1992).
22. J. G. Ackenhusen and others, "ATCURE: A Heterogeneous Processor for Image Recognition," *Proc. IEEE Inter. Conf. Acoustics, Speech, Signal,* Detroit, MI, May, 1995, pp. 2679–2682.
23. Russell Epstein and Alan Yuille, "Training a General Purpose Deformable Template," *Proc. IEEE Inter. Conference on Image Processing*, Austin, Texas, October, 1994, pp. 203–207.
24. Esther Levin and Robert Pieraccini, "Dynamic Planar Warping for Optical Character Recognition," *Proc. IEEE Inter. Conf. Acoustics, Speech, Signal Proc.* San Francisco, CA, March, 1992, pp. III-149–III-152.
25. Dane P. Kottke, Jong-Kao Fu, and Kathy Brown, "Hidden Markov Modeling for Automatic Target Recognition," *Proc 31st Asilomar Conference on Signals, Systems, and Computing,* Pacific Grove, CA, November, 1997, pp. 859–863.
26. ARPA/SAIC System Architecture Study Group, "Model-Driven Automatic Target Recognition," (Washington, DC: Advanced Research Projects Agency, October 14, 1994).
27. Rajesh Sharma and Nikola S. Subotic, "Construction of Hybrid Templates from Collected and Simulated Data for SAR ATR Algorithms," *Proc. SPIE 3371, Automatic Target Recognition VIII, 1998* pp. 480–486.

11

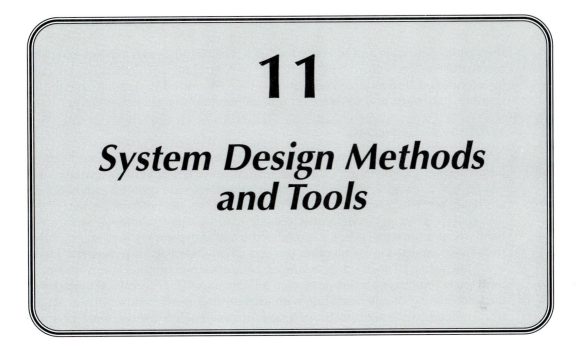

System Design Methods and Tools

A real time signal processing implementation consists of many parts operating in a coordinated manner, thereby constituting a *system*. Having described the underlying principles of real time signal processing and having presented a wide variety of examples that map architecture and algorithm, it is now timely to discuss the integration of these separate entities into a functioning whole.

This description first enumerates the design steps that are needed to develop a signal processing system and describes various ways by which these steps may be ordered into a development process. It then focuses on each of the steps in more detail, beginning with systems engineering and proceeding through hardware design (for both electronics and packaging), software design, integration, and testing. The section then concludes with the case study of a multiprocessor for image recognition, thereby describing by example the steps of system design.

11.1 DESIGN PROCESSES

The design process of signal processing systems begins with verification of the algorithm, placing emphasis on achieving complete validation over all types of expected operating data. At this point, the effort does not yet address algorithm speed of execution. A verified algorithm is then used to synthesize a combination of hardware and software, and portions of algorithm execution are allocated to the various hardware and software modules. The circuit design of processing elements then occurs, followed by the design of control logic. The modules are next interconnected, and module verification is then conducted to ascertain that the algorithm, hardware, and software is implementing the algorithm correctly.

11.1.1 Waterfall Process

A traditional system design process arranges its steps of systems engineering, hardware design, physical design/packaging, software design, integration, and testing into a *waterfall* (Fig. 11–1) by which the stages are ordered as mentioned above and one stage leads directly to the next. The sequential arrangement of the waterfall process provides a disciplined, easy-to-follow approach with clear handoffs. It allows the use of a specialized team of experts for each stage, and it frees those experts for use on other projects when their stage is completed. However, the waterfall method discourages cross-group communications and confines the execution of each stage to a single time interval. This restriction prevents the interrelating of one discipline with another and restricts downstream stages from influencing design decisions made upstream. Every stage must be completed correctly the first time—there is no second chance. The waterfall method also postpones the completion of a working item to the very end of the process.

11.1.2 Spiral Process

The design process may be reordered to provide a crude version of the entire system at the earliest possible time and then refine this version in a series of iterations. This method, known as the *spiral design process*,[1] was originally proposed for software development and is more recently used for the design of entire systems. Fig. 11–2 shows the spiral progression from the center outward.[2] It quickly produces a performance and behavioral simulation as shown at the boundary of Phase II and Phase III, then develops a full virtual (software) prototype and a processor prototype prior to the final production version. Because it quickly produces an end-to-end framework for the system, the spiral development process allows early identification and rectification of cross-discipline or system-wide is-

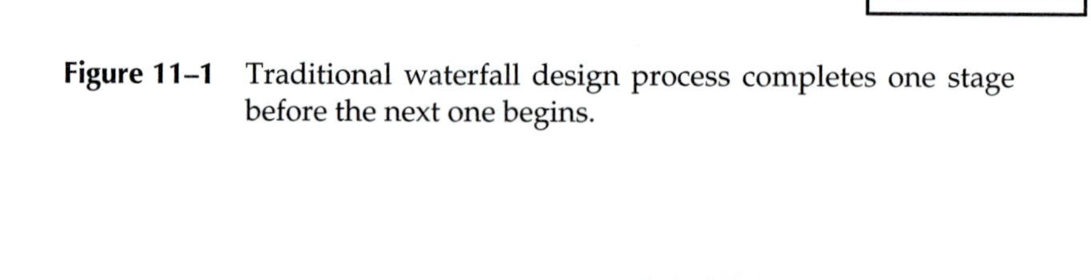

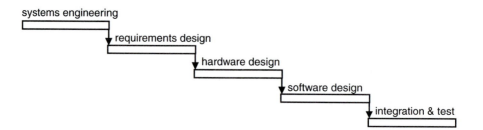

Figure 11–1 Traditional waterfall design process completes one stage before the next one begins.

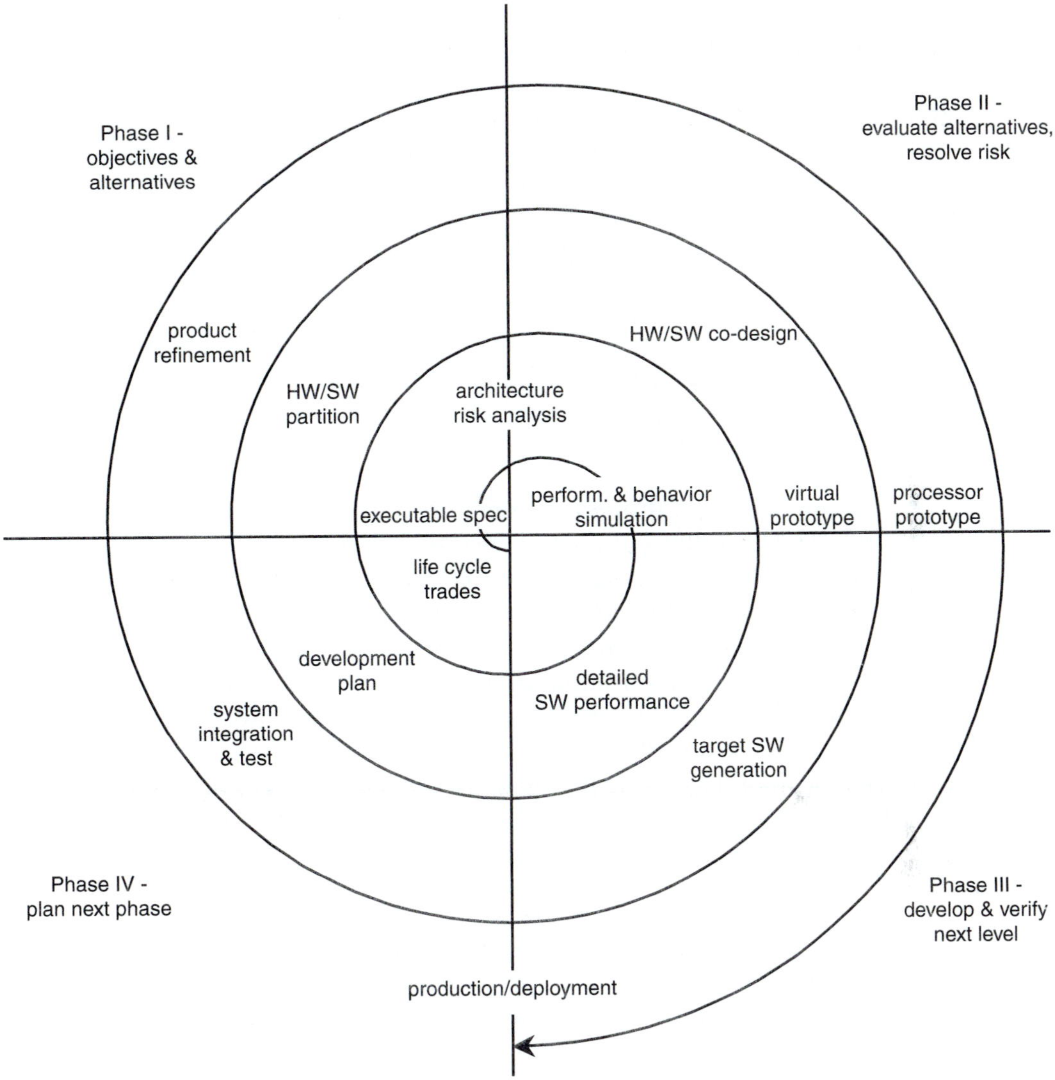

Figure 11–2 Spiral development process provides a total system realization at the end of each revolution (between Phase II and Phase III).[2]

sues. By operating in the mode of successive refinement, it allows a return and refinement to areas needing improvement.

Concurrent engineering[3] completes this trend of overlapping design steps. Concurrent engineering is defined as a systematic approach to creating a product design that simultaneously considers all elements of the product life cycle from conception through disposal at end of life. It includes consideration of manufacturing processes, transportation, maintenance, processing, and the like.[4] A team of designers across all disciplines is established at the start of the design and stays with the design as it evolves. Instead of a design passing from one engineering group to the next, representatives from each engineering group accompany the design as it progresses. Through concurrent engineering, manufacturing issues such as parts inventory and chip placement for automatic assembly can be included in up-front circuit design. Concurrent engineering has been shown to reduce the design time of systems while increasing their quality and manufacturability.

Design timelines have required compression as a result of acceleration of technology development and the increasing degrees to which signal processing systems have been approaching the limits of technology. This trend has resulted in shorter product lifecycles and associated shorter product design cycles. In response, design processes have evolved in several ways:

- overlapped design processes, exemplified by spiral development and concurrent engineering, have supplanted the serial waterfall process;
- programmable circuits such as digital signal processors have been increasingly replacing custom integrated circuits, which take longer to design;
- highly automated design tools that tailor a function-specific architecture to best fit a specific application are being used. In one example,[5] a function-specific silicon compiler generates a range of high-performance customized implementations of the radix-4 decimation-in-frequency FFT using submicron bi-CMOS technology. The user specifies the desired FFT parameters such as FFT size and word size of input, output, and intermediate data. The compiler then automatically generates netlist and floorplan information for fabricating the chip, and the design/verification time is shortened from a few months to less than one week;
- the use of a top-down hardware-less design method that simultaneously operates in the hardware and software domains (*hardware/software codesign*). This method, which has been advanced under a program funded by the U.S. government known as Rapid Prototyping of Application-Specific Signal Processors (RASSP), is now described.

11.1.3 Hardware/Software Codesign

A problem with signal processing system design has been that while technology capabilities advance to the next generation in about two years, a system design requires around four years to complete. As a result, designs are released with one- to two-generation old technology. In one survey, prototyping time from system requirements definition to production and deployment of multi-board signal processors was found to be 37 to 83 months.[6] Of that, 25 to 49 months was devoted to detailed hardware/software design and integration, with 10 to 24 months devoted to integration alone. To speed system design, the use of a *virtual prototype*, realized and modified entirely in software, and the description of the system in a language common to all levels of abstraction, is used. The virtual prototype is an executable description of an embedded system and its stimulus and describes its operation at multiple layers of abstraction. The language, Very High Speed Integrated Circuit Hardware Description Language (VHDL), is governed by an IEEE standard[7] and is subject to continual updates and improvements. It has the ability to describe systems and circuits at multiple levels of detail, and it is suited for logic synthesis as well as simulation. When used at all levels of the design process, VHDL provides *executable* documentation of systems throughout the process.

Fig. 11–3 describes the top-down design process of the VHDL-based rapid prototyping design process.[6] Requirements capture in VHDL provides an executable requirement that includes all types of signal transformations, data formats, timing, and control. The *executable requirement* moves beyond the traditional paper-only requirements description to provide an executable software model, jointly developed between customer and designer, that captures the design intent in an unambiguous software routine. The next step of algorithm and functional level design results in an *executable specification*. The executable specification is independent of implementation information, yet captures data precedence and timing relationships, including system timing, performance-related input/output, latency timing, task execution order information, and an initial functional breakdown of the task into signal processing elements or primitives. The executable specification also reflects such constraints as size, weight, power, and cost.

Between each pair of stages is a verification/validation method that provides a model for the higher stage of the design hierarchy to verify the subsequent more detailed design, represented by the upward and downward arrows. Thus, each higher level includes a *testbench* for verifying that the next lower level properly implements its predecessor.

Data/control flow design captures the concurrency information and data dependencies that are inherent in the algorithm. In this stage, multiple implementation-independent representations of the algorithm are generated to capture potential parallelism at the level of primitives, where primitives are defined as a set of functions used at multiple points of the processing flow and are placed as entries in a design library. At this stage, implementation choices between hardware and software libraries are made and new library elements, based on the recognition of key computational kernels, are created. The next stage, hardware/software architectural design, provides performance-level modeling in support of hardware/software tradeoffs, incorporating the performance and time aspects of each proposed design architecture (latency, system throughput, fraction of computing resources used, as well as power dissipated for power-limited designs). It completes the partitioning of modules between hardware and software implementations.

The hardware virtual prototype establishes the system structure, defining each processing element and associated execution time, communication protocols

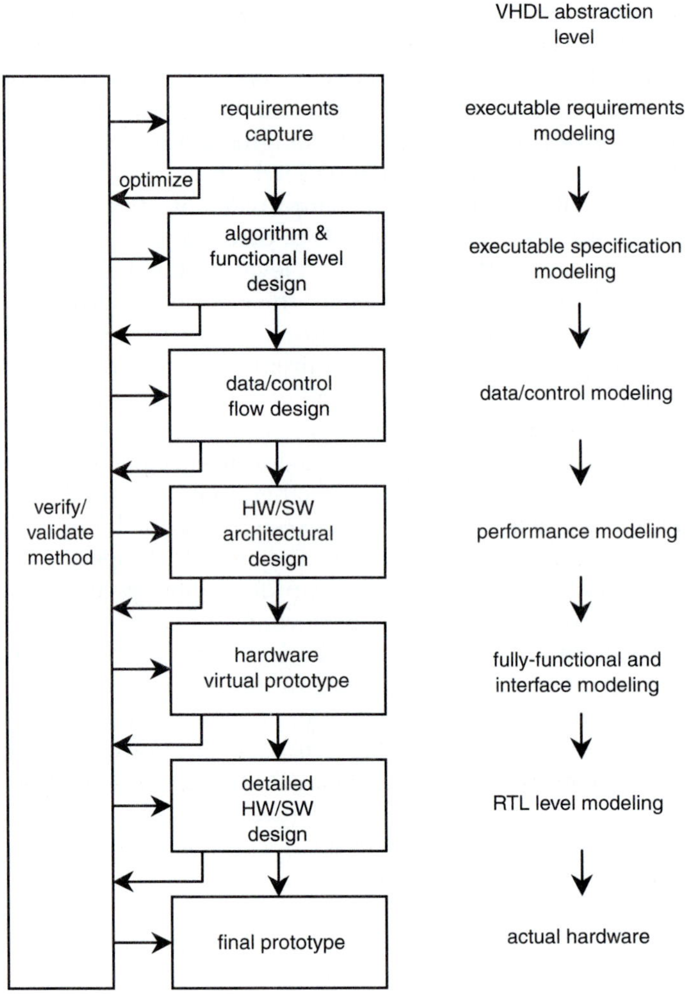

Figure 11–3 VHDL-based rapid prototyping design flow provides an executable description to each stage and a testbench to assure that the next lower level properly implements its parameters.[6] © 1996 Kluwer Academic.

and timing, and input/output timing and formats. It provides complete modeling of all functions and interfaces. The detailed hardware/software codesign culminates in the final register-transfer-level description of the system and precedes the generation of actual hardware.

An example of a commercial design toolset that supports the above design process combines the VHDL description of hardware with high-level language description of software.[8] This tool can start from either the hardware or the software domain and provides a design of a complete hardware/software system. Given a high-level computer program in a language such as Ada or C, the tool can synthesize a machine, expressed in VHDL, that can run the high-level code, providing a VHDL machine description, a VHDL simulator, and a software compiler.[9] Conversely, given a machine description in VHDL, the tool can create both Ada- or C-to-microcode compiler and a hardware simulator for evaluating the execution of Ada/C programs on that machine (Fig. 11–4).

11.1.4 Formal Review Points

The U.S. government specifies design reviews at which the evolving system design is reviewed by the customer.[10] Key customer reviews include the following:

- *System Requirement Review (SRR)*: executes review of the detailed functional requirements that the design will meet;
- *Preliminary Design Review (PDR)*: generally planned for about six months after the SRR, this review examines the functional breakdown of the system into modules and submodules that will be implemented, including the requirements for each module, based upon a top-down allocation of system requirements, and the intermodule communication timing and throughput;

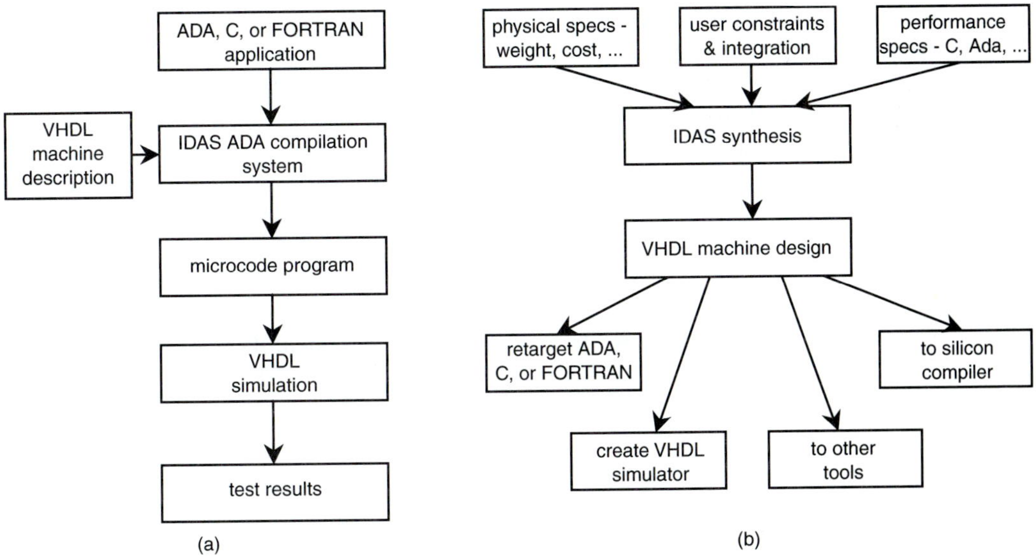

Figure 11–4 IDAS, a design tool from JRS, Inc., can both produce a VHDL machine design from a high-level application program (a) and produce a microcode compiler from a VHDL description (b).[9]

- *Critical Design Review (CDR)*: may occur 6 to 18 months after the PDR, depending upon system complexity—includes all detailed design information, including methods and results of design trade studies, detailed timing specifications, ability of each design modules to meet its allocated requirements, and testability;
- *In Process Review (IPR)*: generally held at regular intervals (e.g., quarterly) to provide customer insight between major reviews described above.

In addition, certain design-specific reviews, such as test readiness review, functional configuration audits, etc. may be held to focus on a particular topic.

These external (customer) reviews are simply samples along the course of the design path. Internal reviews and meetings of the design team, held both as needed and at regular frequent intervals, assure the success of the design and effectiveness of the customer reviews.

11.1.5 Quality Initiatives

Often, a design group finds a need to improve the quality of its results, perhaps by changing its processes. *Process Quality Management and Improvement (PQMI)*[11] is a systematic, customer-driven approach to analyzing processes and identifying defects and their root causes. PQMI can produce a list in order of importance of which process failures, if addressed, will yield the most return, as well as quantifying the amount of improvement. In its later stages, it uses a techniques known as *root cause analysis*, which is an approach for studying and evaluating errors and inefficiencies during the product realization process and identifying changes to the process based on analysis of these errors.

Table 11–1 displays the four steps and seven stages of the PQMI method. Significantly, before introducing any improvements, the PQMI method first insists upon measuring the current state. The first stage, ownership, is a leadership function that identifies and empowers the owner of the process to improve it. It includes upper management providing adequate resources, priority, and expec-

Table 11–1 Steps of the Product Quality Management and Improvement (PQMI) process.

Stage	#	Step
Ownership	1.	Establish process management responsibilities
Assessment	2.	Define process and identify customer requirements
	3.	Define and establish measures
	4.	Assess conformance to customer requirements
Opportunity Selection	5.	Investigate process to identify improvement opportunities
	6.	Rank improvement opportunities and set objectives
Improvement	7.	Improve process quality

tation for quality improvement. The assessment stage begins with defining the current process. This stage usually begins with the development of a detailed flow diagram, including handoffs from one function to the next. Measures of success are then established at points along the chain. These measures are related to the quality of the final results and are selected for ease in quantitative, not qualitative, measurement. The process is then measured in its current (unmodified) form over a period of time. From the measurement results, defects are found, accumulated, sorted by type, and displayed by frequency of occurrence using a histogram. Customer requirements are traced back to set necessary levels of performance against these measures, and the actual measurements are compared to the required thresholds.

At this point, several shortfalls have usually been identified and a histogram of defects has been developed. The next stage, Opportunity Selection, traces back defects to a root cause, determining why each error was introduced, where in the process the errors occurred, where in the process they were detected, and identifying possible areas of the process diagram for process improvements. After performing these activities for all errors, a histogram of root causes can be generated, and root causes can be ranked according to the number and severity of errors that they introduce. This analysis allows the improvement opportunities to be ranked and new objectives to be set for associated measures.

Finally, a few points (or a single point) in the process diagram are selected for improvement. Possible solutions may be brainstormed, and a trial implementation may be made to assess effectiveness of the most promising improvement ideas. Improvement is implemented and the results of the improvement are assessed by continuing the measurement process. As a final step, holding the gain is assured by extending the continuation of the measurement monitoring process.

A key element of process instrumentation, measurement, and improvement is the peer inspection. Applicable at every stage of the design process, from requirements through hardware/software design to test and integration, the peer inspection is an in-process review with specific entry and exit criteria. An inspection team is composed of peers, rather than managers, to reduce the tension associated with finding the design defects that the review is intended to find. Leading the inspection team is a *moderator*, who sets up the team, assures that material is distributed for review prior to the meeting, and coordinates the inspection of a typical design. The moderator leads a *reader*, who reads out loud and interprets the design for other members of the inspection team during the inspection meeting. A *recorder* documents the defects that were discussed at the inspection. In addition are several inspectors who are *subject-matter experts*. The *designer* is also present during the inspection. Typical entrance criteria for an inspection include preparing the inspectors with pre-meeting material to be inspected, the rules against which it is to be inspected, and the assurance that at the start of the meeting, all inspectors have read all pre-inspection material. The moderator postpones the inspection if the team is incomplete or unprepared. Exit criteria that determine whether the inspection is complete include a list of design defects, archived metrics on defects found, and plans for follow-up, which often includes re-inspection of the corrected designs.

11.2 SYSTEMS ENGINEERING

When those tasks that are associated with bringing a complicated contrivance into being are spoken for by the design engineering disciplines (electrical, physical, software design), it is *systems engineering* that claims the remainder. Indeed, one definition of systems engineering includes "guaranteeing that the sum of parts functions as a whole." Systems engineering is the glue that fills the interstices between the design engineering disciplines and binds them into a functioning whole.

11.2.1 Definitions

According to MIL-STD 499A:

> Systems engineering is the application of scientific and engineering efforts to transform an operational need into a description of system performance parameters and a system configuration through the use of an iterative process of definition, synthesis, analysis, design, test, and evaluation;
>
> integrate related technical parameters and ensure compatibility of all physical, functional, and program interfaces in a manner that optimizes the total system definition and design;
>
> integrate reliability, maintainability, safety, survivability, human, and other such factors into the total engineering effort to meet cost, schedule, and technical performance objectives.[12]

The Defense Systems Management College is more succinct:[13]

> Systems engineering is the management function which controls the total system development effort for the purpose of achieving an optimum balance of all system elements. It is a process which transforms an operational need into a description of system parameters and integrates those parameters to optimize the overall system effectiveness.[13]

11.2.2 Translation between User Space and Design Space

To understand the stages of systems engineering, it is useful to regard systems engineering as a translation process by which a system is translated from user space, in which it is characterized by features, to designer space, in which it is characterized by requirements, and back. User needs consist of features, mission requirements, and operational capabilities, and are expressed in the language of the user. Designer requirements are derived from user needs and consist of specifications on performance, system, and detailed design. They are expressed in language suitable for driving electrical, physical, optical, and software design. The process of requirement development performs the translation from user features to design requirements; the process of system test verifies the translation of the requirements-based design back into the user domain. Between these two translations, the design disciplines and their associated processes are at work.

A failure of a system to meet user needs is usually traced to an error in translating those user needs into design requirements and allocating those requirements into lower-level requirements upon system components. Several techniques are used to minimize the chance of error in mapping to and from design space. First, both users and designers can be included in the review of the development specifications that seek to capture user needs. Second, progress on the design and the current view of its performance may be periodically translated back into user space, using systems engineering methods, throughout the course of the design. Third, any evolution of user needs during the design interval can be monitored and validated, using configuration management techniques. Fourth, the final results of the design are tested against both the development requirements and the user needs using a group of testers that is independent of the design team. Finally, to the extent that user requirements can be captured by executable specifications, a software performance model can be executed and maintained to foresee final system performance.

11.2.3 Lifecycle Systems Engineering

Systems engineering for signal processing development may be divided into the four major phases of advanced, development, system test, and user support (Fig. 11–5). The input to the stage of advanced systems engineering is customer require-

Stage	*Advanced*	*Development*	*System Test*	*User*
Activities	Research & development Initial requirements allocation Initial development specifications	Maintain requirements allocation Develop/maintain view of system performance Maintain risk assessment Assess design modifications and workarounds Project management	Plan and witness execution tests Negotiate acceptance criteria Provide early user view	Map user needs to system features Accelerate user learning Compile trouble and bug reports
Input	Customer requirements	Development specifications	Completed system design	Acceptance-tested system design
Output	Development specifications	Completed system design	Acceptance-tested system design	Successful customer

Figure 11–5 Inputs, outputs, and processes for the four stages of life-cycle systems engineering.

ments; its output is design specifications. The advanced systems engineering activity includes the development or identification of key technologies to meet demanding requirements, the decomposition of higher-level requirements onto lower-level modules, and the development of initial requirements for those modules.

One demand of advanced systems engineering for signal processing is determining the size of machine required to run an algorithmic application. The fundamental analysis of determining whether a particular computing architecture can run a selected application in real time is very difficult to perform with high precision for all but the most regular of algorithms and architectures. Simple counts of multiply operations, matched with the appropriately-derated throughput of a multiprocessor, may give a general idea for regular computing processes, but it is not sufficient for decision-dependent data-driven operations, nor does it assure that such irregular operations will not outstrip available computing resources. More detailed sizing estimates may be made, with more difficulty, by fully decomposing the application into a set of primitive library routines, the performance of which is already known. At the most accurate level of detail, a full cycle-accurate timing model of the architecture may be developed. This simulation may either be a stand-alone timing emulator or may derive from an executable specification of the machine. In any event, it is the user application set that drives the machine, determines its cost of design production and maintenance, and determines the degrees of reserve processing and memory resources available for lifecycle upgrades. Indeed, one Department of Defense directive, known as TADSTAND-D, requires a 50 percent reserve capacity in processing as measured as a percentage of the available capacity at full operational loading. A similar requirement exists for memory reserve. Thus, processor sizing is a challenge central to signal processor systems engineering.

Development systems engineering receives as its input the development specifications that are output from advanced systems engineering activities. The development phase of systems engineering maintains the requirements allocation as the design progresses, assuring that if one submodule performance specification must be relaxed, then the others are tightened so that the allocated performance of the group of modules continues to be met.

Development systems engineering requires developing and maintaining a current view of end-to-end system performance and the performance allocated to each of its constituent modules. Performance modeling monitors such timing aspects as latency, throughput, and bandwidth as well as power consumption, size, and weight as the design progresses. During the development interval, system performance follows a path that may be envisioned as a checkmark. Moving from left to right through the checkmark (from early time to later time), a level of performance is set from requirements based on user needs. Developers then add engineering margin to these requirements, placing the goal a bit higher (comprising the small uptick at the left of the checkmark,√). When the design reaches early stages of integration, the first measurements of performance often fall far short of the goal, resulting in the drop to the bottom of the checkmark. Subsequent efforts in completing the integration and optimizing the result cause the performance to increase, rapidly at first and then asymptotically approaching a level at or above the original design goal.

Another function of development systems engineering is continuous risk identification, assessment, and mitigation. Risk assessment begins by identifying all areas where things may go wrong. For each item, a likelihood of occurrence is estimated, based on design experience and historical records. The goal is to identify and manage every risk that could reasonably occur without unduly dissipating energy on unlikely risks. For each reasonable risk, the severity of impact on performance, cost, or schedule is estimated and a mitigation strategy is identified to either remove the risk early in the design (for example, through some parallel backup design or experiment) or work around the failure should the risk event occur. A cost of mitigation is estimated. Fig. 11–6 shows a small section of a risk analysis and indicates a tracking number, risk description, likelihood, severity, and mitigation approach. The list of risks is tracked as the design progresses, and old ones are removed from the list when their danger has passed, while new ones are added when identified. The risk list is best generated by analyzing the list of assumptions that were made during the initial planning of the project. A list of critical dependencies in the design process is developed from the list of risks, and these are discussed at periodic program meetings. Designers who are downstream in the development process identify critical items that they need to start their tasks. As a global measure, an overall tally of total risk, measured by likelihood-weighted cost and/or schedule impact, may be computed, reviewed, and reported to the customer at regular intervals throughout the design cycle.

The development stage of systems engineering also assesses the impact of any in-process design modifications and workarounds. This assessment is per-

#	*Risk*	*Likelihood*	*Mitigation*	*Cost*
461	Chip layout error increases parasitics, decreases speed	M	• Build test chip • Improve manual inspections & redlining • Simulate all paths	• $100K • $50K • $25K
462	Chip fab house will slip schedule	H	• Include them in design reviews • Apply incentive payments	• $10K • $50K

Figure 11–6 Example fragment of risk tracking list for hardware/software design project.

formed against design requirements and translated back into characterization of the system design's ability to continue to meet user needs. Systems engineers may lead the project effort in brainstorming solutions to a cross-discipline design problem to generate possible solutions.

Indeed, the final area of development systems engineering is management of the program or project. Project management is inextricably linked with systems engineering. For example, it is only systems engineering that spans the entire project in the manner necessary for program management. Systems engineering continuously balances the three factors of technical performance, cost, and schedule that are essential to managing a large program.

The output of a successful development systems engineering stage is a completed system design, which then passes to the third phase of systems engineering, the stage of system test.

System test, best performed by an organization separate from the design developers, is often performed by systems engineering, which developed the requirements based on user needs that drove the design. *Requirements testability*, or how the meeting of each requirement can be tested in the design, is a part of the requirements design process and naturally leads to the system test plan. The planning, execution, and customer witnessing of execution tests is a major function of this phase of systems engineering. System test engineering also negotiates the customer acceptance criteria as well as entrance criteria of the system into the system test stage. Here is provided the early user view of system performance. The output of successful system test engineering is implemented in system design that has been accepted by the customer.

Continuing beyond the formation and breakup of the large design team, systems engineering supports the user in effectively applying its newly-received system. User support accelerates user learning through hands-on training and properly-written, inviting user manuals. Systems engineers performing user support become expert in mapping user needs to system features and thereby serve the function of application engineers. In doing this, they identify new features and improvements that may be incorporated into future system designs. As the system is used, bug and trouble reports are compiled and provided to the product support design team, and repairs are tested and installed on customer systems when completed. Indeed, the output of this final stage of systems engineering is the successful customer.

11.2.4 Scope/Schedule/Cost Balance

As mentioned earlier, the continual tradeoff among technical scope (or performance), schedule, and cost is a systems engineering activity essential to program management. Specifying two of these three factors usually determines the third, and different development projects place different factors as the highest priority, as agreed to by both customer and developer. When realism overcomes optimism in the course of system design, everyone must relax requirements in the domain of the lowest-priority factor while seeking to preserve the requirements for the highest priority factor. Similarly, the three areas of scope, schedule, and cost must be kept consistent throughout the design project. Traditionally, the systems engineer generally determines the technical scope as the highest priority, then seeks to minimize cost and lets the results determine schedule. More recently, customers specify cost as an independent variable, with technical scope depending on cost. In all cases, as the design progresses, various factors change the current estimate of each factor. For example, electrical noise may be found to limit the speed of a circuit and therefore reduce performance. Recovering that performance may require adding a second parallel processing unit, thus increasing cost and possibly schedule. With each change, the scope, cost, and schedule must be rebalanced to update the project view. Keeping the rebalancing time small is important to maintain a constant project view for the user.

The scope of a project and its technical performance requirement are set by the initial Statement of Work (SOW) that defines the project. The SOW often contains a *user-level* specification, known as an A-Spec as described below, to define the project. This A-Spec may be in the form of an executable specification. As the project progresses, the customer may change scope, based on evolving needs during the design interval, or the designers may seek to change scope based on design difficulty. Again, such changes must involve both customer and developer, must balance with cost and schedule, and must be configuration managed. Configuration management involves keeping a traceable record of changes to allow reconstruction of the entire SOW that was in effect at any particular instance through the design period.

The schedule for developing a signal processing system must be predicted and enforced, and it occasionally becomes the top priority driving the design. The schedule for a project may be developed by following a progression of steps:

1. Identify the first step of the development (or the next step, if the development is already in progress).
2. Identify all subsequent steps—these steps become the tasks of the schedule.
3. Order these tasks based on time and identify any task-to-task dependencies by which one task output becomes the input of another task.
4. Estimate the duration of effort necessary for each task (and if preparing a cost estimate, also estimate the cost of labor and cost other than labor, such as material, travel, subcontractors, consultants, etc.).
5. Determine which tasks can be overlapped and which must be performed sequentially.
6. Map each task to available resources.
7. Document the result as the project schedule.

Iteration is usually necessary, and for schedule-driven development for which the schedule is fixed, a top-down allocation of schedule to various groups of tasks is performed.

Similarly, many development projects are driven by a particular program cost that cannot be exceeded. To structure a program to meet a not-to-exceed cost, about 10 to 15 percent of the total resources are first segregated for unexpected contingencies that only become known in the midst of the project. Then top-down allocation of remaining costs are made to each major component of the system. Allocation to system components, or configuration items, is preferable to allocating to discipline (software, hardware, . . .), because each configuration item will make use of most of the design disciplines, and tradeoffs to determine whether a module will be implemented by software or hardware have not yet been conducted. Developing a repository of historical cost data from past programs is extremely useful, but rather rare. Such historical records, if available, may be combined with cost estimating rules of thumb to arrive at a more accurate cost estimate.

Another method to influence project costs is to exploit the nature of the spiral design process to produce an early functional system prototype, which may then be used to build advocacy, garner additional funds, or prove feasibility. However, the development and advocacy of a prototype can communicate a false sense of progress, give the mistaken impression that the development is nearly completed, and actually increase the total cost of development. This is because a prototype only meets the subset of total system requirements that can be absorbed in a brief demonstration, and it ignores the requirements that are important at points beyond the initial demonstration (Table 11–2). Worse still, a successful prototype may cause the expectation that the prototype (or a small variant of it) may be produced in quantity, deployed, and maintained. This illusion may extend from the customer to the developer, particularly in programs for which the development of the prototype, rather than a mass-producible system, was the

Table 11–2 Prototype "cream-skims" the eye-catching subset of desired product features.

Feature Class	Feature Example	Prototype	Product
Functionality	Instruction set	√	√
	Memory size	√	√
Performance	Compute speed	X	√
	Memory bandwidth	X	√
Physical	Size	X	√
	Weight	X	√
Manufacturability	Parts selection	X	√
	Assembly cost	X	√
Reliability	Mean time to fault	X	√
	Mean time to repair	X	√

√ - included

X—not included

intent. In such prototype-directed programs, the effort should start over if the goal unexpectedly becomes a fieldable product. In a fresh start, the function of the prototype is captured as a user-level specification and as a set of inputs and corresponding outputs (*test vectors*) to be met by the final product.

Another cost related to development is the cost of lost market opportunity. If the product development takes so long that the product is not available until well after the point of market need, significant revenue will be lost (Fig. 11–7). This loss, coupled with the fact that rapid advances in underlying technology outstrip the pace of system realization and cause the system to be released with older technology, places an emphasis on rapid product realization through concurrent engineering, virtual prototyping, and design reuse. It also indicates a premium on advanced planning, market prediction, and even the speculative development of a design of an item before a market need is identified.

Discussions so far have focused on the cost of development. However, development and design cost is a relatively minor portion of the total lifecycle cost of a system, ranging from 3 to 10 percent. Although the cost of development may be trivial in the lifecycle cost of a product, the resulting design drives the entire lifecycle cost. One source[14] provides a specific example of the general fact that over 80 percent of the lifecycle cost of a product is determined in the first 20 percent of its lifecycle, that is, in the phases of concept and design engineering. Fig. 11–8 plots the cumulative lifecycle cost of a system design vs. time (lower curve), showing that only 20 percent of lifecycle cost is expended at the start of production planning. The upper curve plots the impact on life cycle cost versus progress of design—at the point of design engineering, 80 percent of the lifecycle cost has been determined.

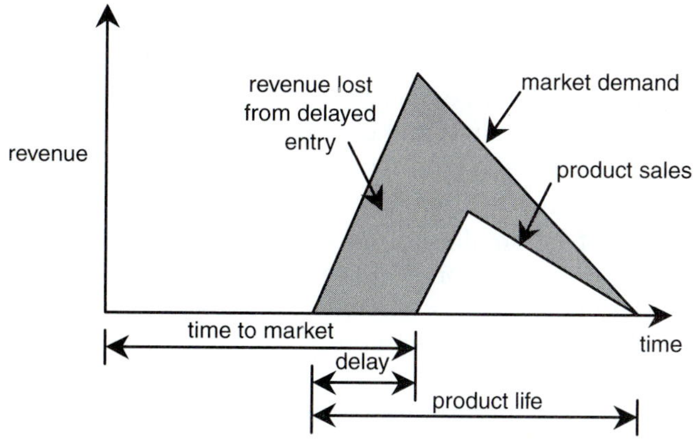

Figure 11–7 A delay in system production and deployment beyond the rise of the market window can result in the loss of a significant fraction of available revenue (shaded area).[6]

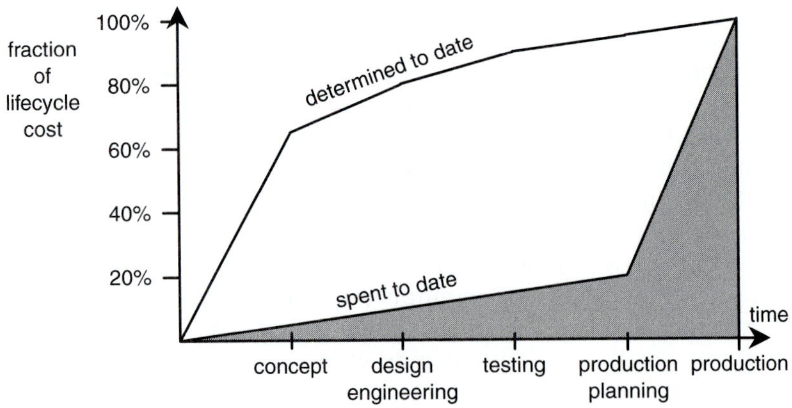

Figure 11–8 Over 80 percent of the lifecycle cost of a product is determined in the first 20 percent of its lifecycle, i.e., in its design stage.[14]

11.2.5 Lifecycle Cost Analysis

The understanding of system lifecycle cost is crucial to end-to-end systems engineering. From this understanding derives various non-intuitive observations, such as:

- Developing a prototype instead of a fully engineered product, while it may reduce by 50 percent the front-end development portion of the system cost, may nearly double lifecycle costs because of lack of optimization to the product factors listed in the lower section of Table 11–2. A 50 percent reduction of design cost that itself is only 10 percent of the lifecycle cost is a relatively small reduction in lifecycle cost.
- Lifecycle cost over several years may be reduced by redesigning the product after a few years to take advantage of more capable electronics, simpler design, and more reliable components.
- The *maintenance philosophy*, which includes the amount of fault tolerance and cost of the smallest replaceable unit, as well as whether repairs are performed by the user or by a separate repair shop, and specialty engineering considerations such as reliability, maintainability, and logistics support can have far more influence on lifecycle cost than the cost of development.

Because of the importance of lifecycle costing in production system design, an extensive but greatly simplified example of lifecycle cost is now presented. A high-speed multiprocessor signal processing system is to be developed which consists of twelve identical boards, each with multiple state-of-the-art signal processing chips, as well as a general purpose computing board with operating system, and a power supply and cabinet. A lifecycle of 10 years is required, with a total of 410 systems being delivered according to the profile of Fig. 11–9. Design begins at the start of Year 1. After a two-year design interval, delivery begins in Year 3. An update to the design that incorporates new technology begins in Year 6, with delivery of the updated system beginning in Year 7 and continuing through Year 10. The systems continue to be operated for two more years, to Year 12. The system is used for continuous 24-hour-a-day surveillance on extended missions upon naval vessels.

The three major components of lifecycle cost are Design and Development, Production, and Operations and Maintenance. The factors of each are now described.

Design and development costs include the development of the signal processor and of any support equipment, such as testing circuits. It is divided into the three components of hardware, software, and specialty engineering (Fig. 11–10).

Hardware costs include the one-time cost of designing the system and the recurring cost of building three prototypes. In systems engineering terminology, one-time design costs are called non-recurring engineering costs, or NRE. The hardware design of this example includes the development of one application-specific integrated circuit (ASIC) and two boards, as well as the design of the cabinet. In Year 6, design of the system upgrade requires the development of one board and one ASIC and the pre-production building of one prototype.

Software development provides for the writing of 20,000 lines of real-time code, which includes operating system and a library of real-time signal processing primitives, as well as test code. A Constructive Cost Model (COCOMO) as described in Section 11.5 is used with complexity factors set rather high to reflect the fact that the code must operate under the tight time constraints of real time and is therefore more difficult and expensive to develop. The software for the

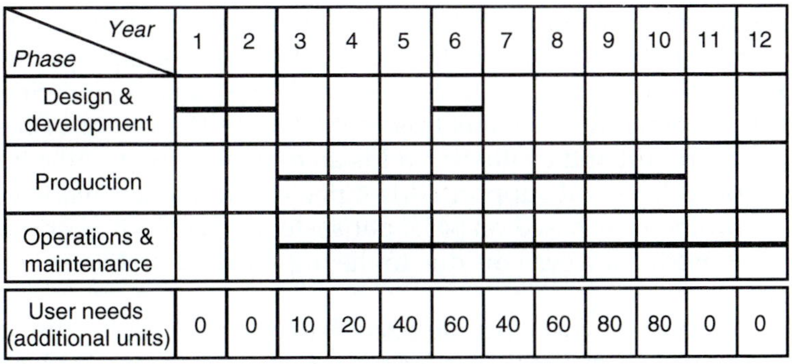

Year
Phase
1 2 3 4 5 6 7 8 9 10 11 12
Design & development
Production
Operations & maintenance
User needs (additional units)
0 0 10 20 40 60 40 60 80 80 0 0

Figure 11–9 User needs dictate a total of 410 systems according to the profile shown. The years for each phase of life cycle cost are shown.

Basic system	*1*	*2*	*3*	*4*	*5*	*6*	*7*	*8*	*9*	*10*	*11*	*12*	*TOT*
Hardware:													
NRE	1000	500											1500
Prototypes (3)		180											180
Software:													
20,000 lines of code	1000	1500											2500
System upgrade													
Hardware:													
NRE (1 ASIC, 1 board)						600							600
Prototypes (1)						90							90
Software:													
50% change to code						1250							1250
HW cost	1000	680				690							2370
HW + SW cost	2000	2180				1940							6120
Specialty engineering:													
Systems eng. (15%)	300	327				291							918
Test & evaluation (8%)	80	54				55							190
Integr. logis. supp. (4%)	40	27				28							95
Installation ($45K)						45							
Total design/develop.	**2420**	**2589**				**2359**							**7367**

Figure 11–10 Design and development cost vs. year consist of a two-year design interval in Years 1 and 2 and a one-year redesign in Year 6 (all costs in $1,000).

system upgrade at Year 6 amounts to a 50 percent change to the code of the basic system.

Specialty engineering, which includes systems engineering, test and evaluation, integrated logistical support (lifecycle support of distribution, maintenance, material flow, inventory management, and customer service), and initial installation and checkout of the prototypes bases its cost as a proportion of hardware or combined hardware and software costs. Systems engineering, which also includes program management, is estimated as 15 percent of hardware and software costs, test and evaluation adds an extra 8 percent of the hardware cost, and integrated logistical support adds 4 percent to the hardware cost. Installation of each prototype is based on $45K per system, which is higher than the installation cost of production systems due to the unpredicted problems of first-time installation.

Production costs include fixed expenditures, production costs for the basic system, and production costs for the upgraded system. Fixed overhead costs provide for the building and equipping of the production facility. These costs are amortized over the entire production run of 410 machines and add a fixed cost to each system (Fig. 11–11).

Fixed cost, amortized over	*1*	*2*	*3*	*4*	*5*	*6*	*7*	*8*	*9*	*10*	*11*	*12*	*TOT*
entire production (per unit)			12	12	12	12	12	12	12	12			
Base system													
Base cost: $150K + fixed													
Less 20%/yr for first 3 yrs			1.00	0.80	0.64	0.51							
-> price per unit			162	132	108	89							
× # units			10	20	40	60							
= cost of units			1622	2644	4328	5340							
Spares: 1 machine to			324	529	866	1068							
2 depots per 10 units													
Upgraded system													
Base cost: $75K + fixed													
-> price per unit							75	60	48	38			
× # units							40	80	80	80			
= cost of units							3000	4800	3840	3072			
Spares: 1 machine to							600	960	768	614			
2 depots per 10 units													
Total hardware costs			1946	3173	5193	6408	3600	5760	4608	3686			
Specialty engineering													
Prog Mang - 11% HW			214	349	571	705	396	634	507	406			
System Eng. - 2% HW			39	63	104	128	72	115	92	74			
Install & check = $5K/unit			50	100	200	300	200	400	400	400			
Total production cost			**2249**	**3685**	**6069**	**7541**	**4268**	**6909**	**5607**	**4566**			**40893**

Figure 11–11 Per-year costs of production include base cost, adjusted for yearly decrease due to learning and parts cost decrease, spares, and installation for basic and upgraded systems.

Production of both the basic and upgraded machine include a cost drop by 20 percent each year that incorporates both the reduction in the cost of electronic parts and a learning factor that reduces the cost to assemble each machine. The cost for the base system products occurs in Years 3 through 6 and produces 170 systems. In addition, an initial set of spare parts, equivalent to one machine, is provided to each of two repair locations ("depots") for every 10 machines that are fielded. Production of the upgraded systems, beginning in Year 7, results in a machine whose unit cost is one half that of the original basic system. It also drops in cost for three years due to parts cost decrease and learning factors, and like the basic system, is equipped with one machine worth of spares for each of the two depots for every 10 machines.

Specialty engineering, amounting to 11 percent of hardware costs for management of the production program, and 2 percent of the hardware costs for systems engineering, are added, as well as an installation and checkout cost of $5K per unit.

The costs of operations and maintenance (Fig. 11–12) dominate the lifecycle costs of this signal processing system. Indeed, average operations and maintenance cost per machine are more than double the initial cost of each machine. Be-

Base system	*1*	*2*	*3*	*4*	*5*	*6*	*7*	*8*	*9*	*10*	*11*	*12*	*TOT*
MTBF = 2000 hrs., 100%													
use => 4.4 fails/machine-yr			4.4	4.4	4.4	4.4	4.4	4.4	4.4	4.4	4.4	4.4	
Cum # machines in field			10	30	70	130	170	170	170	170	170	170	
# maintenance actions			44	132	307	570	745	745	745	745	745	745	
1 failure: 10% of 1 machine-			4	13	31	57	75	75	75	75	75	75	
spares (use machine cost)			711	1739	3321	5073							
+ labor/shipping per action			44	132	307	570	745	745	745	745	745	745	
Spares buyout - #						447							
Spares buyout - cost						39803							
Upgraded system													
MTBF = 3000 hrs													
=> 2.9 fails/machine-yr													
Cum # machines in field							40	120	200	280	280	280	
# maintenance actions							117	351	585	818	818	818	
1 failure: 10% of 1 machine-							12	35	58	82	82	82	
spares (use machine cost)							877	2105	2806	3143			
+ labor/shipping per action							117	351	585	818	818	818	
Spares buyout - #										164			
Spares buyout - cost										6286			
Software maintenance													
1 person/20 KSLOC													
Immaturity factor, yrs 0–3:			1.50	1.33	1.33	1.16	1.50	1.33	1.33	1.16	1.16	1.00	
Effective KLOC:			30.0	26.6	26.6	23.2	30.0	26.6	26.6	23.2	23.2	20.0	
Person yrs to maintain:			1.50	1.33	1.33	1.16	1.50	1.33	1.33	1.16	1.16	1.00	
x Cost per person year =			225	200	200	174	225	200	200	174	174	150	
Total Op. & Maint.			**980**	**2070**	**3827**	**45620**	**1964**	**3400**	**4336**	**11167**	**1738**	**1714**	**76817**

Figure 11–12 Yearly operations and maintenance costs are based on the mean time between failure (MTBF) and the cost of each repair action, as well as software maintenance on maturing software.

cause these machines are used in a continuous naval surveillance operation on extended missions, continuous operation of 24 hrs/day, 365 days/yr is used for projecting maintenance costs. For the base system, a system-wide mean time between fault (MTBF) of 2000 hours is used. A more complete analysis than used in this example would compute the MTBF of the system by combining MTBF of each subsystem and component, such as power supply, signal processing board, and signal processing chip. A system composed of 10 signal processing boards of 20,000 hr MTBF would have an MTBF of 2,000 hours if the MTBF for all other components were very much larger than the 20,000 hrs (and if there were only a limited number of other components). Under continuous operation, a system MTBF of 2,000 hours means that 4.4 repair actions per year are needed for each system.

Again, a more detailed analysis would determine what the most efficient lowest replaceable unit (LRU) would be and whether each repair resulted in the scrapping of an LRU or its repair for re-entry into the spares inventory. As an aside, only through concurrent engineering could the maintenance philosophy, which determines the most efficient LRU, influence the design development as necessary to determine the choice of module size and content that will become the LRU. This simplified analysis merely projects that each action results in replacing one tenth of the system and that costs to localize the fault are negligible. A cost item for replenishing the spares for each maintenance action is therefore included. At the final year of production, sufficient spares are purchased to meet the maintenance needs of all machines for the rest of the lifecycle. This is shown by a substantial cost of spares buyout.

The upgraded system improves the MTBF to 3000 hours, resulting in an average of 2.9 maintenance actions per machine per year. As before, each maintenance action is modeled to require one-tenth of a machine, and again, a spares buyout is executed during the final year of production.

Software maintenance on the 20,000 lines of code is also an operations and maintenance cost factor. A simplified estimate of software maintenance indicates that one engineering year of effort is needed to maintain 20,000 lines of code, once that code has matured and stabilized. Stability is gained after five years of use, and earlier than that, additional efforts per year add to the base in the amounts of 50 percent, 33 percent, 33 percent, 16 percent, and 16 percent for Years 1–5. The software for the base version of the machine has not reached stability before the software for the upgraded system enters the field, and so the software maintenance maturity multiplier decreases for four years from 1.5 and then jumps back to 1.5 for the release of the upgraded system, from which it declines to 1.0.

Fig. 11–13 shows the total lifecycle cost of this signal processing system. Of the $125M lifecycle cost, 6 percent is spent upon design and development, 33 percent is spent on production, and 61 percent is consumed by operations and maintenance. A more thorough analysis would trade off increasing the design effort to improve MTBF and reduce life cycle operation and maintenance costs. For example, an increase of 50 percent in MTBF for each machine type reduces the operations and maintenance costs by $27M, or by over three times the original development cost. Increasing MTBF could require increased design effort to reduce parts

Year:	1	2	3	4	5	6	7	8	9	10	11	12	TOT
Design/develop.	2420	2589				2359							7367
Production			2249	3685	6069	7541	4268	6909	5607	4566			40893
Operations/mainten.			980	2070	3827	45620	1964	3400	4336	11167	1738	1714	76817
Grand Total	2420	2589	3229	5755	9896	55519	6232	10309	9943	15733	1738	1714	125077

Figure 11–13 Total lifecycle cost of 410 parallel signal processing systems, showing yearly cost of development and design, production, and operations and maintenance (O&M).

count, or the use of more reliable parts, or the use of ASIC technology to integrate more standard parts into a single circuit, or the use of improved cooling of components during operation to extend their life. These can all be leveraged into major reductions in lifecycle cost. An actual design would perform trade studies to assess each of these options. It is often found that up-front investment that doubles or triples the development cost and that slightly increases the production costs can reduce the total lifecycle cost by 10 percent to 40 percent. An analogy to the purchase of a home is appropriate—an upfront cash payment on the house will lower the total purchase cost of the house, but the more expensive mortgage opens the house purchase to more buyers by lowering the cost of entry. Similarly, higher lifecycle costs may be accepted if the initial cost of system ownership is lowered to match purchasing power, and a system may be designed to minimize acquisition cost rather than minimizing lifecycle ownership costs.

11.3 REQUIREMENTS DESIGN

One challenge of successfully performing hierarchical requirements design is to distinguish the definition of the requirements from the approach used to achieve them. A user- or system-level requirement may specify:

"... (the system) shall provide spectral energy density content information on analysis frames of 300 msec with a resolution of 50 Hz over a bandwidth of 5000 Hz with a latency not to exceed 300 msec from the end of the analysis window..."

In contrast, a design-level specification would provide more information on how to meet the above requirement. It will describe the specifics of a particular design choice to meet the system specification and may allocate the above requirement to several design specifications:

"... shall perform a fast Fourier transform on 1024 points in 100 msec ..."

and

"... signal processor with a floating-point multiply-and-accumulate throughput of 1 million floating point operations per second..."

and

"... buffer up to 10,000 samples sequentially and access them in random manner with access time not to exceed 500 nsec ...".

11.3.1 Requirements Hierarchy

System requirements are arranged in a hierarchy from most broad and comprehensive specifications that cover the entire system, to detailed specifications and design information for each subsystem. The first specification level is that of the *system* or *segment* specification, also known as the A-level specification, or A-Spec, in military parlance.[10] Table 11–3 identifies the levels of specifications defined in U.S. MIL-STD 490A, "Specification Practices." This military standard also pro-

Table 11–3 Levels of specifications defined by MIL-STD-490A.

Level	Title of Specification
A	System/Segment
B1	Prime Item Development
B2	Critical Item Development
B3	Non-Complex Item Development
B4	Facility or Ship Development
B5	Software Development
C1a	Prime Item Product
C1b	Prime Item Product Fabrication
C2a	Critical Item Product
C2b	Critical Item Product Fabrication
C3	Non Complex Item Fabrication
C4	Inventory Item
C5	Software Product
D	Process
E	Material

vides the paragraph-level outline of what each of these specifications should contain, and it is a necessary guide whenever such specifications are being written.

Preceding these levels of specification may be higher level documents. For example, a user manual may be written prior to the development of any specifications to define the initial system-wide user view. A Principles of Operations (POPS) document may describe the underlying technology in more readable terms than accommodated by the rigid format of MIL-STD 490A. As mentioned in Section 11.1, executable requirements may be present at several of these levels.

Software tools to manage requirements and ease consistency and traceability across various levels of the above hierarchy are becoming increasingly prevalent, especially for large, complex processor designs. The use of such tools may begin with the original specification put forth in the user manual, the customer Request for Proposal, or the A-level specification. The requirements are entered into the toolset, and each paragraph or requirement is identified with a number. The word "shall" is reserved in these documents to identify a requirement—design goals or other less rigid intents are communicated by other words ("will," "as a goal," etc.).

With a top-level set of requirements in place within the tool, requirements analysis can be automatically performed. Analysis operations include grouping all requirements pertaining to a particular user feature, building an alternative grouping of requirements based on which configuration item they govern, simulating the effects of groups of requirements, and decomposing an A-level specification into one or more B-level (development) requirements. In addition, concurrent with the development of each requirement is the identification of a test for that requirement to determine whether the system, once built, meets that requirement. These tests are subsequently reordered into a *test plan* that defines a sequence and set of metrics for verifying that each requirement was met. Finally, tracking of requirement changes over time, known as *configuration management*, is necessary as customer needs change during the design interval or as design limits are discovered that impact requirements.

One example of a commercial tool to assist in the management and traceability of requirements is RDD-100 by Ascent Logic Corporation.[15] RDD-100 (for Requirements Driven Design) provides an integrated environment for full requirements analysis, traceability, behavioral modeling and simulation, component analysis, and tradeoff analysis. It includes a requirements manager module that supports electronic extraction of requirements from source documents, text editing, configuration management, functional flow block diagrams, and graphical views of the hierarchy of requirements. A module for system analysis provides behavior modeling to model operational scenarios, and the tool provides graphical representations of performance and report generation. A "design-to-cost" module incorporates industry-standard models of cost vs. complexity for hardware and software components. It provides the ability to design a system to a particular cost constraint, using parametric cost estimating models.

A tool for signal processing algorithm development known as Signal Processing Worksystem (SPW)[16] can import and export its files from and to RDD-100. SPW provides a block-oriented design, simulation, and implementation environment of signal processing system design, and it may be used for algorithm development, filter design, C-language code generation, hardware/software codesign, and hardware synthesis. The SPW design process begins with the graphical input of signal processing algorithms, using function blocks from any signal processing library. This input allows the specification, capture, and simulation of the end-to-end signal processing system at various levels of abstraction. SPW provides the ability to hierarchically combine lower-level primitive blocks into higher-level functional blocks and to define new blocks as needed. Function blocks are graphically wired together and simulated to realize an algorithm. Test probes may be placed at any point along the way to capture intermediate signals to aid debugging. SPW provides a source of various types of driving signals as inputs.

An earlier graphical signal processing language, developed by the U.S. Navy Research Laboratory, is PGM (Processing Graph Method).[17] PGM was developed in 1980 as a tool for programming the U.S. Navy AN/UYS-1 Advanced Signal Processor (ASP), and later the multiprocessor AN/UYS-2 Enhanced Modular Signal Processor (EMSP). It has since received use in the commercial sector, mainly disseminated by the Rapid Prototyping of Application Specific Signal Processors (RASSP) program, mentioned earlier. PGM tools allow the design of an application to be performed at a high level, expressed in graph notation, and then translated into code strings that are executable within the multiprocessor signal processor system. In the signal processing graph, the nodes represent processing and the connecting lines represent queues that hold intermediate data between nodes. Associated with each queue is a control block structure containing information such as size, current number of data input elements present, and the threshold number of data elements needed as input to the next node to begin processing.

PGM facilitates the programming of multiprocessor machines, providing an architecture-independent method of programming data flow multiple instruction, multiple data (MIMD) machines. More recently, it has been enhanced to extend from regular signal processing to include control and data-dependent general purpose computing. PGM graphs can be translated into the Ada computer language by a tool known as GrTT (for Graph Translation Tool).[18]

11.3.2 Partitioning and Requirements Allocation

Once the application is specified in some language, the system is partitioned into separate blocks, or subsystems, to meet the higher-level requirements. An initial partitioning is developed using one of several techniques, and then several alternative partitionings are developed and all are assessed according to a set of system-wide criteria such as those shown in Table 11–4. To develop an initial partitioning, areas of functionality can be identified by examining the written specifications. For example, a video playback system that receives compressed input of moving images, decompresses them, and routes them to a video monitor might include the functions of memory, control, and video decoding. These blocks are then subdivided based on elements of the functional specification and by examin-

Table 11–4 Example optimization criteria used to selected best partitioning of signal processing system.

Low power, for battery-operated systems
Throughput
Low latency, for real-time response
Low hardware complexity
Small die size for integrated circuits
Minimal amount of interconnections
High integration
Flexibility
Programmability

ing the areas where one subsystem communicates with another. For example, the memory block may be further decomposed into video decoder memory and system memory. Finally, remaining modules not yet allocated to a higher-level system, as well as elements that are needed but not yet provided, are allocated additional blocks in the diagram. In the video player example, a file to hold one frame of information at the input of the decoder is needed as an additional module.

Another method of partitioning to a lower level is to perform the following steps for each block of the higher-level block diagram:

- List the procedures or functions required of the block.
- Decompose each procedure into a lower list of standard functional elements. For software, these elements are routines in a library; for hardware, they are standard cells in a macrocell library.
- Identify new elements that are needed, but that are not in the collection of standard elements, as requiring design and place these new elements appropriately in the circuit.
- For each higher-level module thus decomposed, list the lower-level elements used and compare these lists across modules, identifying smaller or identical elements that are used across modules.
- Exploit the similarity of like computing elements to define and develop reusable design modules.
- Estimate the complexity of each element number, type of operation, and size (number of macrocells for hardware, number of lines of code for software) and allocate a required throughput to meet real-time computing requirements.
- Identify elements that limit computational throughput and pose bottlenecks to achieving real-time operation.
- Apply parallelism or serial pipelining to bottlenecks to speed them.

An example of this final step was given in Section 5.3.2, in which a slower FFT node in a datapath was divided into parallel and serial execution.

As a system is recursively decomposed into ever-smaller computing elements, requirements are allocated from higher levels to lower levels. In this process, a particular higher-level requirement is allocated across several lower-level elements such that if these elements meet the lower-level requirements, then the upper-level requirement is met. Fig. 11–14 provides an example. The system level requirements of weight, range, and mean time between maintenance actions (MTBM) are indicated at the top. Three units support meeting these requirements, and the weight of each of these units is allocated such that their necessary complexity can be accommodated without exceeding total weight. Unlike weight, which combines by addition, MTBM combines by reciprocal: $1/(MTBM)_{system} = 1/(MTBM)_A + 1/(MTBM)_B + 1/(MTBM)_C$. Appropriate allocation of MTBM is

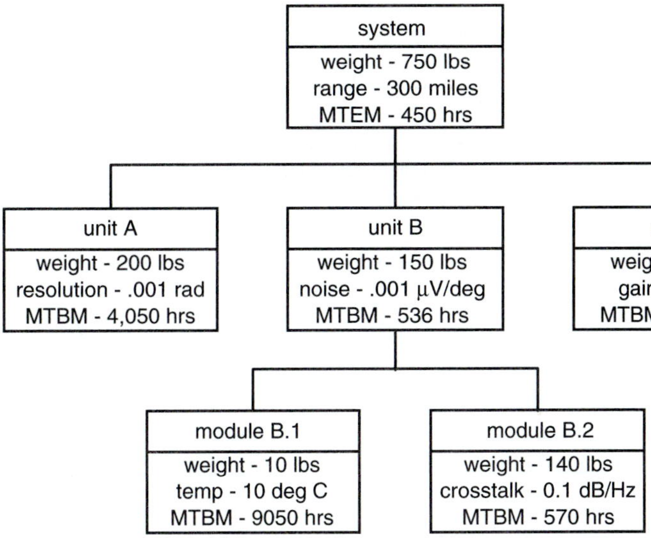

Figure 11–14 Allocation of requirements from system to unit to module is performed in a manner that assures that if lower level requirements are met at the module level, then total system requirements are met.

given to the three units shown. The system level requirement of range is treated as a function of the three variables of resolution, noise, and gain, and one unit is responsible for meeting each of these variables. In like manner, Unit B is shown decomposed into two modules, each of which is allocated low-level requirements that are consistent with its parents.

The effect of changing a requirement for one lower-level module, such as its MTBM, can propagate across the entire system, for example affecting system MTBM. Therefore, requirements traceability and allocation must be maintained, usually with a software tool. In addition, performance models are necessary to combine lower level requirements, such as resolution, noise levels, and gain, to determine the corresponding higher-level performance attribute, such as range.

In practice, each element has many more requirements than the two to three listed in this example—using MIL-SPEC 490A as a guide, as many as several hundred types of requirements may be placed on the lowest level element. Other general requirements include cost, power consumption, heat generation, size, skill level of personnel needed to maintain the unit, mean time between failure, and mean time to repair. Function-specific requirements include input/output interface signal definition, unit bandwidth, storage capacity, compute speed, and repertoire of operations (as well as the timing for each). Actual system designs start the requirements at a much higher level than shown in the example. For instance, a performance-level requirement for the mine detection algorithm de-

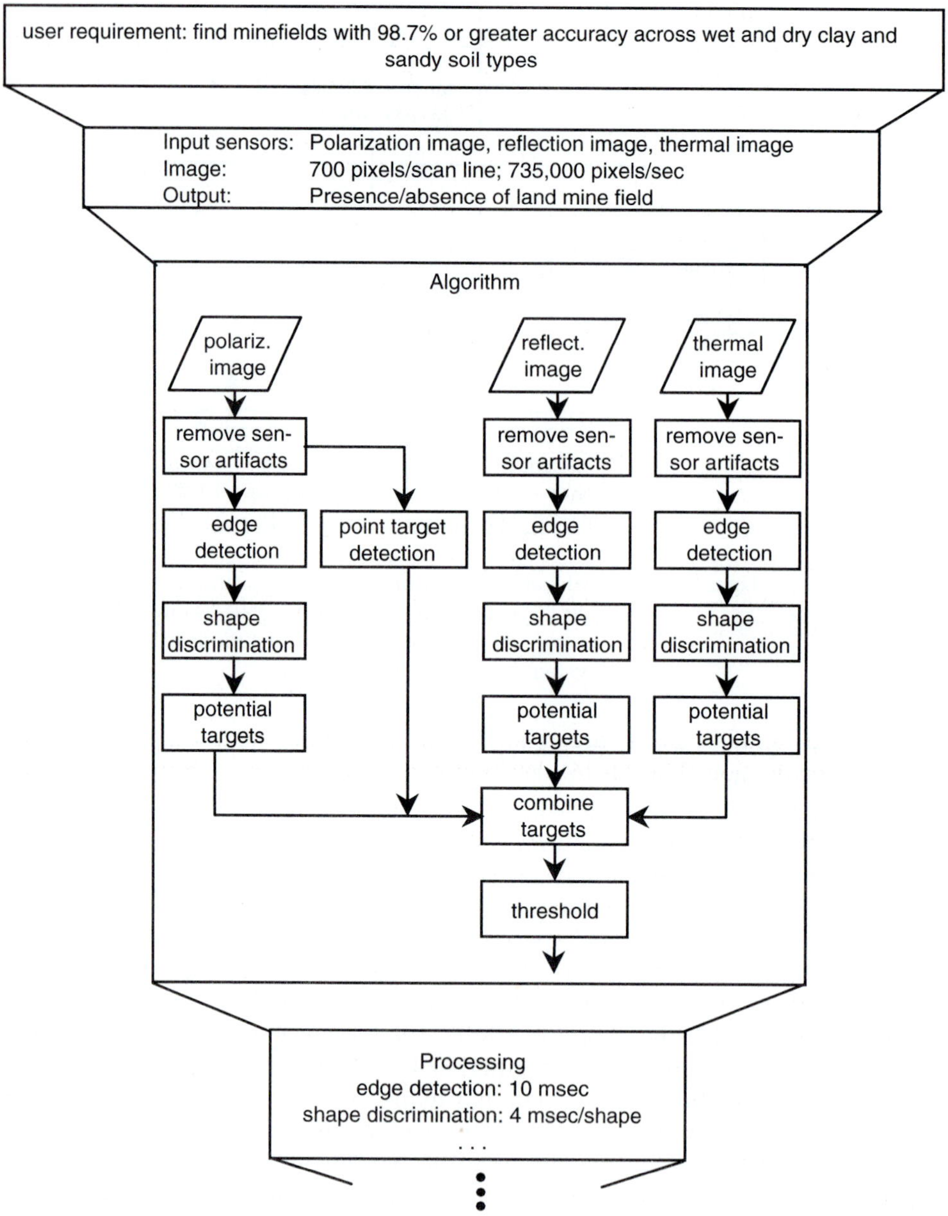

Figure 11–15 Allocation of requirements: translation from user space to design arena.

scribed in Chapter 10 might be to detect mines of a certain type under specific test conditions (soil type, mine depth, mine patterns) with a detection probability of 98.6 percent and a false alarm rate of no more than one per ten square meters. At the lowest end of the requirement hierarchy, an individual integrated circuit is given a data-sheet-like of requirements that specify voltage, current, power supply ripple, and ground bounce, as well as inputs for each pin, definition of expected response and its timing for each input condition, reliability, and many others. Fig. 11–15 illustrates the range of the hierarchy that encompasses the requirements that must be maintained.

Requirements must be designed to be consistent, verified, traceable, and testable across all levels of an expansive hierarchy, ranging from user features at the top to detailed integrated circuit specifications and layout, at the bottom.

11.4 HARDWARE DESIGN

Assessing several trial partitioning and tradeoffs between hardware and software implementations results in identifying the modules (boards), any special-purpose integrated circuits, backplanes, and power systems. An overall system architecture has also been developed and given as input to this stage.

11.4.1 Process

The detailed hardware design process begins with the B-level (Prime Item) specification (Fig. 11–16). Written specifications for each module are next developed that reflect its allocated requirements and the module compatibility with an architecture that will meet these requirements. Peer inspections are conducted for each module in accordance with the methods outlined in Section 11.1.5 and as described in further detail below.

For each module, any integrated circuits that must be developed are identified. A design specification is developed and inspected for each custom integrated circuit. Logic design of integrated circuits and modules then proceeds. In tandem with the design, simulations at increasing levels of detail are used to model and modify the design. At the highest and most all-encompassing level, behavioral simulation models the input/response functions of each block within the module (and across modules for the entire system). This simulation may be at the event level that tracks the inputs that stimulate each block and the associated response. This highest level of simulation models the ordering, but not the timing, of events. At the next level, clock-cycle-accurate register-level simulation tracks each event at the timing resolution of the clock cycle. Register-level simulation preserves the fidelity of events to the nearest clock cycle, but its scope is narrowed from system-wide modeling to module- or chip-wide modeling. Finally, detailed timing simulation models the shape of each switching waveform at the highest frequency to occur. This simulation, usually restricted to critical parts of

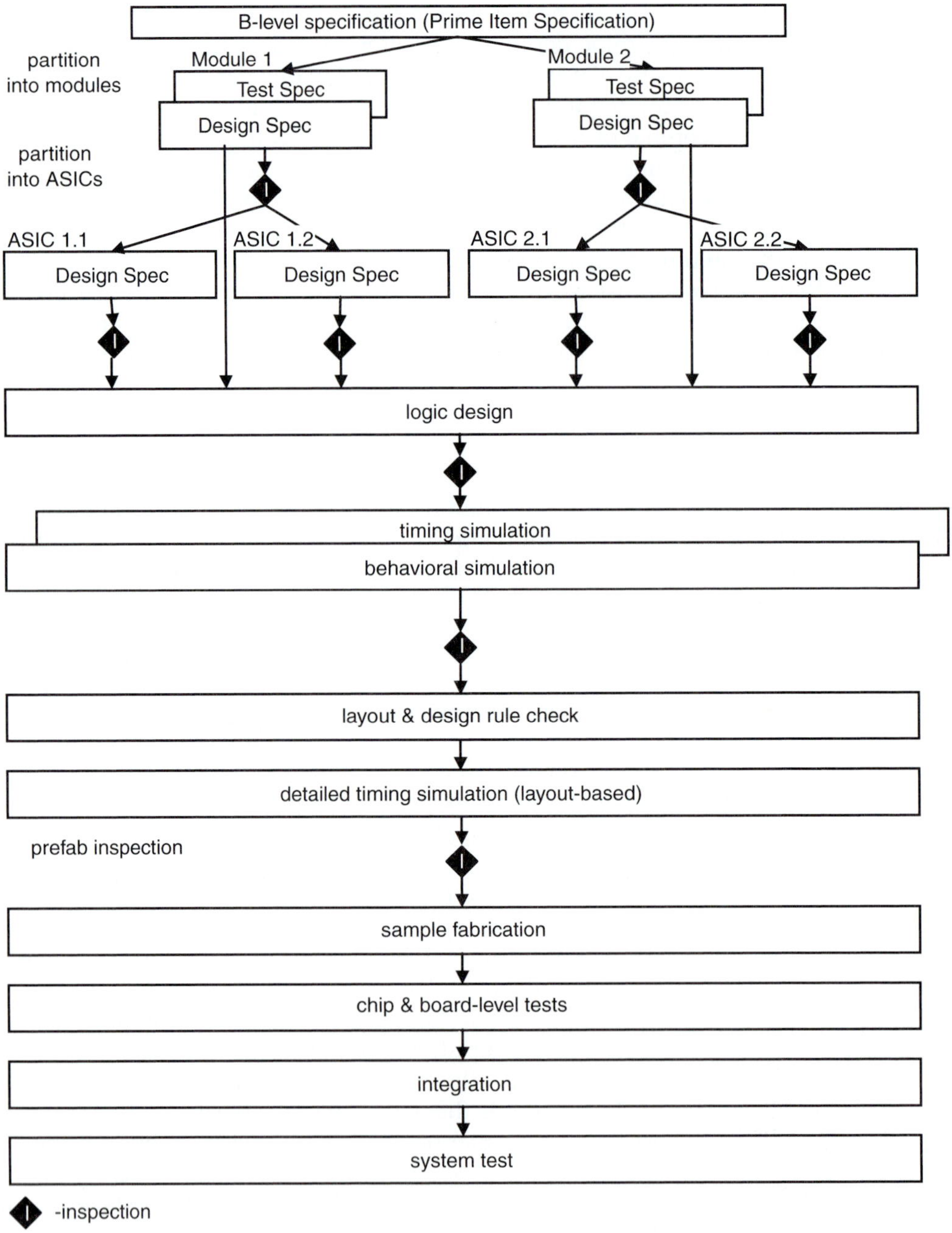

Figure 11–16 Hardware design process begins with B-level specification and proceeds by partitioning the design, completing and inspecting several design steps, and reintegrating the results.

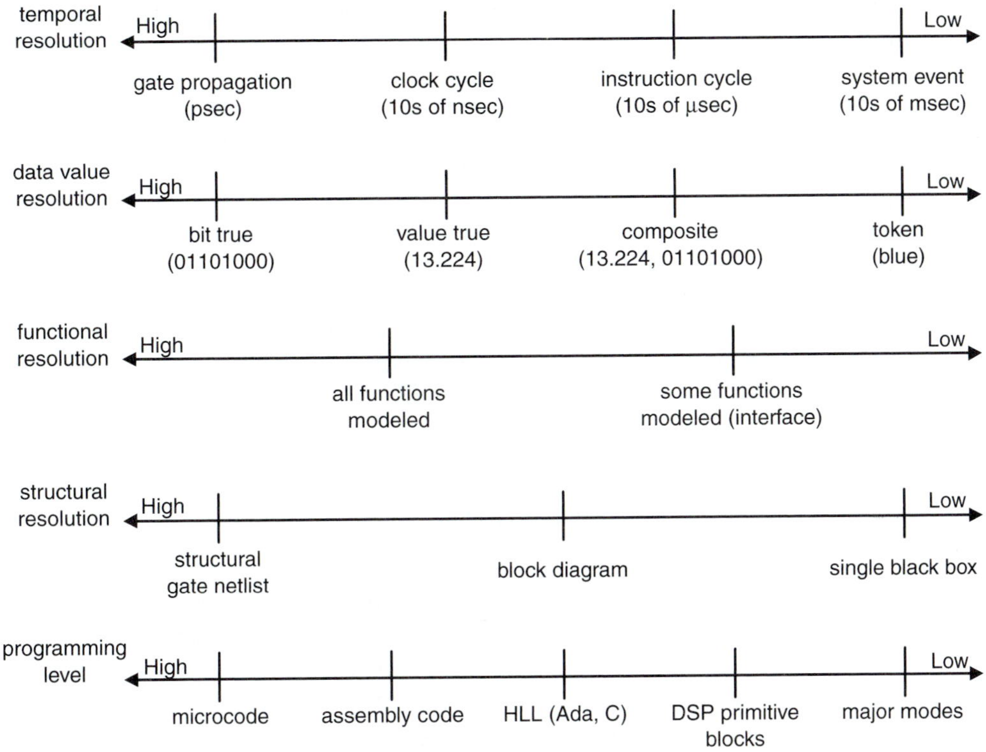

Figure 11–17 Scheme for classifying the fidelity of simulation models.[6] © 1996 Kluwer Academic.

the circuit, models switching noise, spikes, crosstalk, and other events that perturb a waveform between clock cycles. Fig. 11–17[6] shows a scheme for classifying the fidelity of various modeling and simulations.

After correct logic design is achieved and verified for the integrated circuit, logic elements are geometrically arranged and interconnected on the substrate. Based on placement, line length, and proximity, the layout-dependent signal propagation times are simulated. Automatic design rule checks assure that geometric constraints are not violated and that proper allowance is made for variability of feature size during the manufacturing process. The results of the timing simulations are inspected and the design is released for the manufacture of engineering samples. The samples are tested with the test sets developed concurrently with the design, and system integration and testing then begins.

The progression of hardware design, specifying first at the highest levels and partitioning the system into lower-level subsystems with their own specifications, and then implementing from the bottom of the hierarchy upward, suggests that simulations should be linked across levels of the hierarchy. Individual integrated circuit simulations should be combined into multichip simulations, which are then combined into module and system simulations. Linkage of simulations is a characteristic of some advanced toolsets, but manual reformatting and linking of simulation results across different levels is often necessary.

The ability to postpone hardware fabrication until certainty of correct operation, based on simulation, is sought. Chip-level simulations are extremely accurate—one may be confident that the performance that is simulated is what will be obtained. However, errors in chip design can go uncovered because the tests in the simulation domain did not exercise all conditions. The definition of fault coverage, or the fraction of fault conditions that can be tested from a set of tests, determines the chance that certain operating modes remain untested until use.

Combinations of chips operating together are harder to simulate, yet full module simulation requires multichip analysis. Similarly, entire system simulation, consisting of multichip modules, is particularly difficult and is the goal of virtual prototyping activities.

11.4.2 Mixed Analog and Digital Design

One area of particular interest in digital signal processing design is the mixture of analog and digital design techniques as required to build low-noise analog-to-digital (A/D) and digital-to-analog (D/A) conversion circuits. The digital circuitry is switching rapidly as processing proceeds, emitting noise spikes that can be picked up by nearby analog circuits, yet the world around the circuit is often analog, and so conversion from and to the analog domain in close proximity to digital processing circuitry is needed. Several techniques for combined analog/digital design can reduce coupling of digital noise into analog circuits.:

- *Avoid ground loops*: Ground loops are formed when the ground of the analog circuit is not fastened directly to the ground of the analog circuit power supply and the ground of the digital power supply is not directly fastened to digital circuit ground. A loop can be formed when digital ground goes to a wire, and analog ground is connected to another point of the same wire. Ground loops can introduce offset errors and noise into the analog section.
- *Do not mix analog signals with digital signals*: Physical separation of analog and digital sections of the module is necessary to minimize the coupling of digital signals into analog lines.
- *Provide for adequate component cooling*: A thermal gradient between an integrated circuit package and its board can induce errors in an analog-to-digital converter, and the fact that the power dissipation of the complementary metal oxide semiconductor (CMOS) circuits that are typically used depends on switching activity and can affect analog-to-digital converter reference voltages and readings if analog components are heated by digital components.
- *Use the full range of the A/D converter*: The largest signals should activate the most significant bit.

- *Tie down (or up) all digital inputs that are not being used*: This avoids stray noise corrupting A/D operation.
- *Filter digital power in distributed manner*: Filter at the integrated circuit package by providing both a small, fast capacitor for each chip and a larger, slower filtering capacitor for each group of chips.
- *Use ground planes extensively*: Layout that devotes an entire layer or layer per side of a printed circuit board to ground can reduce noise at the surface level, where the chips are located.

11.4.3 Design for Manufacture

In addition to performance, hardware must be design for producibility in manufacture. Indeed, without attention to quantity manufacture during the prototype-building phase, costly delays are necessary as manufacturing is attempted for a design. One of the reasons for the formation of concurrent engineering approaches was to be certain that manufacturing knowledge, normally postponed to a separate phase of production engineering, was included in the upfront system design. For example, minimizing the variety of parts in a design increases its manufacturability by decreasing inventory complexity and allows more economical purchase of parts. This is because quantity purchase is possible for more units of a fewer number of part types. If widely used parts are selected that are manufactured by a variety of sources, system costs are likely to decrease more quickly as competitive pressures lower their cost as their design matures. Furthermore, if the system being designed is only one of many users of a part, greater market stability for the part results than if that part is constructed solely for one system. The physical placement of parts in design layout, such as orienting all integrated circuit packages the same way, can reduce manufacturing costs by allowing all parts to be placed in one run of an automatic pick-and-place machine, rather than requiring repositioning of the board in the middle of the run.

The U.S. military acquisition process provides for particular attention to manufacturability early in the design. A template of guidelines[19] provides a hierarchy of all factors that are addressed during the product acquisition cycle, including design, test, production, facilities, logistics, and funding (Fig. 11–18). For several subareas within each of these factors, a set of common practices are identified and defined as traps that can prevent effective manufacture. The best known practices are then identified. For example (Table 11–5), one trap identified with regard to trade studies in the design phase is that new technology can solve most problems. An alert to such misconception in a design process would be that new technology is used without conducting trade studies, and a consequence of this error is that technology that has not been previously tested in production can cause production schedule delays and cost problems. Instead, trade studies should be required on the benefits and risks of new technology, resulting in the insertion of new technology in the design only where new technology helps.

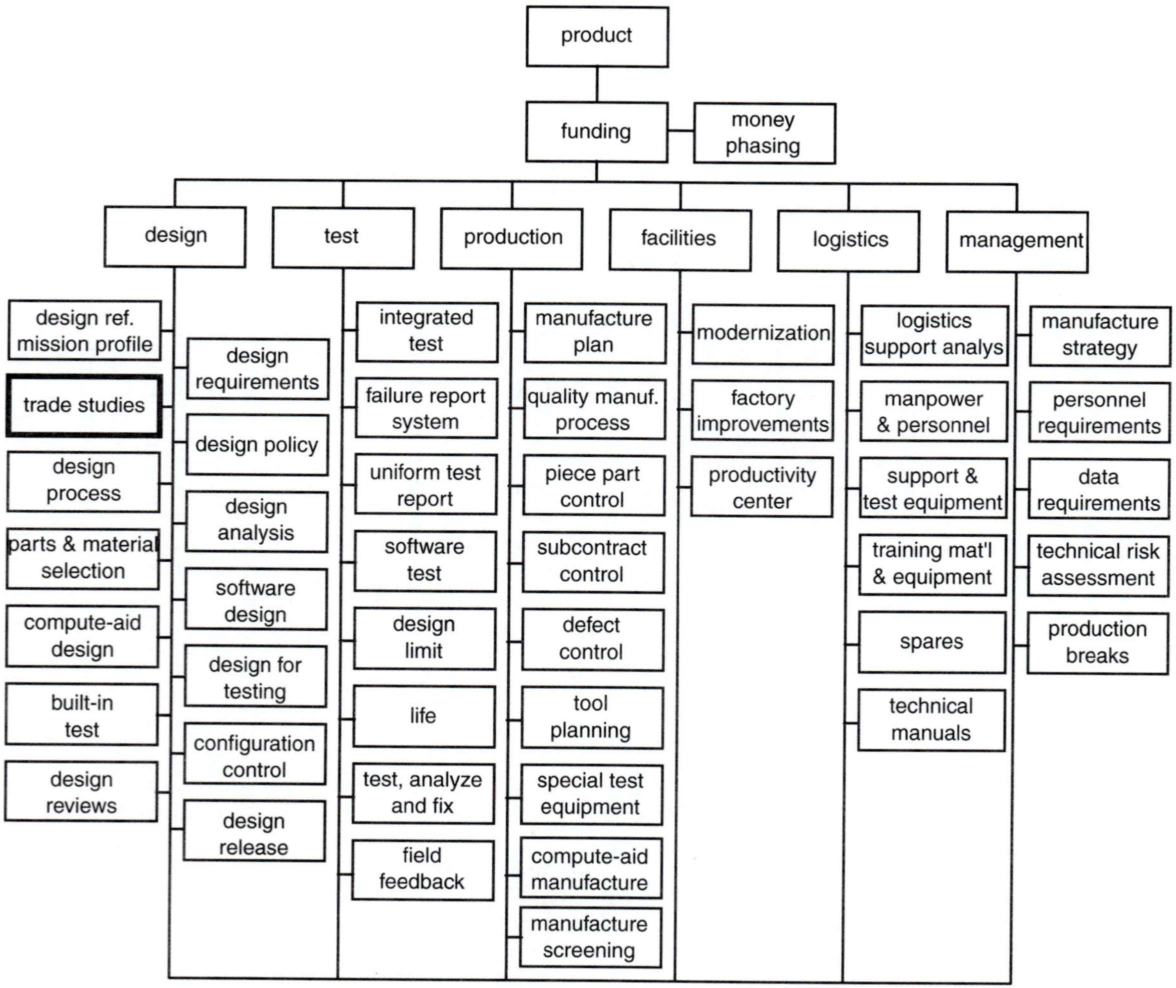

Figure 11–18 Engineering for manufacture considerations expand into several traps for the unwary, alarms to their presence, and suggested process steps to avoid these difficulties;[19] one area, Trade Studies, is expanded in next table.

Table 11–5 **Expansion of Trade Studies box of Figure 11–18 into traps, alarms, consequences, escapes, and benefits.**[19]

		Trade Studies		
Trap	**Alarms**	**Consequences**	**Escapes**	**Benefits**
Conducted as a single event on performance requirements	• Lack of balance exists between effectiveness issues and suitability	• Design surprises will surface during tests	• Verify that trade studies are part of the corporate design policy and process • Ensure that design tradeoff studies continue throughout full-scale development	Best design approach is identified
New technology is the answer	• New technology is used without trade studies being conducted	• Concepts untested in the production environment may cause severe cost and schedule problems	• Use detailed trade studies to identify relative risks of all options associated with new technology	New technology used only when beneficial
Timing and depth of studies are flexible	• Trade studies are not completed prior to Critical Design Review (CDR) • Neither government nor contractor personnel fully understand the process	• Wrong alternative selection could compromise mission effectiveness	• Ensure that trade study procedures establish a specific schedule, identify individuals responsible, and define proper level of reporting	Reduction of repetitive design efforts
Producibility will be considered at start of production contract	• Trade studies during design do not consider alternative manufacturing processes	• Costly redesign for producibility	• Conduct a trade study for each design concept to assess its producibility	Final design concept selected can be efficiently produced

The inspection process is used throughout the hardware design as shown in Fig. 11–16. A hardware design will have a written set of inspection procedures, criteria, and checklists for each type of inspection. Entrance criteria that must be met before a hardware inspection begins include:

- written set of requirements;
- written design specification;
- a list of required simulations and the results of those simulations;
- a schematic diagram;
- identification of critical timing paths;
- list of preferred parts;
- all material assembled into a inspection package that is read and analyzed by all inspectors prior to the inspection.

During the inspection, several activities occur:

- the designated reader presents a page-by-page inspection of the design;
- areas of the circuit are highlighted on the schematic as the inspection proceeds;
- the design checklist is applied to each piece of the design;
- the designated recorder of the inspection team document the result of each checklist question for each stage of the design and compiles defect-based metrics. In addition, the recorder keeps track of time spent in preparation and execution of the inspection;
- plans for correction of defects and the identification of a followup inspection are made.

A typical inspection for a 15-page integrated circuit design requires 18–40 hours of preparation by each inspector and three hours of inspection time. Although this effort is significant, it is far less than the effort to find design defects during the later stages of unit or system test (or worse, after the product is fielded).

11.4.4 Other Hardware Considerations

The previous discussion has focused on circuit design, but the hardware of packaging and the associated physical, mechanical, and thermal design is also a part of system design. System packaging can include a cabinet, backplane, power supply, cooling, and user controls and indicators. A major emphasis in electronic packaging is the ability to cool the integrated circuit during operation. A thermal impedance model that analyzes the thermal transfer from the integrated circuit junction to its package, then to the board, off the board, through the cabinet, and to the ambient surrounding, is conducted. The lifetime of electronic components, and the associated major lifecycle cost of operation and maintenance, depends upon keeping component temperatures as close as possible to ambient. Considerations of using standard parts and reducing the diversity of parts apply in physical design as well as circuit design.

Physical design of a system also is concerned with meeting sometimes demanding requirements on vibration, mechanical shocks, and thermal shock. Means of mounting integrated circuits onto boards and boards into backplane are actively designed rather than passively accepted. One of the design tests for acceptance of the AT&T AN/UYS-2 Navy standard signal processor was the striking of an operating unit with a 2,000-lb hammer several times along each of the three axes while requiring operation during and after the ordeal. The constant vibration of a helicopter can cause components and connections to work loose if not designed against this effect. Signal processors for automotive and military use must operate over a temperature range that extends from between 0° C to above 100° C internal temperature—electronic components must accommodate this temperature range and adjacent dissimilar materials that are fastened together must allow for unequal rates of thermal expansion and relieve the resulting stress.

11.5 SOFTWARE DESIGN

Like hardware design, software design begins with the B-level specification. An analysis of requirements at the system level allocates the meeting of functional requirement to hardware or software, and further analysis decomposes the requirements allocated to software into code modules.

11.5.1 Process

Table 11–6 displays the stages of a typical software development that is part of a larger hardware/software system.[20]

A variety of software processes exists to accomplish the above design stages. The information for Table 11–6 is based on the military standard for software development, following a waterfall process as described in Section 11.1.1. A more recent quality standard for software development from the International Standards Organization (ISO), known as ISO-9000, is more concerned with the presence of and adherence to any reasonable process rather than assessing the specific intricacies of a particular process. The ISO-9000 method requires that the proposed development process be documented in advance. The proposed process is subjected to independent review and must be approved for use. The remainder of the ISO-9000 method is to assure dissemination of the process to all parts of the project and compliance with it. An ISO-9000 audit may perform unannounced spot checks at any time and any place in the development team to measure the degree of com-

Table 11–6 Software design process of MIL-STD 2167A specification for software development.

Stage	Steps	Results
System requirements analysis/ design	• Analyze requirements; allocate requirements to SW • Partition SW into modules (computer software configuration items, or CSCI)	• System Requirements Review (SRR) • System/segment design document • Preliminary Software Requirements Specification (SRS) • Preliminary Interface Requirements Specification (IRS) • Software Development Plan
Software requirements analysis	• Develop requirements for each CSCI • Develop qualification (testing) requirements for each CSCI • Evaluate preliminary SRS and IRS • Initiate configuration management and enter SRS for each CSCI	• Software Specification Review
Preliminary design	• Analyze design alternatives • Update software plan • Develop preliminary design for each CSCI and generate associated software design document (SDD) • Develop preliminary design for each interface and produce Interface Design Document (IDD) • Plan integration, testing, and support programs • Establish quality assurance procedures	• Preliminary Design Review (PDR) • SDD for each CSCI • IDD for each CSCI
Detailed Design	• Develop detailed design and final requirements allocation for each CSCI • Select signal processing algorithms • Develop detailed design and requirements allocation for each interface • Identify test cases for each CSCI and compile for test plan	• Critical Design Review (CDR) • Updated SDDs • Test cases and expected results
Code and unit testing	• Develop source code • Test each separable unit of code	• Source code listings • Test procedures and results
Code integration and testing	• Begin at computer software unit and move up to CSCI level • Conduct each test of each level • Document results • Correct and retest as necessary	• Test Readiness Review
System integration and testing	• Separate development organization (not software team) executes test plan on entire system • User test team conducts tests against functional performance requirements	• Product baseline

pliance to the process, thereby assessing the degree to which the documented process is disseminated in the conduct of the actual design.

One estimate of the relative time intervals of successful software design[21] allocates one-third of the design interval to planning before coding, one-fourth of the design interval to after-coding component- and early system-testing, and one-fourth of the interval to total system test. Software coding itself only requires one-sixth of the design time, if preceded by proper planning and design work.

For systems in which achieving real time operation is a binary success factor (pass/fail), the design and unit testing phases include the running of estimating of execution times in the design phase, running of code modules, profiling their execution time to identify timing bottlenecks, and iterating.

The concept of the virtual prototype, with its associated executable specification, when combined with graphical-oriented signal processing programming languages such as PGM, blends the boundary between requirements design and software design. If a signal processing algorithm has been specified in a signal processing language, and if automatic and efficient conversion of that specification can be made by a software tool, many of the issues of CSCI identification, module interfaces, and timing are addressed in the requirements design phases.

11.5.2 Cost Model

The cost of software consists of development costs and maintenance costs. A commonly-used cost model for software development is the Constructive Cost Model, or COCOMO.[21] The COCOMO model estimates software development effort and time for a development process that includes design, critical design, code and unit test, and integration and test. The basic COCOMO equation is:

$$E = c * a * L^b. \tag{11.1}$$

Here, E is the labor estimate in staff months, L is the number of lines of code produced in thousands, exclusive of comment lines, and a b, and c are parameters that adjust the model to fit the complexity of the code, the demand on computing resources, and the software development environment. The factors a and b reflect whether the code is developed for internal use only, for semi-detached processing (most signal processing applications), or for embedded computing, which is operating with other code in a complex hardware system. The exponent b reflects the inefficiency of scale—larger programs require more effort per line of code to develop than smaller ones, reinforcing the sound software engineering practice of partitioning code into smaller modules. The factor c is decomposed into a product of factors that range from 0.7 to 1.66 as shown in Table 11–7. The factor c can vary from 0.08 to 20 by choice of these factors: for signal processing code, it is generally sufficient to concentrate on c_{time} and c_{stor} and set the remainder to 1. The factor c_{time} is based on how close the program approaches using every clock cycle available in real time. Nominal level, corresponding to $c_{time} = 1$, is set at 50 percent of available clock cycles, with high, very high, and extra high resulting with

Table 11–7 COCOMO cost factors included in multiplier c.[22,a]

Factor Name	Description	Very low	Low	Nominal	High	Very high	Extremely high
c_{rely}	required software reliability	0.75	0.88	1.00	1.15	1.40	
c_{date}	database size		0.94	1.00	1.08	1.16	
c_{time}	execution time constraint			1.00	1.11	1.30	1.66
c_{stor}	storage constraint			1.00	1.06	1.21	1.56
c_{virt}	virtual machine volatility		0.87	1.00	1.15	1.30	
c_{turn}	computer turnaround time		0.87	1.00	1.07	1.15	
c_{acap}	analyst capability	1.46	1.19	1.00	0.86	0.71	
c_{aexp}	application experience	1.29	1.13	1.00	0.91	0.82	
c_{pcap}	programmer capability	1.42	1.17	1.00	0.86	0.70	
c_{vexp}	virtual machine experience	1.21	1.10	1.00	0.90		
c_{pexp}	programming language experience	1.14	1.07	1.00	0.95		
c_{modp}	use of modern programming practices	1.24	1.10	1.00	0.91	0.82	
c_{tool}	use of software tools	1.24	1.10	1.00	0.91	0.80	
c_{sced}	required development schedule	1.23	1.08	1.00	1.04	1.10	
c_{rvol}	requirements volatility		0.91	1.00	1.19	1.38	1.62
c_{ruse}	requirements reusability			1.00	1.10	1.30	1.62
c_{secu}	security requirements			1.00	1.10		
c_{risk}	risk			1.00	1.10	1.20	

[a]*Software Engineering Economics* by B. W. Boehm © 1981. Adapted by permission of Prentice-Hall, Inc., Upper Saddle River, NJ.

c_{time} = 1.11, 1.30, and 1.66 respectively assigned to use of 70 percent, 85 percent, and 95 percent of real time. For c_{stor}, similar utilization fraction of available memory is reflected, with 50 percent use of available memory corresponding to nominal (c_{stor} = 1.0), 70 percent high (1.06), 85 percent very high (1.21), and 95 percent extra high (1.56).

Having a model of software productivity based on the number of lines of code is certainly useful, but how can software development costs be predicted in advance, before the code is written and the number of lines of code is known? Historical factors, represented by code modules that are similar to the ones being designed, can provide one estimate. Certain models, such as Albrecht's function points, may be used to model code module complexity based on numbers of input data, outputs, and types of interactive use. Most often, cost estimates of code module development are made based directly on an experienced assessment of the module function, rather than attempting to estimate lines of code and use code size in a software productivity estimate.

An estimate of software development time in months, T, can be gained from the COCOMO model estimate of the coding effort E:

$$T = 6.3E^{0.32}. \tag{11.2}$$

Software maintenance costs can be modeled by a baseline factor of effort per thousand lines of mature, stabilized code, and a time-dependent multiplier indicating the approach to a stable state of the software:

$$M = d * s(t) * L. \tag{11.3}$$

Here, L is the number of thousands of lines of code, d is a productivity factor in staff months per line of code, and $s(t)$ is a function that asymptotically approaches 1 with time. A typical set of values, used by the U.S. Navy,[23] is $d = .05$ and $s(t) = (1.5, 1.33, 1.33, 1.16, 1.16, 1.00, 1.00, \ldots)$ as the time since release of the code, in years, is given by $t = (1, 2, 3, 4, 5, 6, 7, \ldots)$.

11.5.3 Signal Processing versus Traditional Language

The effect of language type is partially normalized out of the COCOMO model, as the language type determines the power of a line of code, and the COCOMO model is driven by line count rather than by application. For a graphically entered signal processing flow graph, the cost of software development is less than 40 percent of the cost of the same application programmed in a traditional high-order language.[24] Furthermore, the 10-year lifecycle maintenance cost of graphical language software is only 17 to 25 percent of the lifecycle maintenance of software written in a traditional high-order language.

To understand these differences, the two approaches, one using a graph notation (GN) and the other using another high order language (OHOL), are compared. The four phases of traditional software development of preliminary design, detailed design, code and unit test, and developmental test and evaluation occupy about 29 percent, 27 percent, 22 percent, and 22 percent of the design cycle following the completion of software specifications according to the U.S. Navy NAVSEA 06F computer software model.

Preliminary design is the same for either the OHOL or GN approach, so no GN savings is gained at the preliminary design phase (recall that this phase is independent of implementation language, so no difference should be expected).

Based on the analysis of two large signal processing application programs, each written with a version of GN and one in OHOL, the GN code had about one-tenth the number of lines of code of the OHOL. Thus one signal processing graphical node results in 4.5 lines of GN, corresponding to 45 lines of OHOL. For detailed design in GN, only one-tenth the number of code lines are being designed as for OHOL, so the cost of detailed design would be 27 percent of a software development that consists of one-tenth the number of lines of code as

OHOL. In the next design phase of code and unit test, coding for GN is performed by graphically selecting and interconnecting nodes. In productivity studies of GN program for the PGM graphical programming language, an experienced person with a signal processing background can develop PGM code at the rate of 30 nodes/day. Such a person is about 10 percent more expensive than a typical OHOL software engineer, due to the more sophisticated skill needed. A production rate of 30 nodes/day corresponds to 30 × 45 = 1,350 OHOL lines per day, while typical OHOL rates are 100 lines/month to perform all four phases of development, or, adjusting for the fact that the actual coding, performed in Phase 3, occupies 22% of the total effort across all four phases, the OHOL productivity is 100/22 = 454 lines/staff-month, or 24 lines/staff-day. Thus the code and unit test progresses 1350/24 = 56 time more quickly and requires only 22%/56 = 0.4% of the OHOL design interval.

Developmental test and evaluation, which occupies 22 percent of the design interval of OHOL and depends on the number of lines of code, is cut by a factor of 10 because of the ratio of GN to OHOL. As summarized in Table 11–8, this cal-

Table 11–8 Summary comparison of signal processing graph notation (GN) vs. other high-order language (OHOL) development cost.

Phase	Description	OHOL % of OHOL total (based on Navy cost model)	Comment	GN % of OHOL total
1	Preliminary design	29%	Language-independent, no change	29%
2	Detailed design	27%	Depends on number of code statements; OHOL:GN ratio is 10:1	2.7%
3	Code and unit test	22%	Graphical entry provides 4.5x advantage; GN provides another 10x advantage	0.5%
4	Development test and evaluation	22%	Depends on number of code statements	2.2%
Total %		100%		34.4%
Cost factor adjustment for higher skill level for GN		× 1.0		× 1.1
TOTAL COST		100%		37.8%

culation indicates that the resources for a signal processing graph notation development are about 38 percent of these required for OHOL-based development, after adjusting for the more expensive skills needed for GN development.

This example shows a 2.5-fold advantage to developing signal processing applications in high-order graphical signal processing languages. This cost advantage increases when cost models for software maintenance are used to compare GN versus OHOL lifecycle costs.

11.5.4 Software Inspection

As with hardware design, the formal peer inspection process is an essential element of software design. An inspection is held to evaluate the results of the stages of architecture definition, requirements definition, and detailed design. The effort to correct a defect found in the inspection stage is less than one-twentieth of the effort to correct the defect in system and integration test, and 100 times less than correcting the defect once the system is in the hands of the user (Fig. 11–19).

Like other inspections, the software inspection has formal entrance and exit criteria, requires advance preparation by a team consisting of a reader, a coordi-

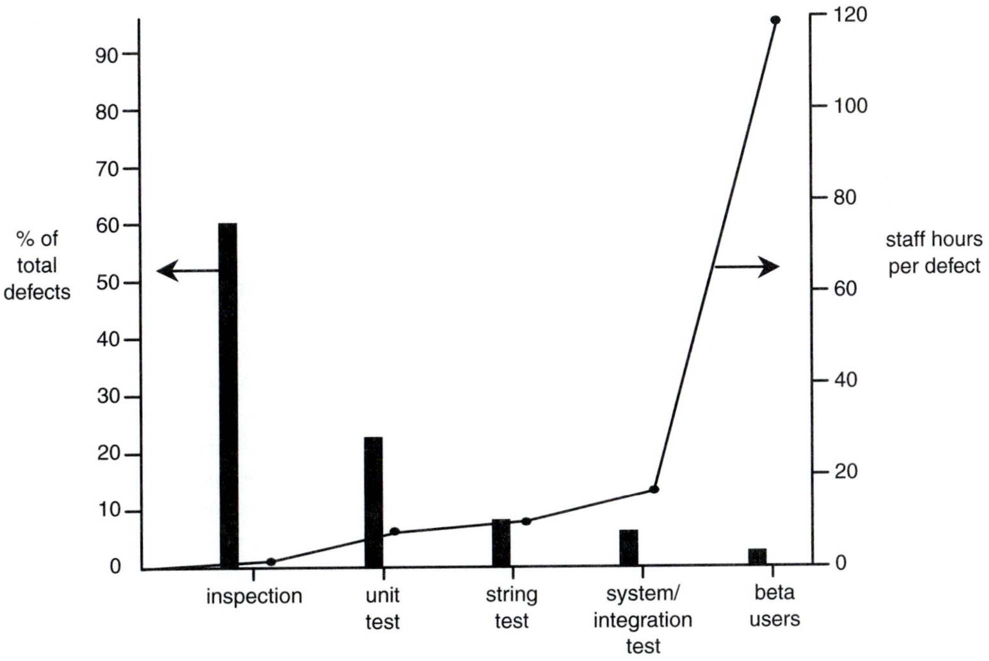

Figure 11–19 Inspections identified most of the errors in a particular software design, and these errors required much less effort to correct.[25]

nator, a recorder, the designer, and several subject-matter experts, and records each defect found, making provision for its correction and reinspection as necessary. A related technique, the code walkthrough, has been traditionally used to obtain high-reliability software designs.

The reuse of proven software from one project to the next can save significantly on future development, but it requires further discipline during the original design process to support this reuse. Similarly, performing *distributed software development*, by which loosely connected team members at different locations and in differing computer environments, is becoming increasingly important with the growth of collaborative research efforts, in which algorithm and application software at one facility is used, improved, and re-released by another facility. Also, orchestrating international software development efforts and recombining the results at an integration point is becoming more common with the growth in world-wide connectivity and numerous international standardization efforts.

The methods to encourage software reuse and portability have been documented in several publications.[26] One element of a successful approach is to develop an overall integration framework in advance and specify the identities, inputs, and outputs of each type of module prior to implementation. Parallel efforts on alternative modules implementing the same function within the framework can be combined and compared if each module adheres to the same input, output, and control structure. An overall software architecture document and an *interface control document* (ICD) are developed to support reuse and portability. Software architecture designs can partition functions into sufficiently small models of component algorithms to support recombining them into other applications. Another way of enforcing portability is in the features of the language. The high-order language Ada has been a standard within the U.S. Department of Defense because it enforces good software engineering as a condition of successful compilation. The Ada language has been shown to result in lower maintenance costs, in large part because its structured approach provides fewer cross-module bugs and offers self-explanatory code easily maintained by others. An analysis[27] has shown that Ada meets compiler efficiency requirements and while it may have a slightly higher development cost for small jobs if training costs are included, its lifecycle cost and its development costs for large embedded software systems is lower than for other high order languages.

For a major system, several types of software are developed. *System software* resides in the system, supporting such fundamental functions as start-up, shutdown, task switching, and fault diagnostics. Although system software is independent of the application, it provides application support such as calling and returning from signal processing library functions. *Application software* consists of both signal processing software libraries and application-specific code. Application libraries provide signal processing primitives that are combined in a particular manner by the application-specific code to implement the desired functions. *Test software* provides a way to exercise each piece of the system in isolation and across its full range of operating conditions, covering many more possible fault areas than typical applications. *Support software* runs on a separate computer connected to the signal processing system, providing data input, control, and diagnostic capability. The software development practices discussed in this section are applied to the development of all four software categories.

11.6 SYSTEM INTEGRATION AND TESTING

At the entrance to this phase of development, all software has been unit tested, all hardware components have been tested as individual entities, and the remaining challenge is to combine and test these items as a whole system. System integration is the application of scientific and engineering efforts to a) combine the assorted components that comprise a system into a self-consistent functional whole, b) integrate related technical elements and ensure compatibility of all physical, electrical, and software interfaces in a manner that optimizes the total system design, and c) incorporate reliability, maintainability, training, field support, and other such factors into the total system to meet the lifecycle cost and performance objectives of the user.

11.6.1 Process

Fig. 11–20 shows the progression from individual pieces to a fully integrated system. A hardware integration framework (HW0) is provided by a backplane; a software framework takes the form of a main control routine (SW0). Moving from left to right, individually-tested hardware modules are separately introduced into the system framework (a backplane or cabinet), and testing and debugging is performed with each new addition, using special test software. Initial software is first run on individual hardware configuration items, as shown by SW1 running on HW1 early in the process of the figure. After a reasonable number of functional hardware components have been integrated, test software is replaced by pieces of actual system and application software. The process of adding hardware, testing, and replacing test software with application software continues until the system is fully integrated.

In some cases, a reference standard system may be available to serve as a basis for comparison. The reference system is usually a previous version of the system or a portion of the new system, such as one processor of a multiprocessor architecture. A known-good reference system can greatly speed integration.

An entire repertory of techniques exists for debugging real time systems during system integration. In some cases, software test routines are sufficient. Usually, they are insufficient, because the test system does not provide software visibility at each interface, real time operation is required, and operation cannot be stopped and captured with a simple software breakpoint. Logic analyzers, processor emulators, and real time signal tracing is necessary. The technique of debugging by symptom[28] proceeds by defining the bug that is to be located. A real time trace is then captured

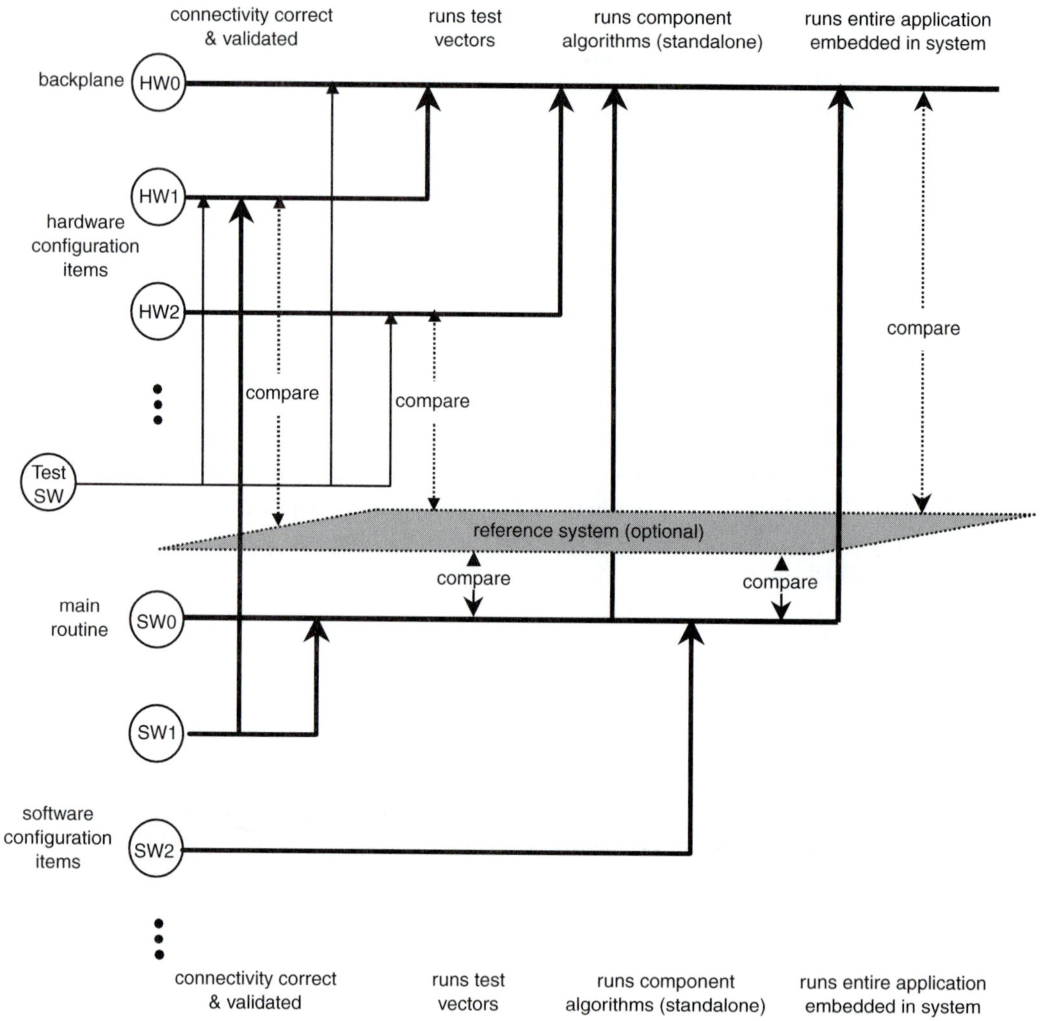

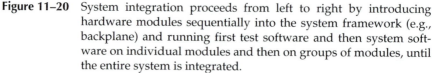

Figure 11–20 System integration proceeds from left to right by introducing hardware modules sequentially into the system framework (e.g., backplane) and running first test software and then system software on individual modules and then on groups of modules, until the entire system is integrated.

to show a listing of the source code that caused the bug. The use of a trigger criterion on a real time logic analyzer provides most of this capability. Other essential operations to using a logic analyzer include start/stop trace collection, cycle counting to prepare for the capture of conditions that lead to the bug, and monitoring the instruction address flow to capture unintended jumps out of the instruction space. Intermittent bugs are often caused by such jumps. In short, scientific analysis and thought on the conditions that lead to a bug, and the translation of those analyses into a set of sophisticated trigger conditions programmed into a real time logic analyzer, are often the fastest way to locate and repair the bug—faster than trial-and-error single stepping or exhaustive search.

System testing is tightly intertwined with system integration but continues after integration is complete. It is conducted after system integration by a group of testers that is independent of the developers. The engineers that specified the system often test the system.

In modern design methods, testing has become a continuous process in parallel with all stages of design, from requirements design through final integration and test. The move toward continuous testing is a direct result of the fact shown in Fig. 11–19—the cost to fix a defect increases as the time of defect detection becomes later. The bottom of Fig. 11–21 shows that the cost to fix a defect increases as the time between defect introduction and its elimination increases—old defects are more expensive to repair than young defects. In the traditional mode of build, then test, shown at the top, the defects found in testing at the end of design were expensive to remove. Incremental testing, shown in the middle, introduces testing at earlier times, and the defects found in these tests are more easily removed. Ultimately, concurrent building and testing removes defects nearly as early as they are introduced. Thus, test has evolved from a one-time event to a continuous activity.

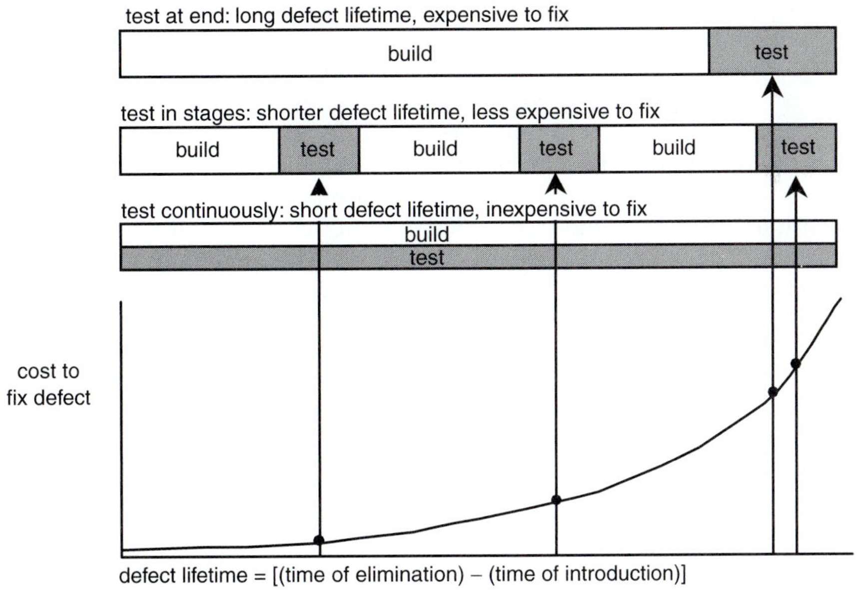

Figure 11–21 In modern design processes, testing is a concurrent activity rather than a one-time event at the end, minimizing the lifetime of defects and reducing the cost to remove them.

11.6.2 Fault Coverage and Test Vectors

Testing proceeds by applying a set of stimulus signals to the inputs of the unit under test, observing the output, and comparing it to the correct response associated with that input. The unit under test is treated as a black box—inputs and outputs are analyzed, but internal operation is not. Indeed, a difference between a measured output and the correct output is known as an *error*—but the incorrect state of the hardware or software, which may or may not be observable as an error, is known as a *fault*.

Attempting to identify all possible faults is necessary to develop tests to detect them. To develop the complete set of faults for requirements, the entire enumerated list of requirements is used. For hardware, it is assumed that the input of each gate is stuck at 0 or 1. For software, each possible path through every code module is used as the fault set. Based on these fault sets, the set of stimulating inputs, known as test vectors, is carefully designed with knowledge of the internal workings of the unit under test to exercise as large a fraction of the set of potential faults as is possible. Fault coverage, based on exhaustive enumeration of all possible faults, is the ratio of faults tested to the total set of possible faults.

Testing is conducted at each stage of the system lifecycle. Hardware, software, and system tests are used during the development stage as described above to verify that the design is correct. During production, quality assurance testing insures against component failure, assembly errors, and system flaws. In the stage of operation and maintenance, diagnostic testing identifies failed components and assures correct repair.

The *test bench* is an important element of testing (Fig. 11–22). The unit under test in the center is driven by a stimulus generator, which also supplies the expected responses to a comparator. Differences between unit response and expected response are collected as errors.

Test vectors may be generated automatically using model-based test generation, which uses the same system model used in simulation to generate the test vectors, including the expected response. The quality of a set of test vectors may be measured by both the percentage of fault coverage and by the number of test

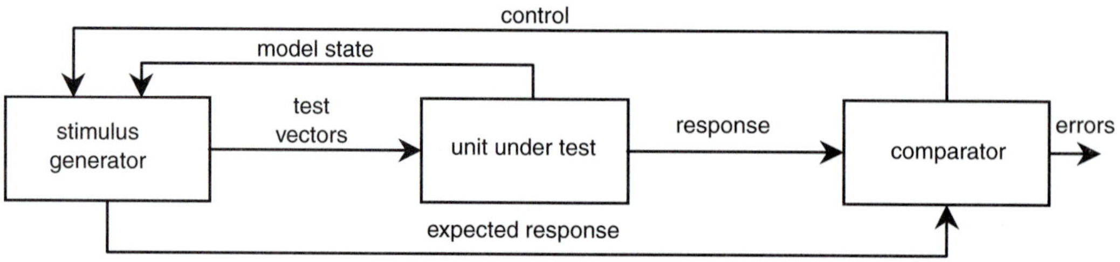

Figure 11–22 Basic test bench stimulates the unit under test and compares its response with known correct results.

vectors required to attain it, or the test vector set efficiency. A major determining factor on the number of test vectors required to achieve a particular fault coverage is the testability of the design. Designs with testability as a design constraint easily admit to testing for each fault with a minimal number of test vectors.

11.6.3 Boundary Scan Testing

As the circuit complexity supported by advancing semiconductor technology increases, access and visibility into the increasing number of modules on a chip becomes less. *Boundary scan testing* addresses the challenges of testing design, providing design for testability as an activity concurrent with logic design. Design of boundary scan testing places a serial scan path at the boundary of the integrated circuit to provide stimulus-and-response testing. An international standard, IEEE Standard 1149.1–1990, "Test Access Port and Boundary-Scan Architecture," known as JTAG, governs the specifics of boundary scan architecture and enables the testing of entire boards to be conducted by daisy-chaining the serial test ports of each integrated circuit. The presence of this international standard means that commercial testing equipment and logic components are available to support the inclusion of boundary scan testing in most circuit designs.

Boundary scan circuitry is added around the perimeter of an integrated circuit as shown in Fig. 11–23. When the chip is placed in test mode by asserting the *test* line, the normally parallel input and output signals are placed in series. The serial signal is stepped through each input; associated output responses are gathered serially at the line labeled "test out." The serial inputs and outputs are related and compared to correct responses to search for errors. The added circuitry for testing consists of switches to place all internal gates in serial connection while in test mode, plus the ability to clock through a serial stimulus signal and

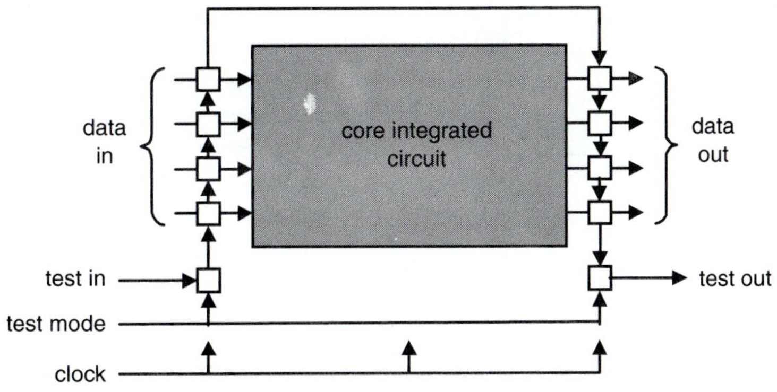

Figure 11–23 Boundary scan testing places inputs and outputs of circuit in serial scan mode.

collect the results. Only a minimal number of additional pins are required—one to place the circuit in test mode, a test data input pin, test data out, and a synchronized clock signal. Serial boundary scan signals from multiple chips on a board can be daisy chained. Design with boundary scan test capability is usually included on advanced logic chips.

11.6.4 Real Time Regression

Boundary scan testing provides precise testing of most logic elements at speeds lower than operating speed and for test data sequences of limited length. For full-speed testing on longer input sequences, a system may be tested and regressed against a standard reference system, which may be either a floating point implementation on a general purpose computer or a known good reference system, such as a previous version of the design. As shown in Fig. 11–24, a common signal input is presented to both systems, and the outputs are compared on a sample-by-sample basis by compiling a histogram of distance scores. To input the same file as is used in the floating point general purpose reference implementation, a memory and clocked test sequencer may be built as a plug-compatible replacement for the analog-to-digital converter chip, and the memory may be filled with the same input file as being used by the general purpose computer reference. This comparison can provide a check for differences that may be introduced by the transition from floating to fixed point or other numeric representation, input signal precision, dynamic range or signal companding limits, or errors that only occur at real time rates.

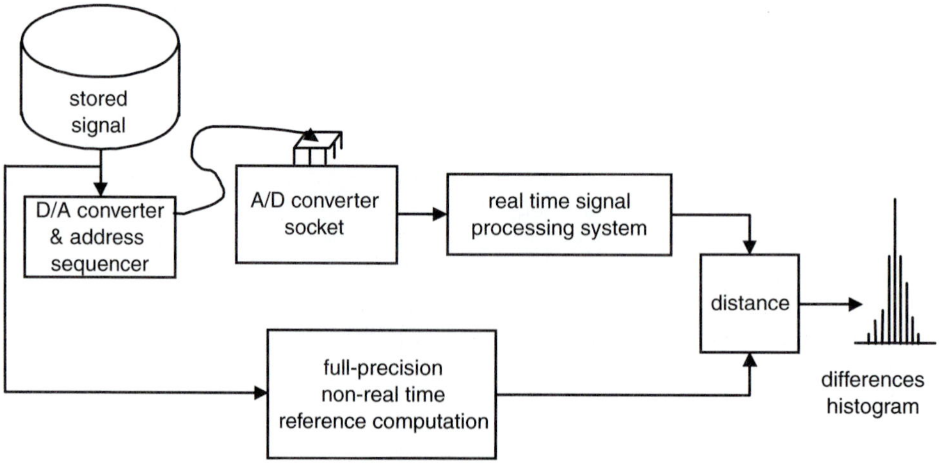

Figure 11–24 Test architecture for real time verification of signal processor against floating point non-real time reference.

11.6.5 Application-Level Testing

At an even higher level, application performance testing conducted in the user domain measures the success of the real time signal processing system in performing its desired function. For speech coders and synthesizers, the metrics may include speech quality and intelligibility. For speech recognizers, the error rate as measured upon a representative variety of talkers, noise conditions, and vocabulary words is used. Image coders can use image quality, image interpretability (measured, for example, by the National Imagery Interpretability Rating Scale, or NIIRS) for human observers, and for machine interpretation, can measure the change in performance of an exploitation algorithm (such as target detection, target or character recognition, resolution enhancement processing) on coded and decoded imagery vs. uncoded imagery. Real time image recognizers may be assessed using metrics of target detection rates versus image complexity, plots of detection rate versus probability of false alarm, and classification accuracy of recognizers. The more accurate a speech or image recognizer, the more tests are required to test it. For example, to obtain a statistically significant set of thirty errors, 300 trials are necessary on a recognizer of 10 percent error rate, but 3,000 trials are needed on a recognizer of 1 percent accuracy. This statistic is really an oversimplification of the number of trials and number of errors needed to evaluate the performance of a recognizer. For a high-performance system, recognition errors are rare events best modeled with Poisson statistics, and typically an extra multiplicative factor of from three to ten (i.e., over 10,000 trials for the recognizer of 1 percent accuracy) is needed to determine an error rate with high confidence.[29]

Testing can be performed to measure quantities that determine lifecycle maintainability. On-line *performance monitoring* refers to the ability of a system to detect nearly all faults while consuming only a negligible fraction of machine resources and while conducting normal operations. On-line performance monitoring is computed with off-line performance diagnostics, which have the purpose of detecting all faults with the machine out of operation and with no constraints on time or computation. Fault monitoring is tested by introducing a fault (for example, a short or stuck-at-1 fault) at a random location of the machine and determining whether on-line performance monitoring detects it and how long it takes to do so. Off-line performance monitoring is tested by measuring its ability to locate the fault.

The mean time between fault (MTBF) is measured by sustained machine operation over several multiples of the MTBF interval. Accelerated aging, brought on by elevated temperatures or by vibration, is occasionally used, but the acceleration factor is often hard to quantify.

Mean time to repair (MTTR) measures the downtime of a machine once a fault has been observed. MTTR is the average time that it takes the typical operator of a machine (not its designer) to diagnose and correct the fault. Achieving a low MTTR requires 1) accurate fault localization software, which narrows down the faulted component to one or two boards, 2) clear maintenance action instructions describing what should be done to localize and repair the fault, and 3) easily accessed system units to allow rapid removal and replacement, including an easily opened cabinet, front-mounted board insertion points, and tool-free capability to remove and replace boards. Lifecycle maintenance, like hardware testability, is most effectively designed concurrently with the other aspects of the system, rather than being added at the end.

All testing is driven by a test plan. The test plan compiles the individual tests that were developed with the design, including the per-requirement tests developed during requirements design, the stuck-at fault tests identified during hardware design, and the test of each path through the software resulting from software design. These tests are organized to group all tests for a particular configuration item together and are ordered to place general tests of broad areas ahead of detailed tests of specific areas. With effective ordering, efficient test sequences may be built that identify most failures as early as possible in the test sequence.

11.7 EXAMPLE SYSTEM DESIGN

To conclude this chapter, the design of a signal processor for real time image recognition is examined.[30] Several unique features applied to this processor:

1. The use of a complete set of documented user requirements of automatic target recognition to derive and validate processor requirements;
2. The use of a heterogeneous architecture that integrates several types of processors, including a pixel-based image processing system (IPS), floating-point DSP-based numeric processing system (NPS), and general-purpose-computer-based symbolic processing system (SPS);
3. The development of custom integrated circuits for image processing operations;
4. The application of wafer-scale multichip module miniaturization to the image processing pipeline;
5. The use of a piece-wise connected hierarchy of simulation tools, providing for connectivity of simulation both vertically (i.e., from chip through boards to subsystem) and horizontally (i.e. board versus multichip module domains).

The development produced four printed circuit board designs, each consisting of up to 800 components and 10 layers, one miniaturized wafer-scale module, consisting of 81 components, three application-specific integrated circuits, with the largest having 132,000 logic gates, and system software, consisting of 100,000 lines of code, as well as application and test software. The project resulted in the completion of two prototype systems, one of which was delivered to the customer and the other kept by the developers.

11.7.1 Systems Engineering

Systems engineering activity included the phases of advanced systems engineering, with requirements development, and developmental systems engineering, with performance monitoring, risk management, and program management. The processor, known as ATCURE (for Advanced Target Cueing and Recognition Engine), was defined by first analyzing the complete set of 125 future systems of the U.S. Army, as documented in the *U.S. Army Next Generation and Future Systems Sourcebook*,[31] and selecting the 39 systems areas that required capabilities in automatic target cueing and recognition (Fig. 11–25).

The concept of a *mission snapshot* was devised to capture the time-critical parameters that pace the real-time execution of image recognition algorithms. Such parameters include frame size (length and width in pixels, frame rate (in frames/sec), pixel depth (in bits per pixel), expected number of targets in a scene, and response time. Representative algorithms from each system were analyzed. The computational burden of each snapshot from each system was assessed for various computational architectures, and an architectural simulation was used to refine estimates of this timing. Iterations were made between architectural alternatives and performance prediction, resulting in the choice of a heterogeneous architecture. The analysis of a snapshot resulted in an algorithm definition, processing timeline, and assignment to a type of processor within the system. From these

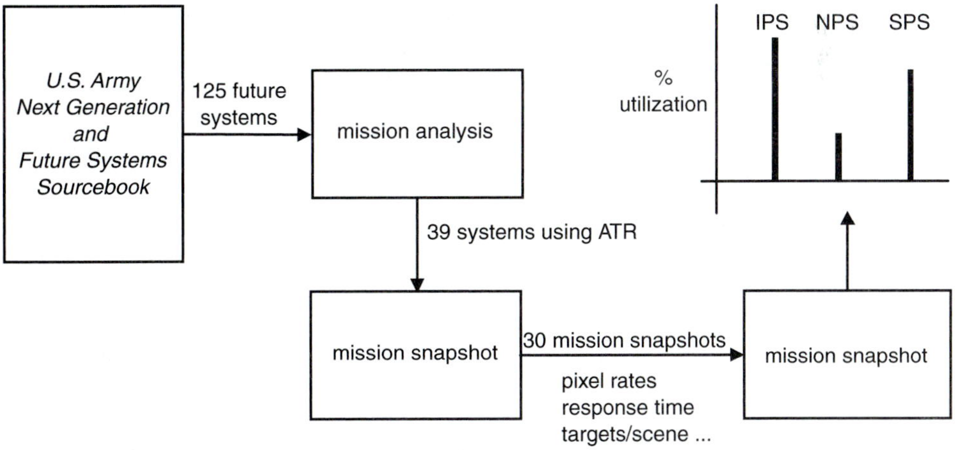

Figure 11–25 Method for requirements development captured the subset of 125 new Army systems that would use automatic target recognition into a set of 30 mission snapshots that emphasized the real-time processing aspects of the task.

quantities, processor throughput, interprocessor communication bandwidths, and processor utilization were calculated for each architectural alternative.

For each processing system, a trade study was conducted to determine the type of processor to use. Criteria of the trade study included processor speed, languages supported, floating point capability, availability of a real time operating system, and completeness of software development environment. As a result of mission analysis, performance modeling, and architectural trade studies, a heterogeneous architecture consisting of various processor types was selected as preferable over a large array of uniform processors. It was found that to cover the breadth of computing requirements imposed by this application, a single processor type suited to one domain would require extension to other modes of computation for which it was not suited. For example, a digital signal processor, suited for floating point multiply-and-add operations, would also be required to perform two-dimensional image neighborhood operations and data-dependent flowchart-type general computations, neither of which it could perform very well. Fig. 11–26 shows the heterogeneous architecture that resulted, consisting of an image processing subsystem dedicated to pixel neighborhood operations, a numeric processing system devoted to floating-point multiply-and-add vector operations, and a symbolic processing subsystem optimized for if-then-else data-dependent computations.

From these top-level architectural definitions, lower-level specifications of integrated circuits, boards, and interfaces were developed. The IPS was struc-

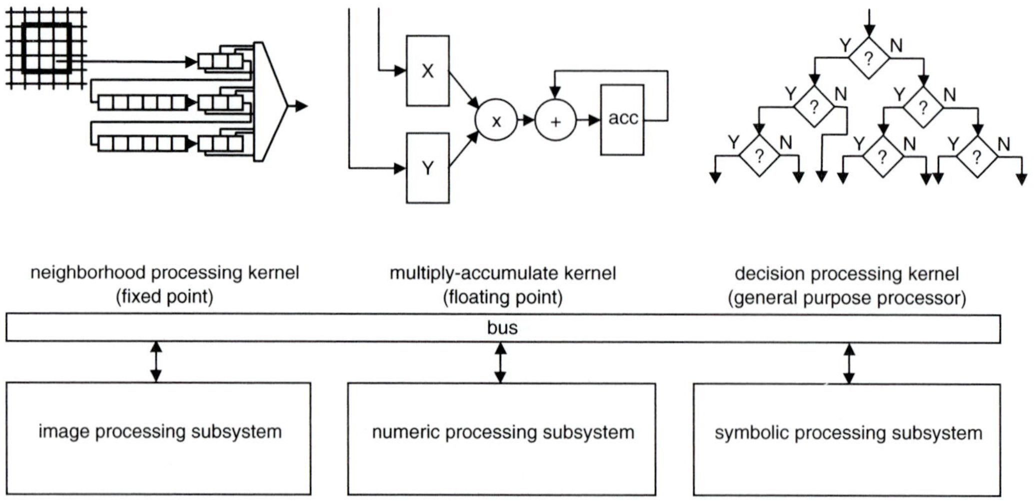

Figure 11–26 Heterogeneous architecture selected for image recognition included subsystems for image, numeric, and symbolic processing, based on kernels for neighborhood, multiply-and-add, and decision operations, respectively.

tured as a cascadable recirculating pipeline of multiple reconfigurable pipeline processing elements (PPEs). The IPS consisted of multiples PPEs with 16 global image buses and several local (direct PPE-to-PPE) buses, each of which transfers 16-bit pixels at a 20 Mpixel/sec rate. Each PPE performs 20 million neighborhood operations/sec, where a neighborhood operation combines a pixel's eight nearest neighbors via simultaneous use of 10 multipliers, 29 adders, 2 arithmetic logic units, and an accumulator, and outputs a pixel that is a linear or non-linear combination of its neighbors. A peak rate of 44 operations/neighborhood pixel × 20 MHz = 880 million operations per PPE is the theoretical limit—in practice, efficiencies of 50 to 70 percent were achieved. An IPS contained a pipeline of 20 PPEs, and software provided a virtual pipeline that allows its effective length to be extended indefinitely through recirculation (Fig. 11–27).

Risk analysis was performed as the design was developed. Table 11–9 shows a tool that was developed for tracking risks on the project. It shows a list of risks, each characterized by a probability of occurrence as of two dates in the program, separated by two months. The cost of recovering from each risk is also shown. The cost and likelihood of each risk change with time: New risks are added at the later date, and some risks decrease or disappear as certain actions become resolved. An aggregate measure of project risk, including total of number of risks, total cost of all, and likelihood-weighted costs, as well as how these totals were changing with time, was tracked and reported both internally to the project and externally to the customer. In addition, a plot of cumulative number of mile-

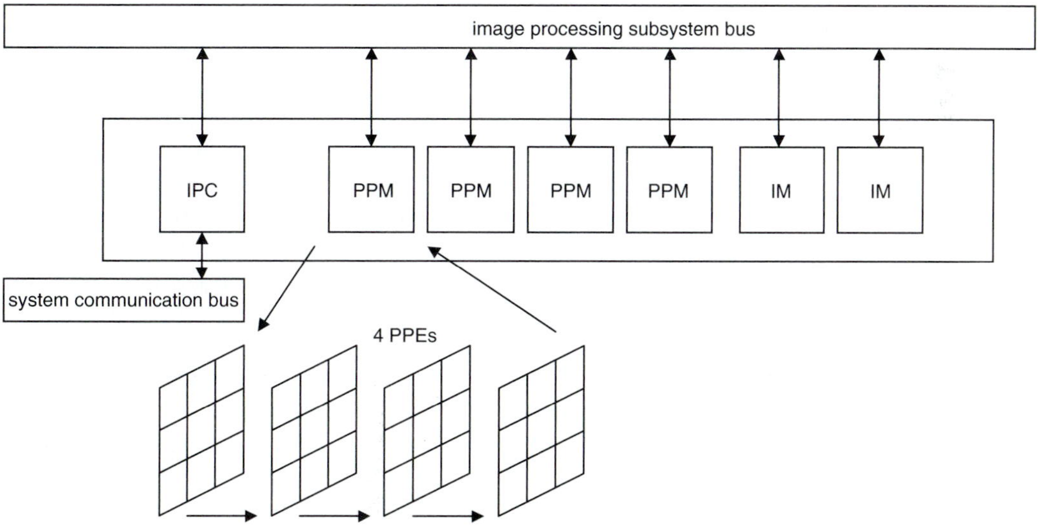

Figure 11–27 Image processing system (IPS) detail: Twenty PPEs, arranged in groups of four on a pipeline processing module (PPM), with intelligent image memories (IM), and crossbar switch.

Table 11–9 Chart of risk tracking and management shows evolution of risk as design progresses.

Risk	11/92 Probability	11/92 Weighted Cost	1/93 Probability	1/93 Weighted Cost	Change in Weighted Cost
Board-level integration takes too long	M	50	M	40	—
IM design late	H	20	L	0	—
PPE design late	M	20	H	30	+
IPSTran debugging on HW takes too long	M	20	M	20	0
Incremental test support SW insufficient			H	30	new
Boundary scan fault isolation insufficient	M	30	0	0	resolved
Bare ASIC die arrive late	M	20	M	20	0
Boundary scan test not ready for first ASICs	M	30	0	0	resolved
TOTAL	**M+**	**190**	**M-**	**140**	—

stones planned to be met, according to the schedule, and the cumulative number actually met, provided an indication of project status versus schedule versus time.

11.7.2 Hardware Design

A formal hardware design process was used to decompose and realize each design into modules and integrated circuits. Fig. 11–28 shows the module design process that was used. A module was decomposed at preliminary design into functional blocks, an overall module behavioral model, test software, and unit simulation models. These were each designed separately, and the results were converged at a pre-fabrication *Engineering Completion Review* (ECR). Upon successful completion of the review, fabrication of integrated circuits and boards proceeded, and system integration began.

Reviews preceding the ECR included the preliminary design review, which determined the feasibility of the partitioning and associated boards and integrated circuits to be developed, and the design description review, which assured that the detailed design for each component was adequate and that a test plan covered all functionality. After the ECR and system integration, a final review, known as the product release review, was held to assure that knowledge gained during system integration and operation of each component was reflected back into its design specification. For each review, a detailed preparation schedule was followed that specified the activities of line management (nominate review team),

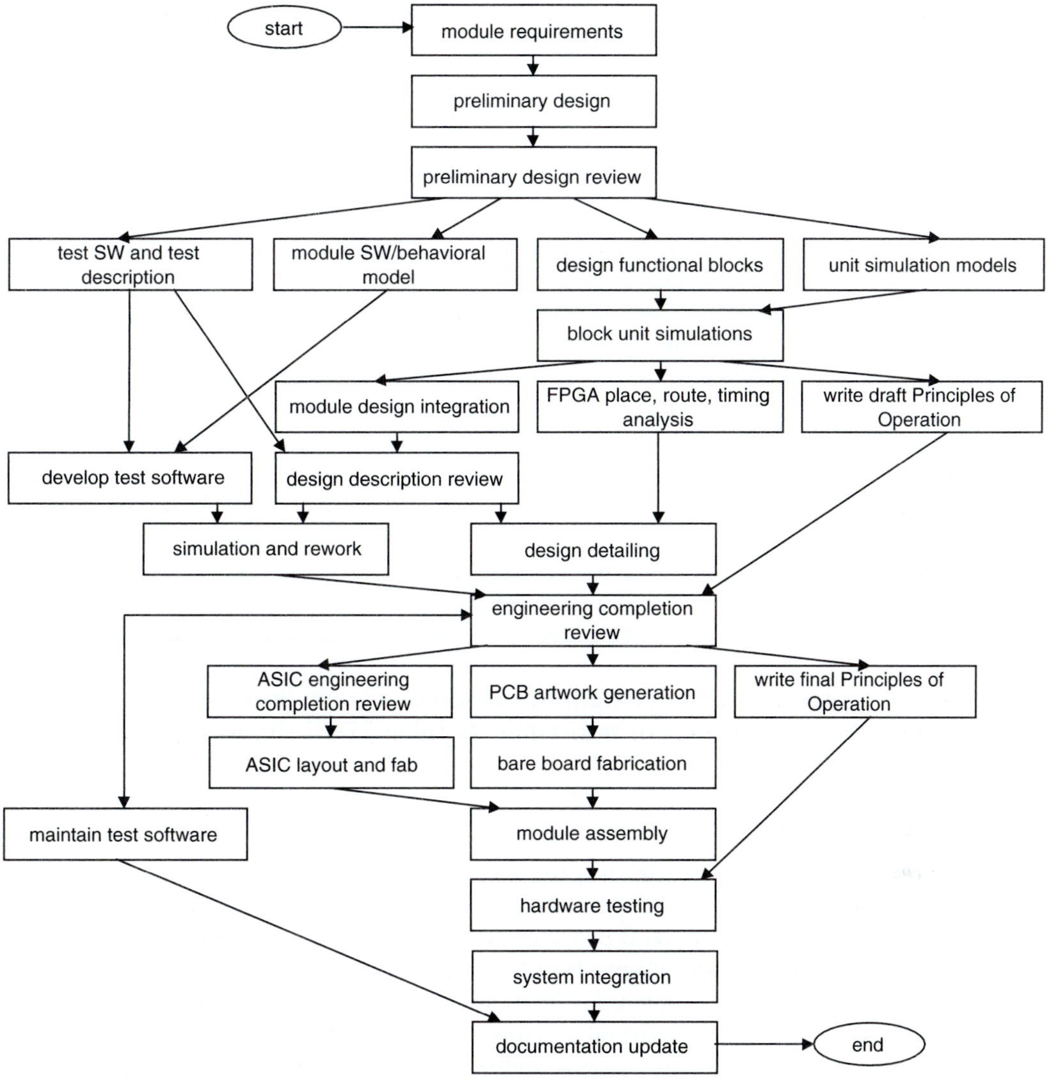

Figure 11–28 Hardware design process used for multiprocessor system combined paths for test software, module design, and application specific integrated circuit (ASIC) design.

lead designer (prepare design), review coordinator (schedule review and ensure its adequacy), and the reviewers. The review team was formed and the review was scheduled two weeks in advance, and a review preparation meeting was held one week prior to the review. The process provided for review checklist, followup, and signoff of the reviewed, revised design.

11.7.3 Software Design

Software consisted of a real time operating system for each subsystem (IPS, NPS, and SPS), upon which was built a support library that was called by the application level of software. The resident software was a mixture of operating system and application software. Off-board software that was developed included test software for integrated circuits, boards, and system, and programming software for the programmable silicon circuit boards used in wafer-scale integration. Both the NPS and SPS were programmed in C. The IPS was programmed in a language designed for the architecture of the PPE, called IPSTran (for IPS Translator).

A *scalable software development process* was used. The scalable software process arose from the need to accommodate development of four types of software: for internal (project) use only versus for outside delivery; small project versus large project. A full development process similar to MIL-STD 2167A was used for the most demanding project scale (large, for external delivery). However, for small test routines or software that would not be delivered, certain parts of the development process were not required. Table 11–10 shows the details of the scalable software process.

11.7.4 Packaging Design

The feasibility of miniaturizing the processor was shown by miniaturizing the image processing pipeline. A novel approach to wafer-scale integration was used that employed a customized programmable silicon circuit board (PSCB) as an interconnecting substrate to wire-bonded bare silicon die.[32] As shown in Fig. 11–29, a PSCB consisted of a layered silicon substrate that provides a level of parallel conductors running orthogonal to a lower level of parallel conductors. At each intersection, an electrically programmable "antifuse" can be permanently transitioned from an insulating state to a conductive state. Because design-specific interconnection information is electrically injected into the PSCB, this technology supports the rapid-prototyping, small-run approach more easily than those multichip module techniques that require that the design-specific interconnection information be imposed as mask layers during fabrication. This analogy is similar to electrically-programmed versus mask-programmed read-only memories.

Using multichip module technology required the consideration of PSCB circuit speed and design constraints early in the design. It also required development of a testing concept called *incremental functional test*. Since the wafer-scale miniaturization technology allowed the construction of a high-value assembly of

Table 11–10 Scalable software process allows rapid prototyping of small modules without the undue burden of process, yet enforces a full process for delivered large software.

		Large		*Small*	
Phase	**Stage**	**External**	**Internal**	**External**	**Internal**
Planning	Project plan	•	•	•	•
	Software development plan	•	•		
	Configuration management plan	•	•		
Requirements	Requirements specification	•	•	•	•
	Functional specifications	•	•	•	
Design	Software system design	•	•		
	Preliminary design	•	•	•	•
	Detailed design	o	o		
	As-built design	o	o		
Test development	System test plan	o	o	o	
	Acceptance test plan	o		o	
	System test report	o		o	
	Acceptance test report	o		o	
	Test stubs	o	o	o	o
	Test driver	o	o	o	o
	Test data	o	o	o	o
Implementation	Software source code	•	•	•	•
	Object files & libraries	•	•	•	•
	Executable files	•	•	•	•
	Build files	•	•	•	•
	Unit testing	•	•	•	•
	Integration testing	•	•		
	System testing	•	•	•	•
	Acceptance testing	o		o	
	Software prep for delivery	•	•	•	•
User documentation	Tutorial	*	*	*	*
	Reference manual	*	*	*	*
	Installation manual	•	o	•	o
	Programmer's manual	o	o		
Training and Installation	Training preparation	o	o		
	Installation procedure	•		•	
Post-project review	Reconciliation	•	•	•	•
	Lessons learned	•	•	•	•
	Project statistics	•	•	•	•
	Product documentation	o	o	o	o

• - required
* - except when exemption provided by recipient
o - as needed

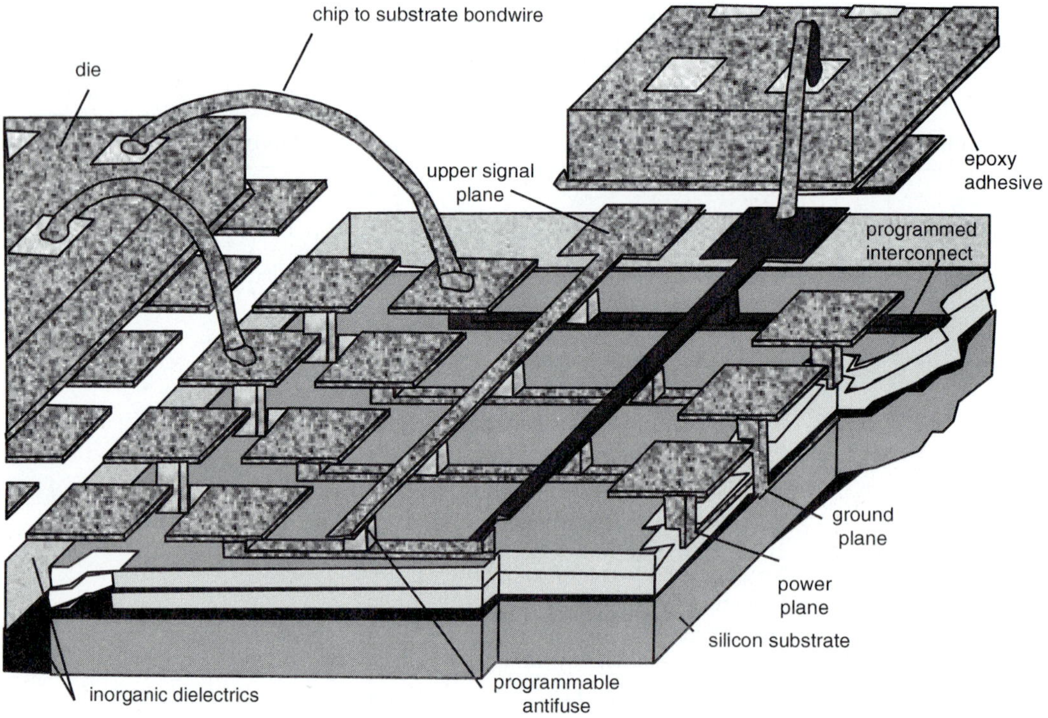

Figure 11–29 Programmable silicon circuit board places electrically-programmable antifuses at the intersections of vertical and horizontal conductors, which are blown into a conducting state to accomplish signal routing.

PSCB and dozens of bare integrated circuits, the traditional "build, then test" approach was replaced by the incremental "build a little, test a little" method, thereby accelerating the detection of defects and reducing the chance of having to scrap an entire finished module. Incremental functional test, based on IEEE 1149.1 boundary scan, embedded a design-for-test capability in the design process. It located faults to a single chip or its connections, thereby allowing the necessary rework to be performed at the smallest subassembly.

As a result of this approach, it was possible to achieve a four-inch-diameter multichip module, allowing an entire 9×13 in. board of conventional circuitry to be collapsed to one multichip module. This correspondence preserved the circuit partitioning between the standard and the miniaturized version of the circuits.

The process for circuit, MCM, and software development were tightly intertwined (Fig. 11–30). For example, PSCB design rules governed preliminary module circuit design, which drove the design and building of test software, detailed designs of modules and ASICs, and the layout of critical nets on the PSCB.

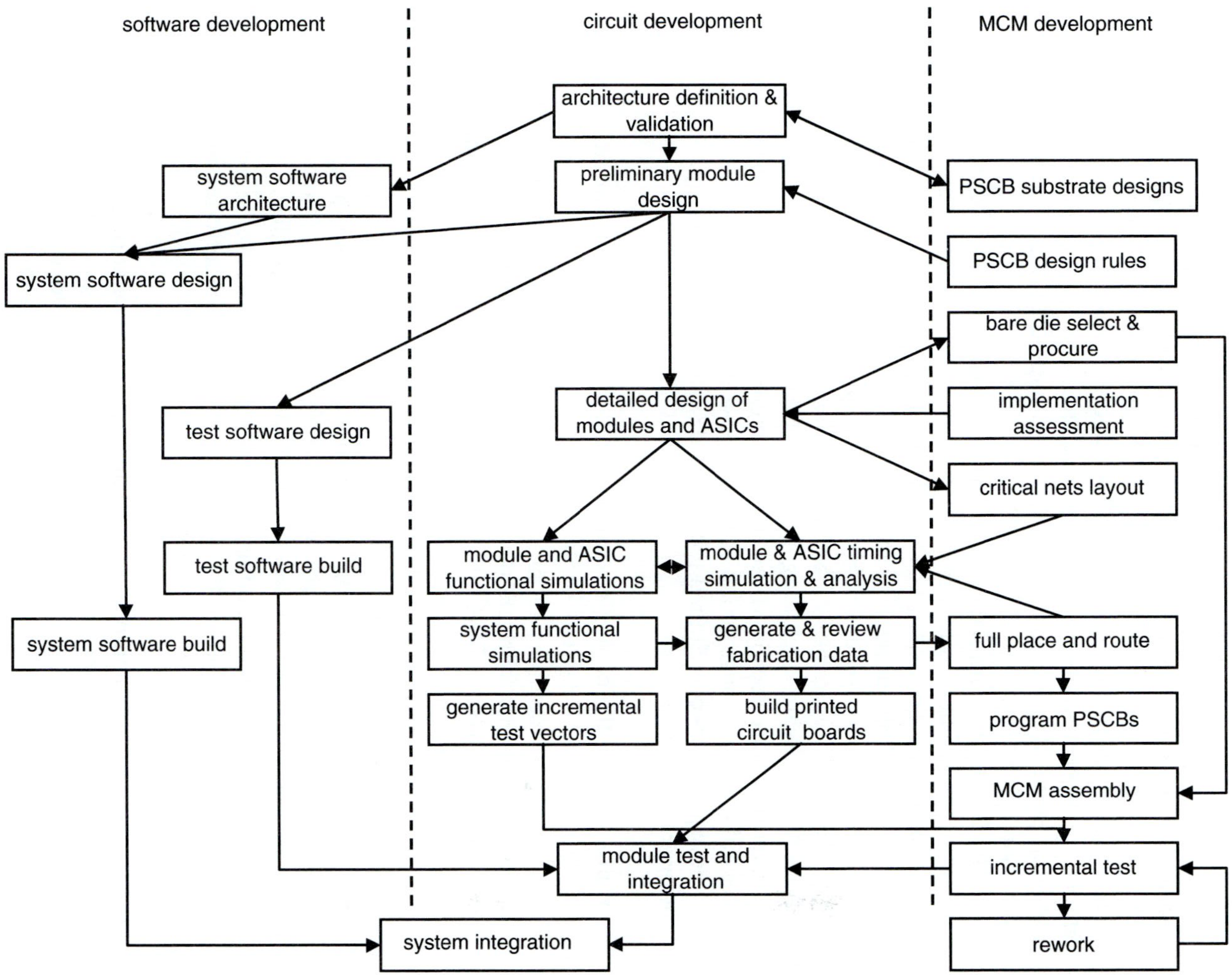

Figure 11–30 Integrated design process connected software, circuit, and multi-chip module design stages.

11.7.5 System Integration

System integration began at the module stage and worked up to the entire system. Starting with the integrated system, integration of application software proceeded up a hierarchy that began with software for basic operations (e.g., image functions of erode, dilate, or convolve), then constructed component algorithms (e.g., edge detection, filtering), and culminated in entire end-to-end object recognition. A reference system, which was a cellular array processor of a generation previous to the IPS, was used to aid the debugging of algorithms on the new processor.

The attention to simulation and rigorous design process paid off in a system integration time that was less than 10 percent of the overall system design and development time. For example, the time from the delivery of the samples of the last of the three ASIC designs to the time that the integrated system was handed off for applications programming was less than one month. Similarly, the time from system handoff to the first application demonstration was also less than one month. All hardware, including ASICs, printed circuit boards, and modules, were first-pass successes, and the use of incremental functional testing enabled successful first-time construction of the wafer-scale IPS.

Several factors contributed to this rapid system integration and correct-first-time operation:

- modeling of the analog circuit parameters of the silicon circuit board and using these in the circuit design;
- a connected hierarchy of simulation provided by the software tools of the commercial chip fabrication house that allowed gate-level simulation of the pipeline processor integrated circuit to be combined with simulation of the image memory custom integrated circuit and the other circuits on a board and the cascading of several boards into a subsystem in the simulation domain;
- use of "virtual prototyping" techniques in which actual image processing component algorithms were run on the above-mentioned hierarchical simulation model well in advance of actual hardware availability;
- use of identical circuit models and test fixtures on the non-miniaturized and miniaturized versions of the circuit, accelerating the introduction of miniaturized modules into the system.

After integration, both component level and application algorithms were running on the system. The power of the IPS as compared with an engineering workstation was measured to provide a 120-fold and 600-fold speedup for a single pipeline element for 3×3 convolution and for a 16-stage morphological filter, respectively, for a single PPE. For a pipeline of 20 PPEs, another factor of 20 speedup was measured for the morphological filter.

Table 11–11 Diverse automatic target cueing and recognition algorithms implemented on ATCURE.

Algorithm	Sensors	Comment
Critical Mobile Target	Synthetic aperture radar (SAR)	Distinguishes certain vehicles from clutter in fine-resolution SAR imagery
Background Adaptation Convexity Operator Region Extraction (BACORE)	Forward-looking infrared (FLIR)	Recognized as indicative of complexity of emerging-generation algorithms
Remote Minefield Detection System (REMIDS)	Three sensor types (polarization, reflectance, thermal infrared)	Sensor fusion combines separate views of same region

11.7.6 Applications

Three diverse end-to-end algorithms were programmed on the ATCURE processor, each using a different input sensor data type and seeking different objects, with one application fusing input from multiple sensors. (Table 11–11).

To summarize, the ATCURE processor requirements were developed from a systematic analysis of a documented set of applications, and the concept of a "mission snapshot" was introduced as a means to capture parameters that would pace real-time processing. As a result of the breadth of computational modes that were identified, a heterogeneous architecture, consisting of three processor types, was used. The mode of pixel-neighborhood operations, prevalent in the image processing front end, was accelerated by developing three custom integrated circuits, one for two-dimensional neighborhood processing, one for two-dimensional image memory addressing, and one to support communication on multiple image buses. Wafer scale multichip module miniaturization placed bare chips on a 4-in. programmable silicon circuit board, attaining a tenfold decrease in size versus the corresponding board-level circuit. The design process used a piecewise-connected hierarchy of tools, connected vertically from the gate level through the multiboard system level, and connected horizontally across the domain of boards, multichip module, and software. All components were operational on first-pass fabrication, and both system integration and the programming of three diverse recognition applications each took less than one month, indicating the power of a front-loaded, simulation-driven design approach in speeding final system completion.

REFERENCES

1. Barry W. Boehm, "A Spiral Model of Software Development and Enhancement," *IEEE Computer* (May, 1986), pp. 61–72.

2. Jeffrey S. Pridmore and W. Bernard Schamming, "RASSP Methodology Overview," *Proc. First Annual RASSP Conference*, Arlington, VA, Aug. 1994, pp. 71–85.
3. A. Rosenblatt and G. F. Watson, "Concurrent Engineering," *IEEE Spectrum* (July, 1991), 22–37.
4. *Defense Acquisition*, Department of Defense Directive 5000.1 (United States Department of Defense, 1991).
5. Y. Wong, W. Gass, T. Yoshino, and L. Johnson, "Silicon Compilation of Fast Fourier Transforms," in *VLSI Signal Processing IV*, H. S. Moscovitz, Kung Yao, and R. Jain, Eds. (Piscataway, NJ: IEEE, 1991), 3–22.
6. T. Egolf and others, "VHDL-Based Rapid System Prototyping," *J. VLSI Signal Processing Systems for Signal, Image, and Video Technology*, 14 (November 1996), 125–156.
7. *IEEE Standard VHDL Language Reference Manual*, ANSI/IEEE Std 1076–1993 (Piscataway, NJ: IEEE, 1993).
8. Integrated Design Automatic System (IDAS) (Santa Clara, CA : JRS Research Laboratories).
9. Deborah W. Runner and Erwin H. Warshawsky, "Synthesizing Ada's Ideal Machine Mate," *VLSI Systems Design* (December 1988).
10. *Specification Practices*, MIL-STD-490A (United States Department of Defense, 1985).
11. AT&T, *Process Quality Management and Improvement Guidelines*, Issue 1.1, Vol 1, AT&T Quality Library Code 500-049 (Indianapolis, IN : AT&T Customer Information Center, 1987).
12. *Engineering Management*, MIL-STD 499A (United States Department of Defense, Washington, DC, 20301, 1986).
13. *Systems Engineering Management Guide*, Supervisor, U.S. Government Printing Office, Washington, DC, 20402 (United States Defense Systems Management College, Jan. 1990).
14. "A Smarter Way to Manufacture: How 'Concurrent Engineering' Can Reinvigorate American Industry," *Business Week* (April 30, 1990), 110.
15. Ascent Logic Corporation, 180 Rose Orchard Way, San Jose, CA.
16. Cadence Design Systems, San Jose, CA.
17. D. J. Kaplan, "An Introduction to the Processing Graph Method," http://ait.nrl.navy.mil/pgraf/Intro_pgm/Final_Intro_pgm.html., August 31, 1998.
18. Management Communications and Control, Inc (MCCI), Arlington, VA.
19. *Best Practices—How to Avoid Surprises in the World's Most Complicated Technical Process*, NAVSO P-6071 (United States Department of the Navy, March, 1986).
20. *Military Standard Defense System Software Development* DOD-STD-2167A (United States Department of Defense, 29 Feb 1988).
21. Frederick P. Brooks, Jr., *The Mythical Man-Month: Essays on Software Engineering* (Reading, MA: Addison-Wesley, 1982).

22. B. W. Boehm, *Software Engineering Economics* (Englewood Cliffs, NJ: Prentice-Hall, 1981).
23. *Comparison of AN/UYS-2 (EMSP) and COTS Implementations of the SURTASS Signal Processor* (Naval Ocean Systems Center, 15 October, 1990).
24. "A Cost Research and Analysis Study to Develop the Ownership Cost of ECOS/ACOS Programs for Use in User Systems Signal Processor Application Software" (Department of the Navy: Naval Research Laboratory, Oct. 13, 1987).
25. Mary S. Albright, "EMSP Software Processing Improvement," *Quality Matters*, AT&T Federal Systems and Advanced Technology (FSAT), June, 1991, p. 5.
26. Edward Yourdon, *Managing the Structured Techniques* (Englewood Cliff, NJ : Prentice Hall, Inc. 1979).
27. Praful V. Bhansali, Bryan K. Pflug, John A. Taylor, and John D. Wooley, "Ada Technology: Current Status and Cost Impact," *Proc. IEEE*, 79 (January 1991), pp. 22–29.
28. Thomas R. Blakesee and Jan Liband, "Real-Time Debugging Techniques: Hardware-Assisted Real-Time Debugging Techniques for Embedded Systems," *Embedded System Programming*, 8 (April, 1995).
29. R. Voles, "Confidence in the Assessment and Uses of Mine Detection System," *Proc. IEE Second Inter. Conf. on Detection of Abandoned Land Mines*, Edinburgh, Scotland, October, 1998, pp. 28–30.
30. J. G. Ackenhusen and others, "ATCURE: A Heterogeneous Processor for Image Recognition" *Proc. IEEE Inter. Conf. Acoustics, Speech, Signal Processing*, Detroit, MI, May, 1995, pp. 2679–2682.
31. *U. S. Army Next Generation and Future Systems Sourcebook* (U.S. Army Materiel Command, Nov. 28, 1989).
32. H. Stopper, "An Advanced Version of the Electrically-Programmable Hybrid-WSI Substrate," *Proc. International Conference on Wafer Scale Integration*, January 1987, pp. 289–298.

Index

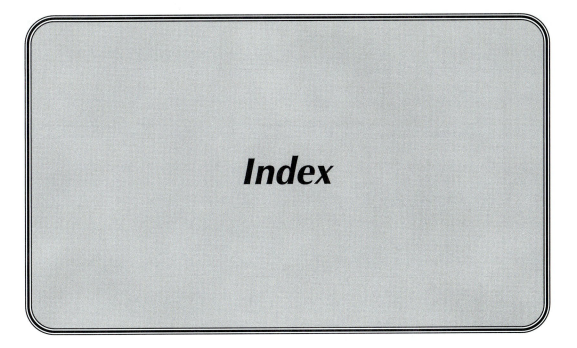

B

C

D

E

J

K

L

Q

R

T

U

V

W

Y

Z

John G. Ackenhusen (BS '75, BSE '75, MS '76, MSE '76, PhD '77, all from The University of Michigan) has performed and directed research and development in real-time signal processing since 1978, both designing computers architectures to accommodate signal processing algorithms and developing algorithms to fit existing computer architectures. He is currently Director of Programs of the Advanced Information Systems Group, ERIM International, Inc., Ann Arbor, MI.

At AT&T Bell Laboratories, he was Head of the Signal Processing Systems Design Department (1988–91), supervisor of the Speech Recognition Group (1981–88), and Member of Technical Staff (1978–81). He developed and patented an early special purpose computer for speech recognition and instigated its evolution into an AT&T product. More recently at AT&T Bell Labs, he led the systems engineering department that defined and directed the design of the U.S. Navy's AN/UYS-2 Enhanced Modular Signal Processor (EMSP).

Dr. Ackenhusen joined the Environmental Research Institute of Michigan (ERIM) in 1991 to lead its Image and Signal Processing Laboratory, a 100-member group devoted to computers and algorithms for real-time image processing and with capabilities in architecture, hardware, software, and multichip module electronic packaging. He served as manager of the Advanced Target Cueing and Recognition Engine (ATCURE), a program that delivered to the U.S. Government a heterogeneous-architecture computer for image recognition that included a wafer-scale multichip module image processing pipeline.

Dr. Ackenhusen was elected a Fellow of the Institute of Electrical and Electronics Engineers (IEEE) for his contributions in real-time signal processing. He served as Technical Program Chair in 1988 for the IEEE International Conference on Acoustics, Speech, and Signal Processing (ICASSP-88), chaired the IEEE Signal Processing Society Conference Board (1986–89), and served as Vice Chair of ICASSP-95. Dr. Ackenhusen served as President of the IEEE Signal Processing Society (1990–1991) and received the 1992 IEEE Signal Processing Society Meritorious Service Award for service as Society President and sustained contributions to the conference operations of the Society. He was elected to the IEEE Board of Directors for 1994–95, leading its Signals and Applications Division.